Thermal Behaviour, Energy Efficiency in Buildings and Sustainable Construction

Thermal Behaviour, Energy Efficiency in Buildings and Sustainable Construction

Editor

Paulo Santos

MDPI • Basel • Beijing • Wuhan • Barcelona • Belgrade • Manchester • Tokyo • Cluj • Tianjin

Editor
Paulo Santos
Department of Civil Engineering, ISISE
University of Coimbra
Coimbra
Portugal

Editorial Office
MDPI
St. Alban-Anlage 66
4052 Basel, Switzerland

This is a reprint of articles from the Special Issue published online in the open access journal *Energies* (ISSN 1996-1073) (available at: www.mdpi.com/journal/energies/special_issues/Buildings_and_Sustainable_Construction).

For citation purposes, cite each article independently as indicated on the article page online and as indicated below:

LastName, A.A.; LastName, B.B.; LastName, C.C. Article Title. *Journal Name* **Year**, *Volume Number*, Page Range.

ISBN 978-3-0365-1175-7 (Hbk)
ISBN 978-3-0365-1174-0 (PDF)

© 2021 by the authors. Articles in this book are Open Access and distributed under the Creative Commons Attribution (CC BY) license, which allows users to download, copy and build upon published articles, as long as the author and publisher are properly credited, which ensures maximum dissemination and a wider impact of our publications.

The book as a whole is distributed by MDPI under the terms and conditions of the Creative Commons license CC BY-NC-ND.

Contents

About the Editor . . . . **vii**

Preface to "Thermal Behaviour, Energy Efficiency in Buildings and Sustainable Construction" **ix**

María José Bastante-Ceca, Alberto Cerezo-Narváez, José-María Piñero-Vilela and Andrés Pastor-Fernández
Determination of the Insulation Solution that Leads to Lower CO_2 Emissions during the Construction Phase of a Building
Reprinted from: *Energies* **2019**, *12*, 2400, doi:10.3390/en12122400 . . . . **1**

Paulo Santos, Gabriela Lemes and Diogo Mateus
Thermal Transmittance of Internal Partition and External Facade LSF Walls: A Parametric Study
Reprinted from: *Energies* **2019**, *12*, 2671, doi:10.3390/en12142671 . . . . **41**

Mingshun Zhang, Xuan Ge, Ya Zhao and Chun Xia-Bauer
Creating Statistics for China's Building Energy Consumption Using an Adapted Energy Balance Sheet
Reprinted from: *Energies* **2019**, *12*, 4293, doi:10.3390/en12224293 . . . . **61**

Aniela Kaminska
Impact of Heating Control Strategy and Occupant Behavior on the Energy Consumption in a Building with Natural Ventilation in Poland
Reprinted from: *Energies* **2019**, *12*, 4304, doi:10.3390/en12224304 . . . . **77**

Lídia Rincón, Ariadna Carrobé, Marc Medrano, Cristian Solé, Albert Castell and Ingrid Martorell
Analysis of the Thermal Behavior of an Earthbag Building in Mediterranean Continental Climate: Monitoring and Simulation
Reprinted from: *Energies* **2019**, *13*, 162, doi:10.3390/en13010162 . . . . **95**

Jorge Fernandes, Raphaele Malheiro, Maria de Fátima Castro, Helena Gervásio, Sandra Monteiro Silva and Ricardo Mateus
Thermal Performance and Comfort Condition Analysis in a Vernacular Building with a Glazed Balcony
Reprinted from: *Energies* **2020**, *13*, 624, doi:10.3390/en13030624 . . . . **115**

Zhiyong Tian, Shicong Zhang, Jie Deng and Bozena Dorota Hrynyszyn
Evaluation on Overheating Risk of a Typical Norwegian Residential Building under Future Extreme Weather Conditions
Reprinted from: *Energies* **2020**, *13*, 658, doi:10.3390/en13030658 . . . . **145**

Paulo Santos, Gabriela Lemes and Diogo Mateus
Analytical Methods to Estimate the Thermal Transmittance of LSF Walls: Calculation Procedures Review and Accuracy Comparison
Reprinted from: *Energies* **2020**, *13*, 840, doi:10.3390/en13040840 . . . . **157**

Sara Elhadad, Chro Hama Radha, István Kistelegdi, Bálint Baranyai and János Gyergyák
Model Simplification on Energy and Comfort Simulation Analysis for Residential Building Design in Hot and Arid Climate
Reprinted from: *Energies* **2020**, *13*, 1876, doi:10.3390/en13081876 . . . . **185**

Victor Lohmann and Paulo Santos
Trombe Wall Thermal Behavior and Energy Efficiency of a Light Steel Frame Compartment: Experimental and Numerical Assessments
Reprinted from: *Energies* **2020**, *13*, 2744, doi:10.3390/en13112744 203

Mirco Andreotti, Marta Calzolari, Pietromaria Davoli, Luisa Dias Pereira, Elena Lucchi and Roberto Malaguti
Design and Construction of a New Metering Hot Box for the In Situ Hygrothermal Measurement in Dynamic Conditions of Historic Masonries
Reprinted from: *Energies* **2020**, *13*, 2950, doi:10.3390/en13112950 229

Elisabete R. Teixeira, Gilberto Machado, Adilson de P. Junior, Christiane Guarnier, Jorge Fernandes, Sandra M. Silva and Ricardo Mateus
Mechanical and Thermal Performance Characterisation of Compressed Earth Blocks
Reprinted from: *Energies* **2020**, *13*, 2978, doi:10.3390/en13112978 251

Helena Monteiro, Fausto Freire and John E. Fernández
Life-Cycle Assessment of Alternative Envelope Construction for a New House in South-Western Europe: Embodied and Operational Magnitude
Reprinted from: *Energies* **2020**, *13*, 4145, doi:10.3390/en13164145 273

Danijela Nikolic, Slobodan Djordjevic, Jasmina Skerlic and Jasna Radulovic
Energy Analyses of Serbian Buildings with Horizontal Overhangs: A Case Study
Reprinted from: *Energies* **2020**, *13*, 4577, doi:10.3390/en13174577 293

Vicente Flores-Alés, Alexis Pérez-Fargallo, Jesús A. Pulido Arcas and Carlos Rubio-Bellido
Effect on the Thermal Properties of Mortar Blocks by Using Recycled Glass and Its Application for Social Dwellings
Reprinted from: *Energies* **2020**, *13*, 5702, doi:10.3390/en13215702 313

Marek Borowski and Klaudia Zwolińska
Prediction of Cooling Energy Consumption in Hotel Building Using Machine Learning Techniques
Reprinted from: *Energies* **2020**, *13*, 6226, doi:10.3390/en13236226 329

Gianmarco Fajilla, Marilena De Simone, Luisa F. Cabeza and Luís Bragança
Assessment of the Impact of Occupants' Behavior and Climate Change on Heating and Cooling Energy Needs of Buildings
Reprinted from: *Energies* **2020**, *13*, 6468, doi:10.3390/en13236468 349

Toba Samuel Olaoye, Mark Dewsbury and Hartwig Kunzel
A Method for Establishing a Hygrothermally Controlled Test Room for Measuring the Water Vapor Resistivity Characteristics of Construction Materials
Reprinted from: *Energies* **2020**, *14*, 4, doi:10.3390/en14010004 367

María José Jiménez, José Alberto Díaz, Antonio Javier Alonso, Sergio Castaño and Manuel Pérez
Non-Intrusive Measurements to Incorporate the Air Renovations in Dynamic Models Assessing the In-Situ Thermal Performance of Buildings
Reprinted from: *Energies* **2020**, *14*, 37, doi:10.3390/en14010037 389

Paola Marrone, Francesco Asdrubali, Daniela Venanzi, Federico Orsini, Luca Evangelisti, Claudia Guattari, Roberto De Lieto Vollaro, Lucia Fontana, Gianluca Grazieschi, Paolo Matteucci and Marta Roncone
On the Retrofit of Existing Buildings with Aerogel Panels: Energy, Environmental and Economic Issues
Reprinted from: *Energies* **2021**, *14*, 1276, doi:10.3390/en14051276 . . . . **405**

About the Editor

Paulo Santos

Paulo Santos [ORCID 0000-0002-0134-6762] is Assist. Prof. in Dep. of Civil Eng. of University of Coimbra, PT. He is a member of the ISISE (Institute for Sustainability and Innovation in Structural Engineering) research centre. His main current scientific research fields are Thermal Behaviour, Energy Efficiency in Buildings and Sustainable Construction. He is author of around 150 scientific publications and has been supervisor of around 50 Doctoral and master's theses. He participated in around 12 funded European and Portuguese research projects, now being the Principal Investigator of the Tyre4BuildIns: "Recycled tyre rubber resin-bonded for building insulation systems towards energy efficiency"research project. Furthermore, he is a member of iiSBE: International Initiative for a Sustainable Built Environment (2011,-) and Technical Committee TC14: Sustainability and Eco-efficiency of Steel Construction of the ECCS: European Convention for Constructional Steelwork (2008,-).

Preface to "Thermal Behaviour, Energy Efficiency in Buildings and Sustainable Construction"

Nowadays, energy and sustainability are two of the major concerns of mankind. Given the relevant energy consumption share of the buildings sector, it is very important to search for innovative design solutions and for the optimal thermal performance of buildings to reduce energy bills and greenhouse gas emissions while maintaining the comfort levels of the occupants. Additionally, given the environmental burdens of the construction sector, seeking more environmentally responsible processes and a more efficient use of resources are currently attracting more attention.

This Special Issue (SI), published in the *Energies* journal, is dedicated to the analysis of recent advances in the following issues: (1) thermal behavior improvement of a building's elements (e.g., walls, floors, roofs, windows, doors, etc.), (2) energy efficiency in buildings, and (3) sustainable construction. The main goal is to compile scientific works within these topics, making use of different possible research approaches, such as experimental, theoretical, numerical, analytical, computational, case studies, and their combinations. This book compiles a set of original research works with academic excellence and scientific soundness.

The guest editor would like to express their sincere and deep gratitude for all of the scientific contributions from the authors among prestigious worldwide scientists as well as to the reviewers who significantly contributed to improving the quality of the manuscripts. Moreover, here, we express our acknowledgments to the research project Tyre4BuildIns—"Recycled tyre rubber resin-bonded for building insulation systems towards energy efficiency", supported by FEDER funds through the Competitivity Factors Operational Programme—COMPETE and by national funds through FCT—Foundation for Science and Technology, within the scope of the project POCI-01-0145-FEDER-032061, which allowed the contribution of three scientific articles to this SI. Additionally, the guest editor also wants to thank the support provided by the following companies, partners of the research project Tyre4BuildIns: Pertecno, Gyptec Ibéria, Volcalis, Sotinco, Kronospan, Hulkseflux, Hilti, and Metabo.

Paulo Santos
Editor

Article

Determination of the Insulation Solution that Leads to Lower CO_2 Emissions during the Construction Phase of a Building

María José Bastante-Ceca [1],*, Alberto Cerezo-Narváez [2], José-María Piñero-Vilela [2] and Andrés Pastor-Fernández [2]

1. Grupo de Investigación en Diseño y Dirección de Proyectos, Universitat Politècnica de València, 46022 Valencia, Spain
2. School of Engineering, University of Cádiz, 11519 Puerto Real, Spain; alberto.cerezo@uca.es (A.C.-N.); josemaria.pinerovilela@mail.uca.es (J.-M.P.-V.); andres.pastor@uca.es (A.P.-F.)
* Correspondence: mabasce1@dpi.upv.es; Tel.: +34-96-387-7000 (ext. 75685)

Received: 29 March 2019; Accepted: 17 June 2019; Published: 21 June 2019

Abstract: The characteristics of the envelope of a building determine, together with other factors, its consumption of energy. Additionally, the climate zone and insulation material may vary the minimum insulation thickness of walls and roofs, making it different, according to cooling down or warming up the home. Spanish legislation establishes different maximum values for energy demand according to different climate area both for heating and for cooling. This paper presents the results of a study that determines the influence of many variables as the climate zone or the orientation, among others, in the optimization of thickness insulation in residential homes in Spain to reduce the CO_2 emissions embodied. To do that, 12 representative cities in Spain corresponding to different climate zones, four orientations, two constructive solutions, and four different configurations of the same house have been combined, for three different hypotheses and four insulation materials, resulting in 4608 cases of study. The results show that, under equal conditions on energy demand, the optimal insulation requirements are determined by heating necessities more than by cooling ones. In addition, a higher insulation thickness need does not necessarily mean more CO_2 emissions, since it can be compensated with a lower Global Warming Potential characterization factor that is associated to the insulation material. The findings of this study can serve to designers and architects to establish the better combination of the variables that are involved in order to minimize the CO_2 emissions embodied during the construction phase of a building, making it more energy efficient.

Keywords: energy demand analysis; insulation materials; climate zones; envelope; CO_2 emissions

1. Introduction

Urban growth following the central years of the "real estate bubble 1998–2007" [1] has produced significant change in Spain in terms of building densities, which fell to substantially below 35 dwellings per hectare [2]. Current legislation, far from restricting the expansion of the urban by occupation of the rural space, promotes it by deregulating the use of undeveloped land [3]. Lower urban densities, high losses of non-urban land covers, the depopulation of metropolitan inner cores, and the expansion of transportation infrastructures confirm the generalization of the dispersed urban model, in which the importance of single housings is highlighted [4,5].

The upward trend in energy prices is growing [6], parallel to this disproportionate development of urban society, which makes it necessary to implement measures that are aimed at optimizing demand and promoting energy saving and efficiency [7]. In this respect, dwellings, like all other buildings, face the challenge of achieving an energy management that allows them to contribute to economic growth, social welfare and sustainability of non-renewable resources, and preservation of the natural environment [8].

Buildings are big consumers of energy and materials and important producers of waste and emissions. Prefabrication presents an opportunity to reduce impacts in the building sector [9]. Among the advantages and benefits that are offered by the prefabricated building systems when compared to conventional construction methods, reductions in cost and time, improved quality, safety, and accuracy in manufacture, speed of installation on-site, and even dismantling and reuse are provided [10,11], as well as customization [12].

Energy consumption in the building sector is gaining increasing interest, as it is directly related to energy economics and sustainable development. The design and the choice of building materials, as well as the energy and thermal systems, evolve very rapidly. In the energy challenge, the building is among the largest consumers of energy in the European Union area [13]. The efficiency and optimization of energy systems remain among the main items that are studied in order to reduce energy consumption and increase system performance. In the area of housing, the cost and optimization of space are the two main reasons that require the decrease of the thickness of walls in new constructions; however, this reduction greatly affects the thermal inertia of the frame and makes it insufficient to effectively damp the oscillations due to the outdoor temperature variation [14]. Under these conditions, the optimization of the thickness insulation plays an important role in reaching a workable compromise between the comfort, the cost of the building, and the consumption of energy (and its corresponding cost during their lifetime).

Spain has generated an intense development of new regulations seeking for better energy performance in buildings in recent years. Thus, it is noteworthy that, as a result of the transposition of Directive 2002/91/CE [15], the Technical Building Code (CTE) is enacted [16], as well as a procedure for energy Certification for Buildings [17] (transposition of Directive 2010/31/EU [18]) and a new Regulation for Thermal Installations in Buildings [19] (transposition of Directive 2012/27/EU [20]).

Many of the potential effects of climate change on the building sector are not well studied, as climate change one of the most important social and environmental concern [21]. At the European level, about 36% of CO_2 emissions are related to buildings. For this reason, the European Union (EU) has identified the building sector as one key area for achieving its objectives for greenhouse gas emission reductions [22].

The EU Directive on the Energy Performance of Buildings [18] specifies that, by the end of 2020, all new buildings shall be nearly Zero Energy Building (nZEB). Directive 2012/27/EU establishes a specific mandatory for member states to draw up national plans to increase the number of nZEB. These plans must include the detailed definition of the nZEB concept in such a way that their national, regional, or local conditions are reflected, and a numerical indicator of the primary energy use must be included and expressed in kWh/m^2 per year.

The Basic Document of Energy Saving (DB-HE) of the CTE [23] is the second revision of the original one dated on 2006 in terms of energy saving (the first revision is dated on 2013). The method of calculation of the characteristic parameters of the elements that compose the thermal envelope of the models is carried out according to the Directives of DB-HE of the CTE. This method consists of the calculation of the thermal transmittance of these elements: enclosures that are in contact with external air, enclosures in contact with the ground, interior partitions in contact with non-habitable spaces and hollows, and skylights considering their modified solar factor.

Usually, the lifetime of the buildings easily reach between 50 and 100 years, so the buildings constructed today need to be resilient to future climates, than can be largely different than the one that we experience today [22]. Pérez-Andreau et al. [24] studied the impacts of climate change on heating and cooling energy demand in a residential building in a Mediterranean climate with two different Global Circulation Models for 2050 and 2100. The authors concluded that climate change has a direct effect on energy demand in homes, and suggested that thermal insulation will have great effect on total energy demand.

Previous studies have analyzed the environmental impact of using different insulation materials [25–30], fixing the rest of parameters (orientation, climate zone, compactness, or constructive solution). This is the case of Braulio-Gonzalo and Bovea [25], which compares eleven insulation materials alternatives for a single-family house that was located in the climate zone B3, with a given orientation, and fixing the envelope description and thermal resistance, in order to see the influence of the insulation material and the thickness on energy demand, to accomplish the Spanish Technical Code. On the other hand, Hill et al. [26] make a review of the different insulation materials environmental information published, with the aim of comparing both the embodied energy and the environmental impact in terms of CO_2 emissions, independently of the rest of variables or the insulation needs. Additionally, Pargana et al. [27] compare the different insulation materials in order to evaluate their environmental impacts, and the consumption of energy on their production. Again, the authors do not consider the needs of insulation materials or the possibility that, although one type of insulation may have a higher environmental impact during its production, this can be compensated with lower insulation thickness needs, resulting in lower CO_2 emissions once placed into the building during its construction phase. Sierra-Pérez et al. [28] analyse different façade-building systems and thermal insulation materials for different climatic conditions, in order to determine their environmental impact. These authors consider five insulation materials, three façade systems, but, as in [25], just consider one climate zone (D), although they perform a sensitivity analysis varying the climate zone, but without varying orientation, compactness, or constructive solutions, variables that also influence the envelope and the insulation thickness needs. The same authors indicate, as one of the weaknesses of their research, that they just consider a unique building façade system in isolation and not as part of an entire building. Asdrubali et al. [29], in line with that indicated for [27], present a report of the state-of-the-art of insulation materials, without going into embodied energy or CO_2 emissions that are associated to its construction, or in the different insulation thickness needs according to variables as orientation, climate zone, and so on. Finally, Schiavoni et al. [30] make a review of the different insulation materials that were used for the building sector, presenting a comparative life cycle assessment between the different insulation materials for four different typical configurations of external walls, in order to compare both the embodied energy and global warming potential in terms of CO_2 emissions, for the same functional unit. Again, the authors do not consider different insulation thickness needs, depending on the climate zone, the orientation of the building, the constructive solution, and the building model, among others, apart from the insulation material.

In addition, different authors have studied the influence of different electricity-to-emissions conversion factors for three different insulation materials into the calculation of lifecycle emissions [13]. Apart from that, other studies [31] have investigated the building energy demand under different climates, or even including variables, such as the configuration of walls [32], but none of them have considered the influence of all the parameters, taken together.

This paper presents the results of a study that determines the influence of different parameters as the climate zone, the compactness of the building and the orientation, as well as the insulation material and the constructive solution in the optimization of thickness insulation in residential prefabricated houses in order to minimize the CO_2 emissions that were embodied during their construction phase.

A series of cases of a single-family semi-detached house is proposed to develop the study. In total, 4608 cases of study have been analyzed, while considering 12 locations according to DB-HE climate zones, four main orientations, two constructive solutions, and four compactnesses, all of them for four insulation materials, under three hypotheses of demand limitation.

The results of this study can help professionals that are involved in the building sector (designers, builders, architects, engineers, and even legislators) to establish the better conditions for minimizing the CO_2 emissions from the insulation during the construction phase for an energy demand fixed for cooling and heating in the use phase. Variables that have been taken into account are the climatic zone, the orientation, the constructive solution for façade and roof, and the compactness of the building, as well as the insulation material and its thickness.

The originality of the research that is presented in this paper consists in the fact that we have considered different variables that have a substantial influence on the determination of the envelope of the building (climate zone, orientation, compactness, constructive solution, insulation material, and energy demand), in order to determine the insulation thickness needs for each case. This way, for a given climate zone, the builders and designers can select the best combination of the variables in order to minimize the embodied CO_2 emissions of the building during its construction phase. Economical aspects are not to be left out of the considerations, since they may affect the final decision. Nevertheless, the difference in cost of implementing the most effective solution in terms of reducing CO_2 emissions and its possible compensation with the savings derived from a minor energy consumption during the use phase of the building is out of the scope of this study and it will be the subject of subsequent research. In addition, the energy requirements for the use phase of the building and the possibility to satisfy them with renewable energies (solar thermal and photovoltaic energies, for example) will also be the subject of further researches.

The paper is structured, as follows. Section 2 presents the method used, establishing the three calculation hypotheses and describing the software used, choosing the location from the climate zones and their orientation, defining the characteristics of the building (compactness and constructive solutions), and selecting the insulation material. Section 3 shows the main results that were obtained of the study, including the thickness of the insulation for each climatic zone, orientation, compactness, constructive solution, and demand hypotheses, as well as their emissions. The major findings are also highlighted and contextualized, discussing them with the literature review made. Section 4 concludes the paper, summarizes the contributions, and proposes further research continuations.

2. Method

2.1. Calculation Procedure and Software Used

The unified tool LIDER-CALENER (HULC) is used in order to assess the energy demand [33]. HULC is the official energy certification tool in Spain, although other homologated tools can also be employed. This tool includes a graphical interface for a three-dimensional (3D) representation of buildings and it performs an hourly simulation considering a transitional regime, while taking into account thermal coupling between adjacent zones and thermal inertia, thanks to its calculation engine, called S3PAS, following the procedure from the ISO 52016-1:2017 standard [34].

There are three demand hypotheses that have been established for each situation (1536 scenarios from 12 climate zones, four orientations, two constructive solutions, and four compactness), making a total of 4608 case studies:

- H1: Compliance with the minimum legal requirements derived from the DB-HE of the CTE.
- H2: Joint (summing heating and cooling up) demand ≤ 30 kWh/m^2 per year.
- H3: Heating demand ≤ 15 kWh/m^2 per year and cooling demand ≤ 15 kWh/m^2 per year.

Hypothesis 1, as shown in Table 1, establishes four different heating demands (a basis of 15 kWh/m^2 per year for climate zones A and B, almost 30 kWh/m^2 for climate zone C, and slightly above 40 and 60 kWh/m^2 for climate zones D and E, respectively, as explained in the next section). Regarding cooling demand, only two requirements are stated (15 kWh/m^2 per year for climate zones 1, 2, and 3, and 20 kWh/m^2 for climate zone 4, as explained in the next section).

Table 1. Maximum heating and cooling demand per climate zone for legal compliance.

Climate Zone	Heating Demand	Cooling Demand
A3	15	15
A4	15	20
B3	15	15
B4	15	20
C1	26.8	15
C2	26.8	15
C3	26.8	15
C4	26.8	20
D1	40.6	15
D2	40.6	15
D3	40.6	15
E1	60.4	15

Units in kilowatts hour per square meter per year (kWh/m^{2}y).

Spanish legal requirements, which fix the maximum energy demand, generate a gap in energy consumption that is faced by final users from some climate zones, especially D and E ones. On the contrary, the hypothesis 3, which is based on the requirements of the Passivhaus standard [35], limits the heating and cooling demand to 15 kWh/m^2 per year each. Given the fact that letter indicates the severity of the winter, whereas the number indicates the severity of summer, for the same winter severity (as explained in the next section), this constraint is detrimental to users in moderate summers as compared to colder ones. Hypothesis 2 is proposed to mitigate this, while considering a joint demand for heating and cooling, aggregating them up to a limit of 30 kWh/m^2 per year.

The procedure has been the following: starting with an initial insulation thickness of 0 mm (both for the façade and for the roof and the ground floor), the energy demand has been calculated and compared to the limits by hypothesis. If the energy demand is under the limits, then an increase in insulation thickness of 5 mm is considered and the process is repeated again. The process continues with an incremental insulation thickness of 5 mm until the limits for each of the hypotheses considered are reached. The incremental insulation thickness of 5 mm has been chosen according to the commercial availability on the market. Other parameters must be taken into account once the insulation thickness for each of the hypotheses considered is fixed, and before the energy demand is determined, according to the characteristics of the building (compactness and constructive solutions), and the other variables considered (orientation, climate zones, block shadows, and so on).

The gains and losses are considered by HULC according to the detailed method of the ISO 52000-1:2017 standard [36], and depend on the type and thickness of insulation, infiltration, orientation, and climate zone, among other variable elements. They also depend on the fenestration, thermal bridges, and ventilation, which remain invariable in this study. Besides, both thermal bridges and ventilation are calculated by the DB-HE of the CTE [22]. Ensuring continuity in the insulation of the constructive elements union solves thermal bridges. In the case of ventilation, the minimum required flow rate is 33 liters per second (intake and extraction), which means 0.27 renovations per hour.

2.2. Climate Zones

The Köppen Climate Classification is chosen in order to identify the climate zones within mainland Spain. This classification, published in 1900, is still one of the most widely classifications systems used for climate studies in the world. According to this, based on the average monthly values for precipitation and air temperature, the climate zones are characterized by a combination of a letter by the climate severity of winter and a number by the climate severity of summer.

For this study, 12 provinces (represented by their capitals) in mainland Spain have been chosen, whose selection is due to its representativeness from their climate zones by their population. Table 2 shows the selected provinces for the study, as well as the climate zone, the altitude of their capitals, their population, and their percentage over the total population of mainland Spain.

Table 2. Characteristics of the cities object of the study [37].

City	Climate Zone	Altitude (m.a.s.l.)	Population [1]	% over Total in Mainland Spain
Cádiz	A3	0	1,238,714	2.86%
Almería	A4	0	709,340	1.64%
Valencia	B3	8	2,547,986	5.89%
Sevilla	B4	9	1,939,887	4.48%
La Coruña	C1	0	1,119,351	2.59%
Barcelona	C2	1	5,609,350	12.96%
Granada	C3	754	912,075	2.11%
Cáceres	C4	385	396,487	0.92%
San Sebastián	D1	5	720,592	1.66%
Gerona	D2	143	761,947	1.76%
Madrid	D3	589	6,578,079	15.19%
Burgos	E1	861	357,070	0.82%

[1] Data at 01/01/2018.

Figure 1 shows the distribution of climate zones for mainland Spain, according to Köppen Climate Classification:

Figure 1. Distribution of climate zones in Spain.

2.3. Orientation

Orientation influences the energy consumption of a building, and the election of an accurate orientation, together with the correct location and landscaping changes, may decrease its energy consumption [38]. For this study, in order to consider different advantage of solar power depending on the orientation of the building due to different shadow, and also to analyze the influence of this parameter on the results of insulation thickness needs, the four cardinal orientations have been selected, following the wind rose: North (N), East (E), South (S), and West (W).

2.4. Characteristics of the Building

All of the buildings considered for this study belong to the category of semidetached houses, joined in a dwelling unit. Each semidetached building consists of three different floors (ground floor, first floor, and roof floor). It can be noted that the same housing units compose all of the studied models). At the ground floor, we can find the dining room, the kitchen, the living room, one bath, and the pantry, apart from the entrance to the house and the ground floor stairs. At the first floor, we can find three bedrooms, two bathrooms, and the first floor stairs. Finally, at the roof floor, there are the roof floor stairs and the access to the deck. Each dwelling unit is made up of three shared median walls and a faade one limiting with the public domain. The block presents a number multiple of four houses. For example, Figure 2 shows a 3D simulation for the models considered, in which the block configuration can be observed.

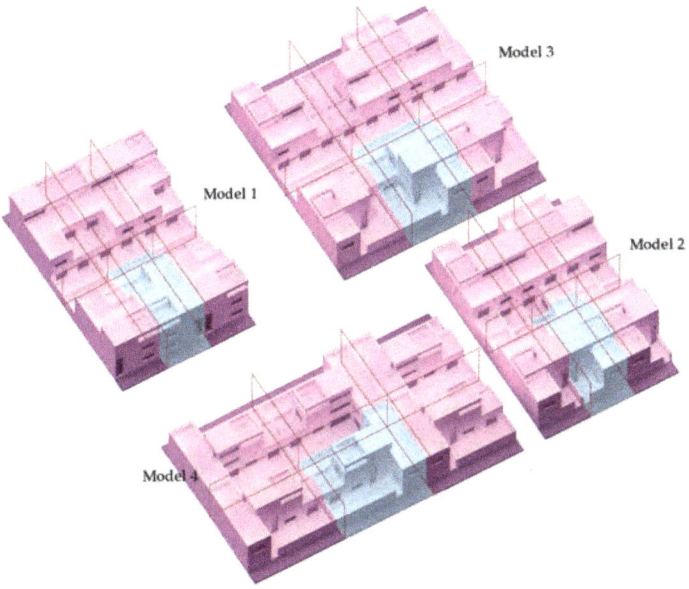

Figure 2. Three-dimensional (3D) Simulation of the block configuration from the dwelling unit for Models 1–4.

Four different building configurations are considered in order to determine the influence of the compactness of the building. For the four models involved, the degree of compactness vary from 1.5 for Model 4 to 2.2 to Model 1. The configurations of the four models studied are shown in Figure 3a–d, in which the green color corresponds to the garden zones (from the ground floor) and the blue color to walkable terraces (from the first and the roof floor).

In addition, the surface of the building is the same (insofar as all the models are made up of the exactly same housing units), but its compactness, which establishes the relationship between the outer shell of the building and its volume, changes. Independent of the orientation, climate zone, and configuration of the elements, the four models studied have the same building characteristics regarding their volume, their built area, their roof, and ground area, but with small differences regarding their opaque façade surface area and their glazed façade surface area, which makes its compactness vary, as can be seen in Table 3.

Figure 3. (**a**) Configuration of the Model 1; (**b**) Configuration of the Model 2; (**c**) Configuration of the Model 3; and, (**d**) Configuration of the Model 4.

Table 3. Characteristics of the building for different models analyzed.

Characteristics	Model 1	Model 2	Model 3	Model 4
Volume (m^3)	441.00	441.00	441.00	441.00
Built area (m^2)	147.00	147.00	147.00	147.00
Roof area (m^2)	73.50	73.50	73.50	73.50
Ground area (m^2)	73.50	73.50	73.50	73.50
Total façade surface area (m^2)	126.00	136.50	199.50	220.50
Opaque façade surface area (m^2)	99.00	107.00	161.50	178.50
Glazed façade surface area (m^2)	27.00	29.50	38.00	42.00
Glazing ratio (%)	21.50	21.50	19.00	19.00
Total insulation surface area (m^2)	246.00	254.00	308.50	325.50
Compactness *	2.20	2.10	1.60	1.50

* Compactness is defined as the 'volume divided by the area exposed to outside air (roof and façades)' ratio.

2.5. Selection of the Insulation Material

The correct choice of the insulation material is relevant when improving the energy-efficiency of the buildings. Different materials can be used to provide similar functions in buildings but the related energy-use and emissions could vary widely [39]. Most commonly used insulation materials in building industry are fiberglass, stone wool (also known as mineral wool or rock wool), glass wool, cellulose fiber, expanded polystyrene (EPS), extruded polystyrene (XPS), polyisocyanurate (PIR), and polyurethane (PUR) [39,40].

For this study, the four commonly insulation materials used have been chosen. The choice has been made according to the state-of-the-art review, where four types of insulation materials have been identified as the most commercialized for building: derived from petroleum (for example, PUR and PIR), polystyrenes (XPS and EPS), minerals (stone wool, glass wool, etcetera), and natural or ecological ones (expanded cork, wood fibreboard, etcetera). According to this, one insulation material of each type has been chosen for this study: Extruded Polystyrene (XPS), Polyurethane foam (PUR), Stone Wool (SW), and Expanded Cork (EC). Table 4 shows the characteristics of insulation materials considered, from Environmental Product Declarations that will be used to determine CO_2 emissions according to their insulation thickness needs.

Table 4. Characteristics of the insulation materials.

Characteristic	XPS Board [41]	PUR Foam [42]	SW Board [43]	EC Board [44]
Thermal conductivity [1]	0.025	0.028	0.031	0.040
Density [2]	32	31	30	115
Global Warming Potential factor [3]	127.35	89.90	64.80	33.30

[1] Data in W/mK. [2] Data in kg/m^3. [3] Data in kg CO_2/m^3 insulation.

As stated before, two different constructive solutions have been considered for the roof and for the façade wall, whereas the intermediate floor, ground floor, medium walls, and partition walls are the same for both cases. The details for their components and layers are shown in Appendix A, Figures A1–A6.

Table 5 includes the data for thermal transmittance (U-value) of the constructive elements detailed. Some of them have a fixed part (because they are invariable) and the others, a variable part, depending on the thickness and the insulation material, as shown in the Figures A7–A9, located in Appendix A.

Table 5. Thermal transmittance (U-value) of different constructive elements.

Elements	Thermal Transmittance (U in W/m^{2}K)
Roof - 1 *	0.77-0.11
Roof - 2 *	0.86-0.11
Intermediate floor	0.45
Ground floor *	0.90-0.11
Façade wall - 1 *	1.45-0.11
Façade wall - 2 *	1.51-0.11
Dry median wall	0.25
Wet median wall	0.27
Dry partition	0.50
Wet/Dry partition	0.51
Wet partition	0.52
Fenestration (windows and exterior doors): Frame: PVC 3 chambers Glass: Low-emissivity double glazing 4/20/4 mm	1.50

* Variable transmittance according to thickness and insulation material.

3. Results and Discussion

Sections 3.1–3.5 present the main results of CO_2 embodied emissions resulting from different insulation requirement needs according to different variable studied: climate zone, insulation material, orientation, constructive solution, and compactness, for the hypotheses H1, H2, and H3, respectively. Finally, a discussion is made in Section 3.6.

Appendix B includes all the results for calculations of different insulation requirement needs for each of the 4608 cases of study in order to reduce the amount of data and extract just the main results obtained from the study, making it more readable and understandable for the reader.

This way, Tables 6–10 show differences between CO_2 emissions in relation to the best possible value for each sequence, according to different variables, in a colour scale varying from blue to red. For each of the hypotheses considered two combinations of different variables have been taken into account: the set up that leads to the lowest CO_2 emissions possible, and the set up that leads to the higher CO_2 emissions possible, in order to analyse the results from both points of view.

Each sequence will be composed by different options, depending on the variable studied. For example, in the case of insulation materials, the options will be EC, SW, PUR, and XPS (as well as for the orientation will be the wind rose, for the compactness will be the four model studied and for the constructive solution will be the two referred in Appendix A). Besides, there will be as sequences as climate zones, set ups, and hypotheses.

For all of the tables, blue colour means situations where no insulation is needed (and consequently no CO_2 emissions derived from insulation is generated). On the other side, grey colour means situations where is not possible to realize this combination of variables due to constructive reasons (and, due to this, the calculation of CO_2 emissions is not applicable). Cells with no background colour indicate the reference value of CO_2 emissions for each sequence, and the rest of the cells will have a different colour, varying from green to red, depending on their difference with the reference value. In this way, the closer the colour of the cell is to light green, the lesser the difference regarding the minimum value of CO_2 emissions; on the other hand, the closer the colour of the cell to dark red, the higher the difference regarding the minimum value of CO_2 emissions.

3.1. Influence of the Climate Zone on CO_2 Emissions

The differences in the insulation needs depend first of all on the climate zone, as can be seen in Table A1a,b, Table A2a,b and Table A3a,b, in Appendix B. The results were shown to correspond to the minimum insulation thicknesses needed (in increments of 5 millimeters, from 0 to 200) to satisfy the energy demands defined in the hypotheses H1, H2, and H3, according to the rest of the variables considered. As the optimal insulation thickness needs are determined more by the needs of heating than for cooling, climate zones where winters are not severe (letters A and B), will need less insulation than climate zones where the winters are colder (letters C, D, and E).

While analyzing the results from the point of view of insulation thickness needs, we can observe that, for a given climate zone (this is the case of someone who wants to build a house in a determined place), XPS material results always in minor insulation material thicknesses than for the rest of materials considered, but in major insulation material emissions, as explained in the next section. These differences between insulation materials needs considerably increase with the degree of compactness, being the lesser compactness the higher differences among the insulation thickness needs. Nevertheless, although these needs also depend on the rest of variables (orientation and constructive solution), analyzing the results from the point of view of CO_2 emissions, the climatic zone is the main factor to be taken into account, as can be understood when analyzing Table 6, which shows that the emissions increased in cold areas, especially for Hypotheses 2 and 3.

In Appendix B, Table A4a–c, Tables A5a–c and A6a–c present the results of CO_2 emissions for Hypotheses H1, H2, and H3, respectively. Expression "n.a" meaning: "not applicable" refers to the situations where the minimum insulation thickness to satisfy energy demand is not possible due to constructive restrictions and, therefore, calculations of CO_2 emissions have no sense.

Table 6. Increase of emissions according to the climatic zone for the best and worst set ups.

Variable	Hypothesis 1 Best	Hypothesis 1 Worst	Hypothesis 2 Best	Hypothesis 2 Worst	Hypothesis 3 Best	Hypothesis 3 Worst
A3	0.00E+00	8.29E+03	0.00E+00	8.29E+03	0.00E+00	8.29E+03
A4	0.00E+00	+25.45%	0.00E+00	+25.45%	0.00E+00	+25.45%
B3	1.23E+03	+175.03%	1.23E+03	+175.03%	1.23E+03	+175.03%
B4	0.00%	+175.03%	0.00%	+175.03%	0.00%	+175.03%
C1	+100.00%	+224.49%	+100.00%	+224.49%	+365.85%	n.a.
C2	+33.33%	+224.49%	+33.33%	+324.61%	+365.85%	n.a.
C3	+33.33%	+200.36%	+133.33%	+475.39%	+332.52%	+725.09%
C4	+33.33%	+224.49%	+233.33%	+750.42%	+332.52%	+799.88%
D1	+100.00%	+275.15%	+332.52%	+900.00%	n.a.	n.a.
D2	+100.00%	+275.15%	+365.85%	n.a.	n.a.	n.a.
D3	+100.00%	+224.49%	+465.85%	n.a.	+1135.77%	n.a.
E1	+33.33%	+149.70%	+665.85%	n.a.	n.a.	n.a.

| 0 | n.a. | +0% | 1–25% | 26–50% | 51–75% | 76–100% | 101–125% | 126–150% | 151–175% | 176–200% | >200% |

3.2. Influence of the Insulation Material on CO_2 Emissions

If we analyze the results in terms of CO_2 emissions, we can observe how, although the recommendations for orientation, compactness, and constructive solution are the same (that is to say, always the combination of North orientation, constructive solution 1, and building Model 1 will result in lower CO_2 emissions; on the other side, the combination of West orientation, constructive solution 2, and building Model 4 will result in more CO_2 emissions, under equal conditions for the rest of variables), the recommendation for the insulation material changes.

The higher insulation thickness that is required to satisfy an energy demand fixed in the case of expanded cork (instead of the minimum thickness need from the extruded polystyrene), as observed in Table A1a,b, Table A2a,b and Table A3a,b, is compensated with its lower Global Warming Potential (GWP) factor, as a result, giving appreciably less CO_2 emissions. This difference increase with the needs of insulation material, so, in order to reduce CO_2 emissions during the construction phase, expanded cork is always preferable, if possible.

Table 7 shows the increase of CO_2 emissions according to the insulation material, for the different climate zones and hypotheses that were considered. The insulation material that generates lower emissions is always the expanded cork. The second one is the stone wool and the third, the polyurethane. The worst is always the extruded polystyrene. However, thanks to its lower thickness needs, it is the most applicable in the cases in which other materials cannot satisfy the demands that are required.

3.3. Influence of the Orientation on CO_2 Emissions

Regarding the orientation, Table A1a,b, Table A2a,b and Table A3a,b in Appendix B show that West orientation is always the most insulation demanding independent of the climate zone, the compactness, the constructive solution, and the insulation material, being the needs higher as long as the compactness of the building decreases. At the same time, the North orientation is also the least insulation demanding.

Table 8 shows the increase of CO_2 emissions according to the orientation, for the different climate zones and hypotheses considered. The orientation that generates lower emissions is always the North. The second one is the East and the third, the South. The worst is always the West orientation. It implies that the North orientation is the most applicable and the West is the orientation in which more cases are not possible. However, sometimes the North and East tie, as well as South and West, due to being included in the same step thickness.

Table 7. Increase of emissions according to the insulation material for the best and worst set ups.

CZ	Variable	Hypothesis 1		Hypothesis 2		Hypothesis 3	
		Best	Worst	Best	Worst	Best	Worst
A3	EC	0.00E+00	4.88E+03	0.00E+00	4.88E+03	0.00E+00	4.88E+03
	SW	0.00E+00	+51.23%	0.00E+00	+51.23%	0.00E+00	+51.23%
	PUR	0.00E+00	+79.92%	0.00E+00	+79.92%	0.00E+00	+79.92%
	XPS	0.00E+00	+113.11%	0.00E+00	+113.11%	0.00E+00	+113.11%
A4	EC	0.00E+00	3.25E+03	0.00E+00	3.25E+03	0.00E+00	3.25E+03
	SW	0.00E+00	+29.85%	0.00E+00	+29.85%	0.00E+00	+29.85%
	PUR	0.00E+00	+80.00%	0.00E+00	+80.00%	0.00E+00	+80.00%
	XPS	0.00E+00	+155.08%	0.00E+00	+155.08%	0.00E+00	+155.08%
B3	EC	1.23E+03	9.21E+03	1.23E+03	9.21E+03	1.23E+03	9.21E+03
	SW	+29.27%	+48.75%	+29.27%	+48.75%	+29.27%	+48.75%
	PUR	+79.67%	+91.10%	+79.67%	+91.10%	+79.67%	+91.10%
	XPS	+154.47%	+147.56%	+154.47%	+147.56%	+154.47%	+147.56%
B4	EC	1.23E+03	9.21E+03	1.23E+03	9.21E+03	1.23E+03	9.21E+03
	SW	+29.27%	+48.75%	+29.27%	+48.75%	+29.27%	+48.75%
	PUR	+79.67%	+91.10%	+79.67%	+91.10%	+79.67%	+91.10%
	XPS	+154.47%	+147.56%	+154.47%	+147.56%	+154.47%	+147.56%
C1	EC	2.46E+03	1.14E+04	2.46E+03	1.14E+04	5.73E+03	n.a.
	SW	+29.67%	+57.02%	+29.67%	+57.02%	+53.05%	n.a.
	PUR	+79.67%	+92.11%	+79.67%	+92.11%	+93.72%	n.a.
	XPS	+154.88%	+135.96%	+154.88%	+135.96%	+146.07%	n.a.
C2	EC	1.64E+03	1.14E+04	1.64E+03	1.46E+04	5.73E+03	n.a.
	SW	45.73%	+57.02%	+45.73%	+51.37%	+53.05%	n.a.
	PUR	102.44%	+92.11%	+102.44%	+90.41%	+93.72%	n.a.
	XPS	186.59%	+135.96%	+186.59%	+141.10%	+146.07%	n.a.
C3	EC	1.64E+03	1.03E+04	2.87E+03	2.01E+04	5.32E+03	n.a.
	SW	+45.73%	+43.69%	+66.55%	+52.24%	+49.81%	n.a.
	PUR	102.44%	+84.47%	+92.68%	+89.05%	+87.03%	5.41E+04
	XPS	186.59%	+141.75%	+118.47%	+137.31%	+134.96%	+26.43%
C4	EC	1.64E+03	1.08E+04	4.10E+03	n.a.	5.32E+03	n.a.
	SW	+45.73%	+56.48%	+55.61%	n.a.	+49.81%	n.a.
	PUR	+102.44%	+89.81%	+88.78%	5.56E+04	+87.03%	5.85E+04
	XPS	186.59%	+149.07%	+129.27%	+26.80%	+134.96%	+27.52%
D1	EC	2.46E+03	1.30E+04	5.32E+03	n.a.	n.a.	n.a.
	SW	+29.67%	+53.85%	+49.81%	n.a.	2.63E+04	n.a.
	PUR	+79.67%	+91.54%	+87.03%	n.a.	+26.24%	n.a.
	XPS	+154.88%	+139.23%	+134.96%	8.29E+04	+60.84%	n.a.
D2	EC	2.46E+03	1.30E+04	5.73E+03	n.a.	n.a.	n.a.
	SW	+29.67%	+53.85%	+53.05%	n.a.	2.55E+04	n.a.
	PUR	+79.67%	+91.54%	+93.72%	n.a.	+25.88%	n.a.
	XPS	+154.88%	+139.23%	+146.07%	n.a.	+59.61%	n.a.
D3	EC	2.46E+03	1.14E+04	6.96E+03	n.a.	1.52E+04	n.a.
	SW	+29.67%	+57.02%	+49.43%	n.a.	+51.97%	n.a.
	PUR	+79.67%	+92.11%	+91.09%	n.a.	+89.47%	n.a.
	XPS	+154.88%	+135.96%	+147.13%	n.a.	+136.84%	n.a.
E1	EC	1.64E+03	8.67E+03	9.42E+03	n.a.	n.a.	n.a.
	SW	+45.73%	+46.48%	+51.80%	n.a.	n.a.	n.a.
	PUR	+102.44%	+85.70%	+87.90%	n.a.	n.a.	n.a.
	XPS	+186.59%	+138.75%	+132.48%	n.a.	n.a.	n.a.

0 | n.a. | +0% | 1–25% | 26–50% | 51–75% | 76–100% | 101–125% | 126–150% | 151–175% | 176–200% | >200%

Best set up: N (Orientation), Model 1 (Compactness), S1 (Constructive Solution). Worst set up: W (Orientation), Model 4 (Compactness), S2 (Constructive Solution).

Table 8. Increase of emissions according to the orientation for the best and worst set ups.

CZ	Variable	Hypothesis 1 Best	Hypothesis 1 Worst	Hypothesis 2 Best	Hypothesis 2 Worst	Hypothesis 3 Best	Hypothesis 3 Worst
A3	N	0.00E+00	8.29E+03	0.00E+00	8.29E+03	0.00E+00	8.29E+03
	E	4.10E+02	0.00%	4.10E+02	0.00%	4.10E+02	0.00%
	S	0.00%	0.00%	0.00%	0.00%	0.00%	0.00%
	W	0.00%	25.45%	0.00%	25.45%	0.00%	25.45%
A4	N	0.00E+00	8.29E+03	0.00E+00	8.29E+03	0.00E+00	8.29E+03
	E	4.10E+02	0.00%	4.10E+02	0.00%	4.10E+02	0.00%
	S	0.00%	0.00%	0.00%	0.00%	0.00%	0.00%
	W	0.00%	0.00%	0.00%	0.00%	0.00%	0.00%
B3	N	1.23E+03	1.45E+04	1.23E+03	1.45E+04	1.23E+03	1.45E+04
	E	+33.33%	0.00%	+33.33%	0.00%	+33.33%	0.00%
	S	+33.33%	+14.48%	+33.33%	+14.48%	+33.33%	+14.48%
	W	+33.33%	+57.24%	+33.33%	+57.24%	+33.33%	+57.24%
B4	N	1.23E+03	1.45E+04	1.23E+03	1.45E+04	1.23E+03	1.45E+04
	E	+33.33%	0.00%	+33.33%	0.00%	+33.33%	0.00%
	S	+33.33%	+14.48%	+33.33%	+14.48%	+33.33%	+14.48%
	W	+33.33%	+57.24%	+33.33%	+57.24%	+33.33%	+57.24%
C1	N	2.46E+03	2.28E+04	2.46E+03	2.28E+04	5.73E+03	6.01E+04
	E	0.00%	0.00%	0.00%	0.00%	0.00%	0.00%
	S	0.00%	+9.21%	0.00%	+9.21%	+14.31%	+10.32%
	W	+16.67%	+17.98%	+16.67%	+17.98%	+21.47%	n.a.
C2	N	1.64E+03	2.07E+04	1.64E+03	2.69E+04	5.32E+03	4.56E+04
	E	+50.00%	0.00%	+50.00%	0.00%	+7.71%	0.00%
	S	+50.00%	+10.14%	+50.00%	+7.81%	+7.71%	+13.60%
	W	+50.00%	+29.95%	+50.00%	+30.86%	+23.12%	+50.00%
C3	N	1.64E+03	1.87E+04	2.87E+03	3.32E+04	5.32E+03	5.39E+04
	E	0.00%	0.00%	0.00%	+6.02%	+7.71%	0.00%
	S	0.00%	+10.70%	+28.57%	+12.35%	+7.71%	+11.50%
	W	+50.00%	+33.16%	+28.57%	+43.67%	+23.12%	+38.40%
C4	N	1.64E+03	2.07E+04	4.10E+03	4.97E+04	n.a.	n.a.
	E	0.00%	0.00%	+10.00%	+8.45%	n.a.	n.a.
	S	0.00%	+10.14%	+29.76%	+12.68%	n.a.	n.a.
	W	+50.00%	+29.95%	+39.76%	+41.85%	n.a.	n.a.
D1	N	2.46E+03	2.69E+04	5.32E+03	7.05E+04	n.a.	n.a.
	E	0.00%	0.00%	0.00%	0.00%	n.a.	n.a.
	S	+16.67%	0.00%	+7.71%	0.00%	n.a.	n.a.
	W	+16.67%	+15.61%	+7.71%	+17.59%	n.a.	n.a.
D2	N	2.46E+03	2.49E+04	5.73E+03	8.29E+04	n.a.	n.a.
	E	0.00%	+8.03%	+14.31%	n.a.	n.a.	n.a.
	S	+16.67%	+8.03%	+21.47%	n.a.	n.a.	n.a.
	W	+16.67%	+24.90%	+35.78%	n.a.	n.a.	n.a.
D3	N	2.46E+03	2.28E+04	6.96E+03	n.a.	1.52E+04	n.a.
	E	0.00%	0.00%	+11.78%	n.a.	+5.26%	n.a.
	S	0.00%	+9.21%	+17.67%	n.a.	+7.89%	n.a.
	W	0.00%	+17.98%	+23.56%	n.a.	n.a.	n.a.
E1	N	1.64E+03	1.87E+04	9.42E+03	n.a.	n.a.	n.a.
	E	0.00%	0.00%	+4.35%	n.a.	n.a.	n.a.
	S	0.00%	+10.70%	+12.53%	n.a.	n.a.	n.a.
	W	0.00%	+10.70%	+17.83%	n.a.	n.a.	n.a.

| 0 | n.a. | +0% | 1–25% | 26–50% | 51–75% | 76–100% | 101–125% | 126–150% | 151–175% | 176–200% | >200% |

Best set up: Model 1 (Compactness), Expanded Cork (EC) (Insulation Material), S1 (Constructive Solution). Worst set up: Model 4 (Compactness), Extruded Polystyrene (XPS) (Insulation Material), S2 (Constructive Solution).

3.4. Influence of the Constructive Solution on CO_2 Emissions

Constructive solution for the roof and façade wall also has an influence on the CO_2 emissions, always being preferable the constructive solution 1, under equal conditions of the rest of variables, since the needs of insulation are lower. It can be noted that the constructive solution 1, as can be checked in the Figures A1a and A4a, presents a more modern solution both for the façade and for the roof (ventilated faade and floating roof) than the traditional ones that are represented in the constructive solution 2 (as shown in Figures A1b and A4b). Table 9 shows the increase of CO_2 emissions, according to the constructive solution, for the different climate zones and hypotheses considered.

Table 9. Increase of emissions due to the constructive solution for the best and worst set ups.

CZ	Variable	Hypothesis 1 Best	Hypothesis 1 Worst	Hypothesis 2 Best	Hypothesis 2 Worst	Hypothesis 3 Best	Hypothesis 3 Worst
A3	S1	0.00E+00	8.29E+03	0.00E+00	8.29E+03	0.00E+00	8.29E+03
	S2	4.10E+02	+25.45%	4.10E+02	+25.45%	4.10E+02	+25.45%
A4	S1	0.00E+00	8.29E+03	0.00E+00	8.29E+03	0.00E+00	8.29E+03
	S2	4.10E+02	0.00%	4.10E+02	0.00%	4.10E+02	0.00%
B3	S1	1.23E+03	2.07E+04	1.23E+03	2.07E+04	1.23E+03	2.07E+04
	S2	+33.33%	+10.14%	+33.33%	+10.14%	+33.33%	+10.14%
B4	S1	1.23E+03	2.07E+04	1.23E+03	2.07E+04	1.23E+03	2.07E+04
	S2	+33.33%	+10.14%	+33.33%	+10.14%	+33.33%	+10.14%
C1	S1	2.46E+03	2.49E+04	2.46E+03	2.49E+04	5.73E+03	7.05E+04
	S2	+16.67%	+8.03%	+16.67%	+8.03%	+21.47%	0.00%
C2	S1	1.64E+03	2.69E+04	1.64E+03	3.32E+04	5.73E+03	7.05E+04
	S2	+75.00%	0.00%	+75.00%	+6.02%	+21.47%	0.00%
C3	S1	1.64E+03	2.07E+04	2.87E+03	4.35E+04	5.32E+03	5.60E+04
	S2	+50.00%	+20.29%	+28.57%	+9.66%	+7.71%	+22.14%
C4	S1	1.64E+03	2.28E+04	4.10E+03	6.01E+04	5.32E+03	6.01E+04
	S2	+50.00%	+17.98%	+29.76%	+17.30%	+23.12%	+24.13%
D1	S1	2.46E+03	2.69E+04	5.32E+03	6.63E+04	n.a.	n.a.
	S2	+16.67%	+15.61%	+7.71%	+25.04%	n.a.	n.a.
D2	S1	2.46E+03	2.69E+04	5.73E+03	n.a.	n.a.	n.a.
	S2	+16.67%	+15.61%	+14.31%	n.a.	n.a.	n.a.
D3	S1	2.46E+03	2.69E+04	6.96E+03	n.a.	1.52E+04	n.a.
	S2	+16.67%	0.00%	+17.67%	n.a.	+7.89%	n.a.
E1	S1	1.64E+03	2.07E+04	9.42E+03	n.a.	n.a.	n.a.
	S2	+50.00%	0.00%	+12.53%	n.a.	n.a.	n.a.

| 0 | n.a. | +0% | 1–25% | 26–50% | 51–75% | 76–100% | 101–125% | 126–150% | 151–175% | 176–200% | >200% |

Best set up: N (Orientation), Model 1 (Compactness), EC (Insulation Material). Worst set up: W (Orientation), Model 4 (Compactness), XPS (Insulation Material).

3.5. Influence of the Compactness on CO_2 Emissions

As observed in Table A2a,b and Table A3a,b in Appendix B, as the compactness of the building diminish, and, depending of the hypotheses considered, it could be possible that the maximum insulation thickness cannot be enough to satisfy the energy demand in those climate zones where the winter is extreme. The situation arrives to that point that, for the hypotheses 3 (Passivhauss Standard), it is not possible to satisfy energy demand in any of the 128 cases that were analyzed for the climate zone E1.

Table 10 shows the increase of CO_2 emissions according to the compactness, for the different climate zones and hypotheses considered. Model 1 generates, in all of the climate zones and for the three hypotheses considered, lower emissions than the other configuration models. This can be noted, since it is the reference base to calculate the differences with the rest of the models, except in those cases where it is not possible to build that configuration due to constructive reasons.

Table 10. Increase of emissions due to the compactness for the best and worst set ups.

CZ	Variable	Hypothesis 1 Best	Hypothesis 1 Worst	Hypothesis 2 Best	Hypothesis 2 Worst	Hypothesis 3 Best	Hypothesis 3 Worst
A3	Model 1	0.00E+00	3.13E+03	0.00E+00	3.13E+03	0.00E+00	3.13E+03
A3	Model 2	0.00E+00	+3.19%	0.00E+00	+3.19%	0.00E+00	+3.19%
A3	Model 3	0.00E+00	+151.12%	0.00E+00	+151.12%	0.00E+00	+151.12%
A3	Model 4	0.00E+00	+232.27%	0.00E+00	+232.27%	0.00E+00	+232.27%
A4	Model 1	0.00E+00	3.13E+03	0.00E+00	3.13E+03	0.00E+00	3.13E+03
A4	Model 2	0.00E+00	+3.19%	0.00E+00	+3.19%	0.00E+00	+3.19%
A4	Model 3	0.00E+00	+151.12%	0.00E+00	+151.12%	0.00E+00	+151.12%
A4	Model 4	0.00E+00	+164.86%	0.00E+00	+164.86%	0.00E+00	+164.86%
B3	Model 1	1.23E+03	6.27E+03	1.23E+03	6.27E+03	1.23E+03	6.27E+03
B3	Model 2	+37.40%	+3.19%	+37.40%	+3.19%	+37.40%	+3.19%
B3	Model 3	+275.61%	+182.30%	+275.61%	+182.30%	+275.61%	+182.30%
B3	Model 4	+340.65%	+263.64%	+340.65%	+263.64%	+340.65%	+263.64%
B4	Model 1	1.23E+03	9.21E+03	1.23E+03	9.21E+03	1.23E+03	9.21E+03
B4	Model 2	+37.40%	+3.19%	+37.40%	+3.19%	+37.40%	+3.19%
B4	Model 3	+275.61%	+182.30%	+275.61%	+182.30%	+275.61%	+182.30%
B4	Model 4	+340.65%	+263.64%	+340.65%	+263.64%	+340.65%	+263.64%
C1	Model 1	2.46E+03	7.83E+03	2.46E+03	7.83E+03	5.73E+03	2.04E+04
C1	Model 2	+20.33%	+3.32%	+20.33%	+3.32%	+25.48%	+10.78%
C1	Model 3	+171.54%	+201.40%	+171.54%	+201.40%	+196.68%	+188.73%
C1	Model 4	+252.44%	+243.55%	+252.44%	+243.55%	n.a.	n.a.
C2	Model 1	1.64E+03	6.27E+03	1.64E+03	6.27E+03	5.73E+03	2.04E+04
C2	Model 2	+54.88%	+29.03%	+80.49%	+54.70%	+25.48%	+2.94%
C2	Model 3	+307.32%	+244.50%	+338.41%	+306.70%	+196.68%	+188.73%
C2	Model 4	+362.80%	+329.03%	+528.05%	+461.40%	n.a.	n.a.
C3	Model 1	1.64E+03	6.27E+03	2.87E+03	1.10E+04	5.32E+03	+1.72E+04
C3	Model 2	+54.88%	+3.19%	+32.75%	+2.73%	+11.28%	+12.79%
C3	Model 3	+244.51%	+212.60%	+240.07%	+203.64%	+161.28%	+185.47%
C3	Model 4	+329.88%	+297.13%	+408.71%	+333.64%	+245.86%	+297.67%
C4	Model 1	1.64E+03	6.27E+03	4.10E+03	1.72E+04	5.32E+03	1.88E+04
C4	Model 2	+54.88%	+29.03%	+34.15%	+12.79%	+11.28%	+11.70%
C4	Model 3	+244.51%	+212.60%	+200.00%	+151.16%	+161.28%	+181.91%
C4	Model 4	+329.88%	+329.03%	+375.61%	+309.88%	+266.54%	+296.81%
D1	Model 1	2.46E+03	7.83E+03	5.32E+03	1.72E+04	n.a.	n.a.
D1	Model 2	+20.33%	+23.88%	+11.28%	+12.79%	n.a.	n.a.
D1	Model 3	+192.28%	+225.67%	+189.47%	+219.77%	n.a.	n.a.
D1	Model 4	+274.39%	+297.19%	n.a.	+381.98%	n.a.	n.a.
D2	Model 1	2.46E+03	7.83E+03	5.73E+03	2.04E+04	n.a.	n.a.
D2	Model 2	+20.33%	+23.88%	+25.48%	10.78%	n.a.	n.a.
D2	Model 3	+192.28%	+201.40%	+249.04%	n.a.	n.a.	n.a.
D2	Model 4	+274.39%	+297.19%	n.a.	n.a.	n.a.	n.a.
D3	Model 1	2.46E+03	6.27E+03	6.96E+03	2.66E+04	1.52E+04	5.17E+04
D3	Model 2	+3.25%	+29.03%	+21.55%	+9.40%	+11.18%	+6.38%
D3	Model 3	+171.54%	+244.50%	n.a.	n.a.	n.a.	n.a.
D3	Model 4	+252.44%	+329.03%	n.a.	n.a.	n.a.	n.a.
E1	Model 1	1.64E+03	6.27E+03	9.42E+03	3.13E+04	n.a.	n.a.
E1	Model 2	+54.88%	+3.19%	+16.77%	+8.63%	n.a.	n.a.
E1	Model 3	+244.51%	+150.40%	n.a.	n.a.	n.a.	n.a.
E1	Model 4	+329.88%	+230.14%	n.a.	n.a.	n.a.	n.a.

Legend: 0 | n.a. | +0% | 1–25% | 26–50% | 51–75% | 76–100% | 101–125% | 126–150% | 151–175% | 176–200% | >200%

Best set up: N (Orientation), EC (Insulation Material), S1 (Constructive Solution). Worst set up: W (Orientation), XPS (Insulation Material), S2 (Constructive Solution).

3.6. Discussion

It is useful to present an overview of buildings' thermal balance with respect to energy gains and losses, checking ventilation and infiltration, heat gains, and transmission through the envelope before discussing the results of insulation thicknesses and CO_2 emissions. Among the 4608 study cases, two from 4008 applicable cases are shown in Tables 11 and 12 as an example (600 of them are not possible due to constructive limitations in which insulation thicknesses are not enough), corresponding to the Hypothesis 2, from Madrid (D3) and Barcelona (C2).

Table 11. Thermal balance. Example 1: Hypothesis 2, Model 1, Constructive solution 1, Climate Zone D3, Orientation North. Thickness 85mm, Insulation Material Expanded Cork.

Elements	Heating *				Cooling *			
	Losses		Gains		Losses		Gains	
Faade	−13.04	20.88%	0.01	0.02%	−0.83	04.27%	1.55	5.72%
Fenestration (windows and doors)	−11.64	18.64%	1.48	3.69%	−1.14	05.86%	3.60	13.27%
Roof	−9.10	14.57%	0.01	0.02%	−0.92	04.73%	0.91	3.36%
Ground floor	−4.30	6.89%	0.08	0.20%	−2.95	15.17%	0.97	3.58%
Thermal bridges	−0.62	1.00%	0.01	0.02%	−0.26	01.34%	0.52	1.92%
Solar heat gains			13.47	33.57%			5.12	18.88%
Internal heat gains			25.07	62.46%			12.80	47.20%
Ventilation and infiltration	−23.74	38.02%	0.01	0.02%	−13.34	68.62%	1.65	6.08%
Sum	−62.44	100%	40.14	100%	−19.44	100%	27.12	100%
Total demand	−22.30						7.68	

* Units in kilowatts hour per square meter per year (kWh/m^2y).

Table 12. Thermal balance. Example 2: Hypothesis 2, Model 4, Constructive solution 2, Climate Zone C2, Orientation West. Thickness 80mm, Insulation Material XPS.

Elements	Heating *				Cooling *			
	Losses		Gains		Losses		Gains	
Faade	−15.37	22.63%	0.01	0.02%	−2.53	9.88%	1.06	3.31%
Fenestration (windows and doors)	−15.74	23.17%	0.17	0.38%	−1.58	6.17%	3.87	12.10%
Roof	−5.87	8.64%	0.01	0.02%	−0.73	2.85%	1.04	3.25%
Ground floor	−3.98	5.86%	0.13	0.29%	−2.42	9.45%	0.85	2.66%
Thermal bridges	−0.87	1.28%	0.02	0.02%	−0.37	1.45%	0.74	2.31%
Solar heat gains			18.79	42.33%			10.49	32.79%
Internal heat gains			25.24	56.86%			12.60	39.39%
Ventilation and infiltration	−26.09	38.41%	0.02	0.02%	−17.97	70.20%	1.34	4.19%
Sum	−67.92	100%	44.39	100%	−25.60	100%	31.99	100%
Total demand	−23.53						6.39	

* Units in kilowatts hour per square meter per year (kWh/m^2y).

In total, we have analyzed 4608 cases of study (1536 cases by hypothesis), corresponding to 12 climatic zones, four main orientations, four models of construction, two constructive solutions, four insulation materials, and three energy demand limitation hypothesis. The results show that just 4008 case studies could really run, from the constructive point of view, given that the 600 remaining cases would require thickness insulation that is incompatible with the constructive characteristics of the building. All of the 600 cases where it was not possible to meet energy demand requirements correspond to the hypothesis H2 (162), and especially to hypothesis H3 (438 cases). However, in many of those cases it would be enough with a small adjustment that allowed a few extra millimeters of insulation in certain cases, in order to achieve compliance with the requirements.

Table 6 has shown the variability of the emissions that were generated to satisfy a specific heating and cooling demand (hypotheses H1–H3), according to the climate zone in which the building is located. For the H1 scenario, these emissions are doubled in the best scenario and tripled in the worst scenario. However, for hypotheses H2 and H3, the differences increase a lot (almost multiplied by ten times). Subsequently, Tables 7–10 show the contribution of the other factors, once a location is fixed. The compactness and the insulation material also have a major influence on the amount of emissions generated. Next, orientation and the constructive solution for the envelope exert a minor but significance influence.

4. Conclusions

In general, it is concluded that the optimal insulation thickness are determined more by the needs of heating than for cooling, even in the most severe summer climates needs. On the demand for energy, in the case of H1, values established by CTE result in similar thicknesses independently of the climate zone, and therefore the costs due to insulation during the construction phase are similar. Nevertheless, this will increase the costs of energy during the use phase of the building, punishing the inhabitants of cold spots due to its higher energy demand for heating. On the contrary, while considering the H3, the users of temperate zones are penalized, given that energy demand for cooling in cold areas is very low. Here follows that the intermediate hypothesis, H2, which tries to balance the joint demand during the phase of use of the building, may be the most optimal when regular energy demand limitations, given that these, and therefore, consumption (and their associated costs), they are similar, both in temperate and in cold-zones. For this case, it would be interesting to determine the satisfaction of the energy demand exclusively with renewable energies.

With regard to CO_2 emissions, and analyzing the results according to the compactness of the building primarily, it is observed that the model 1, regardless of the climatic zone, the orientation and the scenario, always generates less emissions than the rest of the models, for all cases. In terms of the influence of the orientation, regardless of the climatic zone, compactness of the building, constructive solution, and scenario, the orientation W is always that generates a greater number of emissions. These differences can reach up to 57% for the same climatic zone. This can be taken into account by the designers and builders in order to minimize the emissions from the stages of design and construction of the buildings due to the insulation of the envelope. Additionally, the material has influence on the amount of CO_2 emissions, since, as stated before, using expanded cork instead of XPS can reduce the total amount of CO_2 emissions during the construction phase of the building, although the needs for this material are higher, due to its lower GWP factor.

It must be recalled that increased consumption means, not only an increase in CO_2 emissions during the phase of use of the building, but also an increase of the costs for the users of the same, due to the increase in their electric bills. From this point of view, other future research can be done in order to incorporate a cost analysis to determine the influence of the different variables that are considered into the final cost of the electricity, with the aim of minimizing it. It will be also interesting to analyze, from an eco-efficiency point of view, the costs of fabrication, installation, and maintenance for different materials, which will be material for further research. Other research include the extension of the scope in order to include lighting requirements, and the inclusion of active measures, such as the use of photovoltaic and/or solar thermal energy.

Author Contributions: Conceptualization, M.J.B.-C. and A.C.-N.; Methodology, M.J.B.-C. and A.C.-N.; Data Curation, A.C.-N. and J.-M.P.-V.; Formal analysis, A.C.-N. and J.-M.P.-V.; Writing—original draft preparation, M.J.B.-C. and A.C.-N.; Writing—review and editing, M.J.B.-C., A.C.-N. and A.P.-F.

Funding: This research received no external funding.

Acknowledgments: The authors would like to thank to the "Promotion and Support of the Research Activity Program" of University of Cádiz by their support during this research.

Conflicts of Interest: The authors declare no conflict of interest.

Appendix A

Appendix A includes the constructive description of the different solutions, both variable and permanent, for roofs, ground and intermediate floors, faades, median walls and partitions. The end of the appendix present the thermal transmittance of the variable elements.

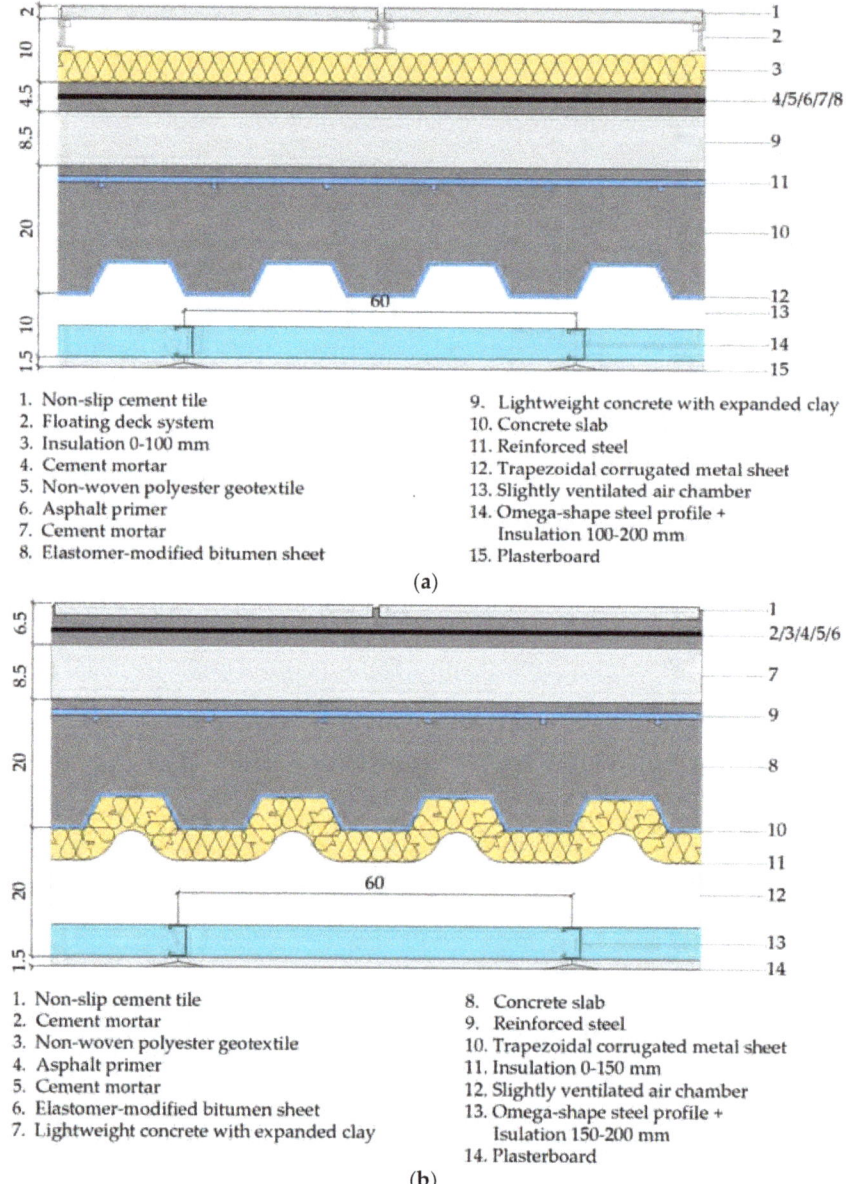

1. Non-slip cement tile
2. Floating deck system
3. Insulation 0-100 mm
4. Cement mortar
5. Non-woven polyester geotextile
6. Asphalt primer
7. Cement mortar
8. Elastomer-modified bitumen sheet
9. Lightweight concrete with expanded clay
10. Concrete slab
11. Reinforced steel
12. Trapezoidal corrugated metal sheet
13. Slightly ventilated air chamber
14. Omega-shape steel profile + Insulation 100-200 mm
15. Plasterboard

(a)

1. Non-slip cement tile
2. Cement mortar
3. Non-woven polyester geotextile
4. Asphalt primer
5. Cement mortar
6. Elastomer-modified bitumen sheet
7. Lightweight concrete with expanded clay
8. Concrete slab
9. Reinforced steel
10. Trapezoidal corrugated metal sheet
11. Insulation 0-150 mm
12. Slightly ventilated air chamber
13. Omega-shape steel profile + Isulation 150-200 mm
14. Plasterboard

(b)

Figure A1. (**a**) Roof floor components detail for constructive solution 1. (**b**) Roof floor components detail for constructive solution 2.

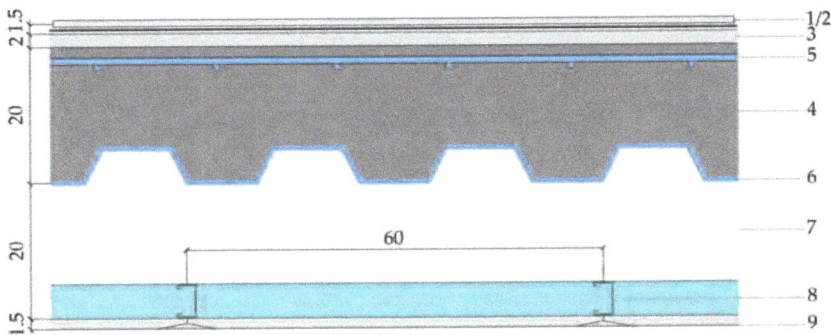

1. Floating laminate floor
2. Foam underlayment
3. Self-leveling mortar
4. Concrete slab
5. Reinforced steel
6. Trapezoidal corrugated metal sheet
7. Slightly ventilated air chamber
8. Omega-shape steel profile + Glass Wool (acoustic insulation)
9. Laminated plasterboard

Figure A2. Intermediate floor components detail.

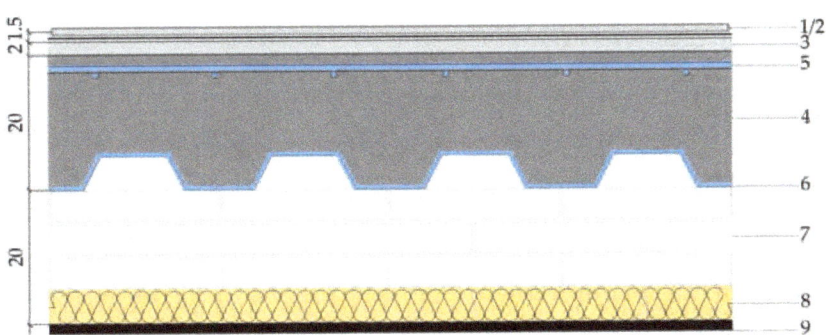

1. Floating laminate floor
2. Foam underlayment
3. Self-leveling mortar
4. Concrete slab
5. Reinforced steel
6. Trapezoidal corrugated metal sheet
7. Slightly ventilated air chamber + Perimeter wall
8. Insulation 0-200mm
9. Insitu foundation

Figure A3. Ground floor components detail.

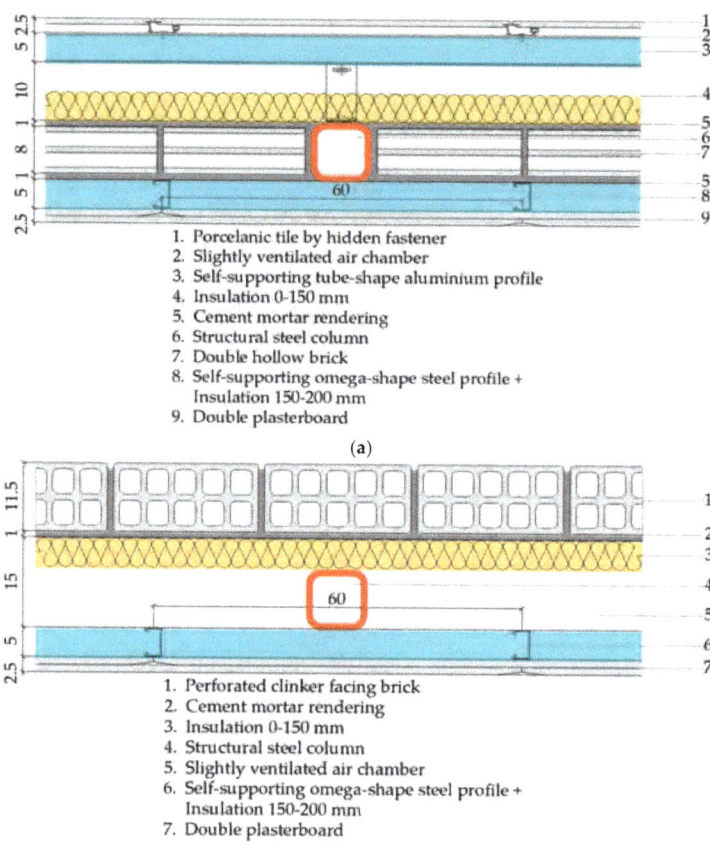

Figure A4. (**a**) Façade wall components detail for constructive solution 1. (**b**) Façade wall components detail for constructive solution 2.

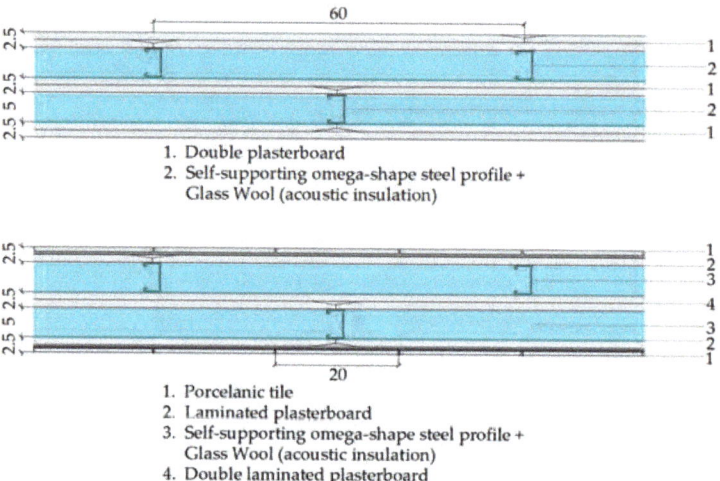

Figure A5. Median walls components detail.

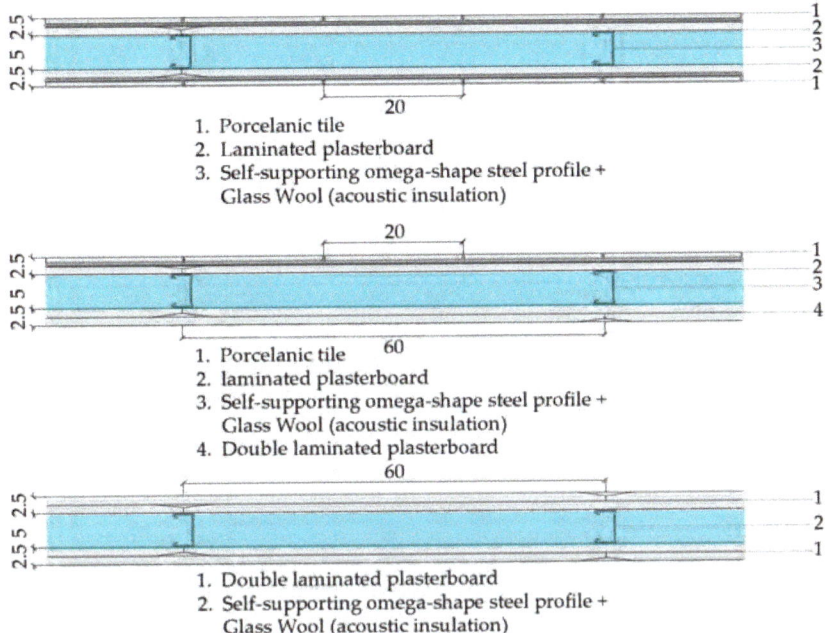

Figure A6. Partition walls component details.

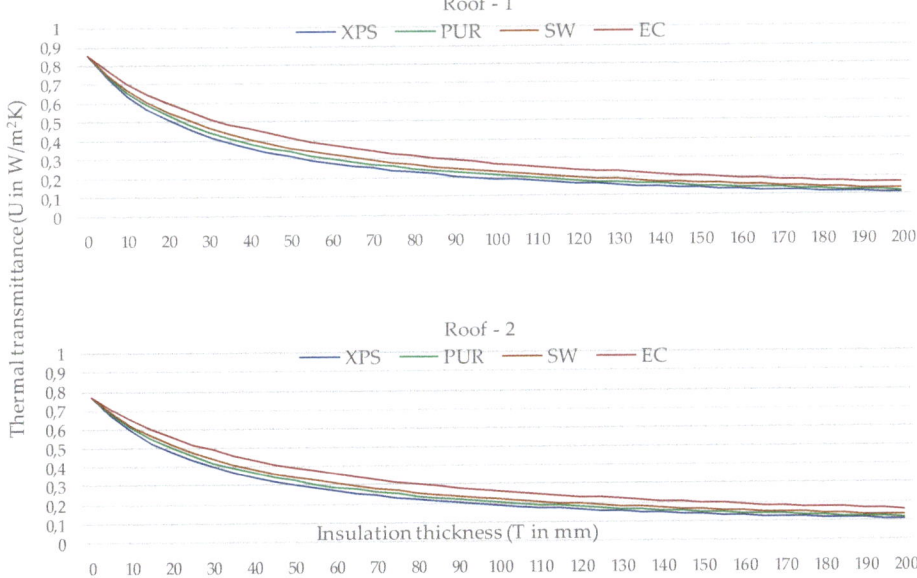

Figure A7. Thermal transmittance (U-value) according to insulation thickness, for roofs.

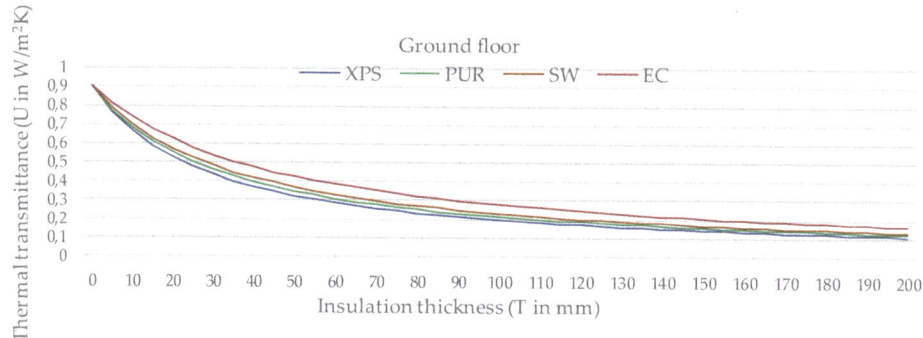

Figure A8. Thermal transmittance (U-value) according to insulation thickness, for ground floor.

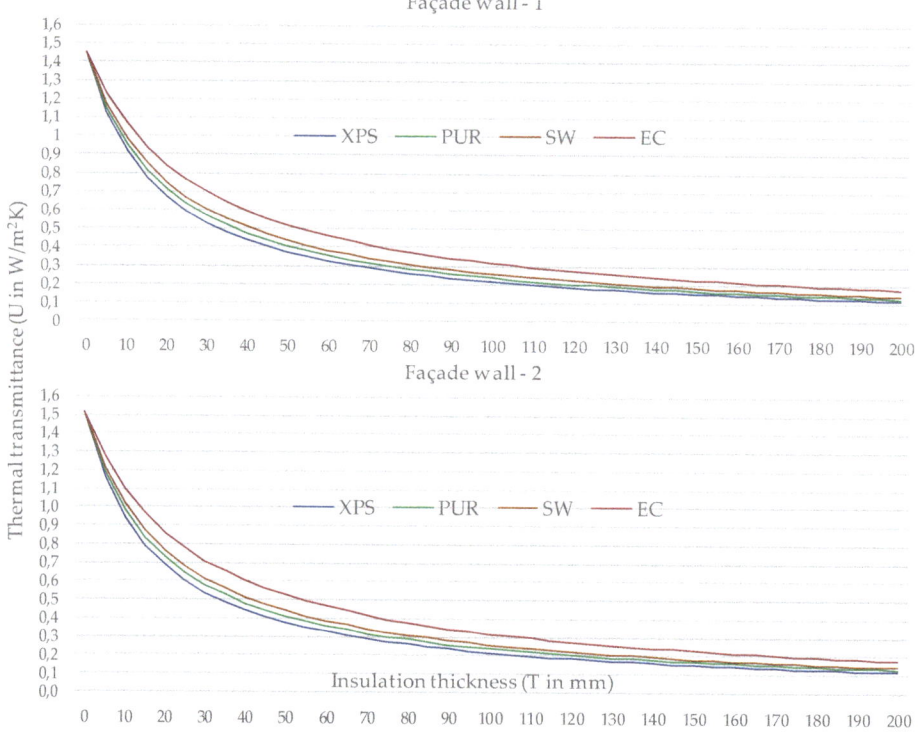

Figure A9. Thermal transmittance (U-value) according to insulation thickness, for façade walls.

Appendix B

Appendix B includes all the results from the 4608 cases studied, both for insulation thickness requirements and for embodied CO_2 emissions.

Table A1. Insulation thicknesses (in mm), for Hypothesis 1: Legal Minimum Compliance.

(a)

CZ	O	S	Model 1				Model 2				Model 3				Model 4			
			XPS	PUR	SW	EC	XPS	PUR	SW	EC	XPS	PUR	SW	EC	XPS	PUR	SW	EC
A3	N	S1	0	0	0	0	0	0	0	0	10	10	10	15	10	10	10	15
		S2	5	5	5	5	10	10	10	15	15	15	15	20	20	20	20	30
	E	S1	5	5	5	5	5	5	5	5	10	10	10	15	10	10	10	15
		S2	5	5	5	5	10	10	10	15	15	15	15	20	20	20	20	30
	S	S1	5	5	5	5	5	5	5	5	15	15	15	20	10	10	10	15
		S2	5	5	5	5	10	10	10	15	20	20	20	30	20	20	20	30
	W	S1	5	5	5	5	5	5	5	5	15	15	15	20	20	25	30	35
		S2	10	10	10	15	10	10	10	15	20	25	30	35	20	20	20	30
A4	N	S1	0	0	0	0	0	0	0	0	10	10	10	15	10	10	10	15
		S2	5	5	5	5	10	10	10	15	15	15	15	20	20	20	20	30
	E	S1	5	5	5	5	5	5	5	5	10	10	10	15	10	10	10	15
		S2	5	5	5	5	10	10	10	15	15	15	15	20	20	20	20	30
	S	S1	5	5	5	5	5	5	5	5	15	15	15	20	10	10	10	15
		S2	5	5	5	5	10	10	10	15	20	20	20	30	20	20	20	30
	W	S1	5	5	5	5	5	5	5	5	15	15	15	20	20	25	30	35
		S2	10	10	10	15	10	10	10	15	20	25	30	35	25	30	35	45
B3	N	S1	10	10	10	15	15	15	15	20	25	30	35	45	30	35	40	50
		S2	15	15	15	20	20	20	20	30	30	35	40	50	35	40	45	55
	E	S1	15	15	15	20	15	15	15	20	30	35	40	50	30	35	40	50
		S2	20	20	20	30	20	25	30	35	30	35	40	50	35	40	45	55
	S	S1	15	15	15	20	15	15	15	20	35	40	45	55	35	40	45	55
		S2	20	20	20	30	20	25	30	35	40	45	50	65	40	45	50	65
	W	S1	15	15	15	20	15	15	15	20	40	45	50	65	50	55	60	80
		S2	20	20	20	30	20	25	30	35	45	50	55	70	55	60	65	85
B4	N	S1	10	10	10	15	15	15	15	20	25	30	35	45	30	35	40	50
		S2	15	15	15	20	20	20	20	30	30	35	40	50	35	40	45	55
	E	S1	15	15	15	20	15	15	15	20	30	35	40	50	30	35	40	50
		S2	20	20	20	30	20	25	30	35	30	35	40	50	40	45	50	65
	S	S1	15	15	15	20	15	15	15	20	35	40	45	55	35	40	45	55
		S2	20	20	20	30	20	25	30	35	35	40	45	55	40	45	50	65
	W	S1	15	15	15	20	15	15	15	20	40	45	50	65	50	55	60	80
		S2	20	20	20	30	20	25	30	35	45	50	55	70	55	60	65	85
C1	N	S1	20	20	20	30	20	25	30	35	40	45	50	65	50	55	60	80
		S2	20	25	30	35	20	25	30	35	45	50	55	70	55	60	65	85
	E	S1	20	20	20	30	20	25	30	35	40	45	50	65	50	55	60	80
		S2	20	25	30	35	25	30	35	45	45	50	55	70	55	60	65	85
	S	S1	20	20	20	30	20	25	30	35	50	55	60	80	55	60	65	85
		S2	20	25	30	35	25	30	35	45	55	60	65	85	60	65	70	95
	W	S1	20	25	30	35	20	25	30	35	55	60	65	85	60	65	70	95
		S2	25	30	35	45	25	30	35	45	60	65	70	95	65	75	85	105
C2	N	S1	15	15	15	20	20	20	20	30	40	45	50	65	45	50	55	70
		S2	20	25	30	35	20	25	30	35	45	50	55	70	50	55	60	80
	E	S1	20	20	20	30	20	25	30	35	40	45	50	65	45	50	55	70
		S2	20	25	30	35	25	30	35	45	45	50	55	70	50	55	60	80
	S	S1	20	20	20	30	20	25	30	35	45	50	55	70	50	55	60	80
		S2	20	25	30	35	25	30	35	45	50	55	60	80	55	60	65	85
	W	S1	20	20	20	30	20	25	30	35	50	55	60	80	65	70	80	100
		S2	20	25	30	35	25	30	35	45	55	60	65	85	65	75	85	105

Table A1. Cont.

(b)

CZ	O	S	Model 1				Model 2				Model 3				Model 4			
			XPS	PUR	SW	EC	XPS	PUR	SW	EC	XPS	PUR	SW	EC	XPS	PUR	SW	EC
C3	N	S1	15	15	15	20	20	20	20	30	35	40	45	55	40	45	50	65
		S2	20	20	20	30	20	25	30	35	40	45	50	65	45	50	55	70
	E	S1	15	15	15	20	20	20	20	30	35	40	45	55	40	45	50	65
		S2	20	20	20	30	20	25	30	35	40	45	50	65	45	50	55	70
	S	S1	15	15	15	20	20	20	20	30	40	45	50	65	45	50	55	70
		S2	20	20	20	30	20	25	30	35	45	50	55	70	50	55	60	80
	W	S1	20	20	20	30	20	25	30	35	45	50	55	70	50	55	60	80
		S2	20	25	30	35	20	25	30	35	50	55	60	80	60	65	70	95
C4	N	S1	15	15	15	20	20	20	20	30	35	40	45	55	40	45	50	65
		S2	20	20	20	30	20	25	30	35	40	45	50	65	50	55	60	80
	E	S1	15	15	15	20	20	20	20	30	35	40	45	55	40	45	50	65
		S2	20	20	20	30	20	25	30	35	40	45	50	65	50	55	60	80
	S	S1	15	15	15	20	20	20	20	30	40	45	50	65	45	50	55	70
		S2	20	25	30	35	20	25	30	35	45	50	55	70	55	60	65	85
	W	S1	20	20	20	30	20	25	30	35	45	50	55	70	55	60	65	85
		S2	20	25	30	35	25	30	35	45	50	55	60	80	65	70	80	100
D1	N	S1	20	20	20	30	20	25	30	35	45	50	55	70	55	60	65	85
		S2	20	25	30	35	25	30	35	45	50	55	60	80	65	70	80	100
	E	S1	20	20	20	30	25	30	35	45	45	50	55	70	55	60	65	85
		S2	20	25	30	35	25	30	35	45	50	55	60	80	65	70	80	100
	S	S1	20	25	30	35	25	30	35	45	55	60	65	85	60	65	70	95
		S2	25	30	35	45	30	35	40	50	60	65	70	95	65	75	85	105
	W	S1	20	25	30	35	25	30	35	45	60	65	70	95	65	75	85	105
		S2	25	30	35	45	30	35	40	50	65	70	80	100	75	85	95	120
D2	N	S1	20	20	20	30	20	25	30	35	45	50	55	70	55	60	65	85
		S2	20	25	30	35	25	30	35	45	50	55	60	80	65	70	80	100
	E	S1	20	20	20	30	20	25	30	35	45	50	55	70	50	55	60	80
		S2	20	25	30	35	25	30	35	45	50	55	60	80	60	65	70	95
	S	S1	20	25	30	35	25	30	35	45	50	55	60	80	55	60	65	85
		S2	20	25	30	35	30	35	40	50	55	60	65	85	65	70	80	100
	W	S1	20	25	30	35	25	30	35	45	55	60	65	85	65	75	85	105
		S2	25	30	35	45	30	35	40	50	60	65	70	95	75	85	95	120
D3	N	S1	20	20	20	30	20	20	20	30	40	45	50	65	50	55	60	80
		S2	20	25	30	35	20	25	30	35	45	50	55	70	55	60	65	85
	E	S1	20	20	20	30	20	25	30	35	40	45	50	65	45	50	55	70
		S2	20	25	30	35	25	30	35	45	45	50	55	70	55	60	65	85
	S	S1	20	20	20	30	20	25	30	35	45	50	55	70	50	55	60	80
		S2	20	25	30	35	25	30	35	45	50	55	60	80	60	65	70	95
	W	S1	20	20	20	30	20	25	30	35	50	55	60	80	65	70	80	100
		S2	20	25	30	35	25	30	35	45	55	60	65	85	65	75	85	105
E1	N	S1	15	15	15	20	20	20	20	30	35	40	45	55	40	45	50	65
		S2	20	20	20	30	20	20	20	30	35	40	45	55	45	50	55	70
	E	S1	15	15	15	20	20	20	20	30	35	40	45	55	40	45	50	65
		S2	20	20	20	30	20	25	30	35	40	45	50	65	45	50	55	70
	S	S1	15	15	15	20	20	20	20	30	40	45	50	65	45	50	55	70
		S2	20	20	20	30	20	25	30	35	40	45	50	65	50	55	60	80
	W	S1	15	15	15	20	20	20	20	30	40	45	50	65	50	55	60	80
		S2	20	20	20	30	20	25	30	35	40	45	50	65	50	55	60	80

CZ (Climate Zone); O (Orientation); S (Constructive Solution).

Table A2. Insulation thicknesses (in mm), for Hypothesis 2: Joint (heating + cooling) demand ≤ 30.

(a)

CZ	O	S	Model 1				Model 2				Model 3				Model 4			
			XPS	PUR	SW	EC	XPS	PUR	SW	EC	XPS	PUR	SW	EC	XPS	PUR	SW	EC
A3	N	S1	0	0	0	0	0	0	0	0	10	10	10	15	10	10	10	15
		S2	5	5	5	5	10	10	10	15	15	15	15	20	20	20	20	30
	E	S1	5	5	5	5	5	5	5	5	10	10	10	15	10	10	10	15
		S2	5	5	5	5	10	10	10	15	15	15	15	20	20	20	20	30
	S	S1	5	5	5	5	5	5	5	5	15	15	15	20	10	10	10	15
		S2	5	5	5	5	10	10	10	15	20	20	20	30	20	20	20	30
	W	S1	5	5	5	5	5	5	5	5	15	15	15	20	20	25	30	35
		S2	10	10	10	15	10	10	10	15	20	25	30	35	25	30	35	45
A4	N	S1	0	0	0	0	0	0	0	0	10	10	10	15	10	10	10	15
		S2	5	5	5	5	10	10	10	15	15	15	15	20	20	20	20	30
	E	S1	5	5	5	5	5	5	5	5	10	10	10	15	10	10	10	15
		S2	5	5	5	5	10	10	10	15	15	15	15	20	20	20	20	30
	S	S1	5	5	5	5	5	5	5	5	15	15	15	20	10	10	10	15
		S2	5	5	5	5	10	10	10	15	20	20	20	30	20	20	20	30
	W	S1	5	5	5	5	5	5	5	5	15	15	15	20	20	25	30	35
		S2	10	10	10	15	10	10	10	15	20	25	30	35	20	20	20	30
B3	N	S1	10	10	10	15	15	15	15	20	25	30	35	45	30	35	40	50
		S2	15	15	15	20	20	20	20	30	30	35	40	50	35	40	45	55
	E	S1	15	15	15	20	15	15	15	20	30	35	40	50	30	35	40	50
		S2	20	20	20	30	20	25	30	35	30	35	40	50	35	40	45	55
	S	S1	15	15	15	20	15	15	15	20	35	40	45	55	35	40	45	55
		S2	20	20	20	30	20	25	30	35	40	45	50	65	40	45	50	65
	W	S1	15	15	15	20	15	15	15	20	40	45	50	65	50	55	60	80
		S2	20	20	20	30	20	25	30	35	45	50	55	70	55	60	65	85
B4	N	S1	10	10	10	15	15	15	15	20	25	30	35	45	30	35	40	50
		S2	15	15	15	20	20	20	20	30	30	35	40	50	35	40	45	55
	E	S1	15	15	15	20	15	15	15	20	30	35	40	50	30	35	40	50
		S2	20	20	20	30	20	25	30	35	30	35	40	50	40	45	50	65
	S	S1	15	15	15	20	15	15	15	20	35	40	45	55	35	40	45	55
		S2	20	20	20	30	20	25	30	35	35	40	45	55	40	45	50	65
	W	S1	15	15	15	20	15	15	15	20	40	45	50	65	50	55	60	80
		S2	20	20	20	30	20	25	30	35	45	50	55	70	55	60	65	85
C1	N	S1	20	20	20	30	20	25	30	35	40	45	50	65	50	55	60	80
		S2	20	25	30	35	20	25	30	35	45	50	55	70	55	60	65	85
	E	S1	20	20	20	30	20	25	30	35	40	45	50	65	50	55	60	80
		S2	20	25	30	35	25	30	35	45	45	50	55	70	55	60	65	85
	S	S1	20	20	20	30	20	25	30	35	50	55	60	80	55	60	65	85
		S2	20	25	30	35	25	30	35	45	55	60	65	85	60	65	70	95
	W	S1	20	25	30	35	20	25	30	35	55	60	65	85	60	65	70	95
		S2	25	30	35	45	25	30	35	45	60	65	70	95	65	75	85	105
C2	N	S1	15	15	15	20	20	25	30	35	45	50	55	70	60	65	70	95
		S2	20	25	30	35	25	30	35	45	50	55	60	80	65	70	80	100
	E	S1	20	20	20	30	20	25	30	35	55	60	65	85	65	70	80	100
		S2	20	25	30	35	25	30	35	45	60	65	70	95	65	75	85	105
	S	S1	20	20	20	30	20	25	30	35	55	60	65	85	65	70	80	100
		S2	20	25	30	35	25	30	35	45	60	65	70	95	70	80	90	115
	W	S1	20	20	20	30	20	25	30	35	60	65	70	95	80	90	100	130
		S2	20	25	30	35	30	35	40	50	65	70	80	100	85	95	105	135

Table A2. Cont.

(b)

CZ	O	S	Model 1				Model 2				Model 3				Model 4			
			XPS	PUR	SW	EC	XPS	PUR	SW	EC	XPS	PUR	SW	EC	XPS	PUR	SW	EC
C3	N	S1	20	25	30	35	25	30	35	45	60	65	70	95	85	95	105	135
		S2	25	30	35	45	35	40	45	55	65	70	80	100	80	90	100	130
	E	S1	20	25	30	35	25	30	35	45	65	75	85	105	85	95	105	135
		S2	25	30	35	45	35	40	45	55	70	80	90	115	85	95	105	135
	S	S1	25	30	35	45	30	35	40	50	65	75	85	105	90	100	110	145
		S2	35	40	45	55	35	40	45	55	70	80	90	115	90	100	110	145
	W	S1	25	30	35	45	30	35	40	50	80	90	100	130	105	120	135	170
		S2	35	40	45	55	35	40	45	55	85	95	105	135	115	130	145	185
C4	N	S1	30	35	40	50	40	45	50	65	75	85	95	120	110	125	140	180
		S2	40	45	50	65	45	50	55	70	80	90	100	130	120	135	150	195
	E	S1	35	40	45	55	40	45	50	65	90	100	110	145	110	125	140	180
		S2	40	45	50	65	50	55	60	80	95	105	115	150	130	145	160	-
	S	S1	40	45	50	65	45	50	55	70	95	105	115	150	130	145	160	-
		S2	55	60	65	85	55	60	65	85	100	110	120	155	135	150	165	-
	W	S1	45	50	55	70	45	50	55	70	105	120	135	170	145	160	175	-
		S2	55	60	65	85	60	65	70	95	110	125	140	180	170	190	-	-
D1	N	S1	40	45	50	65	45	50	55	70	95	105	115	150	130	145	160	-
		S2	45	50	55	70	50	55	60	80	100	110	120	155	170	190	-	-
	E	S1	40	45	50	65	50	55	60	80	105	120	135	170	135	150	165	-
		S2	45	50	55	70	55	60	65	85	110	125	140	180	170	190	-	-
	S	S1	45	50	55	70	50	55	60	80	120	135	150	195	145	160	175	-
		S2	50	55	60	80	55	60	65	85	125	140	155	200	170	190	-	-
	W	S1	45	50	55	70	55	60	65	85	135	150	165	-	160	180	200	-
		S2	55	60	65	85	60	65	70	95	140	155	170	-	200	-	-	-
D2	N	S1	45	50	55	70	55	60	65	85	120	135	150	195	145	160	175	-
		S2	50	55	60	80	60	65	70	95	125	140	155	200	200	-	-	-
	E	S1	50	55	60	80	60	65	70	95	140	155	170	-	150	170	190	-
		S2	50	55	60	80	65	75	85	105	150	170	190	-	-	-	-	-
	S	S1	55	60	65	85	60	65	70	95	145	160	175	-	160	180	200	-
		S2	65	70	80	100	65	75	85	105	150	170	190	-	-	-	-	-
	W	S1	60	65	70	95	65	70	80	100	145	160	175	-	-	-	-	-
		S2	65	70	80	100	70	80	90	115	-	-	-	-	-	-	-	-
D3	N	S1	55	60	65	85	65	70	80	100	145	160	175	-	-	-	-	-
		S2	65	70	80	100	70	80	90	115	160	180	-	-	-	-	-	-
	E	S1	60	65	70	95	70	80	90	115	150	170	190	-	-	-	-	-
		S2	65	70	80	100	75	85	95	120	-	-	-	-	-	-	-	-
	S	S1	65	70	80	100	70	80	90	115	165	185	-	-	-	-	-	-
		S2	80	90	100	130	80	90	100	130	-	-	-	-	-	-	-	-
	W	S1	65	75	85	105	80	90	100	130	-	-	-	-	-	-	-	-
		S2	85	95	105	135	90	100	110	145	-	-	-	-	-	-	-	-
E1	N	S1	70	80	90	115	80	90	100	130	145	165	185	-	-	-	-	-
		S2	80	90	100	130	90	100	110	145	200	-	-	-	-	-	-	-
	E	S1	75	85	95	120	90	100	110	145	170	190	-	-	-	-	-	-
		S2	80	90	100	130	100	110	120	155	-	-	-	-	-	-	-	-
	S	S1	80	90	100	130	90	100	110	145	200	-	-	-	-	-	-	-
		S2	90	100	110	145	105	115	125	165	-	-	-	-	-	-	-	-
	W	S1	85	95	105	135	100	110	120	155	200	-	-	-	-	-	-	-
		S2	100	110	120	155	105	120	135	170	-	-	-	-	-	-	-	-

CZ (Climate Zone); O (Orientation); S (Constructive Solution); - Thickness not enough to satisfy demand.

Table A3. Insulation thicknesses (in mm), for Hypothesis 3: Both heating and cooling demand ≤ 15.

(a)

CZ	O	S	Model 1				Model 2				Model 3				Model 4			
			XPS	PUR	SW	EC	XPS	PUR	SW	EC	XPS	PUR	SW	EC	XPS	PUR	SW	EC
A3	N	S1	0	0	0	0	0	0	0	0	10	10	10	15	10	10	10	15
		S2	5	5	5	5	10	10	10	15	15	15	15	20	20	20	20	30
	E	S1	5	5	5	5	5	5	5	5	10	10	10	15	10	10	10	15
		S2	5	5	5	5	10	10	10	15	15	15	15	20	20	20	20	30
	S	S1	5	5	5	5	5	5	5	5	15	15	15	20	10	10	10	15
		S2	5	5	5	5	10	10	10	15	20	20	20	30	20	20	20	30
	W	S1	5	5	5	5	5	5	5	5	15	15	15	20	20	25	30	35
		S2	10	10	10	15	10	10	10	15	20	25	30	35	25	30	35	45
A4	N	S1	0	0	0	0	0	0	0	0	10	10	10	15	10	10	10	15
		S2	5	5	5	5	10	10	10	15	15	15	15	20	20	20	20	30
	E	S1	5	5	5	5	5	5	5	5	10	10	10	15	10	10	10	15
		S2	5	5	5	5	10	10	10	15	15	15	15	20	20	20	20	30
	S	S1	5	5	5	5	5	5	5	5	15	15	15	20	10	10	10	15
		S2	5	5	5	5	10	10	10	15	20	20	20	30	20	20	20	30
	W	S1	5	5	5	5	5	5	5	5	15	15	15	20	20	25	30	35
		S2	10	10	10	15	10	10	10	15	20	25	30	35	20	20	20	30
B3	N	S1	10	10	10	15	15	15	15	20	25	30	35	45	30	35	40	50
		S2	15	15	15	20	20	20	20	30	30	35	40	50	35	40	45	55
	E	S1	15	15	15	20	15	15	15	20	30	35	40	50	30	35	40	50
		S2	20	20	20	30	20	25	30	35	30	35	40	50	35	40	45	55
	S	S1	15	15	15	20	15	15	15	20	35	40	45	55	35	40	45	55
		S2	20	20	20	30	20	25	30	35	40	45	50	65	40	45	50	65
	W	S1	15	15	15	20	15	15	15	20	40	45	50	65	50	55	60	80
		S2	20	20	20	30	20	25	30	35	45	50	55	70	55	60	65	85
B4	N	S1	10	10	10	15	15	15	15	20	25	30	35	45	30	35	40	50
		S2	15	15	15	20	20	20	20	30	30	35	40	50	35	40	45	55
	E	S1	15	15	15	20	15	15	15	20	30	35	40	50	30	35	40	50
		S2	20	20	20	30	20	25	30	35	30	35	40	50	40	45	50	65
	S	S1	15	15	15	20	15	15	15	20	35	40	45	55	35	40	45	55
		S2	20	20	20	30	20	25	30	35	35	40	45	55	40	45	50	65
	W	S1	15	15	15	20	15	15	15	20	40	45	50	65	50	55	60	80
		S2	20	20	20	30	20	25	30	35	45	50	55	70	55	60	65	85
C1	N	S1	45	50	55	70	55	60	65	85	105	115	125	165	135	150	165	-
		S2	55	60	65	85	65	70	80	100	105	120	135	170	145	165	185	-
	E	S1	50	55	60	80	60	65	70	95	110	125	140	180	135	150	165	-
		S2	60	65	70	95	65	75	85	105	115	130	145	185	145	160	175	-
	S	S1	60	65	70	95	65	70	80	100	135	150	165	215	145	160	175	-
		S2	60	65	70	95	65	75	85	105	140	155	170	220	165	185	-	-
	W	S1	60	65	70	95	65	70	80	100	145	165	185	235	170	190	-	-
		S2	65	75	85	105	70	80	90	115	150	170	190	245	-	-	-	-
C2	N	S1	45	50	55	70	55	60	65	85	105	115	125	165	135	150	165	-
		S2	55	60	65	85	60	65	70	95	105	120	135	170	145	160	175	-
	E	S1	50	55	60	80	60	65	70	95	105	120	135	170	135	150	165	-
		S2	60	65	70	95	65	75	85	105	110	125	140	180	145	160	175	-
	S	S1	45	50	55	70	60	65	70	95	130	145	160	205	135	150	165	-
		S2	60	65	70	95	65	75	85	105	135	150	165	-	160	180	200	-
	W	S1	55	60	65	85	65	70	80	100	145	165	185	235	170	190	-	-
		S2	65	70	80	100	65	75	85	105	150	170	190	-	-	-	-	-

Table A3. Cont.

(b)

CZ	O	S	Model 1				Model 2				Model 3				Model 4			
			XPS	PUR	SW	EC	XPS	PUR	SW	EC	XPS	PUR	SW	EC	XPS	PUR	SW	EC
C3	N	S1	40	45	50	65	45	50	55	70	85	95	105	135	105	120	135	170
		S2	45	50	55	70	55	60	65	85	90	100	110	145	110	125	140	180
	E	S1	45	50	55	70	50	55	60	80	90	100	110	145	105	120	135	170
		S2	50	55	60	80	60	65	70	95	95	105	115	150	110	125	140	180
	S	S1	45	50	55	70	50	55	60	80	105	120	135	170	115	130	145	185
		S2	50	55	60	80	60	65	70	95	110	125	140	180	125	140	155	200
	W	S1	50	55	60	80	55	60	65	85	120	135	150	195	135	150	165	-
		S2	55	60	65	85	60	65	70	95	125	140	155	200	165	185	-	-
C4	N	S1	40	45	50	65	45	50	55	70	85	95	105	135	110	125	140	180
		S2	50	55	60	80	55	60	65	85	95	105	115	150	130	145	160	-
	E	S1	45	50	55	70	50	55	60	80	90	100	110	145	110	125	140	180
		S2	50	55	60	80	60	65	70	95	105	115	125	165	130	145	160	-
	S	S1	45	50	55	70	55	60	65	85	105	115	125	165	130	145	160	-
		S2	55	60	65	85	60	65	70	95	115	130	145	185	145	165	185	-
	W	S1	50	55	60	80	55	60	65	85	115	130	145	185	145	160	175	-
		S2	60	65	70	95	65	70	80	100	135	150	165	-	180	200	-	-
D1	N	S1	135	150	165	-	145	160	175	-	-	-	-	-	-	-	-	-
		S2	160	180	200	-	175	195	-	-	-	-	-	-	-	-	-	-
	E	S1	140	155	170	-	165	185	-	-	-	-	-	-	-	-	-	-
		S2	165	185	-	-	200	-	-	-	-	-	-	-	-	-	-	-
	S	S1	140	155	170	-	190	-	-	-	-	-	-	-	-	-	-	-
		S2	190	-	-	-	-	-	-	-	-	-	-	-	-	-	-	-
	W	S1	150	170	190	-	195	-	-	-	-	-	-	-	-	-	-	-
		S2	-	-	-	-	-	-	-	-	-	-	-	-	-	-	-	-
D2	N	S1	130	145	160	-	145	160	175	-	-	-	-	-	-	-	-	-
		S2	150	170	190	-	165	185	-	-	-	-	-	-	-	-	-	-
	E	S1	135	150	165	-	160	180	200	-	-	-	-	-	-	-	-	-
		S2	155	175	195	-	200	-	-	-	-	-	-	-	-	-	-	-
	S	S1	145	160	175	-	165	185	-	-	-	-	-	-	-	-	-	-
		S2	180	200	-	-	-	-	-	-	-	-	-	-	-	-	-	-
	W	S1	155	175	195	-	180	200	-	-	-	-	-	-	-	-	-	-
		S2	-	-	-	-	-	-	-	-	-	-	-	-	-	-	-	-
D3	N	S1	115	130	145	185	125	140	155	200	-	-	-	-	-	-	-	-
		S2	125	140	155	200	135	140	165	-	-	-	-	-	-	-	-	-
	E	S1	120	135	150	195	135	150	165	-	-	-	-	-	-	-	-	-
		S2	135	150	165	-	150	170	190	-	-	-	-	-	-	-	-	-
	S	S1	125	140	155	200	145	160	175	-	-	-	-	-	-	-	-	-
		S2	145	165	185	-	160	180	200	-	-	-	-	-	-	-	-	-
	W	S1	140	155	170	-	150	170	190	-	-	-	-	-	-	-	-	-
		S2	165	185	-	-	170	190	-	-	-	-	-	-	-	-	-	-
E1	N	S1	-	-	-	-	-	-	-	-	-	-	-	-	-	-	-	-
		S2	-	-	-	-	-	-	-	-	-	-	-	-	-	-	-	-
	E	S1	-	-	-	-	-	-	-	-	-	-	-	-	-	-	-	-
		S2	-	-	-	-	-	-	-	-	-	-	-	-	-	-	-	-
	S	S1	-	-	-	-	-	-	-	-	-	-	-	-	-	-	-	-
		S2	-	-	-	-	-	-	-	-	-	-	-	-	-	-	-	-
	W	S1	-	-	-	-	-	-	-	-	-	-	-	-	-	-	-	-
		S2	-	-	-	-	-	-	-	-	-	-	-	-	-	-	-	-

CZ (Climate Zone); O (Orientation); S (Constructive Solution); - Thickness not enough to satisfy demand.

Table A4. Emissions of CO_2 according to climate zone (CZ), orientation (O), constructive solution (S), building model and insulation material model (in Kg CO_2), for H1.

(a)

CZ	O	S	Model 1				Model 2				Model 3				Model 4			
			XPS	PUR	SW	EC	XPS	PUR	SW	EC	XPS	PUR	SW	EC	XPS	PUR	SW	EC
A3	N	S1	0.00E+00	0.00E+00	0.00E+00	0.00E+00	0.00E+00	0.00E+00	0.00E+00	0.00E+00	3.93E+03	2.77E+03	2.00E+03	1.54E+03	4.15E+03	2.93E+03	2.11E+03	1.63E+03
		S2	1.57E+03	1.11E+03	7.97E+02	4.10E+02	3.23E+03	2.28E+03	1.65E+03	1.27E+03	5.89E+03	4.16E+03	3.00E+03	2.05E+03	8.29E+03	5.85E+03	4.22E+03	3.25E+03
	E	S1	1.57E+03	1.11E+03	7.97E+02	4.10E+02	1.62E+03	1.14E+03	8.23E+02	4.23E+02	3.93E+03	2.77E+03	2.00E+03	1.54E+03	4.15E+03	2.93E+03	2.11E+03	1.63E+03
		S2	1.57E+03	1.11E+03	7.97E+02	4.10E+02	3.23E+03	2.28E+03	1.65E+03	1.27E+03	5.89E+03	4.16E+03	3.00E+03	2.05E+03	8.29E+03	5.85E+03	4.22E+03	3.25E+03
	S	S1	1.57E+03	1.11E+03	7.97E+02	4.10E+02	1.62E+03	1.14E+03	8.23E+02	4.23E+02	3.93E+03	2.77E+03	2.00E+03	1.54E+03	4.15E+03	2.93E+03	2.11E+03	1.63E+03
		S2	1.57E+03	1.11E+03	7.97E+02	4.10E+02	3.23E+03	2.28E+03	1.65E+03	1.27E+03	7.86E+03	5.55E+03	4.00E+03	3.08E+03	8.29E+03	5.85E+03	4.22E+03	3.25E+03
	W	S1	1.57E+03	1.11E+03	7.97E+02	4.10E+02	1.62E+03	1.14E+03	8.23E+02	4.23E+02	5.89E+03	4.16E+03	3.00E+03	2.05E+03	8.29E+03	7.32E+03	6.33E+03	3.79E+03
		S2	3.13E+03	2.21E+03	1.59E+03	1.23E+03	3.23E+03	2.28E+03	1.65E+03	1.27E+03	7.86E+03	6.93E+03	6.00E+03	3.60E+03	8.29E+03	5.85E+03	4.22E+03	3.25E+03
A4	N	S1	0.00E+00	0.00E+00	0.00E+00	0.00E+00	0.00E+00	0.00E+00	0.00E+00	0.00E+00	3.93E+03	2.77E+03	2.00E+03	1.54E+03	4.15E+03	2.93E+03	2.11E+03	1.63E+03
		S2	1.57E+03	1.11E+03	7.97E+02	4.10E+02	3.23E+03	2.28E+03	1.65E+03	1.27E+03	5.89E+03	4.16E+03	3.00E+03	2.05E+03	8.29E+03	5.85E+03	4.22E+03	3.25E+03
	E	S1	1.57E+03	1.11E+03	7.97E+02	4.10E+02	1.62E+03	1.14E+03	8.23E+02	4.23E+02	3.93E+03	2.77E+03	2.00E+03	1.54E+03	4.15E+03	2.93E+03	2.11E+03	1.63E+03
		S2	1.57E+03	1.11E+03	7.97E+02	4.10E+02	3.23E+03	2.28E+03	1.65E+03	1.27E+03	5.89E+03	4.16E+03	3.00E+03	2.05E+03	8.29E+03	5.85E+03	4.22E+03	3.25E+03
	S	S1	1.57E+03	1.11E+03	7.97E+02	4.10E+02	1.62E+03	1.14E+03	8.23E+02	4.23E+02	3.93E+03	2.77E+03	2.00E+03	1.54E+03	4.15E+03	2.93E+03	2.11E+03	1.63E+03
		S2	1.57E+03	1.11E+03	7.97E+02	4.10E+02	3.23E+03	2.28E+03	1.65E+03	1.27E+03	7.86E+03	5.55E+03	4.00E+03	3.08E+03	8.29E+03	5.85E+03	4.22E+03	3.25E+03
	W	S1	1.57E+03	1.11E+03	7.97E+02	4.10E+02	1.62E+03	1.14E+03	8.23E+02	4.23E+02	5.89E+03	4.16E+03	3.00E+03	2.05E+03	8.29E+03	7.32E+03	6.33E+03	3.79E+03
		S2	3.13E+03	2.21E+03	1.59E+03	1.23E+03	3.23E+03	2.28E+03	1.65E+03	1.27E+03	7.86E+03	6.93E+03	6.00E+03	3.60E+03	1.04E+04	8.78E+03	7.38E+03	4.88E+03
B3	N	S1	3.13E+03	2.21E+03	1.59E+03	1.23E+03	4.85E+03	3.43E+03	2.47E+03	1.69E+03	9.82E+03	8.32E+03	7.00E+03	4.62E+03	1.24E+04	1.02E+04	8.44E+03	5.42E+03
		S2	4.70E+03	3.32E+03	2.39E+03	1.64E+03	6.47E+03	4.57E+03	3.29E+03	2.54E+03	1.18E+04	9.71E+03	8.00E+03	5.14E+03	1.45E+04	1.17E+04	9.49E+03	5.96E+03
	E	S1	4.70E+03	3.32E+03	2.39E+03	1.64E+03	4.85E+03	3.43E+03	2.47E+03	1.69E+03	1.18E+04	9.71E+03	8.00E+03	5.14E+03	1.24E+04	1.02E+04	8.44E+03	5.42E+03
		S2	6.27E+03	4.42E+03	3.19E+03	2.46E+03	6.47E+03	5.71E+03	4.94E+03	2.96E+03	1.18E+04	9.71E+03	8.00E+03	5.14E+03	1.45E+04	1.17E+04	9.49E+03	5.96E+03
	S	S1	4.70E+03	3.32E+03	2.39E+03	1.64E+03	4.85E+03	3.43E+03	2.47E+03	1.69E+03	1.38E+04	1.11E+04	9.00E+03	5.65E+03	1.45E+04	1.32E+04	1.05E+04	7.05E+03
		S2	6.27E+03	4.42E+03	3.19E+03	2.46E+03	6.47E+03	5.71E+03	4.94E+03	2.96E+03	1.57E+04	1.25E+04	1.00E+04	6.68E+03	1.66E+04	1.61E+04	1.27E+04	8.67E+03
	W	S1	4.70E+03	3.32E+03	2.39E+03	1.64E+03	4.85E+03	3.43E+03	2.47E+03	1.69E+03	1.57E+04	1.25E+04	1.00E+04	6.68E+03	2.07E+04	1.61E+04	1.27E+04	8.67E+03
		S2	6.27E+03	4.42E+03	3.19E+03	2.46E+03	6.47E+03	5.71E+03	4.94E+03	2.96E+03	1.77E+04	1.39E+04	1.10E+04	7.19E+03	2.28E+04	1.76E+04	1.37E+04	9.21E+03
B4	N	S1	3.13E+03	2.21E+03	1.59E+03	1.23E+03	4.85E+03	3.43E+03	2.47E+03	1.69E+03	9.82E+03	8.32E+03	7.00E+03	4.62E+03	1.24E+04	1.02E+04	8.44E+03	5.42E+03
		S2	4.70E+03	3.32E+03	2.39E+03	1.64E+03	6.47E+03	4.57E+03	3.29E+03	2.54E+03	1.18E+04	9.71E+03	8.00E+03	5.14E+03	1.45E+04	1.17E+04	9.49E+03	5.96E+03
	E	S1	4.70E+03	3.32E+03	2.39E+03	1.64E+03	4.85E+03	3.43E+03	2.47E+03	1.69E+03	1.18E+04	9.71E+03	8.00E+03	5.14E+03	1.24E+04	1.02E+04	8.44E+03	5.42E+03
		S2	6.27E+03	4.42E+03	3.19E+03	2.46E+03	6.47E+03	5.71E+03	4.94E+03	2.96E+03	1.38E+04	1.11E+04	9.00E+03	5.65E+03	1.66E+04	1.32E+04	1.05E+04	7.05E+03
	S	S1	4.70E+03	3.32E+03	2.39E+03	1.64E+03	4.85E+03	3.43E+03	2.47E+03	1.69E+03	1.38E+04	1.11E+04	9.00E+03	5.65E+03	1.45E+04	1.17E+04	9.49E+03	5.96E+03
		S2	6.27E+03	4.42E+03	3.19E+03	2.46E+03	6.47E+03	5.71E+03	4.94E+03	2.96E+03	1.57E+04	1.25E+04	1.00E+04	6.68E+03	1.66E+04	1.32E+04	1.05E+04	7.05E+03
	W	S1	4.70E+03	3.32E+03	2.39E+03	1.64E+03	4.85E+03	3.43E+03	2.47E+03	1.69E+03	1.57E+04	1.25E+04	1.00E+04	6.68E+03	2.07E+04	1.61E+04	1.27E+04	8.67E+03
		S2	6.27E+03	4.42E+03	3.19E+03	2.46E+03	6.47E+03	5.71E+03	4.94E+03	2.96E+03	1.77E+04	1.39E+04	1.10E+04	7.19E+03	2.28E+04	1.76E+04	1.37E+04	9.21E+03

Table A4. Cont.

(b)

CZ	O	S	Model 1				Model 2				Model 3				Model 4			
			XPS	PUR	SW	EC			XPS	PUR	SW	EC				XPS	PUR
C1	N	S1	6.27E+03	4.42E+03	3.19E+03	2.46E+03	6.47E+03	5.71E+03	4.94E+03	2.96E+03	1.57E+04	1.25E+04	1.00E+04	6.68E+03	2.07E+04	1.61E+04	1.27E+04	8.67E+03
		S2	6.27E+03	5.53E+03	4.78E+03	2.87E+03	6.47E+03	5.71E+03	4.94E+03	2.96E+03	1.77E+04	1.39E+04	1.10E+04	7.19E+03	2.28E+04	1.76E+04	1.37E+04	9.21E+03
	E	S1	6.27E+03	4.42E+03	3.19E+03	2.46E+03	6.47E+03	5.71E+03	4.94E+03	2.96E+03	1.57E+04	1.25E+04	1.00E+04	6.68E+03	2.07E+04	1.61E+04	1.27E+04	8.67E+03
		S2	6.27E+03	5.53E+03	4.78E+03	2.87E+03	8.09E+03	6.85E+03	5.76E+03	3.81E+03	1.77E+04	1.39E+04	1.10E+04	7.19E+03	2.28E+04	1.76E+04	1.37E+04	9.21E+03
	S	S1	6.27E+03	4.42E+03	3.19E+03	2.46E+03	6.47E+03	5.71E+03	4.94E+03	2.96E+03	1.96E+04	1.53E+04	1.20E+04	8.22E+03	2.49E+04	1.90E+04	1.48E+04	1.03E+04
		S2	6.27E+03	5.53E+03	4.78E+03	2.87E+03	8.09E+03	6.85E+03	5.76E+03	3.81E+03	2.16E+04	1.66E+04	1.30E+04	8.73E+03	2.49E+04	1.90E+04	1.48E+04	1.03E+04
	W	S1	6.27E+03	5.53E+03	4.78E+03	2.87E+03	6.47E+03	5.71E+03	4.94E+03	2.96E+03	2.16E+04	1.66E+04	1.30E+04	8.73E+03	2.69E+04	2.19E+04	1.79E+04	1.14E+04
		S2	7.83E+03	6.63E+03	5.58E+03	3.69E+03	8.09E+03	6.85E+03	5.76E+03	3.81E+03	2.36E+04	1.80E+04	1.40E+04	9.76E+03	2.69E+04	2.19E+04	1.79E+04	1.14E+04
C2	N	S1	4.70E+03	3.32E+03	2.39E+03	1.64E+03	6.47E+03	4.57E+03	3.29E+03	2.54E+03	1.57E+04	1.25E+04	1.00E+04	6.68E+03	1.87E+04	1.46E+04	1.16E+04	7.59E+03
		S2	6.27E+03	5.53E+03	4.78E+03	2.87E+03	6.47E+03	5.71E+03	4.94E+03	2.96E+03	1.77E+04	1.39E+04	1.10E+04	7.19E+03	2.07E+04	1.61E+04	1.27E+04	8.67E+03
	E	S1	6.27E+03	4.42E+03	3.19E+03	2.46E+03	6.47E+03	5.71E+03	4.94E+03	2.96E+03	1.57E+04	1.25E+04	1.00E+04	6.68E+03	1.87E+04	1.46E+04	1.16E+04	7.59E+03
		S2	6.27E+03	5.53E+03	4.78E+03	2.87E+03	8.09E+03	6.85E+03	5.76E+03	3.81E+03	1.77E+04	1.39E+04	1.10E+04	7.19E+03	2.07E+04	1.61E+04	1.27E+04	8.67E+03
	S	S1	6.27E+03	4.42E+03	3.19E+03	2.46E+03	6.47E+03	5.71E+03	4.94E+03	2.96E+03	1.77E+04	1.39E+04	1.10E+04	7.19E+03	2.07E+04	1.61E+04	1.27E+04	8.67E+03
		S2	6.27E+03	5.53E+03	4.78E+03	2.87E+03	8.09E+03	6.85E+03	5.76E+03	3.81E+03	1.96E+04	1.53E+04	1.20E+04	8.22E+03	2.28E+04	1.76E+04	1.37E+04	9.21E+03
	W	S1	6.27E+03	4.42E+03	3.19E+03	2.46E+03	6.47E+03	5.71E+03	4.94E+03	2.96E+03	1.96E+04	1.53E+04	1.20E+04	8.22E+03	2.69E+04	2.05E+04	1.69E+04	1.08E+04
		S2	6.27E+03	5.53E+03	4.78E+03	2.87E+03	8.09E+03	6.85E+03	5.76E+03	3.81E+03	2.16E+04	1.66E+04	1.30E+04	8.73E+03	2.69E+04	2.19E+04	1.79E+04	1.14E+04
C3	N	S1	4.70E+03	3.32E+03	2.39E+03	1.64E+03	6.47E+03	4.57E+03	3.29E+03	2.54E+03	1.38E+04	1.11E+04	9.00E+03	5.65E+03	1.66E+04	1.32E+04	1.05E+04	7.05E+03
		S2	6.27E+03	4.42E+03	3.19E+03	2.46E+03	6.47E+03	5.71E+03	4.94E+03	2.96E+03	1.57E+04	1.25E+04	1.00E+04	6.68E+03	1.87E+04	1.46E+04	1.16E+04	7.59E+03
	E	S1	4.70E+03	3.32E+03	2.39E+03	1.64E+03	6.47E+03	4.57E+03	3.29E+03	2.54E+03	1.38E+04	1.11E+04	9.00E+03	5.65E+03	1.66E+04	1.32E+04	1.05E+04	7.05E+03
		S2	6.27E+03	4.42E+03	3.19E+03	2.46E+03	6.47E+03	5.71E+03	4.94E+03	2.96E+03	1.57E+04	1.25E+04	1.00E+04	6.68E+03	1.87E+04	1.46E+04	1.16E+04	7.59E+03
	S	S1	6.27E+03	4.42E+03	3.19E+03	2.46E+03	6.47E+03	5.71E+03	4.94E+03	2.96E+03	1.77E+04	1.39E+04	1.10E+04	7.19E+03	2.07E+04	1.61E+04	1.27E+04	8.67E+03
		S2	6.27E+03	4.42E+03	3.19E+03	2.46E+03	6.47E+03	5.71E+03	4.94E+03	2.96E+03	1.77E+04	1.39E+04	1.10E+04	7.19E+03	2.07E+04	1.61E+04	1.27E+04	8.67E+03
	W	S1	6.27E+03	4.42E+03	3.19E+03	2.46E+03	6.47E+03	5.71E+03	4.94E+03	2.96E+03	1.96E+04	1.53E+04	1.20E+04	8.22E+03	2.49E+04	1.90E+04	1.48E+04	1.03E+04
C4	N	S1	4.70E+03	3.32E+03	2.39E+03	1.64E+03	6.47E+03	4.57E+03	3.29E+03	2.54E+03	1.38E+04	1.11E+04	9.00E+03	5.65E+03	1.66E+04	1.32E+04	1.05E+04	7.05E+03
		S2	6.27E+03	4.42E+03	3.19E+03	2.46E+03	6.47E+03	5.71E+03	4.94E+03	2.96E+03	1.57E+04	1.25E+04	1.00E+04	6.68E+03	2.07E+04	1.61E+04	1.27E+04	8.67E+03
	E	S1	4.70E+03	3.32E+03	2.39E+03	1.64E+03	6.47E+03	4.57E+03	3.29E+03	2.54E+03	1.38E+04	1.11E+04	9.00E+03	5.65E+03	1.66E+04	1.32E+04	1.05E+04	7.05E+03
		S2	6.27E+03	4.42E+03	3.19E+03	2.46E+03	6.47E+03	5.71E+03	4.94E+03	2.96E+03	1.57E+04	1.25E+04	1.00E+04	6.68E+03	2.07E+04	1.61E+04	1.27E+04	8.67E+03
	S	S1	4.70E+03	3.32E+03	2.39E+03	1.64E+03	6.47E+03	4.57E+03	3.29E+03	2.54E+03	1.57E+04	1.25E+04	1.00E+04	6.68E+03	1.87E+04	1.46E+04	1.16E+04	7.59E+03
		S2	6.27E+03	5.53E+03	4.78E+03	2.87E+03	6.47E+03	5.71E+03	4.94E+03	2.96E+03	1.77E+04	1.39E+04	1.10E+04	7.19E+03	2.28E+04	1.76E+04	1.37E+04	9.21E+03
	W	S1	6.27E+03	4.42E+03	3.19E+03	2.46E+03	6.47E+03	5.71E+03	4.94E+03	2.96E+03	1.77E+04	1.39E+04	1.10E+04	7.19E+03	2.28E+04	1.76E+04	1.37E+04	9.21E+03
		S2	6.27E+03	5.53E+03	4.78E+03	2.87E+03	8.09E+03	6.85E+03	5.76E+03	3.81E+03	1.96E+04	1.53E+04	1.20E+04	8.22E+03	2.69E+04	2.05E+04	1.69E+04	1.08E+04

Table A4. Cont.

(c)

CZ	O	S	Model 1						Model 2						Model 3						Model 4			
			XPS	PUR	SW	EC									SW	EC						XPS		PUR
D1	N	S1	6.27E-03	4.42E-03	3.19E-03	2.46E-03	6.47E-03	5.71E-03	4.94E-03	2.96E-03	1.77E-04	1.39E-04	1.10E-04	7.19E-03	2.28E-04	1.76E-04	1.37E-04	9.21E+03
		S2	6.27E-03	5.53E-03	4.78E-03	2.87E-03	8.09E-03	6.85E-03	5.76E-03	3.81E-03	1.96E-04	1.53E-04	1.20E-04	8.22E-03	2.69E-04	2.05E-04	1.69E-04	1.08E-04
	E	S1	6.27E-03	4.42E-03	3.19E-03	2.46E-03	8.09E-03	6.85E-03	5.76E-03	3.81E-03	1.77E-04	1.39E-04	1.10E-04	7.19E-03	2.28E-04	1.76E-04	1.37E-04	9.21E+03
		S2	6.27E-03	5.53E-03	4.78E-03	2.87E-03	8.09E-03	6.85E-03	5.76E-03	3.81E-03	1.96E-04	1.53E-04	1.20E-04	8.22E-03	2.49E-04	2.05E-04	1.69E-04	1.08E-04
	S	S1	7.83E-03	5.53E-03	5.58E-03	3.69E-03	9.70E-03	7.99E-03	6.58E-03	4.23E-03	2.16E-04	1.66E-04	1.30E-04	8.73E-03	2.49E-04	1.90E-04	1.48E-04	1.03E-04
		S2	7.83E-03	6.63E-03	5.58E-03	3.69E-03	9.70E-03	7.99E-03	6.58E-03	4.23E-03	2.36E-04	1.80E-04	1.40E-04	9.76E-03	2.69E-04	2.19E-04	1.79E-04	1.14E-04
	W	S1	6.27E-03	5.53E-03	4.78E-03	2.87E-03	8.09E-03	6.85E-03	5.76E-03	3.81E-03	2.36E-04	1.80E-04	1.40E-04	9.76E-03	2.69E-04	2.19E-04	1.79E-04	1.14E-04
		S2	7.83E-03	6.63E-03	5.58E-03	3.69E-03	9.70E-03	7.99E-03	6.58E-03	4.23E-03	2.55E-04	1.94E-04	1.60E-04	1.03E-04	3.11E-04	2.49E-04	2.00E-04	1.30E-04
D2	N	S1	6.27E-03	4.42E-03	3.19E-03	2.46E-03	6.47E-03	5.71E-03	4.94E-03	2.96E-03	1.77E-04	1.39E-04	1.10E-04	7.19E-03	2.28E-04	1.76E-04	1.37E-04	9.21E+03
		S2	6.27E-03	5.53E-03	4.78E-03	2.87E-03	8.09E-03	6.85E-03	5.76E-03	3.81E-03	1.96E-04	1.53E-04	1.20E-04	8.22E-03	2.69E-04	2.05E-04	1.69E-04	1.08E-04
	E	S1	6.27E-03	4.42E-03	3.19E-03	2.46E-03	6.47E-03	5.71E-03	4.94E-03	2.96E-03	1.77E-04	1.39E-04	1.10E-04	7.19E-03	2.07E-04	1.61E-04	1.27E-04	8.67E-03
		S2	6.27E-03	5.53E-03	4.78E-03	2.87E-03	8.09E-03	6.85E-03	5.76E-03	3.81E-03	1.96E-04	1.53E-04	1.20E-04	8.22E-03	2.49E-04	1.90E-04	1.48E-04	1.03E-04
	S	S1	6.27E-03	5.53E-03	4.78E-03	2.87E-03	8.09E-03	6.85E-03	5.76E-03	3.81E-03	1.96E-04	1.53E-04	1.20E-04	8.22E-03	2.28E-04	1.76E-04	1.37E-04	9.21E+03
		S2	6.27E-03	5.53E-03	4.78E-03	2.87E-03	9.70E-03	7.99E-03	6.58E-03	4.23E-03	2.16E-04	1.66E-04	1.30E-04	8.73E-03	2.69E-04	2.05E-04	1.69E-04	1.08E-04
	W	S1	6.27E-03	4.42E-03	3.19E-03	2.46E-03	6.47E-03	5.71E-03	5.76E-03	3.81E-03	2.16E-04	1.66E-04	1.30E-04	8.73E-03	2.69E-04	2.19E-04	1.79E-04	1.14E-04
		S2	7.83E-03	6.63E-03	5.58E-03	3.69E-03	9.70E-03	7.99E-03	6.58E-03	4.23E-03	2.36E-04	1.80E-04	1.40E-04	9.76E-03	3.11E-04	2.49E-04	2.00E-04	1.30E-04
D3	N	S1	6.27E-03	4.42E-03	3.19E-03	2.46E-03	6.47E-03	4.57E-03	3.29E-03	2.54E-03	1.57E-04	1.25E-04	1.00E-04	6.68E-03	2.07E-04	1.61E-04	1.27E-04	8.67E-03
		S2	6.27E-03	5.53E-03	4.78E-03	2.87E-03	6.47E-03	5.71E-03	4.94E-03	2.96E-03	1.77E-04	1.39E-04	1.10E-04	7.19E-03	2.28E-04	1.76E-04	1.37E-04	9.21E+03
	E	S1	6.27E-03	4.42E-03	3.19E-03	2.46E-03	6.47E-03	4.57E-03	3.29E-03	2.54E-03	1.57E-04	1.25E-04	1.00E-04	6.68E-03	1.87E-04	1.46E-04	1.16E-04	7.59E-03
		S2	6.27E-03	5.53E-03	4.78E-03	2.87E-03	6.47E-03	5.71E-03	4.94E-03	2.96E-03	1.77E-04	1.39E-04	1.10E-04	7.19E-03	2.28E-04	1.76E-04	1.37E-04	9.21E+03
	S	S1	6.27E-03	4.42E-03	3.19E-03	2.46E-03	6.47E-03	5.71E-03	4.94E-03	2.96E-03	1.77E-04	1.39E-04	1.10E-04	7.19E-03	2.07E-04	1.61E-04	1.27E-04	8.67E-03
		S2	6.27E-03	5.53E-03	4.78E-03	2.87E-03	8.09E-03	6.85E-03	5.76E-03	3.81E-03	1.96E-04	1.53E-04	1.20E-04	8.22E-03	2.49E-04	1.90E-04	1.48E-04	1.03E-04
	W	S1	6.27E-03	4.42E-03	3.19E-03	2.46E-03	6.47E-03	5.71E-03	4.94E-03	2.96E-03	1.96E-04	1.53E-04	1.20E-04	8.22E-03	2.69E-04	2.05E-04	1.69E-04	1.08E-04
		S2	6.27E-03	5.53E-03	4.78E-03	2.87E-03	8.09E-03	6.85E-03	5.76E-03	3.81E-03	2.16E-04	1.66E-04	1.30E-04	8.73E-03	2.69E-04	2.19E-04	1.79E-04	1.14E-04
E1	N	S1	4.70E-03	3.32E-03	2.39E-03	1.64E-03	6.47E-03	4.57E-03	3.29E-03	2.54E-03	1.38E-04	1.11E-04	9.00E-03	5.65E-03	1.66E-04	1.32E-04	1.05E-04	7.05E-03
		S2	6.27E-03	4.42E-03	3.19E-03	2.46E-03	6.47E-03	4.57E-03	3.29E-03	2.54E-03	1.38E-04	1.11E-04	9.00E-03	5.65E-03	1.87E-04	1.46E-04	1.16E-04	7.59E-03
	E	S1	4.70E-03	3.32E-03	2.39E-03	1.64E-03	6.47E-03	4.57E-03	3.29E-03	2.54E-03	1.38E-04	1.11E-04	9.00E-03	5.65E-03	1.66E-04	1.32E-04	1.05E-04	7.05E-03
		S2	6.27E-03	4.42E-03	3.19E-03	2.46E-03	6.47E-03	5.71E-03	4.94E-03	2.96E-03	1.57E-04	1.25E-04	1.00E-04	6.68E-03	1.87E-04	1.46E-04	1.16E-04	7.59E-03
	S	S1	4.70E-03	3.32E-03	2.39E-03	1.64E-03	6.47E-03	4.57E-03	3.29E-03	2.54E-03	1.57E-04	1.25E-04	1.00E-04	6.68E-03	1.87E-04	1.46E-04	1.16E-04	7.59E-03
		S2	6.27E-03	4.42E-03	3.19E-03	2.46E-03	6.47E-03	5.71E-03	4.94E-03	2.96E-03	1.57E-04	1.25E-04	1.00E-04	6.68E-03	2.07E-04	1.61E-04	1.27E-04	8.67E-03
	W	S1	4.70E-03	3.32E-03	2.39E-03	1.64E-03	6.47E-03	4.57E-03	3.29E-03	2.54E-03	1.57E-04	1.25E-04	1.00E-04	6.68E-03	2.07E-04	1.61E-04	1.27E-04	8.67E-03
		S2	6.27E-03	4.42E-03	3.19E-03	2.46E-03	6.47E-03	5.71E-03	4.94E-03	2.96E-03	1.57E-04	1.25E-04	1.00E-04	6.68E-03	2.07E-04	1.61E-04	1.27E-04	8.67E-03

Table A5. Emissions of CO_2 according to climate zone (CZ), orientation (O), constructive solution (S), building model and insulation material model (in Kg CO_2), for H2.

(a)

CZ	O	S	Model 1				Model 2				Model 3				Model 4		
			XPS	PUR	SW	EC	XPS	PUR	SW	EC	XPS	PUR	SW	EC	XPS	PUR	
A3	N	S1	0.00E+00	0.00E+00	0.00E+00	0.00E+00	0.00E+00	0.00E+00	0.00E+00	0.00E+00	3.93E+03	2.77E+03	2.00E+03	1.54E+03	4.15E+03	2.93E+03	2.11E+03	1.63E+03
		S2	1.57E+03	1.11E+03	7.97E+02	4.10E+02	3.23E+03	2.28E+03	1.65E+03	1.27E+03	5.89E+03	4.16E+03	3.00E+03	2.05E+03	8.29E+03	5.85E+03	4.22E+03	3.25E+03
	E	S1	1.57E+03	1.11E+03	7.97E+02	4.10E+02	1.62E+03	1.14E+03	8.23E+02	4.23E+02	3.93E+03	2.77E+03	2.00E+03	1.54E+03	4.15E+03	2.93E+03	2.11E+03	1.63E+03
		S2	1.57E+03	1.11E+03	7.97E+02	4.10E+02	3.23E+03	2.28E+03	1.65E+03	1.27E+03	5.89E+03	4.16E+03	3.00E+03	2.05E+03	8.29E+03	5.85E+03	4.22E+03	3.25E+03
	S	S1	1.57E+03	1.11E+03	7.97E+02	4.10E+02	1.62E+03	1.14E+03	8.23E+02	4.23E+02	3.93E+03	2.77E+03	2.00E+03	1.54E+03	4.15E+03	2.93E+03	2.11E+03	1.63E+03
		S2	1.57E+03	1.11E+03	7.97E+02	4.10E+02	3.23E+03	2.28E+03	1.65E+03	1.27E+03	5.89E+03	4.16E+03	3.00E+03	2.05E+03	8.29E+03	5.85E+03	4.22E+03	3.25E+03
	W	S1	1.57E+03	1.11E+03	7.97E+02	4.10E+02	1.62E+03	1.14E+03	8.23E+02	4.23E+02	7.86E+03	5.55E+03	4.00E+03	3.08E+03	8.29E+03	7.32E+03	6.33E+03	3.79E+03
		S2	3.13E+03	2.21E+03	1.59E+03	1.23E+03	3.23E+03	2.28E+03	1.65E+03	1.27E+03	7.86E+03	6.93E+03	6.00E+03	3.60E+03	1.04E+04	8.78E+03	7.38E+03	4.88E+03
A4	N	S1	0.00E+00	0.00E+00	0.00E+00	0.00E+00	0.00E+00	0.00E+00	0.00E+00	0.00E+00	3.93E+03	2.77E+03	2.00E+03	1.54E+03	4.15E+03	2.93E+03	2.11E+03	1.63E+03
		S2	1.57E+03	1.11E+03	7.97E+02	4.10E+02	3.23E+03	2.28E+03	1.65E+03	1.27E+03	5.89E+03	4.16E+03	3.00E+03	2.05E+03	8.29E+03	5.85E+03	4.22E+03	3.25E+03
	E	S1	1.57E+03	1.11E+03	7.97E+02	4.10E+02	1.62E+03	1.14E+03	8.23E+02	4.23E+02	3.93E+03	2.77E+03	2.00E+03	1.54E+03	4.15E+03	2.93E+03	2.11E+03	1.63E+03
		S2	1.57E+03	1.11E+03	7.97E+02	4.10E+02	3.23E+03	2.28E+03	1.65E+03	1.27E+03	5.89E+03	4.16E+03	3.00E+03	2.05E+03	8.29E+03	5.85E+03	4.22E+03	3.25E+03
	S	S1	1.57E+03	1.11E+03	7.97E+02	4.10E+02	1.62E+03	1.14E+03	8.23E+02	4.23E+02	3.93E+03	2.77E+03	2.00E+03	1.54E+03	4.15E+03	2.93E+03	2.11E+03	1.63E+03
		S2	1.57E+03	1.11E+03	7.97E+02	4.10E+02	3.23E+03	2.28E+03	1.65E+03	1.27E+03	5.89E+03	4.16E+03	3.00E+03	3.08E+03	8.29E+03	5.85E+03	4.22E+03	3.25E+03
	W	S1	1.57E+03	1.11E+03	7.97E+02	4.10E+02	1.62E+03	1.14E+03	8.23E+02	4.23E+02	7.86E+03	5.55E+03	4.00E+03	3.08E+03	8.29E+03	7.32E+03	6.33E+03	3.79E+03
		S2	3.13E+03	2.21E+03	1.59E+03	1.23E+03	3.23E+03	2.28E+03	1.65E+03	1.27E+03	7.86E+03	6.93E+03	6.00E+03	3.60E+03	8.29E+03	5.85E+03	4.22E+03	3.25E+03
B3	N	S1	3.13E+03	2.21E+03	1.59E+03	1.23E+03	4.85E+03	3.43E+03	2.47E+03	1.69E+03	9.82E+03	8.32E+03	7.00E+03	4.62E+03	1.24E+04	1.02E+04	8.44E+03	5.42E+03
		S2	4.70E+03	3.32E+03	2.39E+03	1.64E+03	6.47E+03	4.57E+03	3.29E+03	2.54E+03	1.18E+04	9.71E+03	8.00E+03	5.14E+03	1.45E+04	1.17E+04	9.49E+03	5.96E+03
	E	S1	4.70E+03	3.32E+03	2.39E+03	1.64E+03	4.85E+03	3.43E+03	2.47E+03	1.69E+03	1.18E+04	9.71E+03	8.00E+03	5.14E+03	1.24E+04	1.02E+04	8.44E+03	5.42E+03
		S2	6.27E+03	4.42E+03	3.19E+03	2.46E+03	6.47E+03	5.71E+03	4.94E+03	2.96E+03	1.18E+04	9.71E+03	8.00E+03	5.14E+03	1.45E+04	1.17E+04	9.49E+03	5.96E+03
	S	S1	4.70E+03	3.32E+03	2.39E+03	1.64E+03	4.85E+03	3.43E+03	2.47E+03	1.69E+03	1.38E+04	1.11E+04	9.00E+03	5.65E+03	1.45E+04	1.17E+04	9.49E+03	5.96E+03
		S2	6.27E+03	4.42E+03	3.19E+03	2.46E+03	6.47E+03	5.71E+03	4.94E+03	2.96E+03	1.57E+04	1.25E+04	1.00E+04	6.68E+03	1.66E+04	1.32E+04	1.05E+04	7.05E+03
	W	S1	4.70E+03	3.32E+03	2.39E+03	1.64E+03	4.85E+03	3.43E+03	2.47E+03	1.69E+03	1.57E+04	1.25E+04	1.00E+04	6.68E+03	2.07E+04	1.61E+04	1.27E+04	8.67E+03
		S2	6.27E+03	4.42E+03	3.19E+03	2.46E+03	6.47E+03	5.71E+03	4.94E+03	2.96E+03	1.77E+04	1.39E+04	1.10E+04	7.19E+03	2.28E+04	1.76E+04	1.37E+04	9.21E+03
B4	N	S1	3.13E+03	2.21E+03	1.59E+03	1.23E+03	4.85E+03	3.43E+03	2.47E+03	1.69E+03	9.82E+03	8.32E+03	7.00E+03	4.62E+03	1.24E+04	1.02E+04	8.44E+03	5.42E+03
		S2	4.70E+03	3.32E+03	2.39E+03	1.64E+03	6.47E+03	4.57E+03	3.29E+03	2.54E+03	1.18E+04	9.71E+03	8.00E+03	5.14E+03	1.45E+04	1.17E+04	9.49E+03	5.96E+03
	E	S1	4.70E+03	3.32E+03	2.39E+03	1.64E+03	4.85E+03	3.43E+03	2.47E+03	1.69E+03	1.18E+04	9.71E+03	8.00E+03	5.14E+03	1.24E+04	1.02E+04	8.44E+03	5.42E+03
		S2	6.27E+03	4.42E+03	3.19E+03	2.46E+03	6.47E+03	5.71E+03	4.94E+03	2.96E+03	1.18E+04	9.71E+03	8.00E+03	5.14E+03	1.66E+04	1.32E+04	1.05E+04	7.05E+03
	S	S1	4.70E+03	3.32E+03	2.39E+03	1.64E+03	4.85E+03	3.43E+03	2.47E+03	1.69E+03	1.38E+04	1.11E+04	9.00E+03	5.65E+03	1.45E+04	1.17E+04	9.49E+03	5.96E+03
		S2	6.27E+03	4.42E+03	3.19E+03	2.46E+03	6.47E+03	5.71E+03	4.94E+03	2.96E+03	1.57E+04	1.25E+04	1.00E+04	6.68E+03	1.66E+04	1.32E+04	1.05E+04	7.05E+03
	W	S1	4.70E+03	3.32E+03	2.39E+03	1.64E+03	4.85E+03	3.43E+03	2.47E+03	1.69E+03	1.57E+04	1.25E+04	1.00E+04	6.68E+03	2.07E+04	1.61E+04	1.27E+04	8.67E+03
		S2	6.27E+03	4.42E+03	3.19E+03	2.46E+03	6.47E+03	5.71E+03	4.94E+03	2.96E+03	1.77E+04	1.39E+04	1.10E+04	7.19E+03	2.28E+04	1.76E+04	1.37E+04	9.21E+03

Table A5. Cont.

(b)

CZ	O	S	Model 1					Model 2					Model 3					Model 4			
			XPS	PUR	SW	EC		XPS	PUR	SW	EC		XPS	PUR	SW	EC		XPS	PUR		
C1	N	S1	6.27E+03	4.42E+03	3.19E+03	2.46E+03	6.47E+03	5.71E+03	4.94E+03	2.96E+03	1.57E+04	1.25E+04	1.00E+04	6.68E+03	2.07E+04	1.61E+04	1.27E+04	8.67E+03
		S2	6.27E+03	5.53E+03	4.78E+03	2.87E+03	6.47E+03	5.71E+03	4.94E+03	2.96E+03	1.77E+04	1.39E+04	1.10E+04	7.19E+03	2.28E+04	1.76E+04	1.37E+04	9.21E+03
	E	S1	6.27E+03	4.42E+03	3.19E+03	2.46E+03	6.47E+03	5.71E+03	4.94E+03	2.96E+03	1.57E+04	1.25E+04	1.00E+04	6.68E+03	2.07E+04	1.61E+04	1.27E+04	8.67E+03
		S2	6.27E+03	5.53E+03	4.78E+03	2.87E+03	8.09E+03	6.85E+03	5.76E+03	3.81E+03	1.77E+04	1.39E+04	1.10E+04	7.19E+03	2.28E+04	1.76E+04	1.37E+04	9.21E+03
	S	S1	6.27E+03	4.42E+03	3.19E+03	2.46E+03	6.47E+03	5.71E+03	4.94E+03	2.96E+03	1.96E+04	1.53E+04	1.20E+04	8.22E+03	2.28E+04	1.76E+04	1.37E+04	9.21E+03
		S2	6.27E+03	5.53E+03	4.78E+03	2.87E+03	8.09E+03	6.85E+03	5.76E+03	3.81E+03	2.16E+04	1.66E+04	1.30E+04	8.73E+03	2.49E+04	1.90E+04	1.48E+04	1.03E+04
	W	S1	6.27E+03	5.53E+03	4.78E+03	2.87E+03	6.47E+03	5.71E+03	4.94E+03	2.96E+03	2.16E+04	1.66E+04	1.30E+04	8.73E+03	2.49E+04	1.90E+04	1.48E+04	1.03E+04
		S2	7.83E+03	6.63E+03	5.58E+03	3.69E+03	8.09E+03	6.85E+03	5.76E+03	3.81E+03	2.36E+04	1.80E+04	1.40E+04	9.76E+03	2.69E+04	2.19E+04	1.79E+04	1.14E+04
C2	N	S1	4.70E+03	3.32E+03	2.39E+03	1.64E+03	6.47E+03	5.71E+03	4.94E+03	2.96E+03	1.77E+04	1.39E+04	1.10E+04	7.19E+03	2.49E+04	1.90E+04	1.48E+04	1.03E+04
		S2	6.27E+03	5.53E+03	4.78E+03	2.87E+03	8.09E+03	6.85E+03	5.76E+03	3.81E+03	1.96E+04	1.53E+04	1.20E+04	8.22E+03	2.69E+04	2.05E+04	1.69E+04	1.08E+04
	E	S1	6.27E+03	4.42E+03	3.19E+03	2.46E+03	6.47E+03	5.71E+03	4.94E+03	2.96E+03	2.16E+04	1.66E+04	1.30E+04	8.73E+03	2.69E+04	2.05E+04	1.69E+04	1.08E+04
		S2	6.27E+03	5.53E+03	4.78E+03	2.87E+03	8.09E+03	6.85E+03	5.76E+03	3.81E+03	2.36E+04	1.80E+04	1.40E+04	9.76E+03	2.69E+04	2.19E+04	1.79E+04	1.14E+04
	S	S1	6.27E+03	4.42E+03	3.19E+03	2.46E+03	6.47E+03	5.71E+03	4.94E+03	2.96E+03	2.16E+04	1.66E+04	1.30E+04	8.73E+03	2.69E+04	2.05E+04	1.69E+04	1.08E+04
		S2	6.27E+03	5.53E+03	4.78E+03	2.87E+03	8.09E+03	6.85E+03	5.76E+03	3.81E+03	2.36E+04	1.80E+04	1.40E+04	9.76E+03	2.90E+04	2.34E+04	1.90E+04	1.25E+04
	W	S1	6.27E+03	4.42E+03	3.19E+03	2.46E+03	6.47E+03	5.71E+03	4.94E+03	2.96E+03	2.36E+04	1.80E+04	1.40E+04	9.76E+03	3.32E+04	2.63E+04	2.11E+04	1.41E+04
		S2	6.27E+03	5.53E+03	4.78E+03	2.87E+03	9.70E+03	7.99E+03	6.58E+03	4.23E+03	2.55E+04	1.94E+04	1.60E+04	1.03E+04	3.52E+04	2.78E+04	2.21E+04	1.46E+04
C3	N	S1	6.27E+03	5.53E+03	4.78E+03	2.87E+03	8.09E+03	6.85E+03	5.76E+03	3.81E+03	2.36E+04	1.80E+04	1.40E+04	9.76E+03	3.52E+04	2.78E+04	2.21E+04	1.46E+04
		S2	7.83E+03	6.63E+03	5.58E+03	3.69E+03	8.09E+03	6.85E+03	5.76E+03	3.81E+03	2.36E+04	1.80E+04	1.40E+04	9.76E+03	3.52E+04	2.78E+04	2.21E+04	1.46E+04
		S1	6.27E+03	5.53E+03	4.78E+03	2.87E+03	8.09E+03	6.85E+03	5.76E+03	3.81E+03	2.55E+04	1.94E+04	1.60E+04	1.03E+04	3.32E+04	2.63E+04	2.11E+04	1.41E+04
		S2	7.83E+03	6.63E+03	5.58E+03	3.69E+03	1.13E+04	9.13E+03	7.41E+03	4.65E+03	2.55E+04	1.94E+04	1.60E+04	1.03E+04	3.32E+04	2.63E+04	2.11E+04	1.41E+04
	E	S1	6.27E+03	5.53E+03	4.78E+03	2.87E+03	8.09E+03	6.85E+03	5.76E+03	3.81E+03	2.55E+04	2.08E+04	1.70E+04	1.08E+04	3.52E+04	2.78E+04	2.21E+04	1.46E+04
		S2	7.83E+03	6.63E+03	5.58E+03	3.69E+03	1.13E+04	9.13E+03	7.41E+03	4.65E+03	2.75E+04	2.22E+04	1.80E+04	1.18E+04	3.52E+04	2.78E+04	2.21E+04	1.46E+04
	S	S1	7.83E+03	6.63E+03	5.58E+03	3.69E+03	9.70E+03	7.99E+03	6.58E+03	4.23E+03	2.55E+04	2.08E+04	1.70E+04	1.08E+04	3.73E+04	2.93E+04	2.32E+04	1.57E+04
		S2	1.10E+04	8.85E+03	7.17E+03	4.51E+03	1.13E+04	9.13E+03	7.41E+03	4.65E+03	2.75E+04	2.22E+04	1.80E+04	1.18E+04	3.73E+04	2.93E+04	2.32E+04	1.57E+04
	W	S1	7.83E+03	6.63E+03	5.58E+03	3.69E+03	9.70E+03	7.99E+03	6.58E+03	4.23E+03	3.14E+04	2.50E+04	2.00E+04	1.34E+04	3.51E+04	2.85E+04	1.84E+04	
		S2	1.10E+04	8.85E+03	7.17E+03	4.51E+03	1.13E+04	9.13E+03	7.41E+03	4.65E+03	3.34E+04	2.63E+04	2.10E+04	1.39E+04	4.77E+04	3.80E+04	3.06E+04	2.01E+04
C4	N	S1	9.40E+03	7.74E+03	6.38E+03	4.10E+03	1.29E+04	1.03E+04	8.23E+03	5.50E+03	2.95E+04	2.36E+04	1.90E+04	1.23E+04	4.56E+04	3.66E+04	2.95E+04	1.95E+04
		S2	1.25E+04	9.95E+03	7.97E+03	5.32E+03	1.46E+04	1.14E+04	9.05E+03	5.92E+03	3.14E+04	2.50E+04	2.00E+04	1.34E+04	4.97E+04	3.95E+04	3.16E+04	2.11E+04
	E	S1	1.10E+04	8.85E+03	7.17E+03	4.51E+03	1.29E+04	1.03E+04	8.23E+03	5.50E+03	3.54E+04	2.77E+04	2.20E+04	1.49E+04	4.56E+04	3.66E+04	2.95E+04	1.95E+04
		S2	1.25E+04	9.95E+03	7.97E+03	5.32E+03	1.62E+04	1.26E+04	9.88E+03	6.77E+03	3.73E+04	2.91E+04	2.30E+04	1.54E+04	5.39E+04	4.24E+04	3.37E+04	n.a.
	S	S1	1.25E+04	9.95E+03	7.97E+03	5.32E+03	1.46E+04	1.14E+04	9.05E+03	5.92E+03	3.73E+04	2.91E+04	2.30E+04	1.54E+04	5.39E+04	4.24E+04	3.37E+04	n.a.
		S2	1.72E+04	1.33E+04	1.04E+04	6.96E+03	1.78E+04	1.37E+04	1.07E+04	7.19E+03	3.93E+04	3.05E+04	2.40E+04	1.59E+04	5.60E+04	4.39E+04	3.48E+04	n.a.
	W	S1	1.41E+04	1.11E+04	8.77E+03	5.73E+03	1.46E+04	1.14E+04	9.05E+03	5.92E+03	4.13E+04	3.33E+04	2.70E+04	1.75E+04	6.01E+04	4.68E+04	3.69E+04	n.a.
		S2	1.72E+04	1.33E+04	1.04E+04	6.96E+03	1.94E+04	1.48E+04	1.15E+04	8.04E+03	4.32E+04	3.47E+04	2.80E+04	1.85E+04	7.05E+04	5.56E+04	n.a.	n.a.

Table A5. Cont.

(c)

CZ	O	S	Model 1				Model 2				Model 3				Model 4		
			XPS	PUR	SW	EC	XPS	PUR	SW	EC	XPS	PUR	SW	EC	XPS	PUR	
D1	N	S1	1.25E+04	9.95E+03	7.97E+03	5.32E+03	1.46E+04	1.14E+04	9.05E+03	5.92E+03	3.73E+04	2.91E+04	2.30E+04	1.54E+04	5.39E+04	4.24E+04	3.37E+04	n.a.
		S2	1.41E+04	1.11E+04	8.77E+03	5.73E+03	1.62E+04	1.26E+04	9.88E+03	6.77E+03	3.93E+04	3.05E+04	2.40E+04	1.59E+04	7.05E+04	5.56E+04	n.a.	n.a.
	E	S1	1.25E+04	9.95E+03	7.97E+03	5.32E+03	1.62E+04	1.26E+04	9.88E+03	6.77E+03	4.13E+04	3.33E+04	2.70E+04	1.75E+04	5.60E+04	4.39E+04	3.48E+04	n.a.
		S2	1.41E+04	1.11E+04	8.77E+03	5.73E+03	1.78E+04	1.37E+04	1.07E+04	7.19E+03	4.32E+04	3.47E+04	2.80E+04	1.85E+04	7.05E+04	5.56E+04	n.a.	n.a.
	S	S1	1.41E+04	1.11E+04	8.77E+03	5.73E+03	1.62E+04	1.26E+04	9.88E+03	6.77E+03	4.71E+04	3.74E+04	3.00E+04	2.00E+04	6.01E+04	4.68E+04	3.69E+04	n.a.
		S2	1.57E+04	1.22E+04	9.56E+03	6.55E+03	1.78E+04	1.37E+04	1.07E+04	7.19E+03	4.91E+04	3.88E+04	3.10E+04	2.00E+04	6.01E+04	4.68E+04	n.a.	n.a.
	W	S1	1.41E+04	1.11E+04	8.77E+03	5.73E+03	1.78E+04	1.37E+04	1.07E+04	7.19E+03	5.30E+04	4.16E+04	3.30E+04	2.05E+04	7.05E+04	5.56E+04	n.a.	n.a.
		S2	1.72E+04	1.33E+04	1.04E+04	6.96E+03	1.94E+04	1.48E+04	1.15E+04	8.04E+03	5.50E+04	4.30E+04	3.40E+04	n.a.	6.63E+04	5.27E+04	4.22E+04	n.a.
D2	N	S1	1.41E+04	1.11E+04	8.77E+03	5.73E+03	1.78E+04	1.37E+04	1.07E+04	7.19E+03	4.71E+04	3.74E+04	3.00E+04	2.00E+04	8.29E+04	n.a.	3.69E+04	n.a.
		S2	1.57E+04	1.22E+04	9.56E+03	6.55E+03	1.94E+04	1.48E+04	1.15E+04	8.04E+03	4.91E+04	3.88E+04	3.10E+04	2.05E+04	8.29E+04	n.a.	n.a.	n.a.
	E	S1	1.57E+04	1.22E+04	9.56E+03	6.55E+03	1.94E+04	1.48E+04	1.15E+04	8.04E+03	5.50E+04	4.30E+04	3.40E+04	n.a.	6.22E+04	4.97E+04	4.01E+04	n.a.
		S2	1.57E+04	1.22E+04	9.56E+03	6.55E+03	2.10E+04	1.71E+04	1.40E+04	8.88E+03	5.89E+04	4.71E+04	3.80E+04	n.a.	n.a.	n.a.	n.a.	n.a.
	S	S1	1.72E+04	1.33E+04	1.04E+04	6.96E+03	1.94E+04	1.48E+04	1.15E+04	8.04E+03	5.70E+04	4.44E+04	3.50E+04	n.a.	6.63E+04	5.27E+04	4.22E+04	n.a.
		S2	2.04E+04	1.55E+04	1.28E+04	8.19E+03	2.10E+04	1.71E+04	1.40E+04	8.88E+03	5.89E+04	4.71E+04	3.80E+04	n.a.	n.a.	n.a.	n.a.	n.a.
	W	S1	1.88E+04	1.44E+04	1.12E+04	7.78E+03	2.10E+04	1.60E+04	1.32E+04	8.46E+03	5.70E+04	4.44E+04	3.50E+04	n.a.	n.a.	n.a.	n.a.	n.a.
		S2	2.04E+04	1.55E+04	1.28E+04	8.19E+03	2.26E+04	1.83E+04	1.48E+04	9.73E+03	n.a.	n.a.	n.a.	n.a.	n.a.	n.a.	n.a.	n.a.
D3	N	S1	1.72E+04	1.33E+04	1.04E+04	6.96E+03	2.10E+04	1.60E+04	1.32E+04	8.46E+03	5.70E+04	4.44E+04	3.50E+04	n.a.	n.a.	n.a.	n.a.	n.a.
		S2	2.04E+04	1.55E+04	1.28E+04	8.19E+03	2.26E+04	1.83E+04	1.48E+04	9.73E+03	6.29E+04	4.99E+04	3.80E+04	n.a.	n.a.	n.a.	n.a.	n.a.
	E	S1	1.88E+04	1.44E+04	1.12E+04	7.78E+03	2.26E+04	1.83E+04	1.48E+04	9.73E+03	5.89E+04	4.71E+04	3.80E+04	n.a.	n.a.	n.a.	n.a.	n.a.
		S2	2.04E+04	1.55E+04	1.28E+04	8.19E+03	2.43E+04	1.94E+04	1.56E+04	1.01E+04	n.a.	n.a.	n.a.	n.a.	n.a.	n.a.	n.a.	n.a.
	S	S1	2.04E+04	1.55E+04	1.28E+04	8.19E+03	2.26E+04	1.83E+04	1.48E+04	9.73E+03	6.48E+04	5.13E+04	n.a.	n.a.	n.a.	n.a.	n.a.	n.a.
		S2	2.51E+04	1.99E+04	1.59E+04	1.06E+04	2.59E+04	2.06E+04	1.65E+04	1.10E+04	n.a.	n.a.	n.a.	n.a.	n.a.	n.a.	n.a.	n.a.
	W	S1	2.04E+04	1.66E+04	1.35E+04	8.60E+03	2.59E+04	2.06E+04	1.65E+04	1.10E+04	n.a.	n.a.	n.a.	n.a.	n.a.	n.a.	n.a.	n.a.
		S2	2.66E+04	2.10E+04	1.67E+04	1.11E+04	2.91E+04	2.28E+04	1.81E+04	1.23E+04	n.a.	n.a.	n.a.	n.a.	n.a.	n.a.	n.a.	n.a.
E1	N	S1	2.19E+04	1.77E+04	1.43E+04	9.42E+03	2.59E+04	2.06E+04	1.65E+04	1.10E+04	5.70E+04	4.58E+04	3.70E+04	n.a.	n.a.	n.a.	n.a.	n.a.
		S2	2.51E+04	1.99E+04	1.59E+04	1.06E+04	2.91E+04	2.28E+04	1.81E+04	1.23E+04	7.86E+04	n.a.	n.a.	n.a.	n.a.	n.a.	n.a.	n.a.
	E	S1	2.35E+04	1.88E+04	1.51E+04	9.83E+03	2.91E+04	2.28E+04	1.81E+04	1.23E+04	6.68E+04	5.27E+04	n.a.	n.a.	n.a.	n.a.	n.a.	n.a.
		S2	2.51E+04	1.99E+04	1.59E+04	1.06E+04	3.23E+04	2.51E+04	1.98E+04	1.31E+04	n.a.	n.a.	n.a.	n.a.	n.a.	n.a.	n.a.	n.a.
	S	S1	2.51E+04	1.99E+04	1.59E+04	1.06E+04	2.91E+04	2.28E+04	1.81E+04	1.23E+04	7.86E+04	n.a.	n.a.	n.a.	n.a.	n.a.	n.a.	n.a.
		S2	2.82E+04	2.21E+04	1.75E+04	1.19E+04	3.40E+04	2.63E+04	2.06E+04	1.40E+04	n.a.	n.a.	n.a.	n.a.	n.a.	n.a.	n.a.	n.a.
	W	S1	2.66E+04	2.10E+04	1.67E+04	1.11E+04	3.23E+04	2.51E+04	1.98E+04	1.31E+04	7.86E+04	n.a.	n.a.	n.a.	n.a.	n.a.	n.a.	n.a.
		S2	3.13E+04	2.43E+04	1.91E+04	1.27E+04	3.40E+04	2.74E+04	2.22E+04	1.44E+04	n.a.	n.a.	n.a.	n.a.	n.a.	n.a.	n.a.	n.a.

34

Table A6. Emissions of CO_2 according to climate zone (CZ), orientation (O), constructive solution (S), building model and insulation material model (in Kg CO_2), for H3.

(b)

CZ	O	S	Model 1					Model 2					Model 3					Model 4			
			XPS	PUR	SW	EC	XPS	PUR	SW	EC	XPS	PUR	SW	EC	XPS	PUR	SW	EC	XPS	PUR
C1	N	S1	1.41E+04	1.11E+04	8.77E+03	5.73E+03	1.78E+04	1.37E+04	1.07E+04	7.19E+03	4.13E+04	3.19E+04	2.50E+04	1.70E+04	5.60E+04	4.39E+04	3.48E+04	n.a.
		S2	1.72E+04	1.33E+04	1.04E+04	6.96E+03	2.10E+04	1.60E+04	1.32E+04	8.46E+03	4.13E+04	3.33E+04	2.70E+04	1.75E+04	6.01E+04	4.83E+04	3.90E+04	n.a.
	E	S1	1.57E+04	1.22E+04	9.56E+03	6.55E+03	1.94E+04	1.48E+04	1.15E+04	8.04E+03	4.32E+04	3.47E+04	2.80E+04	1.85E+04	5.60E+04	4.39E+04	3.48E+04	n.a.
		S2	1.88E+04	1.44E+04	1.12E+04	7.78E+03	2.10E+04	1.71E+04	1.40E+04	8.88E+03	4.52E+04	3.61E+04	2.90E+04	1.90E+04	6.01E+04	4.68E+04	3.69E+04	n.a.
	S	S1	1.88E+04	1.44E+04	1.12E+04	7.78E+03	2.10E+04	1.60E+04	1.32E+04	8.46E+03	5.30E+04	4.16E+04	3.30E+04	2.21E+04	6.01E+04	4.68E+04	3.69E+04	n.a.
		S2	1.88E+04	1.44E+04	1.12E+04	7.78E+03	2.10E+04	1.71E+04	1.40E+04	8.88E+03	5.50E+04	4.30E+04	3.40E+04	2.26E+04	6.84E+04	5.41E+04	n.a.	n.a.
	W	S1	1.88E+04	1.44E+04	1.12E+04	7.78E+03	2.10E+04	1.60E+04	1.32E+04	8.46E+03	5.70E+04	4.58E+04	3.70E+04	2.41E+04	7.05E+04	5.56E+04	n.a.	n.a.
		S2	2.04E+04	1.66E+04	1.35E+04	8.60E+03	2.26E+04	1.83E+04	1.48E+04	9.73E+03	5.89E+04	4.71E+04	3.80E+04	2.52E+04	n.a.	n.a.	n.a.	n.a.
C2	N	S1	1.41E+04	1.11E+04	8.77E+03	5.73E+03	1.78E+04	1.37E+04	1.07E+04	7.19E+03	4.13E+04	3.19E+04	2.50E+04	1.70E+04	5.60E+04	4.39E+04	3.48E+04	n.a.
		S2	1.72E+04	1.33E+04	1.04E+04	6.96E+03	1.94E+04	1.48E+04	1.15E+04	8.04E+03	4.13E+04	3.33E+04	2.70E+04	1.75E+04	6.01E+04	4.68E+04	3.69E+04	n.a.
	E	S1	1.41E+04	1.11E+04	8.77E+03	5.73E+03	1.94E+04	1.48E+04	1.15E+04	8.04E+03	4.32E+04	3.33E+04	2.70E+04	1.75E+04	5.60E+04	4.39E+04	3.48E+04	n.a.
		S2	1.88E+04	1.44E+04	1.12E+04	7.78E+03	2.10E+04	1.71E+04	1.40E+04	8.88E+03	4.32E+04	3.47E+04	2.80E+04	1.85E+04	6.01E+04	4.68E+04	3.69E+04	n.a.
	S	S1	1.57E+04	1.22E+04	9.56E+03	6.55E+03	1.94E+04	1.48E+04	1.15E+04	8.04E+03	5.11E+04	4.02E+04	3.20E+04	2.11E+04	5.60E+04	4.39E+04	3.48E+04	n.a.
		S2	1.88E+04	1.44E+04	1.12E+04	7.78E+03	2.10E+04	1.71E+04	1.40E+04	8.88E+03	5.30E+04	4.16E+04	3.30E+04	n.a.	6.63E+04	5.27E+04	4.22E+04	n.a.
	W	S1	1.72E+04	1.33E+04	1.04E+04	6.96E+03	2.10E+04	1.60E+04	1.32E+04	8.46E+03	5.70E+04	4.58E+04	3.70E+04	2.41E+04	7.05E+04	5.56E+04	n.a.	n.a.
		S2	1.55E+04	1.28E+04	8.19E+03	8.19E+03	2.10E+04	1.71E+04	1.40E+04	8.88E+03	5.89E+04	4.71E+04	3.80E+04	n.a.	n.a.	n.a.	n.a.	n.a.
C3	N	S1	1.25E+04	9.95E+03	7.97E+03	5.32E+03	1.46E+04	1.14E+04	9.05E+03	5.92E+03	3.34E+04	2.63E+04	2.10E+04	1.39E+04	4.35E+04	3.51E+04	2.85E+04	1.84E+04
		S2	1.41E+04	1.11E+04	8.77E+03	6.55E+03	1.78E+04	1.37E+04	1.07E+04	7.19E+03	3.54E+04	2.77E+04	2.20E+04	1.49E+04	4.56E+04	3.66E+04	2.95E+04	1.95E+04
	E	S1	1.41E+04	1.11E+04	8.77E+03	5.73E+03	1.62E+04	1.26E+04	9.88E+03	6.77E+03	3.54E+04	2.77E+04	2.20E+04	1.49E+04	4.35E+04	3.51E+04	2.85E+04	1.84E+04
		S2	1.57E+04	1.22E+04	9.56E+03	6.55E+03	1.94E+04	1.48E+04	1.15E+04	8.04E+03	3.73E+04	2.91E+04	2.30E+04	1.54E+04	4.56E+04	3.66E+04	2.95E+04	1.95E+04
	S	S1	1.41E+04	1.11E+04	8.77E+03	5.73E+03	1.62E+04	1.26E+04	9.88E+03	6.77E+03	4.13E+04	3.33E+04	2.70E+04	1.75E+04	4.77E+04	3.80E+04	3.06E+04	2.01E+04
		S2	1.57E+04	1.22E+04	9.56E+03	6.55E+03	1.94E+04	1.48E+04	1.15E+04	8.04E+03	4.32E+04	3.47E+04	2.80E+04	1.85E+04	5.18E+04	4.10E+04	3.27E+04	2.17E+04
	W	S1	1.57E+04	1.22E+04	9.56E+03	6.55E+03	1.78E+04	1.37E+04	1.07E+04	7.19E+03	4.71E+04	3.74E+04	3.00E+04	2.00E+04	5.60E+04	4.39E+04	3.48E+04	n.a.
		S2	1.72E+04	1.33E+04	1.04E+04	6.96E+03	1.94E+04	1.48E+04	1.15E+04	8.04E+03	4.91E+04	3.88E+04	3.10E+04	2.05E+04	6.84E+04	5.41E+04	3.48E+04	n.a.
C4	N	S1	1.25E+04	9.95E+03	7.97E+03	5.32E+03	1.46E+04	1.14E+04	9.05E+03	5.92E+03	3.34E+04	2.63E+04	2.10E+04	1.39E+04	4.56E+04	3.66E+04	2.95E+04	1.95E+04
		S2	1.57E+04	1.22E+04	9.56E+03	6.55E+03	1.78E+04	1.37E+04	1.07E+04	7.19E+03	3.73E+04	2.91E+04	2.30E+04	1.54E+04	5.39E+04	4.24E+04	3.37E+04	n.a.
	E	S1	1.41E+04	1.11E+04	8.77E+03	5.73E+03	1.62E+04	1.26E+04	9.88E+03	6.77E+03	3.54E+04	2.77E+04	2.20E+04	1.49E+04	4.56E+04	3.66E+04	2.95E+04	1.95E+04
		S2	1.57E+04	1.22E+04	9.56E+03	6.55E+03	1.94E+04	1.48E+04	1.15E+04	8.04E+03	4.13E+04	3.19E+04	2.50E+04	1.70E+04	5.39E+04	4.24E+04	3.37E+04	n.a.
	S	S1	1.41E+04	1.11E+04	8.77E+03	5.73E+03	1.78E+04	1.37E+04	1.07E+04	7.19E+03	4.13E+04	3.19E+04	2.50E+04	1.70E+04	5.39E+04	4.24E+04	3.37E+04	n.a.
		S2	1.72E+04	1.33E+04	1.04E+04	6.96E+03	1.94E+04	1.48E+04	1.15E+04	7.19E+04	4.13E+04	3.61E+04	2.90E+04	1.90E+04	6.01E+04	4.83E+04	3.90E+04	n.a.
	W	S1	1.57E+04	1.22E+04	9.56E+03	6.55E+03	1.78E+04	1.37E+04	1.07E+04	7.19E+03	4.52E+04	3.61E+04	2.90E+04	1.90E+04	6.01E+04	4.68E+04	3.69E+04	n.a.
		S2	1.88E+04	1.44E+04	1.12E+04	7.78E+03	2.10E+04	1.60E+04	1.32E+04	8.46E+03	5.30E+04	4.16E+04	3.30E+04	n.a.	7.46E+04	5.85E+04	n.a.	n.a.

35

Table A6. *Cont.*

(b)

CZ	O	S	Model 1					Model 2					Model 3					Model 4				
			XPS	PUR	SW	EC		XPS	PUR	SW	EC		XPS	PUR	SW	EC		XPS	PUR	SW	EC	
C1	N	S1	1.41E+04	1.11E+04	8.77E+03	5.73E+03	1.78E+04	1.37E+04	1.07E+04	7.19E+03	4.13E+04	3.19E+04	2.50E+04	1.70E+04	5.60E+04	4.39E+04	3.48E+04	n.a.
		S2	1.72E+04	1.33E+04	1.04E+04	6.96E+03	2.10E+04	1.60E+04	1.32E+04	8.46E+03	4.13E+04	3.33E+04	2.70E+04	1.75E+04	6.01E+04	4.83E+04	3.90E+04	n.a.
	E	S1	1.57E+04	1.22E+04	9.56E+03	6.55E+03	1.94E+04	1.48E+04	1.15E+04	8.04E+03	4.32E+04	3.47E+04	2.80E+04	1.85E+04	5.60E+04	4.39E+04	3.48E+04	n.a.
		S2	1.88E+04	1.44E+04	1.12E+04	7.78E+03	2.10E+04	1.71E+04	1.40E+04	8.88E+03	4.52E+04	3.61E+04	2.90E+04	1.90E+04	6.01E+04	4.68E+04	3.69E+04	n.a.
	S	S1	1.88E+04	1.44E+04	1.12E+04	7.78E+03	2.10E+04	1.60E+04	1.32E+04	8.46E+03	5.30E+04	4.16E+04	3.30E+04	2.21E+04	6.01E+04	4.68E+04	3.69E+04	n.a.
		S2	1.88E+04	1.44E+04	1.12E+04	7.78E+03	2.10E+04	1.71E+04	1.40E+04	8.88E+03	5.50E+04	4.30E+04	3.40E+04	2.26E+04	6.84E+04	5.41E+04	n.a.	n.a.
	W	S1	1.88E+04	1.44E+04	1.12E+04	7.78E+03	2.10E+04	1.60E+04	1.32E+04	8.46E+03	5.70E+04	4.58E+04	3.70E+04	2.41E+04	7.05E+04	5.56E+04	n.a.	n.a.
		S2	2.04E+04	1.66E+04	1.35E+04	8.60E+03	2.26E+04	1.83E+04	1.48E+04	9.73E+03	5.89E+04	4.71E+04	3.80E+04	2.52E+04	n.a.	n.a.	n.a.	n.a.
C2	N	S1	1.41E+04	1.11E+04	8.77E+03	5.73E+03	1.78E+04	1.37E+04	1.07E+04	7.19E+03	4.13E+04	3.19E+04	2.50E+04	1.70E+04	5.60E+04	4.39E+04	3.48E+04	n.a.
		S2	1.72E+04	1.33E+04	1.04E+04	6.96E+03	1.94E+04	1.48E+04	1.15E+04	8.04E+03	4.13E+04	3.33E+04	2.70E+04	1.75E+04	6.01E+04	4.68E+04	3.69E+04	n.a.
	E	S1	1.41E+04	1.11E+04	8.77E+03	5.73E+03	1.94E+04	1.48E+04	1.15E+04	8.04E+03	4.13E+04	3.33E+04	2.70E+04	1.75E+04	5.60E+04	4.39E+04	3.48E+04	n.a.
		S2	1.88E+04	1.44E+04	1.12E+04	7.78E+03	2.10E+04	1.71E+04	1.40E+04	8.88E+03	4.32E+04	3.47E+04	2.80E+04	1.85E+04	6.01E+04	4.68E+04	3.69E+04	n.a.
	S	S1	1.57E+04	1.22E+04	9.56E+03	6.55E+03	1.94E+04	1.48E+04	1.15E+04	8.04E+03	5.11E+04	4.02E+04	3.20E+04	2.11E+04	5.60E+04	4.39E+04	3.48E+04	n.a.
		S2	1.88E+04	1.44E+04	1.12E+04	7.78E+03	1.94E+04	1.71E+04	1.40E+04	8.88E+03	5.30E+04	4.16E+04	3.30E+04	n.a.	6.63E+04	5.27E+04	4.22E+04	n.a.
	W	S1	1.72E+04	1.33E+04	1.04E+04	6.96E+03	2.10E+04	1.60E+04	1.32E+04	8.46E+03	5.70E+04	4.58E+04	3.70E+04	2.41E+04	7.05E+04	5.56E+04	n.a.	n.a.
		S2	2.04E+04	1.55E+04	1.28E+04	8.19E+03	2.10E+04	1.71E+04	1.40E+04	8.88E+03	5.89E+04	4.71E+04	3.80E+04	n.a.	n.a.	n.a.	n.a.	n.a.
C3	N	S1	1.25E+04	9.95E+03	7.97E+03	5.32E+03	1.46E+04	1.14E+04	9.05E+03	5.92E+03	3.34E+04	2.63E+04	2.10E+04	1.39E+04	4.35E+04	3.51E+04	2.85E+04	1.84E+04
		S2	1.41E+04	1.11E+04	8.77E+03	5.73E+03	1.78E+04	1.37E+04	1.07E+04	7.19E+03	3.54E+04	2.77E+04	2.20E+04	1.49E+04	4.56E+04	3.66E+04	2.95E+04	1.95E+04
	E	S1	1.41E+04	1.11E+04	8.77E+03	5.73E+03	1.62E+04	1.26E+04	9.88E+03	6.77E+03	3.54E+04	2.77E+04	2.20E+04	1.49E+04	4.35E+04	3.51E+04	2.85E+04	1.84E+04
		S2	1.57E+04	1.22E+04	9.56E+03	6.55E+03	1.94E+04	1.48E+04	1.15E+04	8.04E+03	3.73E+04	2.91E+04	2.30E+04	1.54E+04	4.56E+04	3.66E+04	2.95E+04	1.95E+04
	S	S1	1.41E+04	1.11E+04	8.77E+03	5.73E+03	1.62E+04	1.26E+04	9.88E+03	6.77E+03	4.13E+04	3.33E+04	2.70E+04	1.75E+04	4.77E+04	3.80E+04	3.06E+04	2.01E+04
		S2	1.57E+04	1.22E+04	9.56E+03	6.55E+03	1.94E+04	1.48E+04	1.15E+04	8.04E+03	4.32E+04	3.47E+04	2.80E+04	1.85E+04	5.18E+04	4.10E+04	3.27E+04	2.17E+04
	W	S1	1.57E+04	1.22E+04	9.56E+03	6.55E+03	1.78E+04	1.37E+04	1.07E+04	7.19E+03	4.71E+04	3.74E+04	3.00E+04	2.00E+04	5.60E+04	4.39E+04	3.48E+04	n.a.
		S2	1.72E+04	1.33E+04	1.04E+04	6.96E+03	1.94E+04	1.48E+04	1.15E+04	8.04E+03	4.91E+04	3.88E+04	3.10E+04	2.05E+04	6.84E+04	5.41E+04	n.a.	n.a.
C4	N	S1	1.25E+04	9.95E+03	7.97E+03	5.32E+03	1.46E+04	1.14E+04	9.05E+03	5.92E+03	3.34E+04	2.63E+04	2.10E+04	1.39E+04	4.56E+04	3.66E+04	2.95E+04	1.95E+04
		S2	1.57E+04	1.22E+04	9.56E+03	6.55E+03	1.78E+04	1.37E+04	1.07E+04	7.19E+03	3.73E+04	2.91E+04	2.30E+04	1.54E+04	5.39E+04	4.24E+04	3.37E+04	n.a.
	E	S1	1.41E+04	1.11E+04	8.77E+03	5.73E+03	1.62E+04	1.26E+04	9.88E+03	6.77E+03	3.54E+04	2.77E+04	2.20E+04	1.49E+04	4.56E+04	3.66E+04	2.95E+04	1.95E+04
		S2	1.57E+04	1.22E+04	9.56E+03	6.55E+03	1.94E+04	1.48E+04	1.15E+04	8.04E+03	4.13E+04	3.19E+04	2.50E+04	1.70E+04	5.39E+04	4.24E+04	3.37E+04	n.a.
	S	S1	1.41E+04	1.11E+04	8.77E+03	5.73E+03	1.78E+04	1.37E+04	1.07E+04	7.19E+03	4.13E+04	3.19E+04	2.50E+04	1.70E+04	5.39E+04	4.24E+04	3.37E+04	n.a.
		S2	1.72E+04	1.33E+04	1.04E+04	6.96E+03	1.94E+04	1.48E+04	1.07E+04	8.04E+03	4.52E+04	3.61E+04	2.90E+04	1.90E+04	6.01E+04	4.83E+04	3.90E+04	n.a.
	W	S1	1.57E+04	1.22E+04	9.56E+03	6.55E+03	1.94E+04	1.48E+04	1.07E+04	7.19E+03	4.52E+04	3.61E+04	2.90E+04	1.90E+04	6.01E+04	4.68E+04	3.69E+04	n.a.
		S2	1.88E+04	1.44E+04	1.12E+04	7.78E+03	2.10E+04	1.60E+04	1.32E+04	8.46E+03	5.30E+04	4.16E+04	3.30E+04	n.a.	7.46E+04	5.85E+04	n.a.	n.a.

Table A6. Cont.

(c)

CZ	O	s	Model 1				Model 2				Model 3				Model 4			
			XPS	PUR	SW	EC	XPS	PUR	SW	EC	XPS	PUR	SW	EC	XPS	PUR	SW	EC
D1	N	S1	4.23E+04	3.32E+04	2.63E+04	n.a.	4.69E+04	3.65E+04	2.88E+04	n.a.	n.a.	n.a.	n.a.	n.a.	n.a.	n.a.	n.a.	n.a.
		S2	5.01E+04	3.98E+04	3.19E+04	n.a.	5.66E+04	4.45E+04	n.a.	n.a.	n.a.	n.a.	n.a.	n.a.	n.a.	n.a.	n.a.	n.a.
	E	S1	4.39E+04	3.43E+04	2.71E+04	n.a.	5.34E+04	4.22E+04	n.a.	n.a.	n.a.	n.a.	n.a.	n.a.	n.a.	n.a.	n.a.	n.a.
		S2	5.17E+04	4.09E+04	n.a.	n.a.	6.47E+04	n.a.	n.a.	n.a.	n.a.	n.a.	n.a.	n.a.	n.a.	n.a.	n.a.	n.a.
	S	S1	4.39E+04	3.43E+04	2.71E+04	n.a.	6.15E+04	n.a.	n.a.	n.a.	n.a.	n.a.	n.a.	n.a.	n.a.	n.a.	n.a.	n.a.
		S2	5.95E+04	n.a.	n.a.	n.a.	n.a.	n.a.	n.a.	n.a.	n.a.	n.a.	n.a.	n.a.	n.a.	n.a.	n.a.	n.a.
	W	S1	4.70E+04	3.76E+04	3.03E+04	n.a.	6.31E+04	n.a.	n.a.	n.a.	n.a.	n.a.	n.a.	n.a.	n.a.	n.a.	n.a.	n.a.
		S2	4.86E+04	n.a.	n.a.	n.a.	n.a.	n.a.	n.a.	n.a.	n.a.	n.a.	n.a.	n.a.	n.a.	n.a.	n.a.	n.a.
D2	N	S1	4.07E+04	3.21E+04	2.55E+04	n.a.	4.69E+04	3.65E+04	2.88E+04	n.a.	n.a.	n.a.	n.a.	n.a.	n.a.	n.a.	n.a.	n.a.
		S2	4.70E+04	3.76E+04	3.03E+04	n.a.	5.34E+04	4.22E+04	n.a.	n.a.	n.a.	n.a.	n.a.	n.a.	n.a.	n.a.	n.a.	n.a.
	E	S1	4.23E+04	3.32E+04	2.63E+04	n.a.	5.18E+04	4.11E+04	3.29E+04	n.a.	n.a.	n.a.	n.a.	n.a.	n.a.	n.a.	n.a.	n.a.
		S2	4.86E+04	3.87E+04	3.11E+04	n.a.	6.47E+04	n.a.	n.a.	n.a.	n.a.	n.a.	n.a.	n.a.	n.a.	n.a.	n.a.	n.a.
	S	S1	4.54E+04	3.54E+04	2.79E+04	n.a.	5.34E+04	4.22E+04	n.a.	n.a.	n.a.	n.a.	n.a.	n.a.	n.a.	n.a.	n.a.	n.a.
		S2	5.64E+04	4.42E+04	n.a.	n.a.	n.a.	n.a.	n.a.	n.a.	n.a.	n.a.	n.a.	n.a.	n.a.	n.a.	n.a.	n.a.
	W	S1	4.86E+04	3.87E+04	3.11E+04	n.a.	5.82E+04	4.57E+04	n.a.	n.a.	n.a.	n.a.	n.a.	n.a.	n.a.	n.a.	n.a.	n.a.
		S2	n.a.	n.a.	n.a.	n.a.	n.a.	n.a.	n.a.	n.a.	n.a.	n.a.	n.a.	n.a.	n.a.	n.a.	n.a.	n.a.
D3	N	S1	3.60E+04	2.88E+04	2.31E+04	1.52E+04	4.04E+04	3.20E+04	2.55E+04	1.69E+04	n.a.	n.a.	n.a.	n.a.	n.a.	n.a.	n.a.	n.a.
		S2	3.92E+04	3.10E+04	2.47E+04	1.64E+04	4.37E+04	3.43E+04	2.72E+04	n.a.	n.a.	n.a.	n.a.	n.a.	n.a.	n.a.	n.a.	n.a.
	E	S1	3.76E+04	2.99E+04	2.39E+04	1.60E+04	4.37E+04	3.43E+04	2.72E+04	n.a.	n.a.	n.a.	n.a.	n.a.	n.a.	n.a.	n.a.	n.a.
		S2	4.23E+04	3.32E+04	2.63E+04	n.a.	4.85E+04	3.88E+04	3.13E+04	n.a.	n.a.	n.a.	n.a.	n.a.	n.a.	n.a.	n.a.	n.a.
	S	S1	3.92E+04	3.10E+04	2.47E+04	1.64E+04	4.69E+04	3.65E+04	2.88E+04	n.a.	n.a.	n.a.	n.a.	n.a.	n.a.	n.a.	n.a.	n.a.
		S2	4.54E+04	3.65E+04	2.95E+04	n.a.	5.18E+04	4.11E+04	3.29E+04	n.a.	n.a.	n.a.	n.a.	n.a.	n.a.	n.a.	n.a.	n.a.
	W	S1	4.39E+04	3.43E+04	2.71E+04	n.a.	4.85E+04	3.88E+04	3.13E+04	n.a.	n.a.	n.a.	n.a.	n.a.	n.a.	n.a.	n.a.	n.a.
		S2	5.17E+04	4.09E+04	n.a.	n.a.	5.50E+04	4.34E+04	n.a.	n.a.	n.a.	n.a.	n.a.	n.a.	n.a.	n.a.	n.a.	n.a.
E1	N	S1	n.a.	n.a.	n.a.	n.a.	n.a.	n.a.	n.a.	n.a.	n.a.	n.a.	n.a.	n.a.	n.a.	n.a.	n.a.	n.a.
		S2	n.a.	n.a.	n.a.	n.a.	n.a.	n.a.	n.a.	n.a.	n.a.	n.a.	n.a.	n.a.	n.a.	n.a.	n.a.	n.a.
	E	S1	n.a.	n.a.	n.a.	n.a.	n.a.	n.a.	n.a.	n.a.	n.a.	n.a.	n.a.	n.a.	n.a.	n.a.	n.a.	n.a.
		S2	n.a.	n.a.	n.a.	n.a.	n.a.	n.a.	n.a.	n.a.	n.a.	n.a.	n.a.	n.a.	n.a.	n.a.	n.a.	n.a.
	S	S1	n.a.	n.a.	n.a.	n.a.	n.a.	n.a.	n.a.	n.a.	n.a.	n.a.	n.a.	n.a.	n.a.	n.a.	n.a.	n.a.
		S2	n.a.	n.a.	n.a.	n.a.	n.a.	n.a.	n.a.	n.a.	n.a.	n.a.	n.a.	n.a.	n.a.	n.a.	n.a.	n.a.
	W	S1	n.a.	n.a.	n.a.	n.a.	n.a.	n.a.	n.a.	n.a.	n.a.	n.a.	n.a.	n.a.	n.a.	n.a.	n.a.	n.a.
		S2	n.a.	n.a.	n.a.	n.a.	n.a.	n.a.	n.a.	n.a.	n.a.	n.a.	n.a.	n.a.	n.a.	n.a.	n.a.	n.a.

References

1. Díaz, J.M.C.; Araujo, J.M. Historic Urbanization Process in Spain (1746–2013): From the Fall of the American Empire to the Real Estate Bubble. *J. Urban Hist.* **2017**, *43*, 33–52. [CrossRef]
2. Pozueta Echávarri, J. Rasgos urbanísticos del crecimiento residencial asociado a la burbuja inmobiliaria. 1995–2006. *Cuad. Investig. Urbanística* **2015**, *100*, 87–94. [CrossRef]
3. Jiménez, V.; Hidalgo, R.; Campesino, A.-J.; Alvarado, V. Normalización del modelo neoliberal de expansión residencial más allá del límite urbano en Chile y España. *EURE* **2018**, *44*, 27–46. [CrossRef]
4. Catalán, B.; Saurí, D.; Serra, P. Urban sprawl in the Mediterranean? *Landsc. Urban Plan.* **2008**, *85*, 174–184. [CrossRef]
5. Gil-Alonso, F.; Bayona-i-Carrasco, J.; Pujadas-i-Rúbies, I. From boom to crash: Spanish urban areas in a decade of change (2001–2011). *Eur. Urban Reg. Stud.* **2013**, *23*, 198–216. [CrossRef]
6. Faiella, I.; Mistretta, A. Energy Costs and Competitiveness in Europe Preliminary Draft. In Proceedings of the Sixth IAERE (Italian Association of Environmental and Resource Economists) Annual Conference, Rome, Italy, 15–16 February 2018.
7. Merini, I.; Molina-García, A.; García-Cascales, M.S.; Ahachad, M. *Energy Efficiency Regulation and Requirements: Comparison Between Morocco and Spain*; Springer: Cham, Switzerland, 2019; Volume 914, pp. 197–209. ISBN 978-3-030-11883-9.
8. Duarte, R.; Sánchez-Chóliz, J.; Sarasa, C. Consumer-side actions in a low-carbon economy: A dynamic CGE analysis for Spain. *Energy Policy* **2018**, *118*, 199–210. [CrossRef]
9. Tavares, V.; Lacerda, N.; Freire, F. Embodied energy and greenhouse gas emissions analysis of a prefabricated modular house: The "Moby" case study. *J. Clean. Prod.* **2019**, *212*, 1044–1053. [CrossRef]
10. Lawson, R.M.; Ogden, R.G.; Bergin, R. Application of Modular Construction in High-Rise Buildings. *J. Archit. Eng.* **2011**, *18*, 148–154. [CrossRef]
11. Mostafa, S.; Tam, V.W.; Dumrak, J.; Mohamed, S. Leagile Strategies for Optimizing the Delivery of Prefabricated House Building Projects. *Int. J. Constr. Manag.* **2018**, 1–15. [CrossRef]
12. Park, J. Prefabricated House: Defining Architectural Quality and Identity through the Innovation of Prefab Tectonics. Ph.D. Thesis, University of Hawaii at Manoa, Honolulu, HI, USA, 2017.
13. Lolli, N.; Hestness, A.G. The influence of different electricity-to-emissions conversion factors on the choice of insulation materials. *Energy Build.* **2014**, *85*, 362–373. [CrossRef]
14. Laaouatni, A.; Martaj, N.; Bennacer, R.; Lachi, M.; El Omari, M.; El Ganaoui, M. Thermal building control using active ventilated block integrating phase change material. *Energy Build.* **2019**, *187*, 50–63. [CrossRef]
15. The European Parliament and the Council of the European Union. Directive 2002/91/EC on the Energy Performance of Buildings. *Off. J. Eur. Communities* **2003**, *L001*, 65–71.
16. Spanish Ministry of the Presidency Royal. Decree 314/2006 for the approval of the Technical Building Code. *Off. State Gaz. Gov. Spain* **2006**, *74*, 11816–11831.
17. Spanish Ministry of the Presidency Royal. Decree 235/2013 for the approval of the basic procedure for the Certification of the Energy Efficiency of Buildings. *Off. State Gaz. Gov. Spain* **2013**, *89*, 27548–27562.
18. The European Parliament and the Council of the European Union. Directive 2010/31/EU on the Energy Performance of Buildings. *Off. J. Eur. Union* **2010**, *L153*, 13–35.
19. Spanish Ministry of the Presidency Royal. Decree 238/2013 amending the Regulation for Thermal Installations in Buildings. *Off. State Gaz. Gov. Spain* **2013**, *89*, 27563–27593.
20. The European Parliament and the Council of the European Union. Directive 2012/27/EU on Energy Efficiency. *Off. J. Eur. Union* **2012**, *L315*, 1–56.
21. Zhai, Z.J.; Helman, J.M. Implications of climate changes to building energy and design. *Sustain. Cities Soc.* **2019**, *44*, 511–519. [CrossRef]
22. Cellura, M.; Guarino, F.; Longo, S.; Tumminia, G. Climate change and the building sector: Modelling and energy implications to an office building in southern Europe. *Energy Sustain. Dev.* **2018**, *45*, 46–65. [CrossRef]
23. *Spanish Ministry of Development Basic Document of Energy Saving of the Technical Building Code*; Ministry of Development: Madrid, Spain, 2017; ISBN 978-9504628675.
24. Pérez-Andreu, V.; Aparicio-Fernández, C.; Martínez-Ibernón, A.; Vivancos, J.L. Impact of climate change on heating and cooling energy demand in a residential building in a Mediterranean climate. *Energy* **2018**, *165*, 63–74. [CrossRef]

25. Braulio-Gonzalo, M.; Bovea, M.D. Environmental and cost performance of building's envelope insulation materials to reduce energy demand: Thickness optimization. *Energy Build.* **2017**, *150*, 527–545. [CrossRef]
26. Hill, C.; Norton, A.; Dibdiakova, J. A comparison of the environmental impacts of different categories of insulation materials. *Energy Build.* **2018**, *162*, 12–20. [CrossRef]
27. Pargana, N.; Duarte Pinheiro, M.; Dinis Silvestre, J.; de Brito, J. Comparative environmental life cycle assessment of thermal insulation materials of buildings. *Energy Build.* **2014**, *82*, 466–481. [CrossRef]
28. Sierra-Pérez, J.; Boschmonart-Rives, J.; Gabarrell, X. Environmental assessment of façade-building systems and thermal insulation materials for different climatic conditions. *J. Clean. Prod.* **2016**, *113*, 102–113. [CrossRef]
29. Asdrubali, F.; D'Alessandro, F.; Schiavoni, S. A review of unconventional sustainable building insulation materials. *Sustain. Mater. Technol.* **2015**, *4*, 1–17. [CrossRef]
30. Schiavoni, S.; D'Alessandro, F.; Bianchi, F.; Asdrubali, F. Insulation materials for the building sector: A review and comparative analysis. *Renew. Sustain. Energy Rev.* **2016**, *62*, 988–1011. [CrossRef]
31. Kabanshi, A.; Ameen, A.; Hayati, A.; Yang, B. Cooling energy simulation and analysis of an intermittent ventilation strategy under different climates. *Energy* **2018**, *156*, 84–94. [CrossRef]
32. Salah-Eldin Imbabi, M. A passive-active dynamic insulation system for all climates. *Int. J. Sustain. Built Environ.* **2012**, *1*, 247–258. [CrossRef]
33. International Organization for Standardization ISO 52016-1. *Energy Performance of Buildings. Energy Needs for Heating and Cooling, Internal Temperatures and Sensible and Latent Heat Loads. Part 1: Calculation Procedures*; ISO: Geneva, Switzerland, 2017.
34. Spanish Ministry of Development. *Unified Tool LIDER CALENER 2017*. Available online: https://www.codigotecnico.org/index.php/menu-recursos/menu-aplicaciones/282-herramienta-unificada-lider-calener.html (accessed on 24 March 2019).
35. International Organization for Standardization ISO 52000-1. *Energy Performance of Buildings. Overarching EPB Assessment. Part 1: General Framework and Procedures*; ISO: Geneva, Switzerland, 2017.
36. Passive House Institute Passive House requirements. Available online: https://passivehouse.com/02_informations/02_passive-house-requirements/02_passive-house-requirements.htm (accessed on 30 March 2019).
37. Spanish Statistical Office. Available online: https://www.ine.es/ (accessed on 23 February 2019).
38. Spanos, I.; Simons, M.; Holmes, K.L. Cost savings by application of passive solar heating. *Struct. Surv.* **2005**, *23*, 111–130. [CrossRef]
39. Tettey, U.Y.A.; Dodoo, A.; Gustavsson, L. Effects of different insulation materials on primary energy and CO_2 emission of a multi-storey residential building. *Energy Build.* **2014**, *82*, 369–377. [CrossRef]
40. Aditya, L.; Mahlia, T.M.I.; Rismanchi, B.; Ng, H.M.; Hasan, M.H.; Metselaar, H.S.C.; Muraza, O.; Aditya, H.B. A review on insulation materials for energy conservation in buildings. *Renew. Sustain. Energy Rev.* **2017**, *73*, 1352–1365. [CrossRef]
41. Environmental Product Declaration (EPD) for XPS Insulation Board. Available online: https://gryphon4.environdec.com/system/data/files/6/12369/epd501enDANOPREN.pdf (accessed on 9 May 2019).
42. Environmental Product Declaration (EPD) for PU (PUR/PIR) Thermal Insulation Boards and Energy Saving Potential. Available online: http://highperformanceinsulation.eu/wp-content/uploads/2016/08/Factsheet_13-1_Environmental_product_declaration__EPD__for_PU__PUR-PIR__thermal_insulation_boards_and_energy_saving_potential__updated_12-12-14_.pdf (accessed on 24 March 2019).
43. Declaración Ambiental de Producto (DAP) de Panel Lana de Roca. DAPc.001.003. Available online: http://download.rockwool.es/media/215213/dapc001_003_ROCKWOOL.pdf (accessed on 9 May 2019).
44. Environmental Product Declaration (EPD) for Insulation Cork Board (ICB)/Thermal Insulation. Available online: https://daphabitat.pt/assets/Uploads/dap/pdfs/76ad43077d/EPD_Solfalca_EN.pdf (accessed on 9 May 2019).

© 2019 by the authors. Licensee MDPI, Basel, Switzerland. This article is an open access article distributed under the terms and conditions of the Creative Commons Attribution (CC BY) license (http://creativecommons.org/licenses/by/4.0/).

Article

Thermal Transmittance of Internal Partition and External Facade LSF Walls: A Parametric Study

Paulo Santos *, Gabriela Lemes and Diogo Mateus

ISISE, Department of Civil Engineering, University of Coimbra, Pólo II, Rua Luís Reis Santos, 3030-788 Coimbra, Portugal
* Correspondence: pfsantos@dec.uc.pt; Tel.: +351-239-797-199

Received: 22 June 2019; Accepted: 10 July 2019; Published: 11 July 2019

Abstract: Light steel framed (LSF) construction is becoming widespread as a quick, clean and flexible construction system. However, these LSF elements need to be well designed and protected against undesired thermal bridges caused by the steel high thermal conductivity. To reduce energy consumption in buildings it is necessary to understand how heat transfer happens in all kinds of walls and their configurations, and to adequately reduce the heat loss through them by decreasing its thermal transmittance (U-value). In this work, numerical simulations are performed to assess different setups for two kinds of LSF walls: an interior partition wall and an exterior facade wall. Several parameters were evaluated separately to measure their influence on the wall U-value, and the addition of other elements was tested (e.g., thermal break strips) with the aim of achieving better thermal performances. The simulation modeling of a LSF interior partition with thermal break strips indicated a 24% U-value reduction in comparison with the reference case of using the LSF alone ($U = 0.449$ W/(m^2.K)). However, when the clearance between the steel studs was simulated with only 300 mm there was a 29% increase, due to the increase of steel material within the wall structure. For exterior facade walls ($U = 0.276$ W/(m^2.K)), the model with 80 mm of expanded polystyrene (EPS) in the exterior thermal insulation composite system (ETICS) reduced the thermal transmittance by 19%. Moreover, when the EPS was removed the U-value increased by 79%.

Keywords: LSF construction; facade wall; partition wall; thermal transmittance; thermal bridges; parametric study; numerical simulations

1. Introduction

Buildings account for around 40% of the total energy consumption and about 36% of CO2 emissions in Europe [1]. The main factors of building energy consumption are the properties and design of the building envelope, the operation of building services, the occupants' behavior and the climate/location [2–5]. Most of this energy, ranging from nearly 50% [6] up to 60% [7] depending on climate, design, use type and occupational patterns, is used by air-conditioning systems to achieve thermal comfort inside the buildings. Energy in the form of heat is dissipated to the environment at different rates according to the ventilation and building elements' characteristics (e.g., thermal transmittance U-value). The rate of these losses/gains is important because it directly affects the operation and maintenance costs of mechanically ventilated and/or air conditioned buildings [8].

Usually a wall element is composed of several layers, such as internal and external cladding (e.g., cement mortar), one or two supporting panes (e.g., ceramic brick masonry), air cavity, and thermal and acoustic insulation (e.g., expanded polystyrene (EPS) or mineral wool). Typically lightweight steel framed (LSF) walls are made of the following main types of materials [9]: (1) supporting steel frame, which is constituted of cold formed profiles; (2) sheathing panels, such as inner gypsum plasterboard and outer oriented strand board (OSB), and; (3) insulation materials, such as mineral wool filling the air cavity between steel studs (which besides thermal insulation, has also an important acoustic insulation role [10]) and the exterior thermal insulation composite system (ETICS), where the thermal insulation material could be EPS (expanded polystyrene), XPS (extruded polystyrene), mineral wool or other.

The U-value of an opaque building element (e.g., facade LSF wall) depends on several factors, such as the thickness of each layer, the number of layers, the thermal conductivity of each layer material, the existence of thermal bridges due to the presence of an inhomogeneous thermal layer (e.g., a steel stud), the existence of air voids in the insulation, and the external and internal surface thermal resistances [11]. Perhaps the most relevant parameters regarding the thermal transmittance of a LSF building element are the level of insulation (i.e., its thickness), material properties (e.g., thermal conductivity) and positioning of insulation, and the amount of steel frame material [7,12].

In colder climates to reduce the U-value, and consequently the heat transmission losses, the level of thermal insulation is increased to diminish the heating energy demand [13]. While in warmer climates this level of thermal insulation could be reduced, reducing energy consumption for space heating/cooling as well as the embodied energy related with the insulation materials [14]. In these warmer climates the outdoor temperatures are often higher than the indoor temperatures, which could significantly increase the heat gains. Thus, the use of passive cooling strategies, such as natural ventilation [15], phase change materials [16], free cooling [17] and ground ventilation using an earth-to-air heat exchanger [18] becomes more relevant. In order to predict the energy consumption it is usual to perform advanced dynamic simulations of the entire building [19,20] or make use of more simplified approaches [21].

Apart from the level of thermal insulation (i.e., the thickness of thermal insulation layer(s)), in LSF elements, the position in the building element influences the effectiveness of this insulation (i.e., its U-value or thermal transmittance), and is thus very relevant [12]. Notice, that the thermal insulation positioning is also relevant to the effective thermal inertia/mass of the building, but this was not evaluated in the present paper, neither in reference [12] work. Moreover, this insulation, mainly the LSF batt insulation (e.g., mineral wool), is relevant not only for thermal purposes but also for acoustic insulation [10]. A typical interior partition and exterior facade LSF wall cross-sections will be studied in this paper, as presented later in Sections 2.1 and 3.1, respectively.

At the design stage there are several ways to compute the U-value of a building element [11]. The detailed calculation method based on numerical simulations (e.g., finite element method (FEM)) should be performed using the modeling rules prescribed in standard ISO 10211 [22]. The most simple approach, applicable for homogeneous thermal layers, which may contain air layers up to 300 mm thick, is to consider the thermal resistance of each layer (depending on the thickness of the layer and on the thermal conductivity of the material) and to compute the reciprocal of the sum of all these thermal resistances, including both internal and external surface resistances [23]. Notice that the external thermal surface resistance mainly depends on the wind direction and velocity, as well as on the surface roughness [24].

The standard ISO 6946 [11] also prescribes an approximate method, known as the 'Combined Method', for building elements containing homogenous and inhomogeneous layers, including the effect of metal fasteners, by means of a U-value correction term. However, this methodology is not applicable for LSF elements, where the thermal insulation is bridged by metal (cold and hybrid frame construction), making this type of construction even more challenging in order to obtain an accurate and reliable U-value [23].

Several researchers devoted their attention to the thermal behavior and energy efficiency of LSF construction [5,9,23,25,26]. Soares et al. [26] performed a scientific bibliographic review about this kind of research. The first main driving research topic identified in the previous cited work was: "the development of single and combined strategies to reduce thermal bridges and to improve the thermal resistance of LSF envelope elements". The present work deals with this suggested main research issue. Recently Santos et al. [23] accomplished a comparison between experimental measurements in LSF walls' thermal transmittance and numerical simulations (2D and 3D FEM models) and analytical approach (ISO 6946 combined method). It was concluded that for the LSF wall with a simpler frame (i.e., only vertical steel studs) the analytical ISO 6946 and the 2D FEM numerical approaches provide quite good accuracy in the U-value estimation.

Since the ISO 6946 combined method is not applicable for LSF elements where the thermal insulation is bridged by the steel frames, some researchers developed some alternative analytical methods for this type of structure, such as Gorgolewsky [27] who developed a simplified analytical method for calculating U-values in LSF cold and hybrid construction. This method was based on the principles provided by ISO 6946, but adapted to consider the increased thermal effect of the steel frame, increasing the accuracy of the proposed methodology.

Given the high level of heterogeneity regarding the thermal conductivities of the materials composing the LSF elements, namely the steel frame and the thermal insulation, it is very challenging not only to accurately compute its thermal transmittance, but also to perform accurate and reliable measurements, both in-situ and in laboratory [8]. Regarding the experimental approach there are several methods for the thermal characterization of building elements, such as the heat flow meter (HFM) method, the guarded hot plate (GHP) method, the hot box (HB) method (which could be calibrated (CHB) or guarded (GHB)) and the infrared thermography (IRT) method. For LSF elements the most suitable experimental method, given its large heterogeneity in its component materials' thermal conductivity (e.g., steel and thermal insulation), is the hot box apparatus, since the measurements are not local, but instead in a representative wall area [28].

Recently, Atsonios et al. [29] developed two experimental methods for in-situ measurement of the overall thermal transmittance of cold frame LSF walls, namely the representative points method (RPM) and weighted area method (WAM). These methods make use of the analysis of the examined wall using thermal IR images with the recording and processing of indoor/outdoor air temperature and heat flux. Figure 1 displays an infrared thermal image of an LSF wall, where the thermal bridge's effect due to the high thermal conductivity of the vertical steel studs is quite visible. The vertical red lines denote higher surface temperatures due to an increased heat flow in the vicinity of each vertical steel profile, clearly identifying the position of them in the exterior colder surface of the LSF wall.

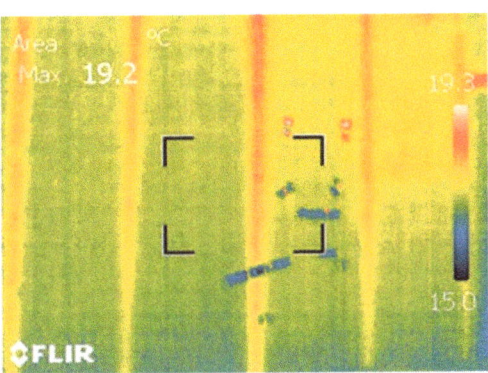

Figure 1. Thermal bridge's effect due to vertical steel studs in a light steel framed (LSF) wall captured in an infrared thermal image [30].

In fact, due to high thermal conductivity of steel in LSF structures, thermal bridges inspired many researchers to investigate the related thermal performance issues. De Angelis and Serra [31] evaluated the thermal insulation performance of metal framed lightweight walls and concluded that the correct evaluation of LSF walls' thermal performance requires more complex and detailed analysis than the ones necessary for traditional reinforced concrete and masonry constructions.

Also in 2014, Santos et al. [30] evaluated the importance of flanking thermal losses of LSF walls using a 3D FEM model validated by comparison with experimental laboratory measurements.

They found heat flux variations from −22% (external surface) to +50% (internal surface) when flanking heat loss was set to zero as a reference case for a LSF wall with a thermal transmittance equal to 0.30 W/(m².K). Later, in 2016, Martins et al. [32] performed a parametric study in order to evaluate the effectiveness of some thermal bridges mitigation strategies in LSF walls, allowing the improvement of thermal performance and reducing energy consumption by air-conditioning systems. A reduction of 8.3% in the U-value was found, comparatively to the reference LSF wall, due to these thermal bridges' mitigation strategies. Additionally, the use of new insulation materials (e.g., aerogel and vacuum insulation panels (VIPs)), which were combined with the mitigation approaches, led to a 68% decrease in the U-value.

In previous research works there was a lack of research on both interior partitions and exterior facade LSF walls, as well as the thermal performance comparison between them. In this work the thermal transmittance (U-value) of LSF walls is evaluated by means of a parametric study related with the wall typology (internal partition and external facade) and its composition. The main objective of this study is to quantify the relevance of several parameters in the U-value of LSF partition and facade walls. The evaluated parameters were selected among the most relevant ones and could be easily implemented in practice with used materials available in the market (e.g., recycled rubber, extruded polystyrene (XPS) and aerogel thermal break strips). Moreover, the analyzed LSF wall configurations were newly implemented for this study (i.e., are different from the ones evaluated before by other researchers).

The simulations were performed bi-dimensionally, and the results could be of interest to building developers and researchers, helping them to mitigate thermal bridges and achieving energy savings, whenever an LSF construction system is used. In Portugal (but probably also in other countries) most of the building designers neglect the effect of repetitive thermal bridges due to the steel frame on the thermal transmittance calculations of LSF elements, leading to lower and erroneous U-values. Consequently, the real building energy consumption will be higher than the predicted one in these cases and there is a higher probability of building pathologies related with the occurrence of interstitial condensations.

After this introduction, the evaluated interior and exterior LSF walls are presented, including the reference partition and facade LSF walls and the parameters used in the sensitivity analysis are described. Next, the accuracy of the used 2D FEM algorithm is verified by means of a comparison with ISO 10211 [22] test cases and with the analytical approach, defined in ISO 6946 [11], for a simplified model assuming no steel frame and homogeneous layers. Then, the 2D FEM simulations are explained, including the used boundary conditions and how the air layers were addressed in these simulations. After, the obtained results are presented and discussed for the two LSF wall typologies evaluated. Finally, the main conclusions of this work are presented.

2. Characterization of LSF Interior Walls

2.1. Reference LSF Interior Partition Wall

The reference interior wall is a configuration of an LSF wall normally used as an internal partition within the same dwelling. As illustrated in Figure 2 and listed in Table 1, this LSF internal partition is constituted by two gypsum plasterboards (12.5 mm thick each) on each side of the steel frame (made with steel studs C90, 90 mm wide, and 0.6 mm of steel sheet thickness) and the air cavity is fully filled with mineral wool batt insulation (90 mm). The distance between vertical profiles for internal reference walls was set on 600 mm. The total thickness of this partition wall is 140 mm.

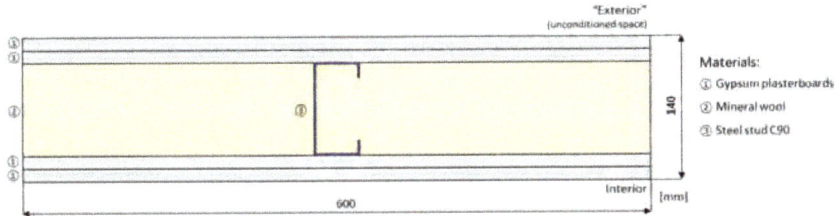

Figure 2. Cross-section of an interior LSF reference partition wall modeled on THERM software.

Table 1. Materials, thicknesses (*d*) and thermal conductivities (λ) of the LSF interior reference partition wall.

Material (From Outer to Innermost Layer)	*d* [mm]	λ [W/(m.K)]	Ref.
GPB [1] (2 × 12.5 mm)	25	0.175	[33]
Mineral wool	90	0.035	[34]
Steel stud (C90 × 43 × 15 × 0.6 mm)	90	50.000	[35]
GPB [1] (2 × 12.5 mm)	25	0.175	[33]
Total Thickness	140	-	-

[1] GPB—gypsum plasterboard.

Notice that, even being an internal partition, this LSF wall can separate a conditioned space from an unconditioned space (e.g., a garage), with lower temperature. Therefore, this internal partition also has thermal requirements. Table 1 also displays the thickness (*d*) of each material layer, as well as the thermal conductivity (λ) of each material. Usually the sheathing panels (e.g., gypsum plasterboard) are fixed to the LSF structure with metallic self-drilling screws. These fixing bolts were not considered in the simulations since its number is very reduced and the related punctual thermal bridge effect on the overall wall *U*-value is very reduced and, thus, could be neglected [12].

2.2. Parameters for the Sensitivity Analysis

Table 2 displays the parameters that will be evaluated in the sensitivity analysis, as well as the values to be used for each one. These models and parameters (illustrated in Figure 3) are: the thickness of the steel studs (Model I1); the clearance between steel studs (Model I2); the material and thickness of the thermal break (TB) strips (Model I3); the TB strip materials (Model I4), and; the sheathing panel materials (Model I5). The parameters and values used for each one will be briefly explained in the next paragraphs.

Table 2. Interior partition LSF wall: models and parameter values to be evaluated.

Model	Evaluated Parameter	Ref. Value	Value 1	Value 2	Value 3
I1	Thickness of Steel Studs [mm]	0.6	1.0	1.2	1.5
I2	Clearance Between Steel Studs [mm]	600	300	400	800
I3	Thickness of Aerogel TB [1] Strips [mm]	0.0	2.5	5.0	10.0
I4	Material of TB [1] Strips with 10 mm Sheathing Panels Materials	-	MS-R1 [2]	XPS [3]	CBS [4]
I5	GPB [5] Thickness [mm]	2 × 12.5	12.5	-	12.5
	OSB [6] Thickness [mm]	-	12.0	2 × 12.0	-
	XPS [3] Thickness [mm]	-	-	-	12.0

[1] TB—thermal break; [2] MS-R1—Acousticork (recycled rubber); [3] XPS—extruded polystyrene; [4] CBS—cold break strip (aerogel); [5] GPB—gypsum plasterboard; [6] OSB—oriented strand board.

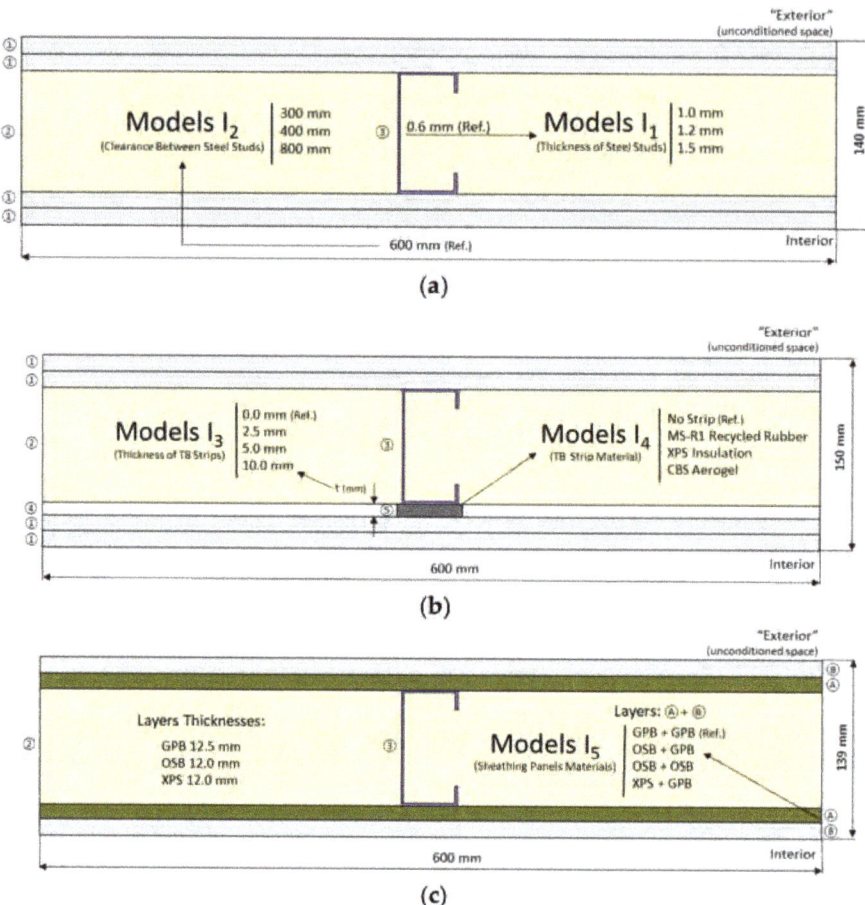

Figure 3. Interior LSF partition cross-sections: (**a**) Models I1 and I2; (**b**) Models I3 and I4; (**c**) Model I5. Layers: ① gypsum plasterboard (GPB); ② mineral wool; ③ steel stud C90; ④ air layer; ⑤ TB strip.

The first parameter to be evaluated was the steel studs thickness used in the wall steel frame (Model I1). The amount of steel inside the wall structure is very relevant because metal has a very high thermal conductivity and its presence in LSF frames create a path that allow the heat to easily cross through the walls, what is known as steel thermal bridges. The reference thickness of the internal partition steel studs is 0.6 mm, which is a usual value for a non-load-bearing partition wall. Steel profiles are also modeled as 1.0, 1.2 and 1.5 mm thick, as this can be found in load-bearing LSF walls (displayed in Table 2 and illustrated in Figure 3a).

The distance between vertical steel studs is another parameter that will be evaluated (Model I2) in order to assess its relevance on the thermal behavior of the LSF internal partitions (Figure 3a). The reference wall has a distance of 600 mm between steel studs (Figure 2), which is the most used clearance given the usual 1.20 m wideness of the sheathing panels. Three more distances will be evaluated in this parametric study, namely 300, 400 and 800 mm (Table 2).

Thermal break is obtained by the insertion of an insulation material (i.e., with a low thermal conductivity), between the steel sections and the innermost layer of the wall, minimizing the heat transfer through the thermal bridges caused by the steel structure and thus, improving/reducing the thermal transmittance (*U*-value) of the wall. In this parametric study three different thicknesses for an aerogel thermal break strip will be evaluated, namely 2.5, 5.0 and 10.0 mm (Model I3 in Figure 3b).

Nowadays, several materials are available to be used as thermal break strips in LSF structures, such as recycled rubber (an environmentally friendly solution), XPS (a cheaper solution) and aerogel (a state-of-the-art insulation material with very low thermal conductivity). In this assessment three different materials were tested as thermal break strips (see Model I4 in Figure 3b), namely: recycled rubber [36], extruded polystyrene (XPS) and cold break strip (CBS) aerogel [37], as displayed in Table 2. The thicknesses of the thermal break strips are 10.0 mm and thermal conductivities are listed in Table 3.

Table 3. Thermal conductivities (λ) of thermal break strips (10.0 mm thick).

Material	λ [W/(m.K)]	Ref.
Recycled Rubber (Acousticork MS-R1)	0.122	[38]
XPS [1] Insulation	0.037	[35]
CBS [2] Aerogel	0.015	[37]

[1] XPS—extruded polystyrene; [2] CBS—cold break strip.

To verify the influence of sheathing panel materials (Model I5), several configurations were modeled for the internal walls as shown in Table 2 and displayed in Figure 3c. The sheathing panels in the reference LSF wall are two gypsum plasterboard panels on each side of the steel structure. On the first parameter variation the inner gypsum plasterboard was replaced by one OSB panel in both sides of the LSF structure. On the second parameter variation, both gypsum plasterboards were replaced by two OSB panels on each side. Regarding the third parameter variation, the inner OSB panel was replaced by one XPS panel with the same thickness (12.0 mm), as illustrated in Figure 3c.

3. Characterization of LSF Exterior Walls

3.1. Reference LSF Exterior Facade Wall

The reference exterior wall is an LSF wall normally used for facades, which means that it is a wall that must be prepared to handle high gradients of environment temperature. Therefore, it has an extra thermal insulation layer which was placed on its outside surface. In this case, ETICS (external thermal insulation composite system) using EPS (expanded polystyrene) was chosen as the main insulation material (50 mm thick).

The steel structure that forms the wall frame is made of galvanized cold-formed steel studs and, different for internal walls, the thickness of the steel profile sheet is now 1.5 mm; since this kind of wall is very often a load bearing wall, C90 vertical studs were adopted. Similar to the interior LSF walls, the distance between the vertical profiles for the reference wall is 600 mm. The horizontal cross-section that shows all the layers of the reference exterior LSF wall is illustrated in Figure 4 and the specifications and characteristics of internal composition materials are detailed in Table 4.

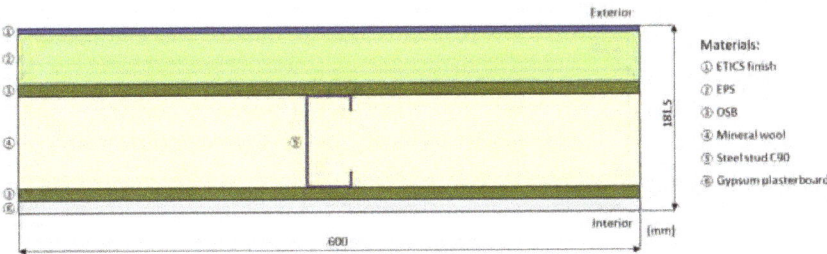

Figure 4. Cross-section of an exterior LSF reference wall modeled on THERM software.

Table 4. Materials, thicknesses (d) and thermal conductivities (λ) of the reference exterior facade wall.

Material (From Outer to Innermost Layer)	d [mm]	λ [W/(m.K)]	Ref.
ETICS [1] finish	5	0.450	[39]
EPS [2]	50	0.036	[40]
OSB [3]	12	0.100	[41]
Mineral wool	90	0.035	[34]
Steel stud (C90 × 43 × 15 × 1.5 mm)	90	50.000	[35]
OSB [3]	12	0.100	[41]
GPB [4]	12.5	0.175	[33]
Total Thickness	181.5	-	-

[1] ETICS—external thermal insulation composite system; [2] EPS—expanded polystyrene; [3] OSB—oriented strand board; [4] GPB—gypsum plasterboard.

3.2. Parameters for Sensitivity Analysis

The parameters and the values that were evaluated in the sensitivity analysis are displayed in Table 5 and illustrated in Figure 5.

Table 5. Exterior facade LSF wall: models and parameters values to be evaluated.

Model	Evaluated Parameter	Ref. Value	Value 1	Value 2	Value 3
E1	Thickness of Steel Studs [mm]	1.5	0.6	1.0	1.2
E2	Clearance Between Steel Studs [mm]	600	300	400	800
E3	Thickness of Aerogel TB [1] Strips [mm]	0.0	2.5	5.0	10.0
E4	Material of TB [1] Strips with 10 mm	-	MS-R1 [2]	XPS [3]	CBS [4]
E5	Inner Sheathing Panels Materials				
	GPB [5] Thickness [mm]	12.5	-	2 × 12.5	12.5
	OSB [6] Thickness [mm]	12.0	2 × 12.0	-	-
	XPS [7] Thickness [mm]	-	-	-	12.0
E6	Thickness of EPS [8] ETICS [9] [mm]	50	0.0	30	80

[1] TB—thermal Break; [2] MS-R1—Acousticork (recycled rubber); [3] XPS—extruded polystyrene; [4] CBS—cold break strip (aerogel); [5] GPB—gypsum plasterboard; [6] OSB—oriented strand board; [7] XPS—extruded polystyrene; [8] EPS—expanded polystyrene; [9] ETICS—external thermal insulation composite system.

The thickness of steel studs used on LSF wall steel frame is the first parameter that will be evaluated (Models E1). The reference value was 1.5 mm and the three additional thicknesses assessed were: 0.6, 1.0 and 1.2 mm (Figure 5a).

Similar to interior partition walls, for exterior facade walls the influence of clearance between the vertical steel studs were also quantified (Models E2). The reference LSF wall has 600 mm of distance between studs and the following clearances were also modeled: 300, 400 and 800 mm (Figure 5a).

Regarding the thermal break strips (Figure 5b), their thickness (Models E3) and materials (Models E4) were the same as for interior partition walls (Figure 3b).

To verify the influence of internal sheathing panels, the exterior wall model was tested in different innermost layer configurations, as shown in Table 5 and illustrated in Figure 5c (Models E5). The reference exterior facade wall has one OSB and one gypsum plasterboard panel as the innermost layer. Notice that these OSB panels are very important in load bearing walls because they give extra resistance to horizontal lateral loads [42]. On the first variation (Value 1), the sheathing panels are composed of two OSBs. For the second variation (Value 2), the internal layers are formed by two gypsum plasterboards (GPBs). In the third variation (Value 3) the OSB panel is replaced by one XPS panel with the same thickness (12.0 mm).

ETICS insulation layer thickness has a great influence on the thermal performance of the external walls. Therefore, this parameter influence will be also evaluated (Models E6). The EPS insulation thickness of the reference exterior LSF wall is 50 mm (Table 5). Three more values will be evaluated, namely: 0.0 mm (i.e., no EPS thermal insulation), 30 and 80 mm (Figure 5c).

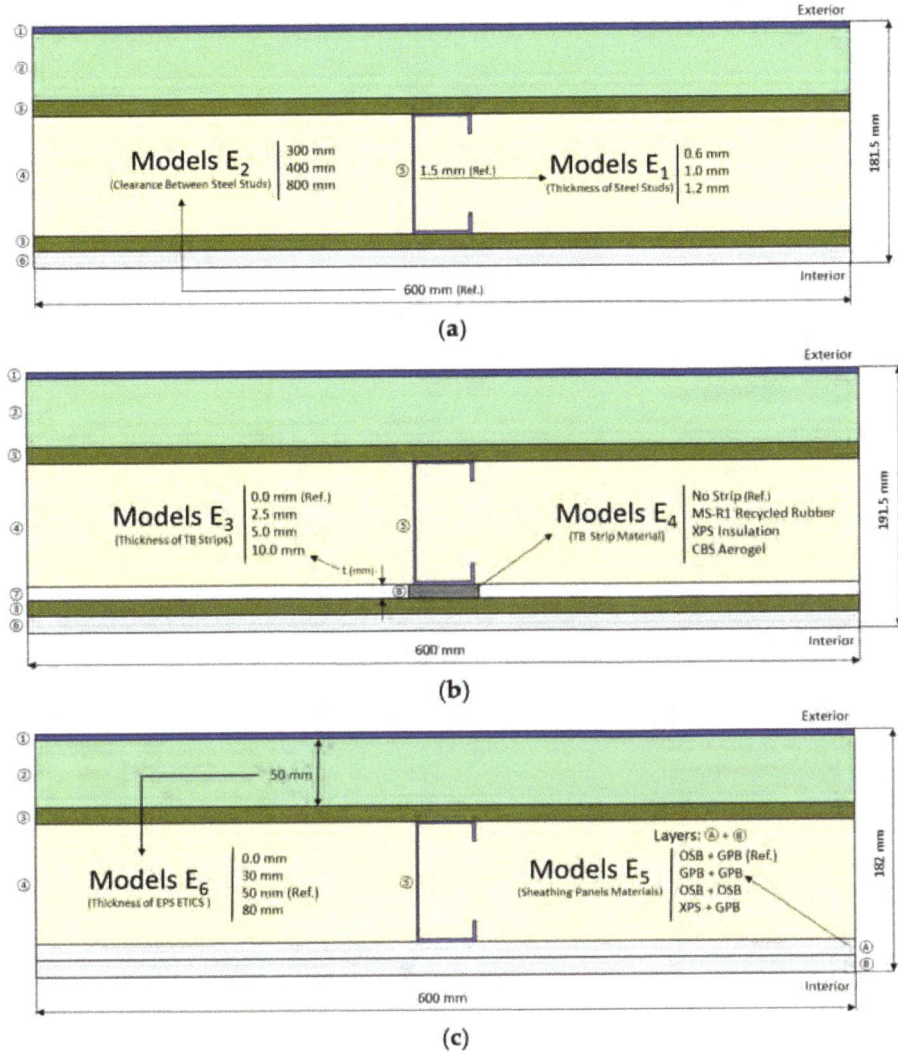

Figure 5. Exterior LSF facade cross-sections: (**a**) Models E1 and E2; (**b**) Models E3 and E4; (**c**) Models E5 and E6. Layers: ① ETICS finish; ② EPS; ③ OSB; ④ mineral wool; ⑤ steel stud C90; ⑥ gypsum plasterboard (GPB); ⑦ air layer; ⑧ TB strip.

4. Verification of 2D FEM Models

In this section the accuracy of the two-dimensional (2D) finite element method (FEM) models used in these computations is verified. First, the numerical results are compared against the two 2D test cases presented in ISO 10211 [22] and implemented by the authors. Then, the numerical 2D results are compared with the analytical solution provided by ISO 6946 [11] for simplified wall models with homogeneous layers (i.e., without LSF structure).

4.1. ISO 10211 Test Cases

To verify the accuracy of two-dimensional calculation algorithms, the ISO 10211 [22] Annex C, provides two test cases reference values (case 1 and 2) that was applied to the 2D FEM THERM software [43] to be classified as a steady-state high precision method.

In the first test case a sketch of a half square column with 28 points placed equidistantly inside the column, for which the corresponding temperatures for each point are known, was provided. The difference between the analytical solution given for each point inside the column and the temperature computed by the algorithm should not exceed 0.1 °C. For all the 28 points provided, the temperatures calculated by THERM (Figure 6a) were the same, with one exception, but stayed below a 0.1 °C difference from the given reference temperature.

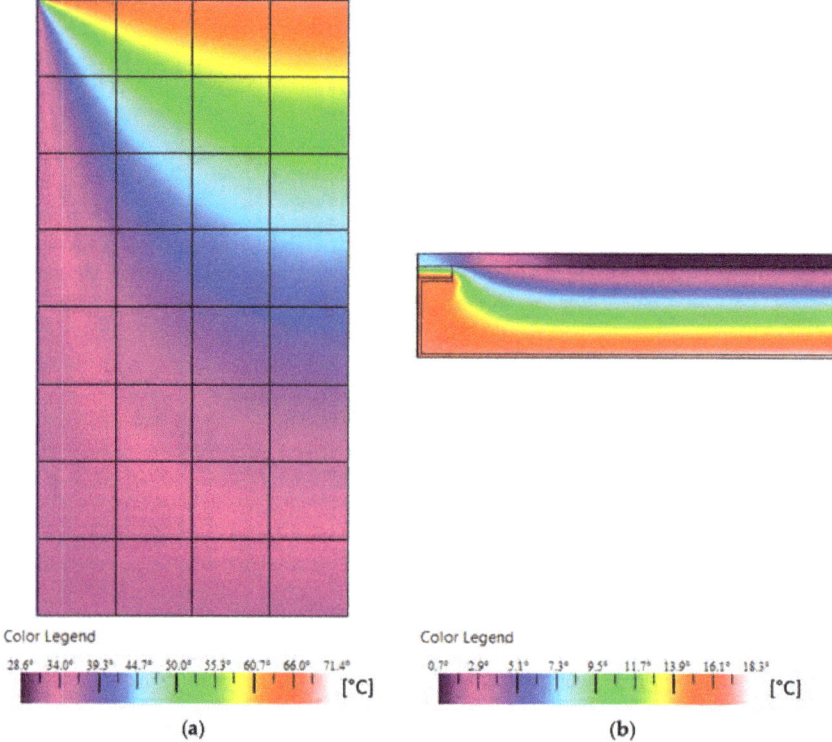

Figure 6. Temperature distribution obtained by the authors for the 2D test cases of ISO 10211 [22]: (a) test case 1; (b) test case 2.

For the second case, ISO 10211 requires that the difference between the temperatures calculated by the method being verified and the reference temperatures listed in the standard shall not exceed 0.1 °C, and the difference between the heat flow calculated and the reference value shall not exceed 0.1 W/m. The temperatures (Figure 6b) and heat flow calculated by THERM for test case 2 were exactly the same as prescribed by ISO 10211 Annex C. Notice that these results ensure not only the precision of the THERM software algorithm [43], but also the authors' expertise to use it.

4.2. ISO 6946 Analytical Approach

Another way to check the reliability of 2D FEM models is to compare the numerical results obtained with a simplified model of the same wall composed only for homogeneous layers (i.e., without the steel frame). For those walls with homogeneous layers, analytical solutions are available in ISO 6946 [11] and easy to calculate based on the thickness of each layer and on the material thermal conductivities. The input values (i.e., materials, layer thicknesses and thermal conductivities) were presented before in Table 1 (reference LSF interior partition wall) and Table 4 (reference LSF exterior facade wall). Regarding surface thermal resistances, the used values were obtained in ISO 6946 [11] for horizontal heat flow, namely 0.13 and 0.04 m².K/W for internal (R_{si}) and external surfaces (R_{se}), respectively.

The obtained thermal transmittance values for the analytical [11] and numerical approach (2D FEM) are displayed in Table 6. These results once again ensure the authors' skills in using THERM software for modeling [43], as well as its high accuracy.

Table 6. Thermal transmittances obtained for simplified wall models with homogeneous layers.

Wall Typology (Without Steel Frame)	U-Value [W/(m².K)]	
	Analytical	2D FEM [1]
Interior Reference Partition Wall	0.321	0.321
Exterior Reference Facade Wall	0.227	0.227

[1] using THERM software [43].

5. Two-Dimensional FEM Simulations

5.1. Boundary Conditions

As a mandatory entry to perform a numerical modeling simulation, it is necessary to define the boundary conditions to be applied on the LSF walls. Regarding temperatures, the interior temperature was set at 20 °C (a usual winter indoor comfort set-point temperature) and the exterior temperature was 0 °C (a usual design outdoor temperature for the winter season in mild climates such as in Portugal). An additional temperature of 10 °C was set for the partition walls 'exterior' unconditioned space; this value was considered an intermediate temperature between the adopted indoor (20 °C) and outdoor (0 °C) temperatures. Notice, that the obtained U-values do not depend on the chosen temperature difference between the interior and exterior environments, since this value is computed for a unitary temperature difference (i.e., per degree Celsius (°C) or, according to international standard units, per Kelvin (K).

Regarding surface thermal resistances, the values set on ISO 6946 [11] for horizontal heat flow were used (i.e., 0.13 and 0.04 m².K/W for internal (R_{si}) and external resistance (R_{se}), respectively). Notice that for the interior partition walls, internal surface resistances were used in both sides of the partition (i.e., 0.13 m².K/W).

5.2. Modeling Air Layers

The air layers inside the walls were modeled with a solid-equivalent thermal conductivity. The thermal resistance for these unventilated air-gaps were obtained in the ISO 6946 [11]. Knowing the thickness of the air-gap and dividing by its tabulated thermal resistance, the solid-equivalent thermal conductivity used in the 2D FEM numerical simulations was obtained, as displayed in Table 7.

Table 7. Thermal resistance and solid-equivalent thermal conductivity of air layers.

d_{air}^1 [mm]	R_{air}^2 [m².K/W]	λ_{eq}^3 [W/(m.K)]
2.5	0.055	0.045
5.0	0.11	0.045
10.0	0.15	0.067
90.0	0.18	0.500

[1] d_{air}—thickness of air layer; [2] R_{air}—thermal resistance of air layer (from ISO 6946); [3] λ_{eq}—solid-equivalent thermal conductivity.

6. Results and Discussion

6.1. Interior LSF Partition Walls

Table 8 displays the obtained thermal transmittances values for interior LSF partition walls, as well as the differences in relation to the reference LSF partition wall. To facilitate the quick analysis, the same results are illustrated graphically in Figure 7.

Table 8. Thermal transmittance obtained for interior LSF partition walls.

Model	Evaluated Parameter	Ref. Value	Value 1	Value 2	Value 3
I1	Thickness of Steel Studs [mm]	0.6	1.0	1.2	1.5
	U-value [W/(m^2.K)]	0.449	0.474	0.482	0.491
	Absolute difference	-	+0.025	+0.033	+0.042
	Percentage difference	-	+5.6%	+7.3%	+9.4%
I2	Clearance Between Steel Studs [mm]	600	300	400	800
	U-value [W/(m^2.K)]	0.449	0.580	0.515	0.420
	Absolute difference	-	+0.131	+0.066	−0.029
	Percentage difference	-	+29.2%	+14.7%	−6.5%
I3	Thickness of Aerogel TB [1] Strips [mm]	0.0	2.5	5.0	10.0
	U-value [W/(m^2.K)]	0.449	0.415	0.392	0.374
	Absolute difference	-	−0.034	−0.057	−0.075
	Percentage difference	-	−7.6%	−12.7%	−16.7%
I4	TB [1] Strips Materials [10 mm]	-	MS-R1 [2]	XPS [3]	CBS [4]
	U-value [W/(m^2.K)]	0.449	0.421	0.396	0.374
	Absolute difference	-	−0.028	−0.053	−0.075
	Percentage difference	-	−6.2%	−11.8%	−16.7%
	Sheathing Panels				
I5	GPB [5] Thickness [mm]	2 × 12.5	12.5	-	12.5
	OSB [6] Thickness [mm]	-	12.0	2 × 12.0	-
	XPS [3] Thickness [mm]	-	-	-	12.0
	U-value [W/(m^2.K)]	0.449	0.419	0.397	0.338
	Absolute difference	-	−0.030	−0.052	−0.111
	Percentage difference	-	−6.7%	−11.6%	−24.7%

[1] TB—thermal Break; [2] MS-R1—Acousticork (rubber); [3] XPS—extruded polystyrene; [4] CBS—cold break strip (aerogel); [5] GPB—gypsum plasterboard; [6] OSB—oriented strand board.

Comparing the obtained thermal transmittance value for the interior reference partition wall without steel frame (Table 6, 0.321 W/(m^2.K)) and the calculated value for the reference interior LSF partition wall (Table 8, 0.449 W/(m^2.K)) it is possible to verify that the LSF metallic structure increases the thermal transmittance value by about 40% (i.e., +0.128 W/(m^2.K)). Notice that this large increase in the U-value is due to the high thermal conductivity of steel (see Table 1)—even for a very small steel thickness (only 0.6 mm)—and due to the fact that all thermal insulation (mineral wool) is bridged by the steel studs (i.e., it is not continuous).

The thickness of steel studs (Model I1) was the first parameter to be assessed (Table 8). As expected, given the higher amount of steel, when increasing the thickness from 0.6 mm (reference value) up to 1.0, 1.2 and 1.5 mm, there was an increase in the U-value of 5.6%, 7.3% and 9.4%, respectively.

The second parameter evaluated (Table 8) was the distance between the vertical studs (Model I2), with the reference value equal to 600 mm. The decrease of this distance to 300 and 400 mm brought an increase in the wall U-value of 29.2% and 14.7%, respectively. This was expected given the increased amount of steel per unit area of the LSF wall. On the other hand, the increase of this distance from 600 mm up to 800 mm brought a wall U-value decrease of about 6.5%.

The existence of a thermal break (TB) strip (Model I3) increases the insulation of the steel structure and consequently decreases the thermal transmittance of the wall, as expected (Table 8). This U-value reduction was 7.6%, 12.7% and 16.7% for an aerogel TB strip with a thickness of 2.5, 5.0 and 10.0 mm, respectively.

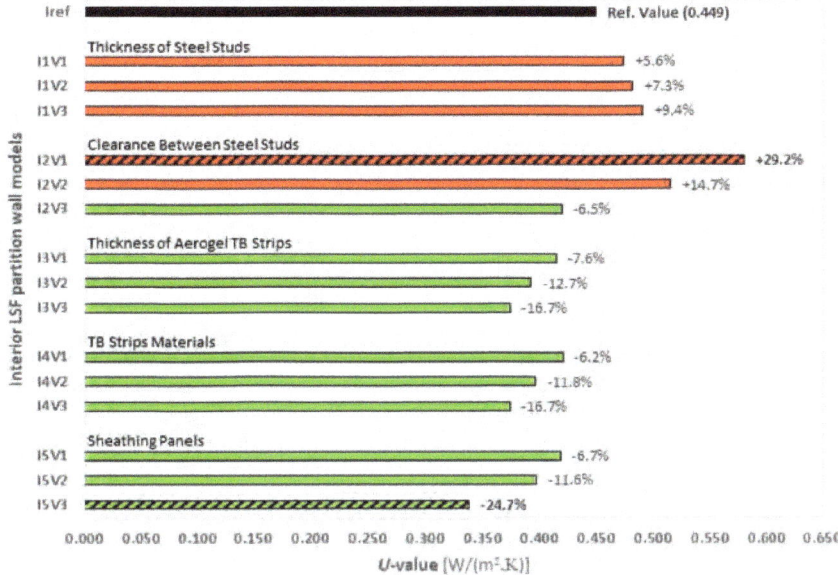

Figure 7. Thermal transmittances obtained for interior LSF partition walls.

The influence of the TB strip material (10 mm thick) was also evaluated (Model I4). Using recycled rubber (MS-R1) as a thermal break material, the U-value reduction was about 6.2% compared with the reference wall model without the TB strip (Table 8). For an XPS TB strip, the U-value decreased 11.8% and when using a material with a lower thermal conductivity (CBS aerogel) the wall thermal transmittance dropped even more (−16.7%). The former material (aerogel) provided the best results but is still quite an expensive material in comparison with the other two (recycled rubber and XPS).

Three variations according to what was previously presented for Model I5 (Table 2) were proposed for the configurations of sheathing panels. All three modeled variations for sheathing panels show better results than the reference interior LSF wall, because gypsum plasterboard has the highest thermal conductivity value, providing the uppermost U-value for the reference interior LSF partition wall (Table 8). The U-value reduction varied from 6.7% for GPB and OSB panels up to 24.7% for GPB and XPS panels. The largest reduction was expected given the very reduced thermal conductivity of XPS material (0.037 W/(m.K)) in comparison with others [i.e., GPB (0.175 W/(m.K)) and OSB (0.100 W/(m.K))].

Looking now to the extreme values obtained (see highlighted values in Table 8 and Figure 7), the highest thermal transmittance increase (+29.2%) was achieved for the Model I2V1, corresponding to a minimum clearance between steel studs (i.e., 300 mm). The lowest thermal transmittance decrease (−24.7%) was achieved for the Model I5V3, corresponding to GPB and XPS sheathing panels. These extreme U-values verify the great relevance of steel inside the LSF wall (Models I2), as well as the importance of providing a continuous thermal insulation layer (Model I5V3), even with a small thickness (only 12.0 mm in each side). Additionally, this XPS sheathing layer has also the advantage of being an affordable solution when compared with more expensive material (e.g., the aerogel TB strips (Models I3)).

In order to visualize and compare the temperature and heat flux distribution for these models, Figure 8 graphically displays this information. The temperature distribution in both LSF wall cross-sections is very similar (Figure 8a), and the influence of the steel stud in the temperature distribution is visible, given the high thermal conductivity from steel and consequently the thermal bridge effect. Analyzing the heat flux images (Figure 8b), the strong concentration of the heat flux around the steel stud is clear. Moreover, there are higher heat flux values for Model I2V1 (i.e., the wall with 300 mm clearance between studs), in comparison to the other model.

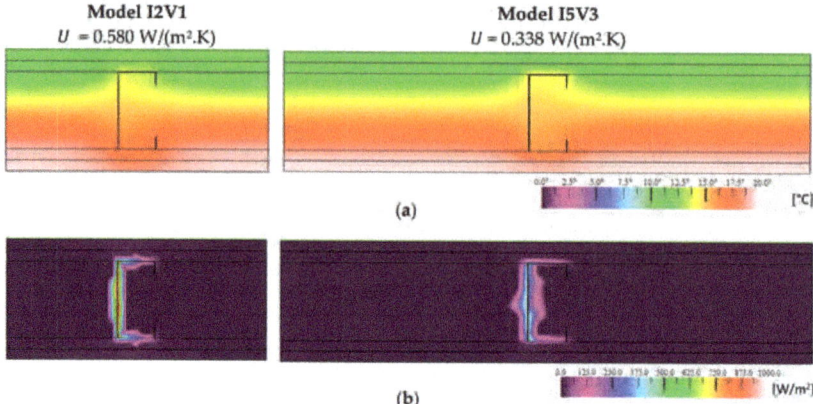

Figure 8. Temperature (**a**) and heat flux (**b**) color distribution for internal LSF wall models with the highest U-value increase (300 mm vertical stud distance) and decrease (XPS + GPB sheathing panels).

6.2. Exterior LSF Facade Walls

On Table 9 are shown the thermal transmittance values obtained for exterior LSF facade walls, as well as the differences between each parameter U-value and the reference LSF exterior wall U-value. For a better visualization and easier analysis for all modeled parameters, the graphic presented in Figure 9 plotted the obtained U-values and percentage differences.

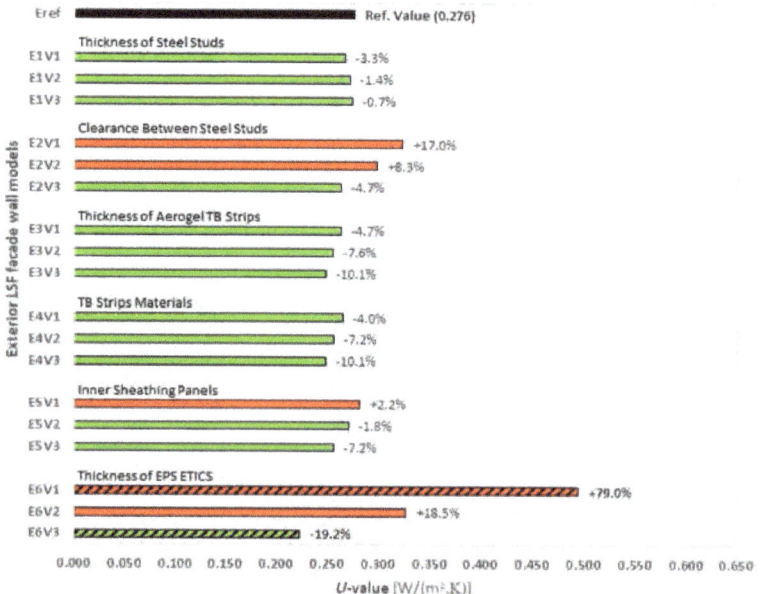

Figure 9. Thermal transmittances obtained for exterior LSF facade walls.

Table 9. Thermal transmittances obtained for exterior LSF facade walls.

Model	Evaluated Parameter	Ref. Value	Value 1	Value 2	Value 3
E1	**Thickness of Steel Studs [mm]**	**1.5**	**0.6**	**1.0**	**1.2**
	U-value [W/(m^2.K)]	0.276	0.267	0.272	0.274
	Absolute difference	-	−0.009	−0.004	−0.002
	Percentage difference	-	−3.3%	−1.4%	−0.7%
E2	**Clearance Between Steel Studs [mm]**	**600**	**300**	**400**	**800**
	U-value [W/(m^2.K)]	0.276	0.323	0.299	0.263
	Absolute difference	-	+0.047	+0.023	−0.013
	Percentage difference	-	+17.0%	+8.3%	−4.7%
E3	**Thickness of Aerogel TB [1] Strips [mm]**	**0.0**	**2.5**	**5.0**	**10.0**
	U-value [W/(m^2.K)]	0.276	0.263	0.255	0.248
	Absolute difference	-	−0.013	−0.021	−0.028
	Percentage difference	-	−4.7%	−7.6%	−10.1%
E4	**TB[1] Strips Materials [10 mm]**	-	**MS-R1 [2]**	**XPS [3]**	**CBS [4]**
	U-value [W/(m^2.K)]	0.276	0.265	0.256	0.248
	Absolute difference	-	−0.011	−0.020	−0.028
	Percentage difference	-	−4.0%	−7.2%	−10.1%
E5	**Inner Sheathing Panels**				
	GPB [5] Thickness [mm]	12.5	2 × 12.5	-	12.5
	OSB [6] Thickness [mm]	12.0	-	2×12.0	-
	XPS [3] Thickness [mm]	-	-	-	12.0
	U-value [W/(m^2.K)]	0.276	0.282	0.271	0.256
	Absolute difference	-	+0.006	−0.005	−0.020
	Percentage difference	-	+2.2%	−1.8%	−7.2%
E6	**Thickness of EPS [7] ETICS [8] [mm]**	**50**	**0**	**30**	**80**
	U-value [W/(m^2.K)]	0.276	0.494	0.327	0.223
	Absolute difference	-	+0.218	+0.051	−0.053
	Percentage difference	-	+79.0%	+18.5%	−19.2%

[1] TB—thermal break; [2] MS-R1—Acousticork (rubber); [3] XPS—extruded polystyrene; [4] CBS—cold break strip (aerogel); [5] GPB—gypsum plasterboard; [6] OSB—oriented strand board; [7] EPS—expanded polystyrene: [8] ETICS—external thermal insulation composite system.

To evaluate the influence of the steel structure the U-value for the exterior wall with homogeneous layers was compared (i.e., without steel frame, from Table 6, 0.227 W/(m^2.K)) with the U-value computed for the complete reference exterior wall (from Table 9, 0.276 W/(m^2.K)). The thermal transmittance increase due to the steel frame was 0.049 W/(m^2.K) (i.e., +22%, or even only 18% for the 0.6 mm thick (Model E1V1)). Notice that this increment in the U-value is much lower when compared with the interior partition wall: +0.128 W/(m^2.K) or +40%. This reduced relevance of the steel structure in the exterior partition wall, even having a steel thickness almost triple from the interior wall (1.5 mm instead of 0.6 mm), could be justified by the continuous thermal insulation in the ETICS (hybrid LSF structure), while in the interior partition wall all the thermal insulation is bridged by the steel frames (cold LSF structure).

Looking to the importance of the steel studs thickness in this exterior facade wall (Model E1), when this thickness is reduced from 1.5 mm to 0.6 mm there is a decrease of only 3.3% in the thermal transmittance (Table 9), while in the interior partition wall the corresponding value when there is an increase in the steel thickness from 0.6 mm up to 1.5 mm is +9.4% (Table 8). This again confirms the higher relevance of the steel structure in the interior partition wall.

The second evaluated parameter is the clearance between the vertical studs (Model E2), where the reference value is 600 mm. When decreasing the distance between the studs—300 and 400 mm—the wall U-value increases 17.0% and 8.3%, respectively. In contrast, when the studs where placed farther apart (800 mm) the U-value decreases 4.7%. As explained before, those thermal transmittance variations are closely linked with the amount of steel inside each wall configuration.

The results of the thickness variation for the CBS aerogel thermal break strip on exterior facade walls were computed using Model E3 (Table 9). As expected, by increasing the TB thickness to 2.5,

5.0 and 10.0 mm, there was a decrease of the wall U-value by 4.7%, 7.6% and 10.1%, respectively. Confronting these results with similar ones for the interior partition wall (7.6%, 12.7% and 16.7% in Table 8), it can be seen that the decrease in U-values is now considerably lower. This could be justified by the reduced importance of the steel frame in the exterior walls and consequently the effect of the TB strips is also reduced.

Evaluating the effectiveness of different materials for the 10 mm thick TB strip (Model E4), as expected, the aerogel (CBS) strip allowed the biggest reduction on wall thermal transmittance (10.1%), followed by the XPS strip (reduction of 7.2%) and recycled rubber (MS-R1) with a 4.0% decrease on the U-value.

Model E5 (Table 9) shows the results of changing the innermost sheathing panels material. Three different configurations were assessed. The first was composed by two panels of GPB and presented a U-value increase of 2.2%. The second configuration used two panels of OSB and it obtained a U-value reduction of 1.8% in comparison with the reference value. For the last variation, the internal layers were composed of a GPB panel and an XPS panel, having the most significant results (i.e., a reduction of 7.2%). Notice, that this last U-value reduction is significantly lower when compared with the one computed for the interior partition wall (−24.7%). Again, this is related with lower relevance of the steel frame thermal bridge transmission due to the existence of the ETICS continuous thermal insulation in the exterior facade wall. Therefore, the relevance of an extra continuous thermal insulation layer is also reduced.

Model E6 evaluates the influence of the EPS thickness in the ETICS. The exterior reference facade wall has 50 mm of EPS, compared with three additional values of 0, 30 and 80 mm. Clearly this was the most relevant evaluated parameter, leading to an increase of 79% in the U-value (Model E6V1) when there is no exterior thermal insulation and a reduction of 19.2% when the EPS thickness was increased to 80 mm (Model E6V3).

Figure 10 displays the color temperature and heat flux distribution for these two models with the most extreme U-value variation. Regarding the temperature distribution (Figure 10a), the influence of the continuous thermal insulation on Model E6V3 (hybrid LSF structure), with a warmer steel frame temperature in comparison with Model E6V1 (cold frame LSF structure) is very visible. Looking at the heat flux distribution (Figure 10b), as expected, the values for Model E6V1 are visually higher than Model E6V3, given the continuous thermal insulation layer in this second model.

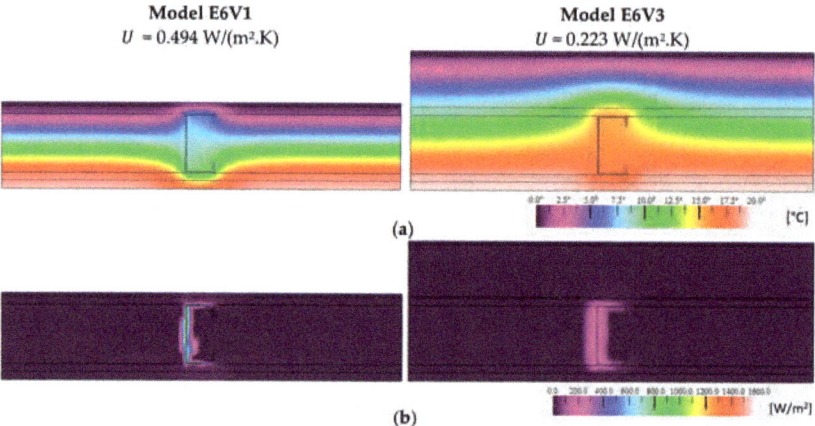

Figure 10. Temperature (**a**) and heat flux (**b**) color distribution for exterior LSF wall models with the highest U-value increase (0 mm EPS ETICS) and decrease (80 mm EPS ETICS).

7. Conclusions

In this work, a sensitivity analysis regarding the thermal transmittance (U-value) was performed for two different types of lightweight steel framed (LSF) walls: interior partition and exterior facade. The numerical results were obtained by using 2D finite element method (FEM) models. The accuracy of these models was verified by comparison with ISO 10211 test cases and with ISO 6946 analytical approach.

The assessed parameters were: (1) thickness of steel studs; (2) clearance between studs; (3) thermal break strips thickness and (4) material; (5) configuration of internal sheathings panels, and; (6) thickness of EPS external thermal insulation composite system (ETICS), only for the external facade wall. The results of this parametric study were compared to a reference interior partition LSF wall, with a U-value equal to 0.449 W/(m^2.K) and to a reference exterior facade LSF wall, with a U-value equal to 0.276 W/(m^2.K). Regarding the obtained results, notice that the percentages of U-value change are high, but the absolute differences are rather small in most cases.

The interior partition LSF wall showed higher U-values and a greater influence of the internal steel structure on the wall thermal transmittance. This was expected given the high thermal conductivity of steel and the absence of a continuous thermal insulation on interior partition walls potentiates the thermal bridges' effects on the LSF structure, resulting in higher U-values. Nevertheless, higher heat flux through the interior walls enables other evaluated parameters to have a greater influence on wall thermal transmittance (e.g., clearance between steel studs (up to +29.2%) and XPS sheathing panel (down to −24.7%)).

The thickness augment of the metallic structure increased the thermal transmittance of the interior wall up to +9.4% (1.5 mm thick). The use of thermal break (TB) strips reduced the U-value of the interior wall down to −16.7% (10 mm thick aerogel strip). The use of different materials in the TB strip was also assessed. The U-value reduction depends on the thermal conductivity of the material used in the TB strip: −6.2% for recycled rubber, −11.8% for XPS and −16.7% for aerogel.

For the exterior facade LSF walls, the existence of an ETICS continuous thermal insulation on the outer side reduces the heat flux through the wall, particularly through the steel frame, resulting in a lower wall U-value and decreasing the importance of other evaluated parameters. In fact, the major and the minor U-value increment changed the thickness of the EPS insulation ETICS layer (i.e., an augment of +79.0% when there is no EPS (0.0 mm thick) and a decrease of −19.2% for 80 mm EPS thickness). Notice that the reference wall has 50 mm of EPS ETICS.

Decreasing the steel thickness (1.5 mm) to 0.6 mm reduced the U-value to only −3.3% (−0.009 W/(m^2.K)). Notice that in the interior partition wall the absolute U-value increased, when the steel thickness changed from 0.6 mm up to 1.5 mm, and was more than four times higher (i.e., +0.042 W/(m^2.K), showing the lower importance of the steel structure in this exterior facade LSF wall.

When changing the distance between the vertical studs from 600 mm to half (300 mm) and doubling the amount of steel, the U-value increased only +17.0% (+0.047 W/(m^2.K)). Notice that in the interior partition wall the absolute U-value increase was almost the triple (i.e., +0.131 W/(m^2.K)).

The use of aerogel thermal break strips with different thicknesses (up to 10 mm) reduced the U-value down to −10.1% (−0.028 W/(m^2.K)). Notice that in the interior wall this absolute U-value reduction was more than double (i.e., −0.075 W/(m^2.K)). Using a 10 mm thick TB strip with different materials (rubber, XPS and aerogel) decreased the U-value to about −4.0% (−0.011 W/(m^2.K)), −7.2% (−0.020 W/(m^2.K)) and −10.1% (−0.028 W/(m^2.K)), respectively. Notice that in the interior wall these U-value reductions were quite higher: −6.2% (−0.028 W/(m^2.K)), −11.8% (−0.053 W/(m^2.K)) and −16.7% (−0.075 W/(m^2.K)), respectively.

The use of different inner sheathing panels (GPB, OSB and XPS) led to a U-value variation down to −7.2% (−0.020 W/(m^2.K)) for the XPS/GPB panels. Notice that in the interior LSF wall this absolute U-value reduction was much more relevant [i.e., more than five times higher (−0.111 W/(m^2.K))]. This was due not only to the absence of any continuous thermal insulation in the reference interior LSF wall, but also to the fact that in this case the two wall sides were updated with an XPS sheathing panel (one in each side), while in the exterior facade only the inner wall surface was updated with an XPS sheathing panel.

For further related research work, the authors intend to perform laboratory experimental measurements in similar interior and exterior LSF walls. These measurements will be useful to ensure the reliability of the numerical simulations and validate the numerical models. In order to consider and evaluate the relevance of some three-dimensional (3D) effects in the thermal performance of these interior and exterior LSF walls, the authors also intend to perform some 3D FEM simulations in a complementary future research work. Another predicted future work is to evaluate the cost-benefit of these thermal performance improvement measures and the provided energy efficiency benefits for an LSF building.

Author Contributions: All the authors participated equally to this work.

Funding: This work was financed by FEDER funds through the Competitivity Factors Operational Programme—COMPETE and by national funds through FCT—Foundation for Science and Technology within the scope of the project POCI-01-0145-FEDER-032061.

Acknowledgments: The authors also want to thank the support provided by the following companies: Pertecno, Gyptec Ibérica, Volcalis, Sotinco, Kronospan, Hukseflux and Hilti.

Conflicts of Interest: The authors declare no conflicts of interest.

References

1. European Union. Directive (EU) 2018/844 of the European Parliament and of the Council of 30 May 2018 amending Directive 2010/31/EU on the energy performance of buildings and Directive 2012/27/EU on energy efficiency. *J. Eur. Union* **2018**, *2018*, 75–91.
2. Aslani, A.; Bakhtiar, A.; Akbarzadeh, M.H. Energy-efficiency technologies in the building envelope: Life cycle and adaptation assessment. *J. Build. Eng.* **2019**, *21*, 55–63. [CrossRef]
3. Ruparathna, R.; Hewage, K.; Sadiq, R. Improving the energy efficiency of the existing building stock: A critical review of commercial and institutional buildings. *Renew. Sustain. Energy Rev.* **2016**, *53*, 1032–1045. [CrossRef]
4. Paone, A.; Bacher, J.P. The impact of building occupant behavior on energy efficiency and methods to influence it: A review of the state of the art. *Energies* **2018**, *11*, 953. [CrossRef]
5. Santos, P.; da Silva, L.S.; Ungureanu, V. *Energy Efficiency of Light-Weight Steel-Framed Buildings*, 1st ed.; Technical Committee 14—Sustainability & Eco-Efficiency of Steel Construction; N. 129; European Convention for Constructional Steelwork (ECCS): Mem Martins, Portugal, 2012; ISBN 978-92-9147-105-8.
6. Pulselli, R.M.; Simoncini, E.; Marchettini, N. Energy and emergy based cost-benefit evaluation of building envelopes relative to geographical location and climate. *Build Environ.* **2009**, *44*, 920–928. [CrossRef]
7. Kaynakli, O. A review of the economical and optimum thermal insulation thickness for building applications. *Renew. Sustain. Energy Rev.* **2012**, *16*, 415–425. [CrossRef]
8. Soares, N.; Martins, C.; Gonçalves, M.; Santos, P.; da Silva, L.S.; Costa, J.J. Laboratory and in-situ non-destructive methods to evaluate the thermal transmittance and behavior of walls, windows, and construction elements with innovative materials: A review. *Energy Build.* **2019**, *182*, 88–110. [CrossRef]
9. Santos, P. Chapter 3—Energy Efficiency of Lightweight Steel-Framed Buildings. In *Energy Efficient Buildings*; Eng Hwa, Y., Ed.; InTech: Rijeka, Croatia, 2017; pp. 35–60.
10. Roque, E.; Santos, P.; Pereira, A.C. Thermal and sound insulation of lightweight steel-framed façade walls. *Sci. Technol. Built Environ.* **2019**, *25*, 156–176. [CrossRef]
11. International Organization for Standardization. *Building Components and Building Elements—Thermal Resistance and Thermal—Calculation Methods, International Organization for Standardization*; ISO 6946; International Organization for Standardization: Geneva, Switzerland, 2017.

12. Roque, E.; Santos, P. The Effectiveness of Thermal Insulation in Lightweight Steel-Framed Walls with Respect to Its Position. *Buildings* **2017**, *7*, 13. [CrossRef]
13. Schiavoni, S.; D'Alessandro, F.; Bianchi, F.; Asdrubali, F. Insulation materials for the building sector: A review and comparative analysis. *Renew. Sustain. Energy Rev.* **2016**, *62*, 988–1011. [CrossRef]
14. Gervásio, H.; Santos, P.; da Silva, L.S.; Lopes, A.M.G. Influence of thermal insulation on the energy balance for cold-formed buildings. *Adv. Steel Constr.* **2010**, *6*, 742–766.
15. Craveiro, A.; Lopes, A.G.; Santos, P.; da Silva, L.S. Natural ventilation potential on thermal comfort of a light-steel framing residential building. In Proceedings of the Green Design, Materials and Manufacturing Processes, Proceedings of the 2nd International Conference on Sustainable Intelligent Manufacturing, SIM 2013, Lisbon, Portugal, 26–29 June 2013.
16. Soares, N.; Gaspar, A.R.; Santos, P.; Costa, J.J. Multi-dimensional optimization of the incorporation of PCM-drywalls in lightweight steel-framed residential buildings in different climates. *Energy Build.* **2014**, *70*, 411–421. [CrossRef]
17. Waqas, A.; Din, Z.U. Phase change material (PCM) storage for free cooling of buildings—A review. *Renew. Sustain. Energy Rev.* **2013**, *18*, 607–625. [CrossRef]
18. Rosa, N.; Santos, P.; Costa, J.; Gervásio, H. Modeling and performance analysis of an earth-to-air heat exchanger in a pilot installation. *J. Build. Phys.* **2018**, *42*, 259–287. [CrossRef]
19. Valdiserri, P.; Biserni, C.; Garai, M. Energy performance of a ventilation system for an apartment according to the Italian regulation. *Int. J. Energy Environ. Eng.* **2016**, *7*, 353–359. [CrossRef]
20. Xu, X.; Feng, G.; Chi, D.; Liu, M.; Dou, B. Optimization of Performance Parameter Design and Energy Use Prediction for Nearly Zero Energy Buildings. *Energies* **2018**, *11*, 3252. [CrossRef]
21. Santos, P.; Martins, R.; Gervásio, H.; Silva, L.S. Assessment of building operational energy at early stages of design—A monthly quasi-steady-state approach. *Energy Build.* **2014**, *79*, 58–73. [CrossRef]
22. International Organization for Standardization. *Thermal Bridges in Building Construction—Heat Flows and Surface Temperatures—Detailed Calculations*; ISO 10211; International Organization for Standardization: Geneva, Switzerland, 2017.
23. Santos, P.; Gonçalves, M.; Martins, C.; Soares, N.; Costa, J.J. Thermal Transmittance of Lightweight Steel Framed Walls: Experimental Versus Numerical and Analytical Approaches. *J. Build. Eng.* **2019**, *25*, 100776. [CrossRef]
24. Evangelisti, L.; Guattari, C.; Gori, P.; Bianchi, F. Heat transfer study of external convective and radiative coefficients for building applications. *Energy Build* **2017**, *151*, 429–438. [CrossRef]
25. Santos, P.; Martins, C.; da Silva, L.S. Thermal performance of lightweight steel-framed construction systems. *Metall. Res. Technol.* **2014**, *111*, 329–338. [CrossRef]
26. Soares, N.; Santos, P.; Gervásio, H.; Costa, J.J.; da Silva, L.S. Energy efficiency and thermal performance of lightweight steel-framed (LSF) construction: A review. *Renew. Sustain. Energy Rev.* **2017**, *78*, 194–209. [CrossRef]
27. Gorgolewski, M. Developing a simplified method of calculating U-values in light steel framing. *Build. Environ.* **2017**, *42*, 230–236. [CrossRef]
28. International Organization for Standardization. *Thermal Insulation—Determination of Steady-State Thermal Transmission Properties—Calibrated and Guarded Hot Box, International Organization for Standardization*; ISO 8990; ISO—International Organization for Standardization: Geneva, Switzerland, 1994.
29. Atsonios, I.A.; Mandilaras, I.D.; Kontogeorgos, D.A.; Founti, M.A. Two new methods for the in-situ measurement of the overall thermal transmittance of cold frame lightweight steel-framed walls. *Energy Build.* **2018**, *170*, 183–194. [CrossRef]
30. Santos, P.; Martins, C.; da Silva, L.S.; Bragança, L. Thermal performance of lightweight steel framed wall: The importance of flanking thermal losses. *J. Build. Phys.* **2014**, *38*, 81–98. [CrossRef]
31. De Angelis, E.; Serra, E. Light steel-frame walls: Thermal insulation performances and thermal bridges. *Energy Procedia* **2014**, *45*, 362–371. [CrossRef]
32. Martins, C.; Santos, P.; da Silva, L.S. Lightweight steel-framed thermal bridges mitigation strategies: A parametric study. *J. Build. Phys.* **2016**, *39*, 342–372. [CrossRef]
33. Gyptec Ibérica. Technical Sheet: Standard Gypsum Plasterboard. 2019. Available online: https://www.gyptec.eu/documentos/Ficha_Tecnica_Gyptec_A.pdf (accessed on 14 March 2019). (In Portuguese)

34. Volcalis. Technical Sheet: Alpha Mineral Wool. 2019. Available online: https://www.volcalis.pt/categoria_file_docs/fichatecnica_volcalis_alpharolo-253.pdf (accessed on 14 March 2019). (In Portuguese)
35. Santos, C.; Matias, L. *ITE50—Coeficientes de Transmissão Térmica de Elementos da Envolvente dos Edifícios*; LNEC—Laboratório Nacional de Engenharia Civil: Lisbon, Portugal, 2006. (In Portuguese)
36. MS-R1. Acousticork MS-R1 Recycled Rubber. 2017. Available online: https://amorimcorkcomposites.com/media/1334/acousticork-book-en.pdf (accessed on 14 March 2019).
37. Spacetherm. Technical Sheet: Cold Bridge Strip. 2018. Available online: https://www.proctorgroup.com/assets/Datasheets/Spacetherm_CBS_Datasheet.pdf (accessed on 14 March 2019).
38. ITeCons. *Test Report HIG 363/12—Determination of Thermal Resistance*; Instituto de Investigação e Desenvolvimento em Ciências da Construção: Coimbra, Portugal, 2012.
39. WeberTherm Uno. Technical Specifications: Weber Saint-Gobain ETICS Finish Mortar. 2018. Available online: https://www.pt.weber/files/pt/2019-04/FichaTecnica_weberthermuno.pdf (accessed on 14 March 2019). (In Portuguese)
40. TincoTerm. Technical Sheet: EPS 100. 2015. Available online: http://www.lnec.pt/fotos/editor2/tincoterm-eps-sistema-co-1.pdf (accessed on 14 March 2019). (In Portuguese)
41. KronoSpan. Technical Sheet: KronoBuild OSB 3. 2019. Available online: https://de.kronospan-express.com/public/files/downloads/kronobuild/kronobuild-en.pdf (accessed on 14 March 2019).
42. Henriques, J.; Rosa, N.; Gervasio, H.; Santos, P.; Simões, L. Structural performance of light steel framing panels using screw connections subjected to lateral loading. *Thin Walled Struct.* **2017**, *121*, 67–88. [CrossRef]
43. THERM. *Software Version 7.6.1*; Lawrence Berkeley National Laboratory, United States Department of Energy: Berkeley, CA, USA, 2017. Available online: https://windows.lbl.gov/software/therm (accessed on 14 February 2019).

© 2019 by the authors. Licensee MDPI, Basel, Switzerland. This article is an open access article distributed under the terms and conditions of the Creative Commons Attribution (CC BY) license (http://creativecommons.org/licenses/by/4.0/).

Article

Creating Statistics for China's Building Energy Consumption Using an Adapted Energy Balance Sheet

Mingshun Zhang [1],*, Xuan Ge [1], Ya Zhao [2] and Chun Xia-Bauer [3]

1. School of Environment and Energy Engineering, Beijing University of Civil Engineering and Architecture, Beijing 100044, China; 210813j117008@stu.bucea.edu.cn
2. School of Construction Management and Real Estate, Chongqing University, Chongqing 400044, China; zyofficial@cqu.edu.cn
3. Wuppertal Institute for Climate, Environment and Energy, 42103 Wuppertal, Germany; chun.xia@wupperinst.org
* Correspondence: zhangmingshun@bucea.edu.cn; Tel.: +86-139-1101-2891

Received: 2 October 2019; Accepted: 8 November 2019; Published: 11 November 2019

Abstract: China's regular energy statistics does not include the building sector, and data on building energy demand is included in other types of energy consumption in the Energy Balance Sheet (EBS). Therefore data on building energy demand is not collected based on statistics, but rather calculated or estimated by various approaches in China. This study aims at developing and testing China's building energy statistics by applying an adapted EBS. The advantage of the adapted EBS is that statistical data is from the regular statistical system and no additional statistical efforts are needed. The research result shows that the adapted EBS can be included in China regular energy statistical system and can be standardized in a transparent way. Testing of the adapted EBS shows that China's building energy demand has shown an annual increase of 7.6% since 2001, and a lower contribution to the total energy demand as compared to the developed world. There is also a close link to lifestyle and living standard while industrial energy demand is mainly driven by economy and decoupling of building energy demand with increasing of building floor area, this is due to a considerable improvement of building energy efficiency. The adapted EBS creates a method for China conducting statistics of building energy consumption at the sector level in a uniform way and serves as the basis for any sound building energy efficiency policy decisions.

Keywords: building energy statistics; building energy consumption; energy balance sheet; building energy efficiency; China

1. Introduction

China's energy consumption has increased dramatically since 1980. Around 2011 China became the largest energy-consuming country, replacing the USA. In 2015, China had a 28% share of the global end-energy demand while USA had a 22% share [1]. Building, transportation and industry are the three key energy demand sectors worldwide. The building sector is responsible for more than 25% of China's total primary energy consumption and this figure will increase to 35% by 2030. The GHG emissions contributed by the building sector are about 25% of China's total emissions [2]. Internationally, buildings consume about 30–45% of the global energy demand [3]. Although China's current building energy consumption share is significantly lower than the international level, the fast urbanization, the fast development of the building sector, rising of living standards and increasing consumption will increase the energy demand in the building sector. China is now facing the challenges of both a fast growing building energy demand and low building energy efficiency. It is estimated that more than 30%–50% of the existing building energy consumption could be saved [4] by adopting various energy efficiency solutions.

Statistics on building energy consumption at the sector level are essential to assess the building energy situation and are indeed used as the basis for any sound building energy efficiency policy decisions. Authorities, building developers, building owners and the public need information about where the building energy demand is and the impact of policy enforcements. The building sector needs information about progress achieved or reasons that prevent progress. The key to meet the information needs mentioned above is to provide complete, timely and reliable data on building energy demand.

In China's national energy statistical system, the final consumption is composed of seven sectors: (1) farming, forestry, animal husbandry, fishery & water conservancy, (2) industrial, (3) construction, (4) transport, storage, postal and telecommunications services, (5) wholesale, retail and catering services, (6) others, and (7) residential consumption. The energy consumption of buildings is mainly included in the consumption sector and also included in other sectors. There are various studies on statistics and monitoring of energy consumption at a single building level, however, there is no specific sectoral building energy statistical system in China and data on building energy consumption at the sector level is calculated or estimated by various approaches [5]. There are many office buildings in industrial sectors, and those buildings' energy consumption is included in the energy demand statistics of the industrial sectors. There are similar cases in other sectors of transportation and construction as well. The lack of reliable and accurate data on building energy consumption at a sector level has been a major barrier for policy making at the national and sectoral levels. The establishment of a national statistical system of building energy consumption in China will create a method for determining the statistics of building energy consumption at the sector level in a uniform way and serve as the basis for any sound building energy efficiency policy decisions.

There are many studies aiming at getting data of building energy demand at the sectoral level. The Building Energy Research Centre of Tsinghua University (THUBERC) has developed the China Building Energy Model, CBEM), based on building energy intensity and building floor area, to estimate China's total building energy consumption [6]. By applying this model, THUBERC publishes an annual report on China's building energy efficiency, which has been one of the key national sources for different stakeholders to get energy consumption data for the building sector. Wang calculated the total building energy demand based on international EBS, surveys and expert workshops and concluded that China's building energy demand was 370 million tons-coal-equivalent (tce) in 2006 [7], which accounts for 21.7% of the total energy demand in China. Long [8] developed a model, based on an analysis of China's energy consumption by sectors and by comparing its industrial structures with USA and Japan, to estimate building energy consumption. According to this model, China's total energy consumption in the building sector was 330 million tce in 2003, which accounts for 20% of the total energy demand of China. The Ministry of Housing and Urban & Rural development of China estimated the existing building energy consumption based on building stocks, climate zone characters and relevant energy efficiency standards, and concluded that China building energy consumption accounts for 27.5% of China total energy demand. From the literature reviews, there are basically four methods to get information on building energy demand at a sector level:

- Sampling surveys [8]. Surveys are widely applied to get detailed information on building energy demand. However, it is costly and impossible for surveys to cover all buildings. Therefore sampling surveys are applied. The scale of the survey depends on the resources and complete data of building energy demand at the sector level is calculated, based on the survey results and estimated total floor areas of all types of buildings.
- Statistics for defined-scale and defined-type buildings [9–12]. This method builds on building energy consumption statistics that are applied for defined-scale buildings (e.g., with a floor area of more than 10.000 m^2) and defined-type buildings (public buildings), and statistics gathered by local statistics departments. Total building energy consumption can therefore be estimated by using the statistical data.

- Modelling [6,13–15]. Modelling is widely used by academic institutes for estimating energy consumption at the sector level. Models are developed based on building types and building stocks, building age, characters of climate zones, and total floor areas of each type of buildings in each of China's climate zones.
- EBS together with expert approach [16–18]. In the EBS of IEA and other developed countries, energy consumption data is collected for the three main sectors of industrial, transportation and buildings, and thus data of building energy consumption is available from the EBS database. However, building energy consumption data in China's EBS is divided into the different sectors of industries, transportation, consumptions and others. Thus getting building energy consumption data from China's EBS needs additional work that may be done by professional energy analysts and therefore the EBS together with expert approach is applied to gather building energy consumption data.

The methods mentioned above have the disadvantages of needing the additional data collection efforts or surveys in addition to regular energy statistics, due to limited samples, using assumptions in modelling, and making estimations by historical experiences. None of the methods mentioned above are standardized and thus it is difficult for China to build up its regional comparisons and benchmarking of building energy performance and compare it to international data.

There have been several studies on China building energy consumption statistics using adapted EBS and those studies are all on a local level [18]. The main disadvantage of these local studies is that the data from EBS is directly used and lacks necessary corrections. This study, built on those local-level studies, focuses on national level and has made necessary amendments to the EBS data. The justified amendments to the data ensure that the study results are closer to the real energy consumption of buildings in China.

EBS is the key resource for all sectors to get energy information. However, building energy consumption is not listed separately in China's EBS and it is divided among the energy consumptions of different sectors. Therefore the first step of this research is to identify and analyze the data sources and statistical definitions of energy consumption of all sectors included in China's EBS. China's EBS has defined the following seven sectors of final energy consumption. This research builds on the following analysis of the final energy consumption of the seven sectors included in China's EBS:

(S1) Farming, forestry, animal husbandry, fishery & water conservancy. This sector is the primary industry and energy consumption of this sector included in EBS covers all production and production related services energy consumption. Thus final energy consumption of this sector does not include building energy consumption. Energy consumption of transportation of this sector is included in EBS.

(S2) Industry. Industries in China's EBS include mining and quarrying, machinery, and power, heating, fuel gas and water production and services. The industrial energy consumption in China's EBS is comparable to but different from the International Energy Agency (IEA) data. There are mainly three differences between China's and IEA's industrial energy consumption: (1) China's industrial energy consumption includes energy consumption of energy industry itself. In IEA energy statistics, energy consumption of energy production is not included in industrial energy consumption but rather listed in the energy loss of energy processing. (2) China's industrial energy consumption includes the transportation energy consumption of this sector. (3) Industrial building energy consumption is included in industrial energy consumption in China's EBS.

(S3) Construction. The energy consumption of construction in China's EBS is mainly the energy consumption of construction process. In IEA's statistics, construction energy consumption is included in industrial energy consumption. China's EBS has separately listed construction energy consumption. Similar to the industrial energy consumption of China, construction energy consumption in China's EBS also includes building energy consumption of construction sector, e.g., office buildings in this sector.

(S4) Transport, storage, postal and telecommunications services. This sector's energy consumption is mainly from transportation enterprises, and it does not include the transportation energy consumption of other sectors as well as citizens' transportation energy consumption. Energy consumption of this sector covers building energy consumption of this sector, e.g., the building energy consumptions of airports, railway stations, bus stations and post office buildings.

(S5) Wholesale, retail trade and catering services. These sectors are mainly tertiary industry and main energy consumption of these sectors are building energy consumption. These sectors' energy consumption is comparable to the IEA's statistics of energy consumption of commercial and public services.

(S6) Others. Energy consumption of "others" in China's EBS is defined as the energy consumption of the tertiary industrial energy consumption excluding the abovementioned sectors (S4) and (S5). "Others" includes software and information technology services, financing, real estate industry, education, science, culture and public health and public administrations. Energy consumption of these sectors are mainly from building performance.

(S7) Residential consumption. Energy demand in residential consumption includes the living energy consumption of citizens and is mainly from building performance. It also includes the transportation energy consumption of citizens.

Another issue is how to amend the data of energy consumption of central heating systems from EBS. The existing data of energy consumption of central heating systems from EBS is significantly low and it is necessary to correct this data. As an example [19], the energy consumption of central heating in the sectors of (S5) Wholesale, retail trade and catering services, (S6) Others and (S7) Residential consumption was 31.11 million tce in 2011 and the floor area of the central heating is 5.18 billion m^2. Thus the energy efficiency of central heating is 6 kg coal-equivalent per M^2, which is incorrect given to the fact that energy efficiency of the most efficient central heating by a large-scale cogeneration project is 9 kg coal-equivalent per m^2, and the national average energy efficiency of central heating by regular boilers is 20 kg coal-equivalent per mM^2 [19]. There are three reasons that data of energy consumption of central heating system from EBS is significantly low:

(1) Heat metering is not well installed in China, in particular there is almost no heat metering in the buildings built before 2010.

(2) China's EBS statistics targets enterprises that are on the scale of 20 million Chinese Yuan turnover or energy consumption of above 10,000 tce. There are many SMEs' heat-generators or heat-suppliers (small and medium enterprises) that do not meet this scale and are therefore excluded from the EBS statistics. Therefore heating generated by those SMEs are not included in the EBS.

(3) Heating energy consumption of cogeneration systems is excluded from the central heating energy consumption, but included in energy consumption of power generation (energy industry energy consumption).

For correcting the data of energy consumption of central heating system from EBS, we use the data of central heating from national and local statistics yearbook. In China, statistics yearbooks include detailed data on central heating and this data covers all heat-generators or heat suppliers regardless of their scale.

Since EBS is recognized as a reliable, timely and complete data source of both energy supply and demand, our focus is to develop and test a method of adapted EBS for providing reliable data of China's energy consumption at sector level. Specifically this paper makes the following contributions to develop China's energy consumption statistics by the adapted EBS:

- Developing and testing a method that is adapted from China's existing EBS. This method is applied to calculate China building energy consumption at the sector level.
- Comparing the calculated results of this research with other sources and evaluating what the differences are between this research and other similar researches.
- Comparing the energy consumption of China's commercial (so-called public buildings in China), urban residential and rural residential buildings by the calculated results of this research for the period of 2001–2015.
- Analyzing building energy demand against economic growth to gain insights on differences of energy demand in industrial and building sectors.

- Comparing China's building energy demand with international practices, in particular with the USA, European Union and Japan.
- Recommending how China could build up its building energy consumption statistical system by applying the adapted EBS.

2. Methods

2.1. Procedures and Methods of This Research

The following Figure 1 presents the procedures and methods applied in this research.

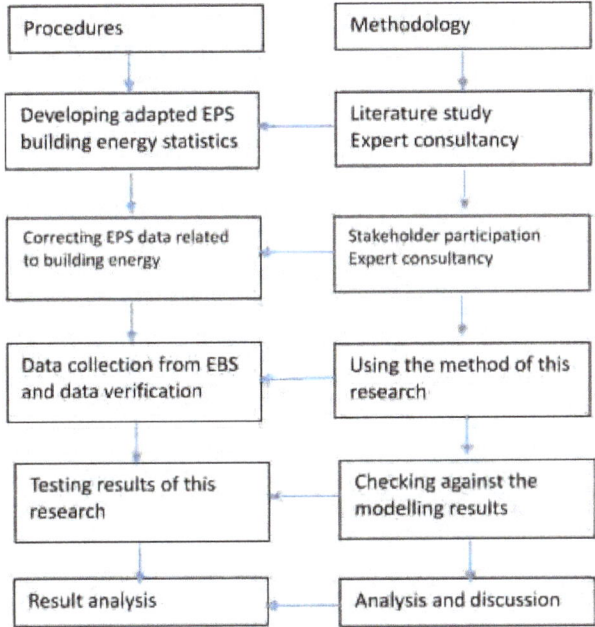

Figure 1. Procedures and methods of this research.

2.2. Data Collection and Analysis

Based on the analysis of the existing EBS of China, we develop the following Formula (1) for statistics of complete building energy consumption in China:

$$E_c = E_b - E_t + E_h + E_o \tag{1}$$

where E_c: complete building energy consumption, E_b: total energy consumption in the sectors of (S5) wholesale, retail trade and catering services, (S6) others and (S7) residential consumption, which are available from the existing EBS, E_t: Transportation energy consumption of (S5) wholesale, retail trade and catering services, (S6) others and (S7) residential consumption, E_h: Energy consumption of corrected central heating and E_o: Building energy consumption from the sectors of (S2) industry, (S3) construction and (S4) transport, storage, postal and telecommunications services.

To gather data on E_t, we conducted a survey aiming at getting information on the consumption of different types of energy among the sectors of (S5) wholesale, retail trade and catering services, (S6) others and (S7) residential consumption, which are available from the existing EBS. The survey conducted by this study took place between October–December 2018 and it included three phases: 1) collecting data of different types of energy consumption of the sectors of S5, S6 and S7 from national

EBS and local EBS of the four megacities of Beijing, Shanghai, Tianjin and Chongqing and seven provinces of Liaoning, Hebei, Shandong, Guangdong, Henan, Sichuan, Ningxia and Gansu. These four megacities and seven provinces are recommended by the China Association of Building Energy Efficiency, given to the facts that: (1) these 11 cities and provinces have quite good energy consumption databases and good EBS, (2) they are good representatives of China's climate zones and (3) they were willing to cooperate with this research and to answer the questionnaire; 2) sending 11 questionnaires to the energy administrations (local Development and Reform Commission) of the four megacities and abovementioned provinces. All 11 questionnaires were answered and have been sent back for this study. The questionnaires aimed at getting information on allocation of energy consumption among different types of energy in the sectors S5, S6 and S7; 3) analyzing the data and providing survey results. The survey results shows that 95% of gasoline and 35% of diesel are used by transportation of the sectors of (S5) wholesale, retail trade and catering services and (S6) others. Almost 100% of gasoline and 95% of diesel consumption are made by transportation in the sector of (S7) residential consumption. This survey results are in line with the study conducted by Wang [7]. Therefore, in this research, we calculated the E_t by using Equation (2):

E_t = (95% gasoline consumption + 35% diesel consumption) of Sector (S5) wholesale, retail trade and catering services and (S6) others + (100% gasoline consumption + 95% diesel consumption) of Sector (S7) residential consumption

(2)

In Equation (2), data on gasoline and diesel consumption in all sectors is available from the existing EBS in China. E_h is calculated by the following Equation (3):

E_h = total energy consumption of central heating − energy consumption of central heating in sector (S5) wholesale, retail trade and catering services, (S6) others and (S7) residential consumption

(3)

The total energy consumption of central heating in Equation (3) is available in the China Statistical Yearbook. Energy consumption of central heating of the sectors (S5) wholesale, retail trade and catering services, (S6) others and (S7) residential consumption is available from EBS.

E_o is calculated by the following Equation (4):

$$E_o = E_{bt} + E_{bi}$$

(4)

E_{bt} is the energy consumption of buildings in the Sector (S4) transport, storage, postal and telecommunications services. An assumption of this research is that coal consumption is only for building energy performance, given to the fact that coal is not used as power fuel in transportation in China. Therefore, E_{bt} is the sum of coal consumption and electricity consumption of buildings in Sector (S4) transport, storage, postal and telecommunications.

E_{bi} is the energy consumption of buildings of the sectors (S2) industry and (S3) construction. Those are mainly office buildings and buildings used for production. Energy consumption of those buildings has a limited contribution to the total energy consumption of buildings. In this research, we organized an expert workshop aiming to estimate E_{bi}. Sixteen experts participated in this workshop. Twelve experts were energy managers from energy-related sectors. Two experts were from Sectors (S4) transport, storage, postal and telecommunications services and the remaining two experts were from universities. Participants were selected based on criteria like: 1) at least seven years of working experience in statistics or estimation of building energy consumption; 2) good knowledge of building energy efficiency; 3) members of the Expert Committee of China Association of Building Energy Efficiency. The workshop, based on the fact that total floor area of buildings of the sectors of (S2) industry and (S3) construction is about same as the floor area of buildings of Sector (S4) transport, storage, postal and telecommunications services, concluded that E_{bi} is comparable to E_{bt}.

3. Results

3.1. China Building Energy Consumption 2001–2015

Figure 2 shows that China's total building energy consumption increased from 310 million tce in 2001 to 860 million tce in 2015, with an average annual increase of 7.6%, which is in line with China's total energy consumption that increased from 1560 million tce to 4300 million tce with an average annual increase of 7.5%. Table 1 shows that the increase of both energy demand and building energy demand is different in the three five-year periods of 2001–2005, 2006–2010 and 2011–2015. During the 2001–2005 period (known as the China National 10th Five-Year Plan), building energy consumption was increasing by 11.9%, while the total energy consumption was increasing by 13.9%. However, the increase of building energy consumption slowed down to 5.3% in the National 11th Five-Year Plan of 2006–2010 and to 5.5% in the National 12th Five Year Plan of 2011–2015. The main reason is that China's national and local governments launched various building energy efficiency initiatives in 2005–2006, and their key initiatives were energy retrofits for existing buildings, promoting green building & low energy building development and compulsorily energy efficiency improvement programme for large-scale public buildings.

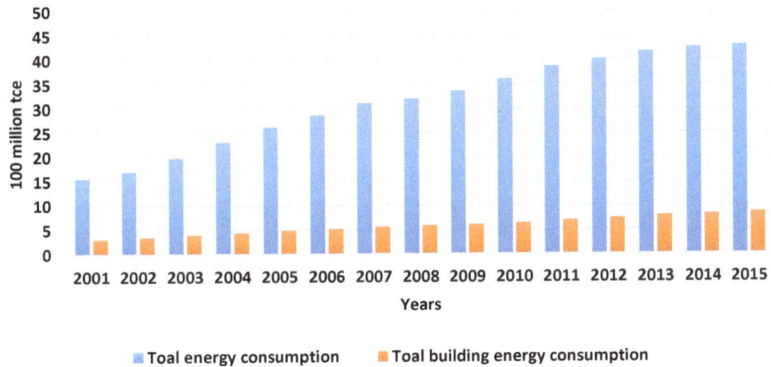

Figure 2. China's Building Energy Consumption 2001–2015. (Source: 1) Data on total energy consumption, China EBS 2001–2015; (Source: 2) Data on total building energy consumption, primary).

Table 1. China Energy Consumption growth rate in the three five-year periods.

Indicator	2001–2005	2006–2010	2011–2015
Average annual growth of China total energy consumption	13.9%	5.9%	2.7%
Average annual growth of China total building energy consumption	11.9%	5.3%	5.5%

Increased building energy consumption is mainly due to the increase of building floor areas and improvement of living standard. However, as shown in Figure 3, the annual growth of building energy consumption has stabilized, while the annual increasing of building floor area has stabilized at about 4%. This decoupling of building energy consumption with increasing building floor area is due to the improvement of building energy efficiency.

Although EBS is recognized as the most reliable energy information source and the original data of this research is from EBS, it is still necessary to compare the result of this research to other sources. Table 2 presents a comparison between this research and building energy consumption data from China Building Energy Model (CBEM). CBEM was developed and being updated by Tsinghua University Building Energy Research Centre (THUBERC). CBEM is based on a sampling of the energy consumption of different types of buildings in different climatic zones. Thus necessary sampling surveys are needed to support CBEM calculation, while the EBS approach of this research is based on official statistics and no additional data collection or survey efforts are needed. Since there is a systematic data verification, data quality control and application of standardized data collection, data from official statistics is highly accepted and well applied by policy makers as well as various stakeholders, while the CBEM approach is widely used by academic institutes and researchers. Table 2 shows a 15-years comparison of the building energy consumption data results from this research and CBEM.

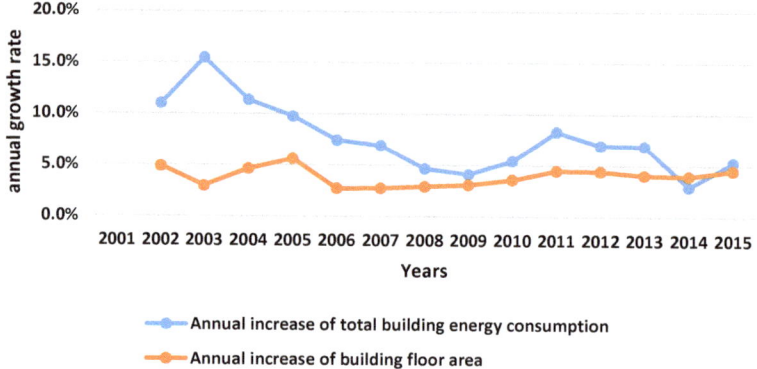

Figure 3. Changing annual growth of building energy consumption 2001–2015. Data source: data on building floor area [19].

Table 2. Data comparison between this research and CBEM, in 100 million tce.

Year	This Research	CBEM [6]	Difference
2001	3.09	3.7	19.7%
2002	3.43	4.1	19.5%
2003	3.96	4.5	13.6%
2004	4.41	5.20	17.9%
2005	4.84	5.50	13.6%
2006	5.20	5.63	8.3%
2007	5.56	5.76	3.6%
2008	5.82	5.93	1.9%
2009	6.06	6.22	2.6%
2010	6.39	6.65	4.1%
2011	6.92	6.95	0.4%
2012	7.40	7.22	-2.4%
2013	7.91	7.81	-1.3%
2014	8.14	8.22	1.0%
2015	8.57	8.64	0.8%

Source: CBEM data [6].

Table 3 presents a comparison of energy consumption of three building types calculated by this research and the CBEM.

Table 3. Data comparison of energy consumption of three building types, in 100 million tce.

Building Types	This Study	CBEM	Difference
Public building	2.83	2.60	−8.8%
Urban residential building	1.85	1.99	7.0%
Rural residential building	1.97	2.13	7.5%
Northern China heating	1.93	1.91	−1%

Source: CBEM data [6].

CBEM has been the most popular tool in China for estimating building energy consumption at the sector level, and this model is revised regularly. Table 2 shows that energy consumption data generated by CBEM in years of 2001 and 2002 is about 20% higher than the data calculated by this research. The difference is getting smaller in 2003–2005, then the difference has been less than 10% since 2006. The reason of the difference that appeared at the earlier stage is the poor-functioning of CBEM and the fact the model needs to be revised to meet the practical conditions. It is interesting that the CBEM results and the results of this research are aligned perfectly after 2007, which could be evidence that CBEM is now working properly and the data calculated by this research is reliable. Table 3 also shows that energy consumptions per category calculated by this study and CBEM are well aligned. We therefore suggest that both methods can be used at the same time for cross-checking and the method developed by this research can be standardized, since it is based on EBS that is dependent on the regular energy statistical system.

3.2. Energy Consumption of Building Types in China

Types of buildings are categorized into public buildings, urban residential buildings and rural residential buildings in China. Floor areas of public buildings and urban residential buildings are available from the statistics yearbooks. However, there are no statistics on the floor area of rural residential buildings. In this study, the expert workshop organized by this study suggested that the floor area of rural residential buildings could be estimated by the average floor area per farmer and the total population of farmers. Average floor area per farmer is available from both national and local housing authorities and the population of farmers is available from statistics yearbooks. The public buildings (so called commercial buildings internationally) are a mix of various buildings like offices, schools and universities, hotels, theaters, warehouses, airports, train stations, retail stores, etc.. Figure 4 provides information on the building energy consumption by the three building types in China, which shows that energy consumptions of all three types are increasing steadily. The main causes of the increasing energy consumption are the increase of building floor areas and improvement of office conditions and living standard. Energy consumption of public buildings represents a 37%–41% share of the total building energy consumption, the energy consumption of urban residential buildings represents a 36%–39% share and the energy consumption of rural residential buildings has a 23%–25% shares.

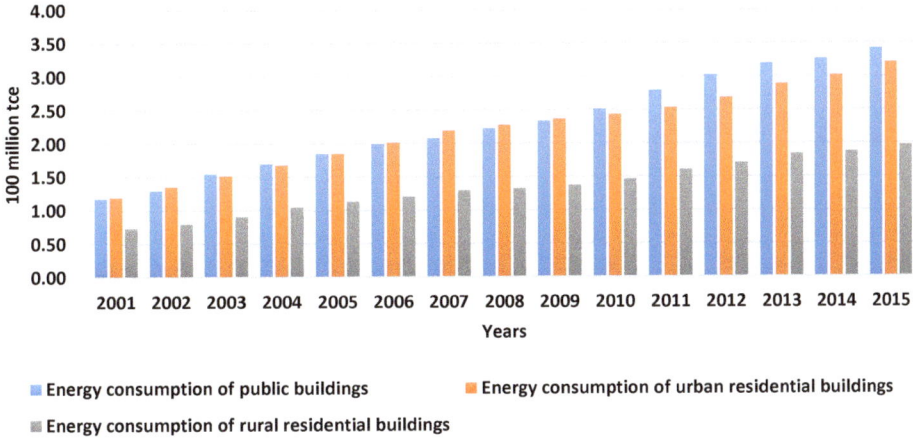

Figure 4. Building Energy Consumption by Categories in China.

As shown in Figure 5, the energy intensity of public buildings is above 30 Kgce/m^2 (Kg coal-equivalent), which is the highest among the three building types. The energy intensity of urban residential buildings is almost double that of rural residential buildings. Among the three building types, the energy intensity of rural residential buildings is increasing slightly, due to the significant improvement of living standards in rural areas. Figure 5 shows that energy intensity of public buildings is the highest. It was increasing in the period of 2001–2005 and then stabilized in the period from 2006–2010. It was decreasing in the period of 2011–2015, given to the successful energy efficiency solutions adopted in this period. The energy intensity of urban residential buildings is almost stabilized and is not increasing in response to the significant improvement of office conditions and living standards. This is also due to the achievements of energy efficiency efforts made by the building sectors. Figure 5 shows that energy intensity of rural residential buildings is increasing significantly, due to the significant improvement of living standards in rural areas.

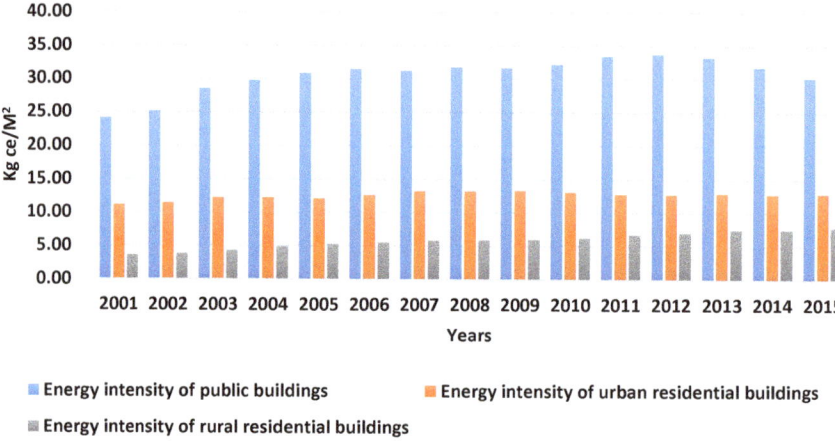

Figure 5. Energy Intensities of Three Types of Buildings in China.

3.3. China Building Energy Consumption Against Economic Growth

As shown in Figure 6, the share of building energy demand in the total energy demand varies from 17% to 21%. The share of building energy consumption is generally decoupled from GDP growth. A lower share of building energy consumption appears when there is higher GDP growth in the period of 2001–2015. During the period of 2002–2007, the GDP growth rate increased annually and reaches its peak of 18.8% in 2007, while the share of building energy consumption decreased from 20.26% in 2002 to 17.86% in 2007. During the period of 2007 to 2014, the GDP growth rate was fluctuating, while the share of building energy consumption is reversely fluctuating. After 2010, the GDP growth is slowing down, while the share of building energy consumption is increasing.

Contrary to the varying share of building energy consumption against GDP growth, the share of industrial energy consumption is coupled with GDP growth, as shown in Figure 7. During the period of 2001 to 2007, the GDP growth rate increased, while the share of industrial energy consumption also increased and reached its peak of 69% in 2007. After 2007, DGP growth is slowing down, and the share of industrial energy consumption is decreasing.

Figures 6 and 7 provide evidence that building energy consumption and industrial energy consumption have different features. Building energy consumption is consumption-related and it is driven by lifestyles and living standards. However, industrial energy consumption is more production-related and it is driven by market and economic activities. Therefore, industrial energy consumption is mainly a consequence of economic development and building energy demand is mainly a consequence of the growing living standards. This can explain why higher GDP growth results in an increasing share of industrial energy demand and a decreasing share of building energy demand in Figures 6 and 7.

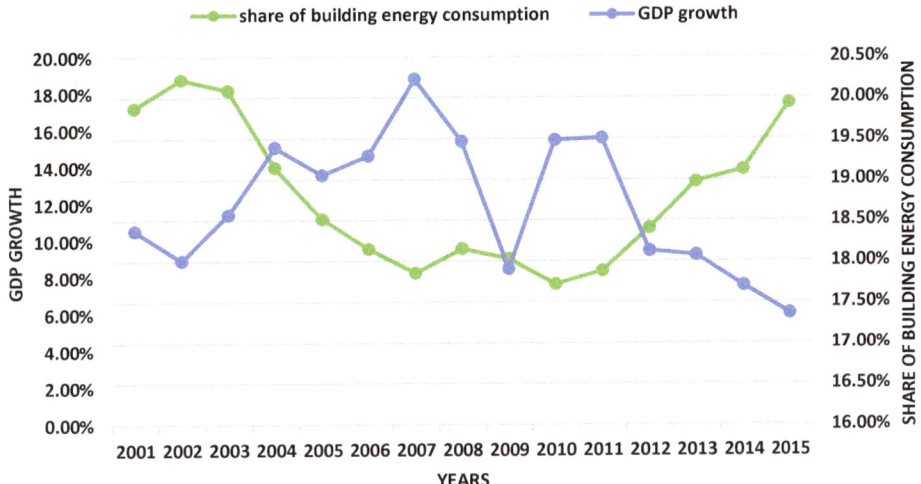

Figure 6. Share of Building Energy Consumption vs. GDP Growth. Data source: GDP data is from [19].

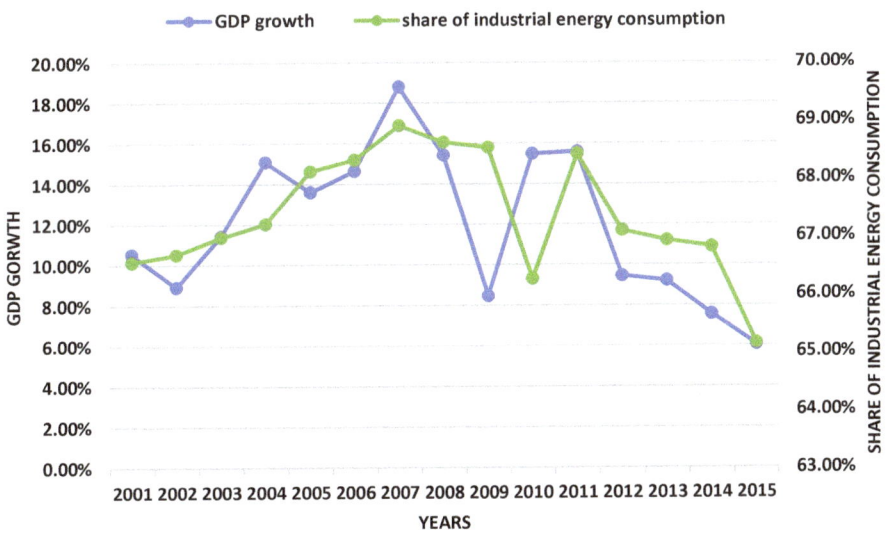

Figure 7. Share of Industrial Energy Consumption vs. GDP Growth.

3.4. China Building Energy Consumption: an International Comparison

After four decades of high economic growth since 1978 when China started its Reforming and Opening Policy, China has been the largest country in terms of total energy consumption and greenhouse gases emissions. China has a share of 28% of the global energy requirements, followed by USA with a share of 22% and by European Union with a share of 15% [1]. However, USA is the largest country in building energy consumption, with a share of 17% of global building energy consumption, while China has a share of 14%. Table 4 provides information that building energy consumption internationally accounts for as much as 30–40% of the global energy requirements. However, China is exceptional and its building energy consumption accounts for only 20.5% of total energy consumption, given the fact that China is still in its high-speed industrializing process and its industrial sectors are more energy-intensive and have a major contribution to total energy consumption, as shown in Figure 2.

Table 4. An international comparison of building energy consumption, Mtoe (million tons-oil-equivalent).

Indicator	China	USA	EU	India	Russia	Japan	Global
Total energy consumption	1913.0	1519.0	1042.5	577.6	456.9	291.3	6929
Building energy consumption	393.0	469.0	364.8	213.2	150.7	95.1	2807
Share of building energy consumption	20.5%	30.9%	35.0%	36.9%	33.0%	32.6%	40.5%

Data source: [1].

4. Discussion

The establishment of building energy consumption statistics by adapted EBS creates a method for China to establish statistics of building energy consumption in a uniform way. An advantage of the adapted EBS is that statistical data comes from the regular statistical system and no additional statistical efforts are needed. However, the adapted EBS method has its limitations. First, all data related to building energy demand is extracted from existing EBS and thus various assumptions are made. Those assumptions are only related to the items (e.g., office building energy consumption in the sectors of transport, storage, postal and telecommunications services, industries and construction) of building energy consumption that have less than 5% contributions to the total energy consumption of buildings. This may cause errors in the final calculation of building energy demand, although we can be sure that those errors are not more than the errors generated by methods of sampling surveys, incomplete statistics, modelling and estimations. Further research is needed for identifying the error percentage and for assuming that the error can be negligible. Second, bio-energy has been widely used in Chinese rural residential buildings and this bio-energy consumption is not included in this adapted EBS system. This means that energy consumption of rural residential buildings calculated by this research is lower or significantly lower than the actual energy consumption of rural buildings. Third, our research finds that data on China building floor areas from various sources are quite different, since not all buildings are registered in building departments. The China Association of Building Energy Efficiency estimates that about 15%-25% of the existing buildings are not registered [19]. Thus building energy intensity calculated by this research is about 20% higher than the actual value. Therefore we suggest that building energy statistics and a complete building registration system should be established together. Lastly, there are more and more clean energy used at a single building scale in China [20,21] at this moment. In the existing China EBS, large-scale renewable energy production (e.g., hydropower) is already included. Clean energy production at a single building level (e.g., solar) is not included in the national EBS and thus clean energy uses at a single building scale is not included in this study. Thus further studies are needed for incorporating clean energy uses into the building energy consumption data. By doing so, the adapted EBS methods can be used to calculate CO_2 emissions. Further research is also needed to test and ascertain whether this adapted EPS approach can truly help China in establishing building energy consumption statistics at both national and local levels.

5. Conclusions

China's energy statistical system is different from the IEA system. Building energy demand is not a statistical sector in China's EBS and thus data on building energy demand is not available directly from China's EBS [22]. To gather data on China's building energy demand at the sector level, various methods have been developed, tested and applied [6,19]. All the methods are based on limited sampling, incomplete statistics, modelling and estimations and thus those methods are not standardized and it is difficult to conduct regional comparisons and benchmarking in the building energy sector. This study explores a possibility where data on energy demand of the building sector can be made available from an adapted China EBS. Since EBS is the most reliable energy data source, our method can be standardized and thus will enable regional and international comparisons and benchmarking. Our study contributes to build up China building energy statistics and our key findings include six perspectives:

1. The building energy statistics by adapted EBS covers all types of buildings (commercial and residential buildings and multi-functions buildings that are used by sectors of industry, transportation and others), and it provides reliable, detailed and complete data on energy demand in China's building sector. Based on EBS data, this method can be included in regular energy statistical system and can be standardized in a transparent way. Thus data generated by this method will ensure the international, national and regional comparisons and benchmarking of building energy demand.
2. Data comparison between this research and the existing most adopted China Building Energy Model (CBEM) shows no significant differences. We therefore suggest that both methods can be used for cross-checking to build up China statistics of building energy consumption.
3. Both total energy consumption and building energy consumption have increased by about 7.6% in China since 2001. However the growth rate of total energy consumption is slowing down and annual growth of building energy consumption has been stabilized, although total building floor area is increasing by about 4%. Decoupling of building energy consumption with increasing of building floor area is due to the improvement of building energy efficiency.
4. The energy consumption of public buildings accounts for a 37%–41% share of the total building energy consumption, residential buildings represent 36%–39% and rural residential buildings have a 23%–25% share in China. The energy intensity of commercial buildings is the highest compared to urban and rural residential buildings and the energy intensity of urban residential buildings is almost double that of rural residential buildings in China. This is completely different from the situation in Europe, where the energy efficiency of rural households is always higher than in urban households [23]. This research concludes that there are tremendous differences in the building energy efficiency of rural buildings between China and Europe, since building energy efficiency solutions are less developed in rural China [24–27].
5. Building energy consumption and industrial energy consumption features are different. Building energy consumption is driven by lifestyle and living standards. However, industrial energy consumption is mainly driven by market and economic activities. Therefore, a higher GDP growth results in an increased share of industrial energy consumption and a decreased share of building energy consumption.
6. China's building energy consumption has less of a contribution to the total energy demand, as compared to that of the developed world. This shows that China industrialization is still playing a more important role in the energy demand, while consumption and transportation sectors have more of a contribution to the total energy demand in the developed world.

The main conclusion from this study is that the existing EBS of China can be adapted to provide a more reliable and complete building energy demand information at the sector level. Thus, the underlying target is not to build up a new building energy statistics that is costly and separated from the existing regular energy statistics, but instead a better option is to apply the adapted EBS developed by this research. In addition, we suggest that the adapted EBS can be used with the support of the existing CBEM to ensure accurate data cross-checking.

Author Contributions: Conceptualization, M.Z., Y.Z. and C.X.-B.; methodology, M.Z. X.G. and C.X.-B.; validation, M.Z. X.G., Y.Z. and C.X.-B.; formal analysis, M.Z. and X.G.; data curation, X.G., Y.Z. and C.X.-B.; writing—original draft preparation, M.Z. X.G. and C.X.-B.; writing—review and editing, M.Z. and X.G.

Funding: This research is funded by the European Commission's Switch Asia II programme: Promoting Sustainable Consumption and Production (contract number: DCI-ASIE/2015/368-399).

Acknowledgments: We would like to thank all the SusBuild project staff involved in this research for their valuable contribution and comments. We would like to give special thanks to China Association of Building Energy Efficiency, who provides us necessary data and supports us for data collection.

Conflicts of Interest: The authors declare no conflict of interest.

References

1. IEA (International Energy Agency). World Energy Balances: Complete Energy Balances for over 150 Countries and Regions. 2018. Available online: https://www.iea.org/statistics/balances/ (accessed on 9 November 2019).
2. Fridley, D.; Zheng, N.; Zhou, N. *Estimating Total Energy Consumption and Emissions of China's Commercial and Office Buildings (LBCN-248E Report)*; Lawrence Berkley National Laboratory: Berkeley, CA, USA, 2008.
3. Asimakopoulos, D.A.; Santamouris, M.; Farrou, I.; Laskari, M.; Saliari, M.; Zanis, G.; Giannakidis, G.; Tigas, K.; Kapsomenakis, J.; Douvis, C.; et al. Modelling the energy demand projection of the building sector in Greece in the 21st century. *Energy Build* **2012**, *49*, 488–498. [CrossRef]
4. Zhang, M.; Wang, M.; Jin, W.; Xia-Bauerc, C. Managing energy efficiency of buildings in China: A survey of energy performance contracting (EPC) in building sector. *Energy Policy* **2018**, *114*, 13–21. [CrossRef]
5. Gu, L.; Yu, C. Data status and statistics of China building energy consumption. *China Energy* **2011**, *33*, 38–41.
6. Building Energy Research Centre of Tsinghua University (THUBERC). *2017 Annual Report on China Building Energy Efficiency*; China Architecture & Building Press: Beijing, China, 2017; pp. 7–9. ISBN 978-7-112-20573-8.
7. Wang, Q. 2007 Research on statistics and calculation of China building energy consumption. *Energy Sav. Environ. Prot.* **2007**, *8*, 9–10.
8. Long, H. Percentage of building energy consumption and building energy efficiency targets. *China Energy* **2005**, *10*, 23–27.
9. Ding, J. Discussions on statistics and methodology of China building energy consumption. *Mod. Econ. Inf.* **2011**, *23*, 191.
10. Chen, S.; Li, N.; Guan, J. Research on statistical methodology to investigate energy consumption in public buildings sector in China. *Energy Convers. Manag.* **2008**, *49*, 2152–2159. [CrossRef]
11. Lombard, P.; Ortiz, J.; Pout, C. A review on buildings energy consumption information. *Energy Build.* **2008**, *40*, 394–398. [CrossRef]
12. Zhao, L.; Liang, R.; Zhang, J.; Ma, L.; Zhao, T. A new method for building energy consumption statistics evaluation: Ratio of real energy consumption expense to energy consumption. *Energy Syst.* **2014**, *5*, 627–642. [CrossRef]
13. Zhou, N.; Lin, J. The reality and future scenarios of commercial building energy consumption in China. *Energy Build.* **2008**, *40*, 2121–2127. [CrossRef]
14. Zhang, Y.; He, C.; Tang, B.; Wei, Y. China's energy consumption in the building sector: A life cycle approach. *Energy Build.* **2015**, *94*, 240–251. [CrossRef]
15. Jiang, Y. Chinese building energy consumption situation and energy efficiency strategy. *New Archit.* **2008**, *2*, 4–8.
16. Chen, S.; Li, N.; Guan, J.; Xie, Y.; Sun, F.; Ni, J. A statistical method to investigate national energy consumption in the residential building sector of China. *Energy Build.* **2008**, *40*, 654–665. [CrossRef]
17. Xiao, H.; Wei, Q.; Jiang, Y. The reality and statistical distribution of energy consumption in office buildings in China. *Energy Build.* **2012**, *50*, 259–265. [CrossRef]
18. Lin, X.; Peng, J.; Jiang, H. Study on statistics and calculation of building energy consumption in Chongqing City. *Build. Energy Effic.* **2008**, *36*, 55–57.
19. China Association of Building Energy Efficiency (CABEE). *2017 Research Report of China Building Energy Consumption*; China Association of Building Energy Efficiency: Beijing, China, 2017; pp. 1–25.
20. Yu, S.; Eom, J.; Evans, M.; Clarke, L. A long-term, integrated impact assessment of alternative building energy code scenarios in China. *Energy Policy* **2014**, *67*, 626–639. [CrossRef]
21. Liu, X.; Wang, C.; Liang, C.; Feng, G.; Yin, Z.; Li, Z. Effect of the energy-saving retrofit on the existing residential buildings in the typical city in northern China. *Energy Build.* **2018**, *177*, 154–172.
22. Wang, J.; Zhang, Y.; Liu, M.; Wang, Y. Baseline comparison of China, USA and UK building energy consumption. *Build. Sci.* **2015**, *31*, 48–51. [CrossRef]
23. Heinonen, J.; Junnila, S. Residential energy consumption patterns and the overall housing energy requirements of urban and rural households in Finland. *Energy Build.* **2014**, *76*, 295–303. [CrossRef]
24. Yang, X. Introduction on building energy consumption statistics of USA. *Build. Sci.* **2010**, *26*, 8–11. [CrossRef]

25. Zhang, M.; Cui, Y.; ter Avest, E.; van Dijk, M.P. Adoption of voluntary approach: Can voluntary approach generate collective impacts for China achieving ambitious energy efficiency targets? *Energy Environ.* **2018**, *29*, 281–299. [CrossRef]
26. Zhang, M.; Li, H.; Jin, W.; ter Avest, E.; van Dijk, M.P. Voluntary agreements to achieve energy efficiency, a comparison between China and The Netherlands. *Energy Environ.* **2018**, *29*, 989–1003. [CrossRef]
27. Lu, N.; Lin, G.; Yu, Y.; Ma, H.; Wang, Q. Analysis on China's residential building energy consumption in 2003–2007. *Constr. Econ.* **2009**, 95–97. [CrossRef]

© 2019 by the authors. Licensee MDPI, Basel, Switzerland. This article is an open access article distributed under the terms and conditions of the Creative Commons Attribution (CC BY) license (http://creativecommons.org/licenses/by/4.0/).

Article

Impact of Heating Control Strategy and Occupant Behavior on the Energy Consumption in a Building with Natural Ventilation in Poland

Aniela Kaminska

Faculty of Electrical Engineering, Poznan University of Technology, ul. Piotrowo 3a, 60-965 Poznań, Poland; aniela.kaminska@put.poznan.pl

Received: 10 October 2019; Accepted: 7 November 2019; Published: 12 November 2019

Abstract: This study aims to provide an experimental assessment of energy consumption in an existing public building in Poland, in order to analyze the impact of occupant behavior on that consumption. The building is naturally ventilated and the occupants have the freedom to change the temperature set point and open or close the windows. The energy consumption is calculated and the calculation results are compared with the experimental data. An analysis of occupants' behavior has revealed that they choose temperature set points in a wide range recognized as thermal comfort, and window opening is accidental and difficult to predict. The implemented heating control algorithms take into account the strong influence of individual occupant preferences on the feeling of comfort. The energy consumption assessment has revealed that the lowering of temperature set point by 1 °C results in an energy saving of about 5%. Comparisons of energy consumption with heating control and without any controls showed that the potential for energy reduction due to heating control reached approximately 10%. The use of windows control, which allows to turn off the heating after opening the window and its impact on energy savings have been discussed as well.

Keywords: building automation systems; building energy efficiency; heating control; energy savings

1. Introduction

Currently, the global building sector has been the main consumer of world energy [1]. Energy consumption in the existing buildings accounts for 40% of the total energy consumption in the United States [2] and in Europe [3] where 75% of buildings are energy inefficient [4]. Therefore, the European Commission has published a series of recommendations on the modernization of buildings including guidance on the automation and controls of buildings [5]. However, despite the large number of building retrofit technologies [6] and the management of heating, ventilation and air conditioning (HVAC) systems, the implementation of these recommendations is a difficult and costly challenge.

In making any decisions regarding the modernization of a building, estimating energy consumption in the building is of key importance. This consumption is influenced by many factors such as ambient weather conditions, building structure and characteristics, the operation of HVAC systems and occupancy. One of the most important factors is climate data, which plays a fundamental role in the building design. Results presented in [7] show that an improvement of around 15% in energy consumption in buildings can be achieved due to changes in building design such as space area, exterior openings and material thickness and the choice of building envelope in all climates. An overview of measures and policies adopted by different countries, allowing the monitoring, management and reduction of energy consumption in buildings is given in [8]. The energy consumption related to HVAC systems in different types of buildings (office, commercial and residential) is analyzed in [9]. It is widely expected that building occupancy is of great importance for energy efficient control of buildings. Therefore, a large number of works have been developed for the estimation and detection of building occupancy. A comprehensive review on this problem is presented in [10]. However it should be noted that new buildings are mostly controlled by a building management system (BMS) where building occupants have minimal access to the controls. In these buildings energy consumption is not strongly correlated with occupancy patterns [11].

Many factors influencing energy consumption mentioned above make the estimation of this consumption very difficult. In [12] recently developed models for solving this problem, including elaborate and simplified engineering methods, statistical methods and artificial intelligence methods are reviewed. Quantitative energy performance assessment methods are described in [13]. To simplify the calculation of energy in the building, a steady-state model was developed as CEN standards, i.e., energy performance of the building—calculation of energy use for space heating and cooling [14]. In this model the predicted energy consumption consists of heat transfer through the building envelope, heat losses for ventilation, heat gain from solar radiation and internal heat gain from people and equipment. In cold climates, such as in Poland, the energy used for heating is predominant, therefore, knowing the thermal characteristic of the building envelope and ventilation is crucial [15]. In old buildings, natural ventilation with operable windows is usually used. In new buildings, this type of ventilation also becomes increasingly popular as a solution with lower energy consumption compared to mechanical ventilation and air conditioning. Over the past decades, the impact of various parameters on the performance of natural ventilation has been studied [16] and many models have been developed. Important natural ventilation models and simulation tools as well as the comparisons of their prediction capabilities are reviewed in [17]. The analysis shows that these models are generally only applicable to specific geometries and driving forces. Furthermore, the most accurate models are developed for cases with small and simple openings. To investigate the air flow pattern inside a building, computational fluid dynamics (CFD) models are developed. The model based on the finite volume numerical solution of the Navier–Stokes equations presented in [18] shows that different positions and shapes of an opening can determine the behavior of the flow stream inside the building. It allows to determine the condition of natural ventilation efficiency of the building. Another fluid dynamics (CFD) model allows to investigate a wind-driven ventilation system in a building with multiple windows [19].

The study mentioned above shows the complexity of a phenomenon that has a decisive influence on thermal comfort and energy consumption. In a naturally ventilated building, thermal comfort can be improved and adapted to individual preferences when occupants have the freedom to change the temperature set points and open or close the windows.

Various case studies [20,21] have shown that occupants tend to adapt to changing environmental conditions in such a way as to achieve their individual comfort. Research on such behavior is called the adaptive approach. The application of this approach to thermal comfort standards is considered in [22] and an equation for naturally ventilated buildings in hot-humid climates is developed in [23]. It was found that acceptable comfort ranges showed asymmetry and leaned towards operative temperatures below thermal neutrality for all climates. However, other results, inter alia in [24], based on the data of surveys conducted in a naturally ventilated building found symmetry of comfort ranges. Many studies also confirm it is difficult to use defined comfort ranges in the real conditions because it depends on the occupants' physiology and subjective perception [22]. The thermal sensations of occupants inside buildings are influenced by many factors such as air temperature and velocity, humidity, concentration of CO_2, building microclimate, as well as age, activities, preferences, etc. [25,26]. Occupants have various means of interacting with the indoor environment: they can interact directly with a given built environment by changing the temperature set points (or adjusting thermostats), operating the windows, shading, or they can adjust themselves to the existing environmental conditions by changing their clothing or activity [27]. As regards the theory of thermal comfort in buildings, a large impact of clothing and activity on the level of comfort is represented by the most extended predicted mean vote (PMV) index [22,25,28]. This index described the statistical response about thermal sensation of a large group of people exposed to specific thermal conditions. Six variables, namely metabolic rate, clothing insulation, air and mean radiant temperatures, air velocity and relative humidity affect the PMV index. Four of them can be recorded during the experiment, while clothing insulation and metabolic rate are not easily measurable and their values are most often taken from [27]. For a typical office the values of clothing insulation are 1.0 and 0.5 clo for winter and summer respectively, whereas a typical value used for metabolic rate is 1.0 met. It is also worth noting that the occupant-building interaction is bidirectional, which means that the building environment and interior also affect the occupants' behavior [25], but this interaction requires additional research to identify and describe.

The behavior of occupants is a key issue in the design of the HVAC system and its integration with other control systems in the building as well as in the assessment of energy efficiency [29]. Various methods of occupant behavior estimation and detection are used in [10] and models of occupant behavior can be an efficient means to be implemented into building energy modeling programs [30]. Detecting the presence and absence of occupants allows to determine the operation time of HVAC systems in the building. Potential annual energy savings are estimated at around 10–40%. It has been shown in [31] that the HVAC system can save up to 9% of energy if occupancy-based HVAC schedules are used. In [32], an algorithm for adjusting temperature set points with various indicators of occupant discomfort tolerance has been proposed and energy savings are estimated at 20% while maintaining the building comfort requirements. In [33], based on the detection of the instantaneous number of occupants in the building and related behaviors, it was demonstrated that the energy consumption of the building could be reduced by 40% without compromising the thermal comfort and air quality. However, although there are many methods for detecting and describing occupant behavior to achieve energy savings, their limitations are revealed when applied to real HVAC systems, and they are mainly related to the difficulty of tracking occupant-provoked changes by the HVAC system.

The use of information about occupant behavior to control the HVAC system and estimate possible energy savings depends on the thermal behavior of the building, which determines the heating and cooling time of the building. Several studies have been carried out to investigate the building thermal behavior and model predictive control (MPC), which allow better tracking of changes in the operating mode and temperature set points [34]. The knowledge of building thermal behavior and the popular gray box model approach are the basis for designing an HVAC control system and estimating the energy savings potential [35,36].

As stated above, because the potential of energy savings depends on various parameters, its estimation shows large discrepancies. This paper deals with the experimental and theoretical evaluation of energy consumption in an existing public building in Poland. The building is naturally ventilated and the occupants have the freedom to change the temperature set point and open or close the windows. The effect of occupant behavior as well as heating control and window operation on energy consumption is investigated. The main purpose of the work is to determine the impact of window opening and the range of temperature set point chosen by the users on energy consumption.

The temperature set points in the heating zones of the building and the outdoor temperature are measured and recorded by the KNX automation system and for these temperatures the energy consumption is calculated taking into account heat transfer through the building envelope and heat losses for ventilation. The calculation results are compared with the experimental data. A heating control strategy has been implemented in the building and the energy saving potential is estimated for this strategy.

2. Methodology

The main purpose of the work was to determine how much energy could be saved in a real building by using heating control. It is also important to determine what factors affect the energy savings in a building. In order to achieve this goal, energy consumption for heating was first calculated. The calculations took into account temperatures outside the building and inside the rooms as they occurred during the one-month period. These temperatures were recorded in the KNX system implemented in the building. It was noted that occupants chose temperature set points corresponding to their thermal comfort, which differed by several degrees. In order to verify the calculations, calculated energy values were compared with measured values. Then, it was assumed that the temperature in the whole building was constant during the analyzed period and that the outside temperature was as in the previous experiment. Two temperature values were selected, namely 20 °C and 21 °C. Energy consumption for these conditions was referred to as "consumption without heating control". The next task was to calculate energy consumption for the same external conditions, but taking into account the control method used in the building. This consumption was marked as "energy consumption with heating control". However, this required determining the temperature changes in the rooms of the building after lowering the temperature set point. This problem was investigated experimentally and discussed. Attention was also paid to the temperature change when the window is tilted from the top by 30° from vertical. This method of window opening is often used by occupants.

3. Building and Experimental Installation

3.1. Construction of the Building

This study deals with the activities of the Laboratory of KNX System and Evolution of Installation Energy Efficiency (SKNX and EIEE Laboratory) at Poznan University of Technology in Poznan, located in the north-western part of Poland (Figure 1). The building was built in the 1980s and is representative of existing Polish buildings from that period considering building envelopes. In 2010 the building was retrofitted and its energy efficiency improved significantly. It is a three-story building with a height of 11.48 m and the external outline surface of 236.8 m^2. On the south the building adjoins another facility up to the level of one story.

Figure 1. External view of the KNX System and Evolution of Installation Energy Efficiency (SKNX and EIEE) Laboratory building.

Figure 2 shows the thickness and the value of the thermal conductivity coefficient of each layer that constitutes part of the building envelope. The thermal conductivity coefficients are taken from PN-EN ISO 6946 [37].

The external walls (Figure 2a) with a thickness of 380 mm were built of full ceramic brick and covered with 15 mm lime and cement-lime plasters. In the ground, the walls were made of cement blocks and covered with two 15 mm layers of cement-lime plasters. As a thermal insulation, a 120 mm layer of styrofoam was used on the external walls. At a height of 50 cm below and above the ground, extruded polystyrene with a thickness of 90 mm was placed.

The roof (Figure 2b) is multi-layered and consists of 240 mm channel slabs, 100 mm layer of Supreme, a void of 210 mm, 20 mm cement plaster and the final layer of 45 mm roofing felt. Thermal isolation was achieved by blowing Rockwool granules into the air void. The laboratory floor was not thermo-modernized, and the layers in contact with the ground in the part corresponding to heating zone 1 are presented in Figure 2c, and those corresponding to zones 2 and 3 are shown in Figure 2d. The main layers of the floor in heating zone 1 are a 150 mm layer of concrete debris and a 300 mm layer of granulated blast-furnace slag. Insulating roofing tar on a layer of waterproof asphalt and cement-bonded wood fiber are used as the insulation. In heating zones 2 and 3 the floor forms layers of concrete debris, leveling concrete and terrazzo. The whole floor in all the zones is covered with floor gres laid on cement-plaster.

The thermal resistance of a component layer i of a building envelope is defined as $R_i = d_i/\lambda_i$, where d_i is the thickness of the layer and λ_i is the thermal conductivity coefficient. The thermal resistance R of a multi-layer building envelope is determined as the sum of the thermal resistance of the component layers and the conventional internal surface thermal resistance R_{si} and the external surface thermal resistance R_{se}. The values of R_{si} and R_{se} resistance depend on the type of building envelope and the direction of heat flow. For external walls and the horizontal direction of heat flow $R_{si} = 0.13$ m^2 K/W and $R_{se} = 0.04$ m^2 K/W, for flat roof $R_{si} = 0.10$ m^2 K/W and $R_{se} = 0.04$ m^2 K/W [38]. The heat transfer coefficient, by definition, is calculated as $U = 1/R$.

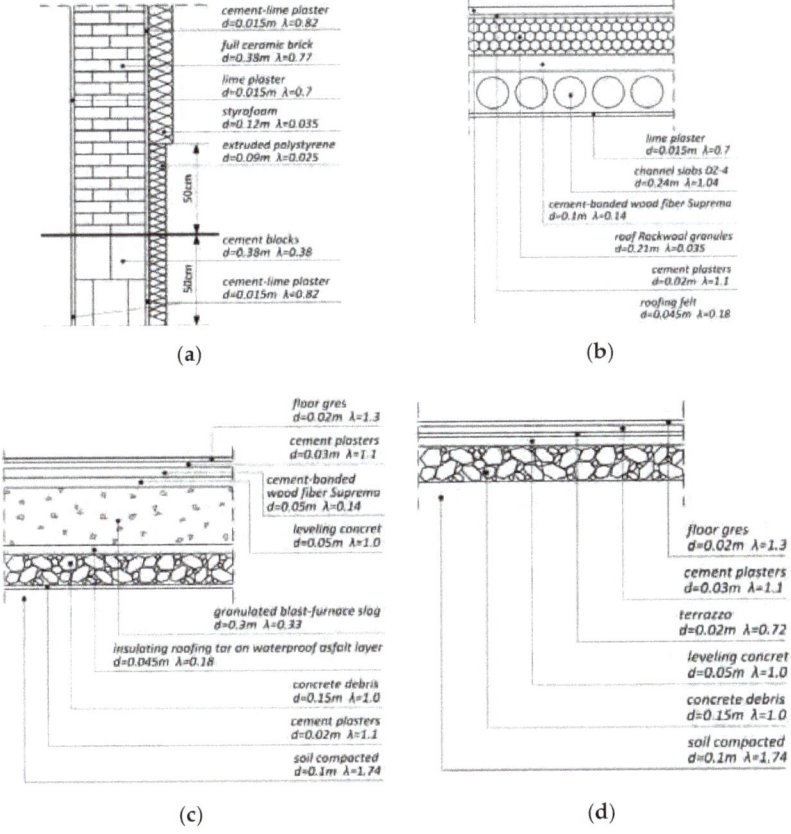

Figure 2. Cross-section of: (**a**) the external wall; (**b**) the roof; (**c**) the floor in heating zone 1 and (**d**) the floor in heating zones 2 and 3.

In the walls, there are window jambs, lintels and wall connections, which result in the formation of thermal bridges that increase heat transfer. They are taken into account by introducing a correction of ΔU. For external walls with windows $\Delta U = 0.05$ W/m² K is assumed.

The heat transfer coefficient for windows is determined as:

$$U_W = \frac{A_g \cdot U_g + A_f \cdot U_f + l_g \cdot \Psi_g}{A_g + A_f}, \qquad (1)$$

where: U_g and U_f are the heat transfer coefficients in the middle part of double glazing and the frame, respectively, A_g and A_f are the surfaces of the glass and the frame, Ψ_g is the linear heat transfer coefficient of the thermal bridge at the interface between the glass and the frame and l_g is the length of the thermal bridge. According to the technical approval for windows $U_g = 0.5$ W/m² K and $U_f = 1.2$ W/m² K. The surface of the glass is 0.4544 m² and that of the frame is 0.7781 m². The length of the thermal bridge amounts to 2.3 m and the linear heat transfer coefficient is taken as 0.06 W/m² K.

The main entrance to the building leads through two doors from the west. The surface of the single door is 3.494 m². There is an additional door with a surface of 3.478 m² on the east of the building, occasionally used for moving heavy equipment. According to the technical approval the heat transfer coefficient is 2.6 W/m² K.

3.2. Heating Zones

The building was divided into heating zones shown in Figure 3, differing in use, size and separation walls. The division into zones determined the pipeline system, in particular the number of heating circuits supplying hot water to panel radiators.

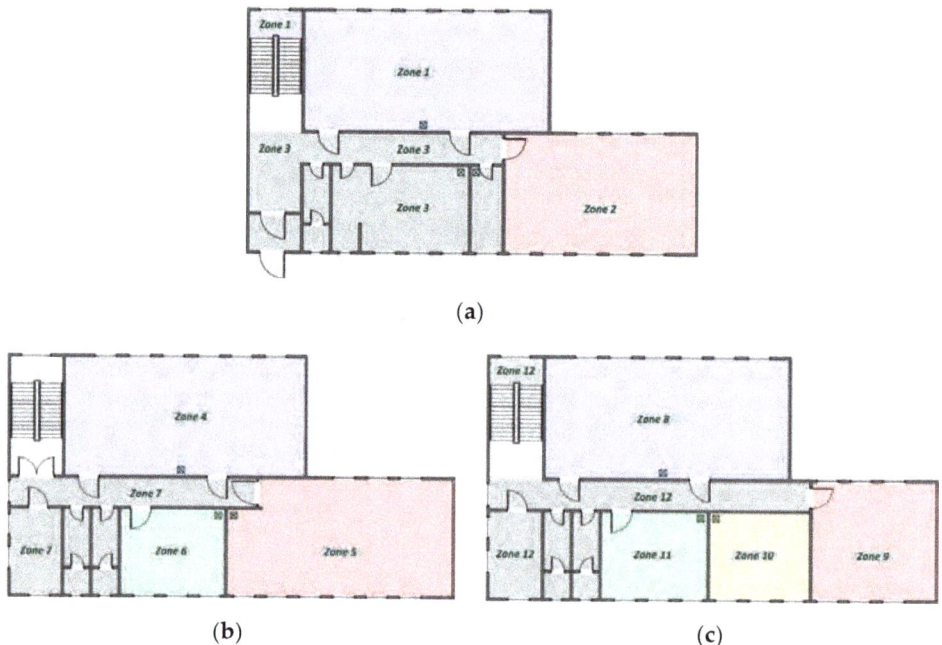

Figure 3. Heating zones in the case study building: (**a**) the ground floor plan; (**b**) the first floor plan and (**c**) the second floor plan.

On the ground floor, there are three heating zones, namely zone 1 and 2 including high-current laboratories and zone 3 including a workshop, sanitary facilities and a corridor. People staying in these rooms do not perform sedentary work and the operation of the devices causes an increase in temperature. The first floor consists of four heating zones. These zones are the most stable in temperature, due to the floor being closed with a staircase door and because of its location between the heated floors of the building. The second floor was divided into five heating zones corresponding to the rooms. The height of all zones is the same and amounts to 2.8 m.

3.3. Control System and Data Acquisition

The heating system in the SKNX and EIEE Laboratory building is designed in such a way that it is possible to estimate the heat consumption in each room and implement various control algorithms as well as to measure, record and visualize useful data [39]. Panel radiators are used as the heating devices. In this system heat is carried by water supplied from the city heating network. The scheme of the pipeline system is shown in Figure 4. In order to force the water flow through the installation, circulation pump (P) is used. At the inflow, a control valve (CV) has been mounted and heating water parameters are measured using a heat meter. Then, the hot water flows into three main circuits assigned to each story and the heating water parameters are also measured at the inflow to each circuit. The water feeds heating circuits assigned to heating zones (Figure 3): on the ground floor—three circuits, on the first floor—four circuits and on the second floor—five circuits. Water from heating devices returns through the pipelines on the stories and then the main pipeline to the city heating network. Each water circuit is equipped with a heat meter and a KNX servo drive. The servo drives are controlled by signals sent directly from the KNX bus. The KNX multi-function push-button with a room temperature control unit is located in each heating zone. In addition, the KNX Laboratory (heating zone 5) is equipped with a KNX touch panel that visualizes the states and parameters of the system. A valve controller at the heating system inflow and heat meters is connected to the ControlMaestro controller with a SCADA (superior control and data acquisition) system using an M-Bus network (Figure 4). This system allows the visualization and acquisition of values measured in the building heating system.

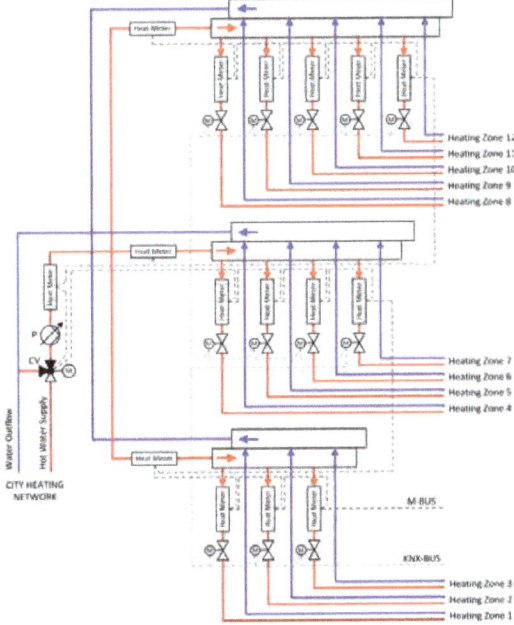

Figure 4. Heating system pipeline scheme in the SKNX and EIEE Laboratory.

To control the heating system KNX devices mentioned above and KNX BACS field network are used. In the KNX system other devices are integrated, including a weather station, brightness and temperature sensor, presence detectors and Gira HomeServer. KNX is an open standard for public, commercial and domestic buildings [40], which allows the integration of many devices from different manufacturers. KNX devices are most often connected by a twisted pair or RF bus and programmed with the use of ETS software. It is worth noting that the system used in the laboratory building can be easily expanded with new devices, and in addition, it allows testing various control algorithms through reprogramming using the available ETS software. Two networks, M-Bus and KNX, are integrated using a M-Bus/KNX converter (Figure 5), which enables the acquisition of all measured values and events in the form of telegrams (standardized KNX messages) by the KNX HomeServer. The HomeServer visualizes the results on-line, archives them and, once a day, sends the results as a csv file to specified e-mail addresses. The recording format allows further processing of the results by external tools and programs.

The following data were recorded by the HomeServer:

- Set point and current temperature in each heating zone from a push-button with room temperature control unit, measured with the accuracy of ±1 °C (logged every 5 min);
- Temperature from the weather station and the external brightness and temperature sensor mounted on the building facades, measured with the accuracy of ±1 °C (logged every 5 min);
- Wind speed from the weather station, measured with the accuracy of ±1.5 m/s (logged every 5 min);
- Occurrence (or absence) of rainfall or snowfall from the weather station (logged every 5 min);
- Illuminance level, from the weather station and the external brightness and temperature sensor, measured with the accuracy of ±5 lux (logged every 5 min);
- Energy from the heat meters, measured with the accuracy of ±5% (logged every 30 min);
- Instantaneous power from the heat meters, measured with the accuracy of ±5% (logged every 5 min);
- Position status of the windows in each room.

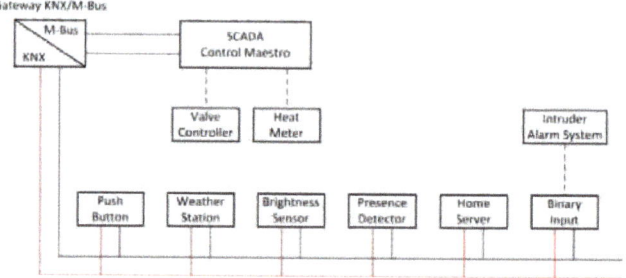

Figure 5. Integration of M-Bus and KNX networks.

In order to determine the position status of the windows and take it into account in the heating control, the intruder alarm system (IAS) in the building was integrated with the KNX system. In window frames, reed switches are mounted and signals from these devices are sent to the alarm control unit, which transmits them to the KNX binary input.

3.4. Temperature Set Point

The temperature set points for the heating seasons are established based on ISO (International Standard Organization) Standard 7730 [41], which defines the comfort ranges according to the specificity of Europe [42]. However, it should be noted that thermal sensations differ between persons sharing the same environment, because there are many factors that affect the perception of human beings [26,28]. The thermal sensations experienced by a human being result mainly from the overall thermal balance of the body. This balance includes two components, namely heat generated by a human being and heat transferred to the environment. The first depends on the physical activity and the second depends on clothing, as well as on environmental parameters such as air temperature, radiant temperature, air velocity and air humidity [43].

The American Society of Heating, Refrigerating and Air-Conditioning Engineers (ASHRAE) standard [44] specifies the conditions in which a fraction of occupants find the environment thermally acceptable. The predicted mean vote (PMV) and the predicted percentage dissatisfied (PPD) are defined in ISO 7730 [41]. The thermal comfort index PMV-PPD reflects the degree of human thermal balance deviation and is a comprehensive comfort indicator that represents the feelings of most people in the same environment. PMV scales constitute seven thermal sensation points ranging from −3 (cold) to +3 (hot), where 0 means a neutral thermal sensation [45]. The PMV index involves activities (expressed through the metabolic rate index), clothing corresponding to the total thermal resistance from the skin to the outer surface of the clothed body and the four environmental parameters mentioned above [41,46].

Depending on the admissible ranges for PMV and PPD, three kinds of comfort zones or categories of thermal requirements are defined by ISO 7730 as: category I (or class A; PPD < 6%, i.e., −0.2 < PMV < 0.2), category II (or class B; PPD < 10%, i.e., −0.5 < PMV < 0.5) and category III (or class C; PPD < 15%, i.e., −0.7 < PMV < 0.7). The ranges of recommended air temperatures for different types of buildings depending on the previous categories are shown in Table 1 [41]. Thus, in the study case building the range of temperature set point was set from 19 to 25°C and the occupant had some freedom to choose the preferred temperature during their presence in the room. It should be noted that this value was a subjective decision of the occupant and the prediction of occupant behavior was a factor of considerable uncertainty in the analysis [47].

Table 1. The range of recommended air temperatures for offices and classrooms, according to ISO7730 [41].

Type of Building	Activity (W/m^2)	Category	Temperature (°C)
Classrooms		A	22.0 ± 1.0
Offices	70	B	22.0 ± 2.0
Conference room		C	22.0 ± 3.0

3.5. Building Use and Heating Control Algorithm

The analyzed information about the occupancy, opening windows, operation mode of the heating system and changing the temperature set point in each room of the building is derived from the data recorded by Gira HomeServer. On weekdays, the building is usually occupied from 8 a.m. to 6 p.m. In this time, the heating system operates in comfort mode with the various temperature set points in the rooms set by the occupants. From 6 p.m. to 8 a.m. the system operates in night mode with the constant temperature of 16 °C. In practice, lowering the temperature set point to 16 °C results in closing the KNX servo drive and switching off the heating system. This control algorithm is considered below and the experimental results were compared with the calculation. To assess the energy saving potential due to heating control the same algorithm was assumed, but the temperature was constant in comfort mode (21 or 20 °C). This case was referred as "with control".

In a real heating control other functions are implemented. One of these functions is the detection of window opening (or tilling from the top by 30° from vertical). This function is essential because the occupants have free and easy access to open the windows in their own office and laboratory rooms. Opening the window by the user in the room results in a transition of the heating system to the anti-frost mode with a temperature of 7 °C. In addition, the heating control system was integrated with the intruder alarm system. It is not possible to arm this system when a window in the building is open. Occupants leaving the building arm the system and they must close all the windows.

Another function is presence detection in the off time, between 6 p.m. and 8 a.m. and on weekends. If users start work earlier, finish later or work on weekends, information about the events is transmitted from the presence sensor to the heating control system, which changes the operating mode to comfort mode in the room where such presence is detected.

4. Calculation of Energy Consumption

The energy consumption Q_{smj} in the time interval Δt_m in the j-th heating zone is estimated taking into account heat transfer through the building envelope and heat losses for ventilation according to the following formula [14,15]:

$$Q_{smj} = \sum_{i=1}^{n} Q_{Tij} + Q_{Vj}, \qquad (2)$$

where: Q_{Tij} is heat losses for transmission through the i-th barrier in the j-th heating zone, Q_{Vj} is heat losses for ventilation in the j-th heating zone and n is the number of partitions.

The heat losses (or gains) for transmission through the i-th barrier are estimated as:

$$Q_{Ti} = U_i \cdot (\vartheta_{im} - \vartheta_{eim}) \cdot A_i \cdot \Delta t_m, \qquad (3)$$

where: U_i is the heat transfer coefficient through the i-th barrier in W/m² K, ϑ_{im} is the air temperature in °C, in the room, in the time interval Δt_m, ϑ_{eim} is the air temperature in °C, outside the i-th barrier, in the time interval Δt_m, A_i is the surface of the i-th barrier in m² and Δt_m is the time interval in hours.

The heat loss for ventilation in the j-th heating zone in Wh is calculated as follows:

$$Q_{Vj} = 0.333 \cdot (\vartheta_{im} - \vartheta_{eim}) \cdot V_j \cdot \Delta t_m, \qquad (4)$$

where: V_j is the ventilation air stream flowing into the j-th heating zone in m³.

Ventilation of the rooms is provided by ventilation ducts (Figure 3) and window ventilators integrated in the frames. Each ventilator is equipped with a regulator allowing different air flow rates. Due to the impact of various parameters on the performance of natural ventilation and the complexity of the phenomenon [16–19], the volume of ventilated air in the room was estimated based on the difference between energy consumption measured with open window ventilators and this energy measured with completely closed ventilators and ventilation duct. This difference determines the heat loss for ventilation and the volume V_j is estimated using Formula (4).

Energy consumption in the analyzed period is estimated as the sum of heat losses calculated in time intervals m in which various temperature increases $\vartheta_{im} - \vartheta_{eim}$ occurred, therefore:

$$Q_{sj} = \sum_{m} Q_{smj}. \qquad (5)$$

5. Results and Discussion

5.1. Ambient Weather Temperature and Daylight Illuminance

The analysis of energy consumption was carried out for the month of January 2017. January is usually the coldest month of the year in Poland. The calculations were performed taking into account the actual ambient temperature measured by the weather station installed on the south-eastern facade of the building. In calculation it is ϑ_{eim} temperature. However, the temperature was also measured by the external brightness and temperature sensor mounted on the northern facade. It should be noted that the values measured by these two sensors on a sunny day differ from each other. Two phenomena are responsible for the measurement discrepancies. The first one is the insolation of the building walls, which is stronger for the south-eastern wall than for the north wall. On cloudy days there is no difference in the heating of the walls by sunlight and the measured temperatures are close to each other. The second is the direct impact of sunlight on the weather station. This effect is mainly observed on a sunny day with high variability of daylight. In this case, variations in illuminance and temperature occur simultaneously. The lowest and highest temperatures on each day of January measured by the weather station and the brightness sensor are shown in Figure 6a. The lowest temperature in the month was about −13 °C and the highest about 4 °C. The temperature difference during the day reached 10 °C.

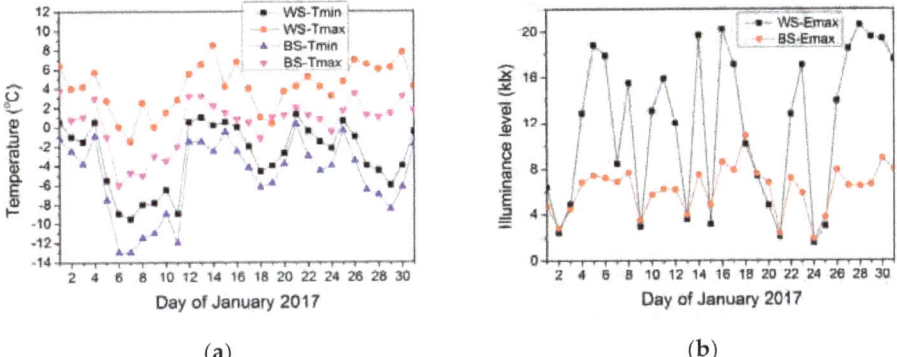

Figure 6. Data measured in January 2017 by the weather station WS and the brightness and temperature sensor BS: (**a**) the lowest and highest temperature and (**b**) the highest daylight illuminance level.

When the illuminance levels measured by the weather station and the brightness sensor are the same (Figure 6b), it means that the day is overcast and the walls are not heated by sunlight. This is from 18 to 21 January. Obviously, there is a time shift between variations in the illuminance level and the temperature. On 18 January, the wall was still warmed up by daylight and there was a difference in the measured temperature values. Due to these temperature differences, on a sunny day, the temperature values measured by the brightness and temperature sensor are represented as ϑ_{eim} temperature in the calculation. Time intervals Δt_m are determined, in which the temperature ϑ_{eim} differs by 1 °C. The air temperature ϑ_{im} is taken as the current temperature in the heating zone, measured by the push-button with room temperature control unit and recorded by the HomeServer.

5.2. Temperature Changes Inside the Building

The implementation of heating control algorithms must take into account temperature changes in the rooms as a result of lowering the temperature set point, switching off the heating, opening the window and other events. Anyway, heating control usually consists in lowering the temperature at night and on weekends and turning off the heating after opening the window. The change in temperature will depend on the thermal properties of the building and the ambient conditions, i.e., temperature, wind speed, rainfall and daylight. Figure 7a shows the temperature inside and Figure 7b the temperature outside the building during three days, i.e., from 0:00 on 10 April to 24:00 on 12 April, which is during 4320 min. To investigate temperature changes in the building, the temperature was first lowered by fully opening (on 9 April) one window in zones 4 and 5. The heating system switched to the anti-frost mode and until 9:40 on 10 April (in 580 min) the temperature in these zones decreased to 20.9 and 21.8 °C, respectively. At that time, the windows were closed and the temperature increased to the temperature set points. Further temperature changes were forced at 18:40 (in 1120 min) by turning the heating system off and then at 9:30 on 12 April (in 3450 min) by turning this system on. Temperature changes in zone 4 prove the high thermal inertia of this zone and it may take several hours to reach a higher temperature set point or comfort mode temperature after earlier turning off the heating. On the other hand, the temperature reduction after switching off the heating is small when the windows are closed. For the considered conditions it was approximately 1 °C for about 39 h. It is worth noting that the temperature changes in zone 4 were even smaller than in zone 5. In zone 3 the window was not open and the temperature increased around 750 and 2100 min as a results of insolation and an increase of temperature outside the building. A slight effect of the outside temperature on the inside temperature could also be seen in zones 4 and 5.

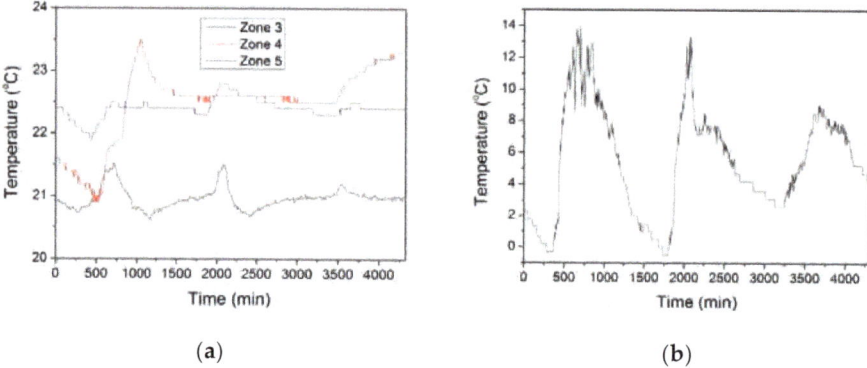

Figure 7. Impact of changing the heating system operating mode on the temperature inside the building. Temperature from 0:00 on 10 April to 24:00 on 12 April: (**a**) in the heating zones and (**b**) outside the building.

In order to estimate the effect of window operation on energy consumption, the temperature changes after tilting the window from the top by 30° from vertical were analyzed. It is worth noting that such window operation was often used by occupants. The window was tilted on 6 April at 16:15, 975 min from 0, which corresponds to 0:00. Figure 8a shows that after tilting the windows the temperature in both zones dropped to about 23 °C, then due to the increase in the outside temperature (Figure 8b) the temperature inside the zones increased too. However, the temperature increase in the two zones was different due to the difference in insolation of these rooms. In zone 4 the windows were located on one wall of the room on the north-east side, while in zone 5, the windows were on two sides of the room, i.e., north-west and north-east. The illuminance level of daylight is shown in Figure 8c.

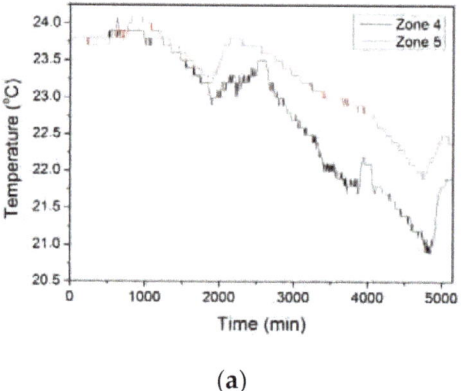

(**a**)

Figure 8. *Cont.*

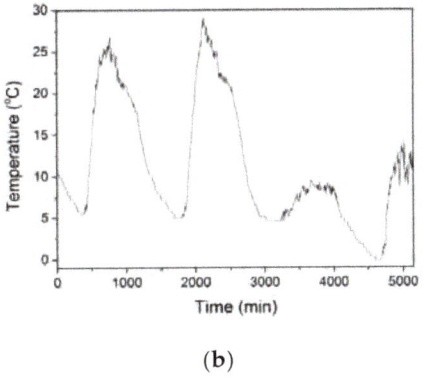
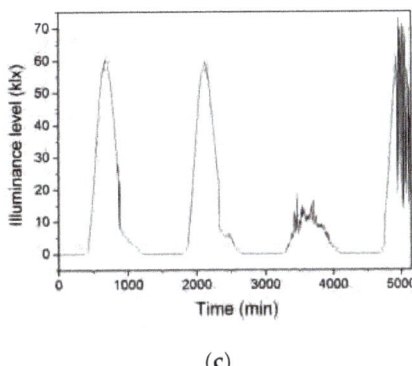

(**b**)   (**c**)

Figure 8. Impact of window operation on the temperature inside the building. (**a**) Temperature change due to one window tilting from the top inside zones 4 and 5; (**b**) temperature outside the building and (**c**) daylight illuminance level.

On 8 April, the daytime temperature dropped below 10 °C and then to around −1 °C at night. This resulted in a lower room temperature, more significant in zone 4. On 9 April at 9:40 (4900 min) the windows were closed in both zones, the heating system turned on and the temperature started increasing to the set point value. It should be noted that tilting of only one window in the room led to a temperature decrease of around 3 °C during the considered time, which corresponds to the weekend time.

5.3. Energy Consumption Experiment and Calculation

Energy consumption in the heating zones on days of January 2017, measured by the heat meters, is shown in Figure 9, and Table 2 presents the measured and calculated energy consumed over the whole month. The results were obtained with no heating control and the actual room temperatures were equal to the temperature set points. The occupants had the freedom to choose the temperature set points and, as can be seen in Table 2, the range of the selected set points was wide: from 19 to 24.5 °C. It reveals a strong influence of individual occupant preferences on the feeling of comfort.

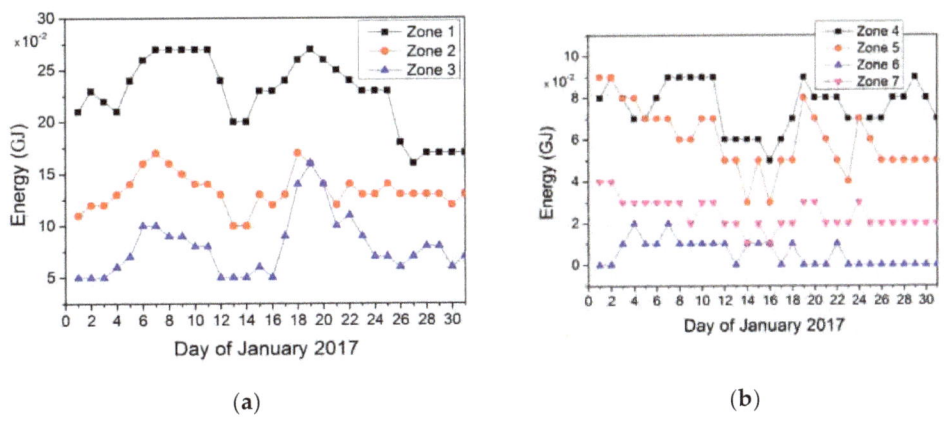

(**a**)   (**b**)

Figure 9. *Cont.*

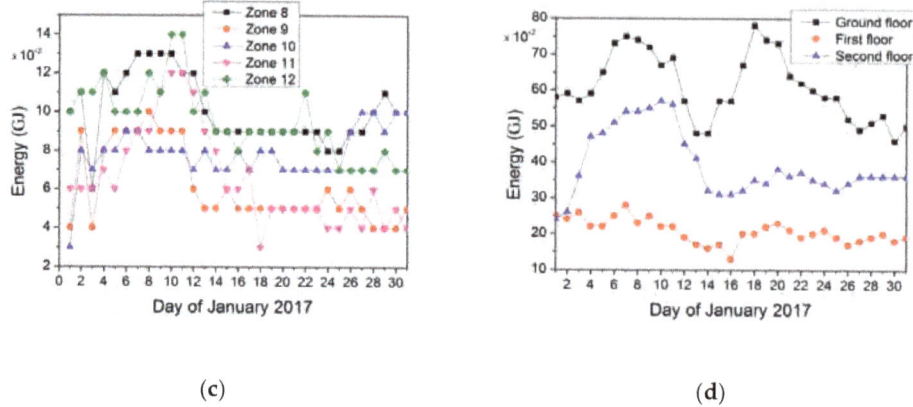

(c)   (d)

Figure 9. Energy consumption on days of January 2017 measured with heat meters: (**a**) in heating zones 1–3; (**b**) in heating zones 4–7; (**c**) in heating zones 8–12 and (**d**) on the stories of the building.

The highest energy consumption (Figure 9a) was in heating zone 1 due to the large volume of air to be heated, which results from the fact that this zone includes not only the laboratory room but also the entrance of the building and the open space of the staircase. The energy consumption in heating zone 2 was higher than in zone 3 due to the heat transfer through the door in zone 2 and a lower temperature in zone 3. The comparison of energy consumption on the three floors of the building (Figure 9d) shows that the highest consumption was on the ground floor due to the poor thermal insulation of the floor and the volume of heated air. The lowest energy consumption occurred in the rooms on the first floor (Figure 9b). On the second floor, where there was heat transfer through the roof, energy consumption was higher than on the first floor.

Table 2. Experimental and calculated energy consumptions in the heating zones, in the month of January 2017.

Heating Zone	Temperature	Energy Consumption—Experiment (GJ)	Energy Consumption per Unit of Room Area (GJ/m^2)	Energy Consumption—Calculation (GJ)	Energy Difference (%)
Ground floor—zone 1	22.0	7.21	0.1105	7.78	7.9
Ground floor—zone 2	22.0	4.25	0.0724	4.62	8.7
Ground floor—zone 3	19.0	2.48	0.0399	2.83	14.1
First floor—zone 4	24.5	2.39	0.0366	2.60	8.8
First floor—zone 5	24.0	1.86	0.0331	2.10	12.9
First floor—zone 6	21.0	0.39	0.0176	0.42	7.7
First floor—zone 7	24.0	1.32	0.0311	1.51	14.4
Second floor—zone 8	20.0	2.96	0.0454	3.31	11.8
Second floor—zone 9	20.0	2.84	0.0752	3.06	7.7
Second floor—zone 10	22.0	1.53	0.0872	1.69	10.4
Second floor—zone 11	21.0	1.78	0.0803	2.01	12.9
Second floor—zone 12	22.0	2.97	0.0648	3.24	9.1

Energy consumption in each room depended on their volume, temperature set point and insolation, therefore it was better to compare the energy consumption per unit of room area, at which the above conclusions were rather obvious. Another good example is the comparison of energy consumption per unit of room area in heating zones 4 and 8 with the same volume, which showed that the consumption in zone 4 was lower despite a higher temperature. It is worth noting that the calculated value was always larger than the measured value, and it seemed to be the case for two reasons. Firstly, heat gains from insolation, people and equipment were not included in the calculations. Secondly, at small energy values measured, heat meter indications were burdened with significant errors, namely the values were underestimated. As the measurement of thermal energy by the heat meter was carried out indirectly, on the basis of measuring the volume of the water and the temperature difference at the inflow and return, the measurement error could be relatively large (±5%). However, the difference

5.4. Effect of Room Temperature and Heating Control on Energy Consumption

In the heating season 2017/2018, between the end of September and the beginning of May, time control of the heating was implemented. Due to different weather conditions, the experimental results of two heating seasons could not be compared in order to estimate the effect of heating control on the reduction of energy consumption. Therefore, the calculations were carried out for the weather conditions in January 2017: first, without heating control, assuming that the room temperature was 21 and 20 °C. The value of 21 °C corresponds to the recommended indoor air temperature in education rooms of category A and 20 °C in rooms of category B (Table 1). Then, based on the observation, it was assumed that after the transition of the heating system to night mode, the temperature dropped in the rooms located on the ground floor by an average of 1.5 °C at night (during 14 h). At weekends (during 62 h), the reduction was about 4 °C. These temperature drops were, respectively, about 0.5 °C and 1.5 °C in the rooms on the first floor and 1 °C and 3 °C on the second floor. The calculation results are given in Table 3. This calculation shows that reducing the temperature set point by 1 °C gives an energy saving of about 5% compared to energy consumption at 21 °C.

Table 3. Energy consumptions in the heating zones without and with control, calculated considering the weather conditions in the month of January 2017. * The energy saving potential is determined in comparison with the energy consumption at the temperature of 21 °C.

Heating Zone	Energy Consumption without Control at 21 °C (GJ)	Energy Consumption without Control at 20 °C (GJ)	Energy Consumption with Control at 21 °C (GJ)	Potential of Energy Saving * (%)
Ground floor—zone 1	7.42	7.06	6.66	10.2
Ground floor—zone 2	4.40	4.19	3.96	10.1
Ground floor—zone 3	3.13	2.98	2.81	10.3
First floor—zone 4	2.23	2.12	2.14	4.0
First floor—zone 5	1.83	1.74	1.76	3.6
First floor—zone 6	0.42	0.40	0.401	3.8
First floor—zone 7	1.41	1.34	1.36	3.8
Second floor—zone 8	3.48	3.31	3.22	3.2
Second floor—zone 9	3.22	3.06	2.98	7.6
Second floor—zone 10	1.62	1.54	1.50	7.6
Second floor—zone 11	2.01	1.91	1.86	7.5
Second floor—zone 12	3.09	2.94	2.86	7.5

The comparison of energy consumption with and without heating control reveals that the energy saving potential mainly depended on the temperature drop after the set point lowering. The greater the decrease, the greater the potential for energy savings. In the study case, in the rooms with a poorly heat-insulated floor, the energy reduction due to heating control reached about 10%. A slightly lower reduction of about 7.5% was estimated for the rooms on the second floor, where heat was transferred through the roof, and the smallest reduction of less than 4% was estimated for the rooms on the first floor. This proves that in well-insulated rooms with a low energy consumption for heating the implementation of the control system gave relatively little benefit.

For energy saving, a very important function was to control the opening of a window. As shown in Figure 8 the tilt of the top of only one window in the room led to a temperature decrease of a few degrees. Leaving the window open before night or weekend would result in a significant increase in energy consumption, by about 5% per 1 °C drop.

6. Conclusions and Future Work

In this paper, the potential of energy savings in an existing public building in Poland was estimated. This estimation includes the most important parameters affecting energy consumption for heating. Experimental verification of the building case study showed that the calculation of energy consumption in a cold climate including the heat transfer through the building envelope and heat losses for ventilation were sufficiently accurate. In such calculations, a good knowledge of the thermal characteristics of the building, the volume of ventilated air and the temperature outside and inside the building is crucial.

Using the KNX system implemented in the building, the behavior of occupants was investigated revealing that occupants choose temperature set points in a wide range recognized as thermal comfort, and window opening was also accidental and difficult to predict. The proposed heating control algorithms took into account the strong influence of individual occupant preferences on the feeling of comfort. However, in order to reduce energy consumption, the anti-frost mode was applied after opening the window, as well as integration with the intruder alarm system. Investigation of temperature changes in the building with changes in the temperature set points and after opening the window showed that from the point of view of energy saving, the most important issue is the window opening control.

Finally, detailed comparisons of energy consumption with heating control and without any controls were performed. It shows that the energy saving potential depended on the temperature drop after lowering the set point, and thus on the dynamics of the thermal behavior of the building. The greater this drop, the greater the potential for energy savings. In the case study, in rooms with poorly heat-insulated floors, the energy reduction potential due to heating control reached about 10%. A slightly lower potential of about 7.5% was estimated for rooms on the second floor, where heat was transferred through the roof, and the smallest potential of less than 4%, for rooms on the first floor. This proved that in a well-insulated room with a low energy consumption for heating, the implementation of the control system gave relatively little benefit.

Future work will include an analysis of information from presence detectors to describe occupant behavior, and the implementation of such information to control heating and estimate energy savings. Research associated with the optimal operation of the heat source will also be undertaken.

Funding: This study is based upon work supported by the National Centre for Research and Development in the context of the Innovative Economy Program under grant No. POIG.02.02.00-00-018/08. This work was also supported by the 2018 Poznan University of Technology funds transferred from the Ministry of Science and Higher Education.

Conflicts of Interest: The author declares no conflict of interest. The funders had no role in the design of the study; in the collection, analyses, or interpretation of data; in the writing of the manuscript, and in the decision to publish the results.

References

1. International Energy Agency (IEA). *Key World Energy Statistics*; International Energy Agency (IEA): Paris, France, 2018.
2. Energy Information Administration (EIA). *Energy Consumption by Sector*; U.S. Department of Energy: Washington, DC, USA, 2019.
3. European Commission. *Consumption of Energy*; European Commission: Brussels, Belgium, 2019.
4. European Commission. *Driving Energy Efficiency in the European Building Stock: New Recommendations on the Modernisation of Buildings*; European Commission: Brussels, Belgium, 2019.
5. European Parliament and of the Council. *Directive EU 2018/844 of the European Parliament and of the Council of 30 May 2018 Amending Directive 2010/31/EU on the Energy Performance of Buildings and Directive 2012/27/EU on Energy Efficiency*; Official Journal of the European Union: Brussels, Belgium, 2018.
6. Ma, Z.; Cooper, P.; Daly, D.; Ledo, L. Existing building retrofits: Methodology and state-of-the-art. *Energy Build.* **2012**, *55*, 889–902. [CrossRef]
7. Najjar, M.K.; Tam, V.W.Y.; Di Gregorio, L.T.; Evangelista, A.C.J.; Hammad, A.W.A.; Haddad, A. Integrating Parametric Analysis with Building Information Modeling to Improve Energy Performance of Construction Projects. *Energies* **2019**, *12*, 1515. [CrossRef]
8. Allouhi, A.; El Fouih, Y.; Kousksou, T.; Jamil, A.; Zeraouli, Y.; Mourad, Y. Energy consumption and efficiency in buildings: current status and future trends. *J. Clean. Prod.* **2015**, *109*, 118–130. [CrossRef]

9. Pérez-Lombard, L.; Ortiz, J.; Pout, C. A review on buildings energy consumption information. *Energy Build.* **2008**, *40*, 394–398. [CrossRef]
10. Chen, Z.; Jiang, C.; Xie, L. Building occupancy estimation and detection: A review. *Energy Build.* **2018**, *169*, 260–270. [CrossRef]
11. Gul, M.S.; Patidar, S. Understanding the energy consumption and occupancy of a multi-purpose academic building. *Energy Build.* **2015**, *87*, 155–165. [CrossRef]
12. Zhao, H.; Magoulès, F. A review on the prediction of building energy consumption. *Renew. Sustain. Energy Rev.* **2012**, *16*, 3586–3592. [CrossRef]
13. Wang, S.; Yan, C.; Xiao, F. Quantitative energy performance assessment methods for existing buildings. *Energy Build.* **2012**, *55*, 873–888. [CrossRef]
14. EN ISO 13790. *Energy Performance of Building—Calculation of Energy Use for Space Heating and Cooling*; European Committee for Standardization (CEN): Brussels, Belgium, 2008.
15. Gładyszewska-Fiedoruk, K.; Krawczyk, D.A. The possibilities of energy consumption reduction and a maintenance of indoor air quality in doctor's offices located in north-eastern Poland. *Energy Build.* **2014**, *85*, 235–245. [CrossRef]
16. Sacht, H.; Lukiantchuki, M.A. Windows Size and the Performance of Natural Ventilation. *Procedia Eng.* **2017**, *196*, 972–979. [CrossRef]
17. Zhai, Z.; El Mankibi, M.; Zoubir, A. Review of Natural Ventilation Models. *Energy Procedia* **2015**, *78*, 2700–2705. [CrossRef]
18. Shetabivash, H. Investigation of opening position and shape on the natural cross ventilation. *Energy Build.* **2015**, *93*, 1–15. [CrossRef]
19. Bangalee, M.Z.I.; Lin, S.Y.; Miau, J.J. Wind driven natural ventilation through multiple windows of a building: A computational approach. *Energy Build.* **2012**, *45*, 317–325. [CrossRef]
20. Gunay, B.; O'Brien, W.; Morrison, I.B. A critical review of observation studies, modelling and simulation of adaptive occupant behaviours in offices. *Build. Environ.* **2013**, *70*, 31–47. [CrossRef]
21. McCartney, K.J.; Nicol, J. Developing an adaptive control algorithm for Europe. *Energy Build.* **2002**, *34*, 623–635. [CrossRef]
22. Nicol, J.F.; Humphreys, M.A. Adaptive thermal comfort and sustainable thermal standards for buildings. *Energy Build.* **2002**, *34*, 563–572. [CrossRef]
23. Toe, D.H.C.; Kubota, T. Development of an adaptive thermal comfort equation for naturally ventilated buildings in hot–humid climates using ASHRAE RP-884 database. *Front. Archit. Res.* **2013**, *2*, 278–291. [CrossRef]
24. Nguyen, A.T.; Singh, M.K.; Reiter, S. An adaptive thermal comfort model for hot humid South-East Asia. *Build. Environ.* **2012**, *56*, 291–300. [CrossRef]
25. Rinaldi, A.; Schweiker, M.; Iannone, F. On uses of energy in buildings: Extracting influencing factors of occupant behaviour by means of a questionnaire survey. *Energy Build.* **2018**, *168*, 298–308. [CrossRef]
26. Rupp, R.F.; Vásquez, N.G.; Lamberts, R. A review of human thermal comfort in the built environment. *Energy Build.* **2015**, *105*, 178–205. [CrossRef]
27. Fanger, P. *Thermal Comfort*; McGraw-Hill: New York, NY, USA, 1973.
28. Robledo-Fava, R.; Hernández-Luna, M.C.; Fernández-de-Córdoba, P.; Michinel, H.; Zaragoza, S.; Castillo-Guzman, A.; Selvas-Aguilar, R. Analysis of the Influence Subjective Human Parameters in the Calculation of Thermal Comfort and Energy Consumption of Buildings. *Energies* **2019**, *12*, 1531. [CrossRef]
29. Yu, Z.; Fung, B.C.M.; Haghighat, F.; Yoshino, H.; Morofsky, E. A systematic procedure to study the influence of occupant behaviour on building energy consumption. *Energy Build.* **2011**, *43*, 1409–1417. [CrossRef]
30. Yan, D.; OBrien, W.; Hong, T.; Feng, X.; Gunay, H.B.; Tahmasebi, F.; Mahdavi, A. Occupant behaviour modeling for building performance simulation: current state and future challenges. *Energy Build.* **2015**, *107*, 264–278. [CrossRef]
31. Shih, H.C. A robust occupancy detection and tracking algorithm for the automatic monitoring and commissioning of a building. *Energy Build.* **2014**, *77*, 270–280. [CrossRef]
32. Dong, J.; Winstead, C.; Nutaro, J.; Kuruganti, T. Occupancy-Based HVAC Control with Short-Term Occupancy Prediction Algorithms for Energy-Efficient Buildings. *Energies* **2018**, *11*, 2427. [CrossRef]

33. Wang, F.; Feng, Q.; Chen, Z.; Zhao, Q.; Cheng, Z.; Zou, J.; Zhang, Y.; Mai, J.; Li, Y.; Reeve, H. Predictive control of indoor environment using occupant number detected by video data and CO_2 concentration. *Energy Build.* **2017**, *145*, 155–162. [CrossRef]
34. Lee, S.; Jung, S.; Lee, J. Prediction Model Based on an Artificial Neural Network for User-Based Building Energy Consumption in South Korea. *Energies* **2019**, *12*, 608. [CrossRef]
35. Brastein, O.M.; Perera, D.W.U.; Pfeifer, C.; Skeie, N.O. Parameter estimation for grey-box models of building thermal behaviour. *Energy Build.* **2018**, *169*, 58–68. [CrossRef]
36. Berthou, T.; Stabat, P.; Salvazet, R.; Marchio, D. Development and validation of a gray box model to predict thermal behaviour of occupied office buildings. *Energy Build.* **2014**, *74*, 91–100. [CrossRef]
37. PN-EN ISO 6946. *Komponenty Budowlane i Elementy Budynku—Opór Cieplny i Współczynnik Przenikania Ciepła—Metoda Obliczania*; Polski Komitet Normalizacyjny: Warsaw, Poland, 2008.
38. PN-EN ISO 10211. *Mostki Cieplne w Budynkach. Strumienie Ciepła i Temperatury Powierzchni. Obliczenia Szczegółowe*; Polski Komitet Normalizacyjny: Warsaw, Poland, 2008.
39. Kaminska, A.; Ożadowicz, A. Lighting Control Including Daylight and Energy Efficiency Improvements Analysis. *Energies* **2018**, *11*, 2166. [CrossRef]
40. KNX Association. *Handbook for Home and Building Control, Basic Principles, 5th ed*; KNX Association: Brussels, Belgium, 2006.
41. Standard 7730. *Ergonomics of the Thermal Environment—Analytical Determination and Interpretation of Thermal Comfort Using Calculation of the PMV and PPD Indices and Local Thermal Comfort Criteria*; International Organization for Standardization: Geneva, Switzerland, 2005.
42. CEN. *15251-Criteria for the Indoor Environment, Including Thermal, Indoor Air Quality (Ventilation), Light And Noise*; CEN: Brussels, Belgium, 2006.
43. Fanger, P.O. Thermal environment—Human requirements. *Environmentalist* **1986**, *6*, 275–278. [CrossRef]
44. ASHRAE. *ANSI/ASHRAE Standard 55-2010: Thermal Environmental Conditions for Human Occupancy*; American Society of Heating, Refrigerating and Air-Conditioning Engineers: Atlanta, GA, USA, 2010.
45. Fanger, P.O. Thermal Comfort. In *Analysis and Applications in Environmental Engineering*; Danish Technical Press: Copenhagen, Denmark, 1970.
46. Antoniadou, P.; Papadopoulos, A.M. Occupants' thermal comfort: State of the art and the prospects of personalized assessment in office buildings. *Energy Build.* **2017**, *153*, 136–149. [CrossRef]
47. Hong, T.; Taylor-Lange, S.C.; D'Oca, S.; Yan, D.; Corgnati, S.P. Advances in research and applications of energy-related occupant behaviour in buildings. *Energy Build.* **2016**, *116*, 694–702. [CrossRef]

© 2019 by the author. Licensee MDPI, Basel, Switzerland. This article is an open access article distributed under the terms and conditions of the Creative Commons Attribution (CC BY) license (http://creativecommons.org/licenses/by/4.0/).

Article

Analysis of the Thermal Behavior of an Earthbag Building in Mediterranean Continental Climate: Monitoring and Simulation

Lídia Rincón, Ariadna Carrobé, Marc Medrano *, Cristian Solé, Albert Castell and Ingrid Martorell †

SEMB Research Group, INSPIRES Research Centre, Universitat de Lleida, Pere de Cabrera s/n, 25001 Lleida, Spain; lrincon@diei.udl.cat (L.R.); acarrobe@diei.udl.cat (A.C.); csole@diei.udl.cat (C.S.); acastell@diei.udl.cat (A.C.); imartore@diei.udl.cat (I.M.)
* Correspondence: mmedrano@diei.udl.cat
† Serra Húnter Fellow, Generalitat de Catalunya.

Received: 7 November 2019; Accepted: 24 December 2019; Published: 30 December 2019

Abstract: Nearly 30% of humanity lives in earthen dwellings. Earthbag is a sustainable, cheap, feasible and comfortable option for emergency housing. A comparative monitoring-simulation analysis of the hygrothermal behavior of an Earthbag dwelling in Mediterranean continental climate, designed under bioclimatic criteria, is presented. The dome shape Earthbag dwelling has a net floor area of 7.07 m^2, a glass door facing south and two confronted windows in the east and west facades. A numerical model (EnergyPlus v8.8) was designed for comparison. Twenty-four hour cross ventilation, night cross ventilation, and no ventilation in free floating mode and a controlled indoor temperature were the tested scenarios. Comparisons between experimental data and simulation show a good match in temperature behavior for the scenarios studied. Reductions of 90% in summer and 88% in winter, in the interior thermal amplitude with respect to exterior temperatures are found. Position of the glazed openings was fundamental in the direct solar gains, contributing to the increase of temperature in 1.31 °C in winter and 1.37 °C in the equinox. Night ventilation in the summer period had a good performance as a passive system. Passive solar gains made a reduction of heating energy consumption of 2.3% in winter and 8.9% in equinox.

Keywords: earth building; thermal comfort; passive design; monitoring and simulation

1. Introduction

Earthen architecture historically has been widely used for wall construction around the world. According to Minke [1] earth construction has been used for more than 10,000 years. Today, it is estimated that nearly 30% of the world's population lives in earthen dwellings, not only in developing countries but also in industrialized countries, where using earth as a construction material has raised interest recently, as it is considered an environmentally friendly solution. Particularly, Earthbag (also called Superadobe) is presented as a sustainable, cheap, feasible and comfortable option to improve thermal comfort. Superadobe is a form of Earthbag construction patented and developed by the Iranian architect Nader Khalili, who proposed fundamental rules for the design and building recommendations [2]. Earthbag and Superadobe are building techniques that consist of the use of earth-filled sandbags in order to build structural walls, usually in a dome shape [3]. The dome shape offers more structural integrity and durability than adobe square buildings [4], but also limits its design to 5 m in diameter and one ground floor. The dome shape allows the foundation, load-bearing walls and roof to be built with the same materials and technique. In low-cost buildings, the roof used to be the part of the building with the highest cost. Thus, this building technique is four times cheaper than conventional techniques [5]. When comparing two low-cost earthen buildings, such as Earthbag dwelling and adobe traditional Burkinabe dwelling, the Earthbag one achieves better thermal performances in hot arid climates. In that case, a combination of night ventilation, roof solar protection, and high-inertia of the Earthbag enclosure lead to an almost total elimination of thermal discomfort during the year [6]. Among their possible uses, the Earthbag building is a good solution to temporary emergency housing, as shown in the construction of 14 Earthbag shelters in the refugee camp of Baninajar [7]. The architect Khalili built them for the displaced Iranians in Iraq after a flood. The project also served to assess the feasibility and cost of building with Earthbag and to evaluate the possibilities of Earthbag shelters in the case of a real emergency [8]. Earthbag building has also been used in cooperation projects, such as the construction of part of the Emsimision Training Medical Center in the Boulmiuogou District, in Ouagadougou, Burkina Faso [9]. After the earthquake of 2010 in Haiti, humanitarian aid of different organizations constructed several buildings with Earthbag and Superadobe techniques, such as numerous houses for those affected inhabitants [10], a medical clinic [11], a school for orphans [12], a community center [13], a shelter for children, and some experimental Earthbag buildings for scientific tests [14]. In Nepal, the Small-Earth partnership built an orphanage with Superadobe [15]. Because of the particular location of the orphanage, in a hilly area with difficult access, the Superadobe system was a good choice, since it allows the use of local resources, such as earth and stones, and it only needed to transport bag rolls to the place.

Previous research includes thermally simulated and monitored raw earthen buildings [16–18], but not Earthbag buildings yet. This research analyzes the hygrothermal performance and comfort of an Earthbag building located in Mediterranean continental climate by experimentation of a real construction and by energy simulation. Passive design strategies are tested, such as the use of high thermal inertia in the enclosure, the collection of direct solar radiation through the glazed openings and the use of natural ventilation. In this research, energy simulation results are also validated with experimental data.

2. Materials and Methods

2.1. Constructive Characteristics of the Monitored Earthbag Building

The experimental Earthbag building was designed under bioclimatic criteria. The building was constructed with a dome shape following the Superadobe technique. The dome shape allows for the reducing of the shape factor, which is a smaller surface of the building envelope per same volume compared to a cubical shape. It has a net floor area of 7.07 m^2, a circular plant of 3 m of diameter, a height of 3.3 m, an envelope surface of 29.96 m^2, an interior volume of 17.67 m^3, and a shape factor of 1.7 (Figure 1). The Earthbag walls are 35 cm thick, but the buttress is formed by a double Earthbag (70 cm thick). The Earthbag dome roof has an average thickness of 28 cm. The continuous polypropylene bag contains an earthen mixture of on-site earth and construction sand in a 1:1 proportion. Slaked lime in water was used as stabilizer, in approximately 10% of the total earthen volume. The sieve analysis showed that the earth mixture contained in weight: 0.80% fine gravel, 92.21% sand, 3.42% slime and 3.57% clay. The earth mixture was manually rammed. The building was exteriorly coated with 4 cm thick lime mortar. The floor is made of lime concrete (9 cm thick) and it is directly in contact with the ground, over a waterproofing plastic layer. The main glass opening is the entrance door, which is facing exactly south (exterior window frame of 0.91 × 2 m, with a glazed surface of 1.09 m^2) to take advantage of the direct solar gain. Two confronted windows in the east (exterior window frame of 0.8 × 0.67 m, 0.25 m^2 of glazed surface) and west (exterior window frame 0.6 × 0.35 m, with a glazed surface of 0.06 m^2) facades allow crossed ventilation. The position of the windows with respect to the walls is in the interior, which produces small solar protection due to the thickness of the walls. Over the square windows, there is a space that has been insulated with polystyrene (6 cm) and covered with a wooden exterior coating (2.2 cm) (Table 1).

Figure 1. Earthbag building, University of Lleida Campus, Spain.

Table 1. Materials properties of the Earthbag building.

System	Material	Thickness (m)	Thermal Conductivity (λ, W/m·K)	Density (ρ, kg/m³)	Specific Heat [a] (Cp, J/kgK)	Glass SHGC [d] (-)
Roof	Earthbag	0.28	2.18 [c]	2190 [b]	1000	-
Walls	Earthbag	0.35	2.18 [c]	2190 [b]	1000	-
	Earthbag (buttress)	0.70	2.18 [c]	2190 [b]	1000	-
	Exterior lime mortar coating	0.04	1 [a]	1700 [a]	1000	-
Floor	Lime concrete	0.09	0.4 [a]	1000 [a]	1000	-
Windows	EXP Insulation	0.060	0.0432 [a]	91 [a]	837	-
	Wooden exterior coating	0.022	0.17 [a]	700 [a]	1600	-
	Wood frame	0.07	0.15 [a]	500 [a]	1600	-
	Double Glazing (6 + 10 + 6)	0.006	0.9 [a]	-	-	0.8662
	Air Chamber	0.01	-	-	-	-

[a] Source: Catálogo de Elementos Constructivos del Código Técnico de la Edificación (2015) [19]. [b] Source: Measured density taken from the building prototype of the Cappont Campus, Lleida [20]. [c] Source: Estimated thermal conductivity from experimental U-value calculation. [d] Solar heat gain coefficient.

2.2. Location and Climate

The prototype is located in the Cappont Campus of the University of Lleida, Spain (41.60N, 0.62E; 167 m above sea level). Lleida has a Mediterranean continental climate, classified as BSk by the Köppen climate classification [21]. It is characterized by hot and dry summers and cold and wet winters due to the presence of fog. Rains are low and irregular, with an annual average of 423 mm. The annual average temperature is 15.2 °C although there are big differences between summer and winter temperatures and between maximum and minimum daily dry air temperatures.

2.3. Instrumentation and Experimental Campaign

For the thermal evaluation of the Earthbag building different experiments were performed.

- Free floating temperature: Internal temperature of the building fluctuates depending on the weather conditions and the thermal behavior of the construction. Temperature oscillations allow evaluating dynamic parameters such as the thermal lag and the decrement factor. No heating or cooling system is used. Two cases for ventilation are considered:

  o No ventilation is provided to the building, just a base level of infiltrations.
  o Natural ventilation is provided to the building. Two different scenarios were tested: all-day long and night ventilation.

- Controlled temperature: The internal temperature of the building is set to a constant value by means of an electric radiator. The energy consumption of the radiator is registered to determine the energy consumption required to maintain a certain level of comfort (set point at 22 °C).

The experimental setup is instrumented with 18 temperature sensors, 2 relative humidity sensors and the control and data acquisition systems, as it is shown in Figure 2. Monitoring consisted in data collection of interior and exterior air temperatures as well as interior surface temperatures in both south and north walls. The interior temperature and humidity sensor was located in the geometrical center of the dome at 1.50 m high (position 2 in Figure 2). Four additional temperature sensors were located in the center of the prototype at different heights, every 0.80 m (positions 3, 4, 5 and 6 in Figure 2). The north surface wall was monitored with 9 temperature sensors, located in a vertical axis every 0.40 m (positions from 8 to 16 in Figure 2). Moreover, 2 temperature sensors (positions 17 and 18 in Figure 2) were located in the north surface wall, next to the sensor in position 12, covering a triangle surface of 300 cm^2. The idea is that sensor 12 follows the 0.4 m distance between all the surface sensors in the north wall. The other two sensors were added next to this 12 sensor drawing a triangle to measure the U-value. The south surface wall temperature sensor was located at 2.10 m (position 7 in Figure 2), above the door. Finally, there was a sensor measuring the external temperature and relative humidity (position 1 in Figure 2), and an energy consumption meter to register the energy consumed by the electric radiator.

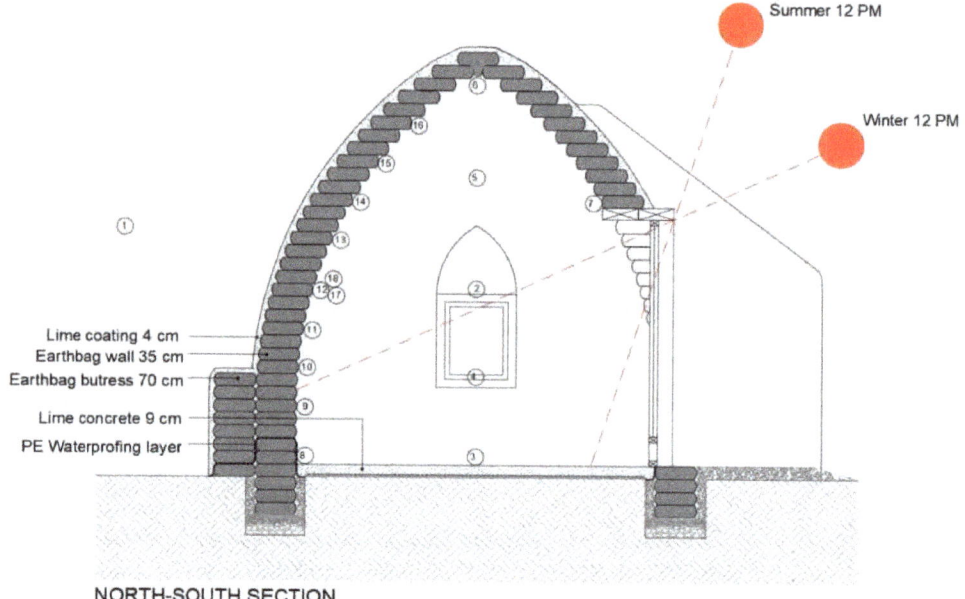

Figure 2. Sensors location in the monitored Earthbag building.

Temperatures were registered every 5 min by means of a data acquisition system connected to a computer. Air temperature and humidity sensors used were Elektronik device model EE210 (±0.1 °C uncertainty for temperature and 0.5% for humidity). PT-100 class B (±0.3 °C uncertainty) sensors were used for surface temperatures. The acquisition data equipment consisted of a data logger (model DIN DL-01-CPU), connected to the adapter data logger-computer (model AC-250). The computer software to compile the data was TCS-01. When controlled temperature experiments were carried out in winter, a 1500 W electric radiator was used and its energy consumption was also measured with a Finder E7energy meter.

2.4. Experimental U-Value Calculation

In order to calculate the U-value of the Earthbag wall, a transmittance test according to [22,23] was performed. The test consists in monitoring indoor, outdoor air and indoor wall temperatures. It is important to locate the surface sensors on the north wall to avoid the solar radiation interfering with the measures, or having a protected sensor. Moreover, for the wall surface temperature reading, three temperature sensors where located in a triangular shape separated about 20 cm from each other, in order to calculate an average temperature to compute the U-value. It is also important that the indoor air temperature and the outdoor air temperature are as constant as possible, as the importance of the calculation is focused on the heat transfer to the wall. To assure this specification, the experimentation was performed during the indoor controlled temperature scenario and during a fog week. The inner temperature was kept constant with a radiator and the external one due to the presence of all-day-long pervasive fog in Lleida. In this situation, a quasi-steady state hypothesis is justified [24] and the expression to obtain the U-value is the following:

$$U = \frac{(T_i - T_{si})}{(T_i - T_e)} * h_{si} \quad (1)$$

where: T_i: Indoor air temperature, °C. T_{si}: Indoor surface temperature, °C. T_e: Outdoor air temperature, °C. h_{si}: Heat transfer coefficiewnt of external envelopes, 7.69 W/m² °C [25].

The experimental U-value will allow calculating the thermal conductivity of the Earthbag wall λ_1, from the equation:

$$U = \frac{1}{R_{si} + \frac{e_1}{\lambda_1} + \frac{e_2}{\lambda_2} + R_{se}} \quad (2)$$

where: R_{si}: Interior surface thermal resistance for a vertical facade, 0.13 m²·°C/W [25]. R_{se}: Exterior surface thermal resistance for a vertical facade, 0.04 m²·°C/W [25]. e_1: Earthbag wall thickness, m. e_2: Lime mortar coating thickness, m. λ_1: Thermal conductivity of the Earthbag wall, W/m²·°C. λ_2: Thermal conductivity of the lime mortar coating, W/m²·°C.

2.5. Thermal Lag and Decrement Factor

The thermal lag (ϕ) represents the time that elapses between the indoor air temperature maximum value and the outdoor maximum value. The decrement factor (μ) is the reduction of the temperature range of both measures (Figure 3).

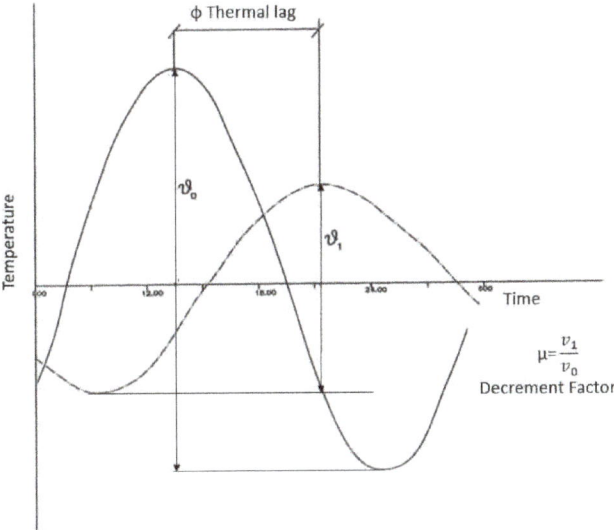

Figure 3. Definition of the thermal lag and the decrement factor of a sinusoidal heat wave. Source: adapted from Yáñez, 2008 [26].

Assuming an external sinusoidal wave temperature, some formulas are presented for the calculation of the thermal lag and the decrement factor in homogeneous walls, knowing the diffusivity of the material (α), the thickness of the wall (l) and the period of the wave.

In this case the heat flow is supposed to be transmitted only in the normal direction to the wall, neglecting the effects of the edge. Likewise, it is assumed that the variation of the temperature inside the wall depends only on the external conditions, neglecting any variation that may be generated inside (semi-rigid solid).

As both parameters depend on the period of the wave, the following formulas correspond to a period of 24 h:

$$\mu = exp.\left(-0.362 * l * \sqrt{1/\alpha}\right) \quad (3)$$

$$\varphi = 1.38 * l * \sqrt{1/\alpha} \quad (4)$$

where: l: Thickness (m). α: Material diffusivity, $\alpha = \frac{\lambda}{\rho * C_p}$ [m^2/s].

2.6. Simulation

A numerical model of the Earthbag prototype designed with EnergyPlus was defined and used for comparisons. Open Studio was used as the graphical user interface. Open Studio does not allow for the creation of dome shapes; this is why a polygonal dome has been drawn for the dome building. Shadow elements were drawn, such as the door buttress and the thickness of the walls producing shadow over the windows. The default heat balance algorithm based on the conduction transfer function (CTF) transformation and 6 time steps per hour for the simulation are applied. CTF is a widely used numerical method to calculate transient heat conduction in Building Energy Simulation tools. It is preferred to the finite difference method thanks to the smaller computational time required [27].

Considerations in the energy simulation:

- The climatic file from EnerygPlus for Lleida has been used as a base in which measured on site temperature and humidity have been incorporated. Solar radiation is also incorporated to this climate file and it is taken from the nearest station located in Raïmat (20 km away).
- In the experiments with no ventilation, a calibration analysis has been performed for different ACH (air changes per hour) values to determine the level of air infiltration. Values of 0, 0.1, 0.3, 0.5 and 0.6 ACH have been considered. The mean absolute error (MAE) of the monitored and simulated indoor air temperature was calculated for the five ACH cases. The best performance is achieved with infiltrations of 0.5 ACH with a MAE value of 1.147 °C.
- In the experiments with ventilation, a ventilation of 10 ACH has been considered to simulate the natural ventilation [28].
- Due to the impossibility of drawing a dome shape with OpenStudio, a polygonal dome has been used. When adapting the geometry, the internal volume of the simulated prototype is 2% larger. Moreover, the roof thickness of the simulated prototype is taken as a mean value of the real roof prototype thickness, which changes slightly with the height.
- The simulation in EnergyPlus gives as a result the total glass solar radiation. In this paper, the solar radiation per square meter entered in each glass opening and the total time of solar radiation per opening are determined based on the hourly sun's path for the latitude and longitude of Lleida, which can be evaluated using the Sketchup's Shadows feature.
- Since the monitored building is not occupied, no internal heat loads due to occupation or electrical devices have been considered in the simulation.

2.7. Initial Hypothesis and Testing Scenarios

The initial hypotheses are:

- The thermal comfort in an Earthbag building can be achieved in Mediterranean continental climate due to the high thermal inertia of the Earthbag walls and a combination of passive strategies for heating and cooling.
- The position and area of the glazed openings can improve the thermal comfort of an Earthbag building in winter conditions due to the direct solar gains, as a passive heating strategy.
- The natural cross ventilation can improve the thermal comfort of the Earthbag building in summer conditions, as a passive cooling strategy.
- The energy simulation can be validated by the experimental monitoring and, therefore, any future design of the Earthbag building could be tested during the design phase in order to improve the thermal comfort of the building.

Different scenarios during summer and winter periods were tested. Each test was design to give an answer to the objectives listed in Table 2. In test #2, the simulation of the Earthbag building has been compared in addition with an equal Earthbag building with no glazed openings.

Table 2. Testing scenarios of monitoring and simulation of the Earthbag prototype.

# Test	Scenario Description		Duration	Data Taken From	Objective
#1	Air stratification		23 March 2018 to 2 April 2018	Monitoring	• Testing the variation of temperature inside the Earthbag dome in function of height
#2.1	Free floating temperature with no ventilation	Winter solstice	12–20 December 2017	Monitoring, simulation and comparison of simulation with no glazed openings.	• Testing thermal inertia of the Earthbag wall
#2.2		Equinox	15–23 March 2018		• Testing the effect of the passive strategy "direct solar gains"
#2.3		Summer solstice	14–21 June 2018		
#3.1	Summer: natural ventilation in free floating mode	24 h cross ventilation	4–11 June 2018	Monitoring and simulation	• Testing the effect of the passive strategy for cooling "natural cross ventilation" in the Earthbag building
#3.2		night cross ventilation	25 July 2017 to 1 August 2017		
#4	Winter: controlled temperature		25 February 2018 to 5 March 2018	Monitoring and simulation	• Calculating the energy consumption for heating
			24–27 December 2017	Monitoring	• Calculating the thermal transmittance (U-value) and the thermal conductivity (λ) of the Earthbag wall

3. Results

In this section, firstly steady-state and dynamic parameters are presented. Experimental data from the monitoring is taken to calculate the thermal transmittance and the conductivity of the Earthbag walls. Secondly, results of the experimental analysis and simulation analysis are presented. The experimental data are presented to analyze the effect of the air stratification. The monitoring and simulation free floating results of temperature and solar radiation data are presented to, in one hand, validate the simulation with the experimental data and, in the other hand, to analyze the thermal inertia and the solar heat gains in winter solstice, equinox and summer solstice. The monitoring and simulation results of power consumption are presented in winter conditions to analyze the energy consumption of the Earthbag building. The monitoring and simulation temperature results of natural ventilation in free floating conditions are presented to analyze its effect and validate the energy simulation.

3.1. Steady-State and Dynamic Thermal Parameters

Figure 4 shows indoor and outdoor air temperatures. For indoor temperatures, air temperature in the geometrical center of the dome (position 2 in Figure 2) is represented with the average value. The indoor north surface temperature monitored (average of sensors 12, 17 and 18 in Figure 2) in quasi-steady state conditions [24] with the average value is also plotted. Moreover, the U-value calculated with the average of the indoor temperature and the uncertainty of this calculation according to the sensors accuracy, are also included. The uncertainty for the U-value was determined to be ±4%, applying the standard method for uncertainty propagation [29].

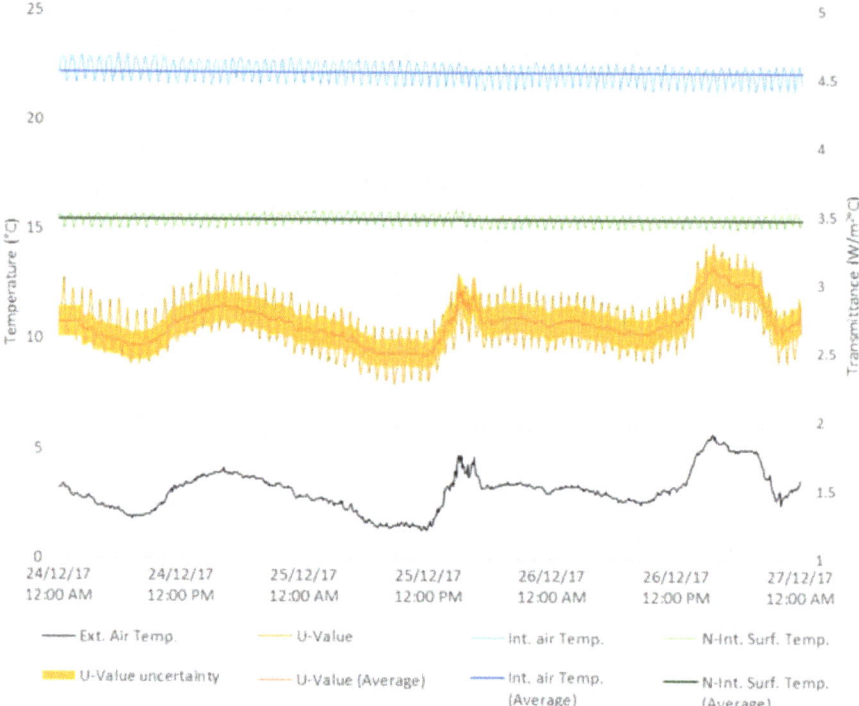

Figure 4. Interior air and surface temperatures, exterior air temperature and U-value calculated of the Earthbag wall. Data taken from experimental monitoring.

As shown in Figure 4, indoor and outdoor temperatures are almost constant, with an indoor temperature average of 22 °C and an indoor north surface temperature average of 15.5 °C. Outdoor temperature is around 3 °C in days of persistent fog, oscillating only 2 °C throughout the day. The U-value calculated for the Earthbag wall of 35 cm with an exterior lime coating of 4 cm has an average value of 2.7 W/m² K. According to this experimentally obtained U-value, and to the Equation (2) described in the methodology section, the thermal conductivity of the Earthbag material is 2.18 W/m·K.

The theoretical thermal lag and decrement factor are calculated considering a homogeneous Earthbag wall of 35 cm (with no exterior coating). According to Equations (1), (3) and (4) and the Earthbag properties (Table 1), the corresponding values are shown in Table 3.

Table 3. Theoretical results of the steady-state and dynamic thermal parameters for the Earthbag wall.

Parameter	Value	Units
Material diffusivity, α	0.00355	m²/h
Decrement factor, μ	0.1194	-
Thermal lag, Φ	8.1	h
Thermal transmittance, U-value	2.7	W/m²·°C

3.2. Experimental and Simulation Results

#1. Air stratification inside the Earthbag dome.

The air stratification testing scenario shows an increase of 1.4 °C from the bottom to the top of the dome in summer and 2.8 °C in the equinox. The surface temperature keeps more stable, oscillating in less than 1 °C (Figure 5).

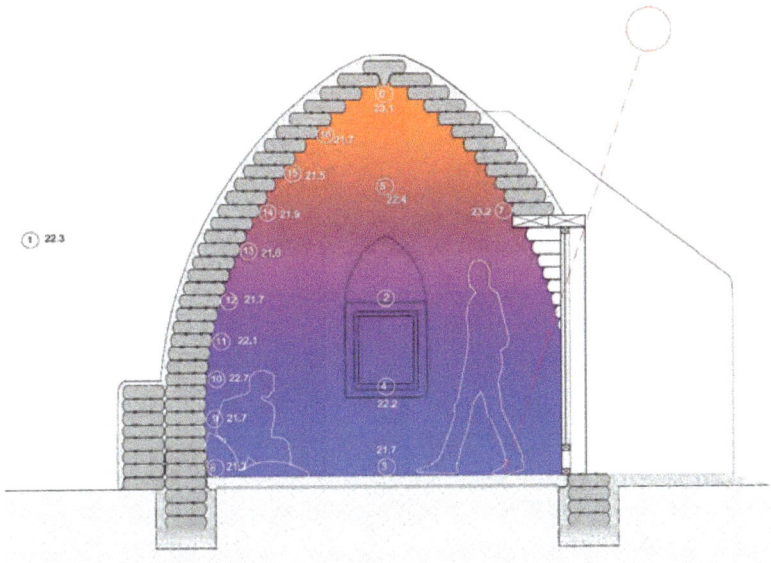

Figure 5. Air stratification inside the Earthbag dome during at solar noon (2 p.m., on 10 June 2018). (Temperature values in °C).

2.1. Winter solstice free floating temperature.

- *Comparison of simulation and monitoring:*

In the winter solstice period simulation and monitoring data follow a very similar trend (Figure 6). The thermal amplitude range is 1.5 °C for simulation and 2.3 °C for monitoring, with some specific days that can increase up to 3.7 °C. While the outdoor maximum temperature is at 3 p.m., inside the Earthbag building the maximum peak of temperature is produced from 1 p.m. to 2 p.m., one hour after the moment of maximum solar radiation. This peak of temperature is produced by the direct solar gain through the south glazed door. In this period of the year, the incident solar radiation in the east and west windows is inexistent and therefore no effect due to these glazed openings is observed. In a cloudy day with significantly less solar radiation, such as 11 December, there is no substantial increase of temperature from 2 p.m. to 3 p.m. In this case, the temperature oscillates moderately as if there were no windows.

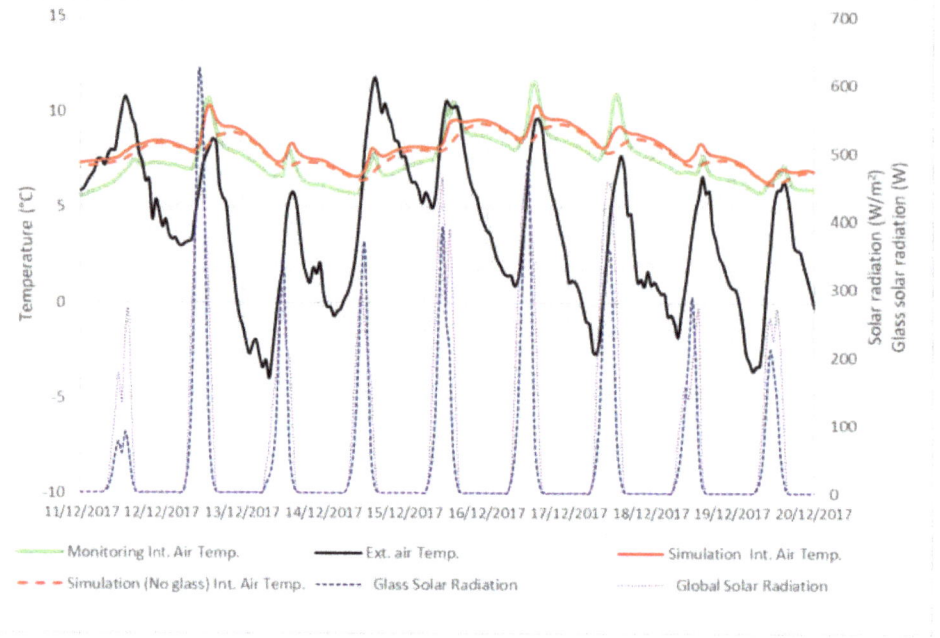

Figure 6. Interior simulated and monitored temperature, exterior temperature, solar radiation and radiation through the glazed surface, during the winter solstice.

- *Comparison between the two simulation cases*:

When comparing the simulated data in Figure 6 with the simulated model with no windows, the effect of the direct solar gains through the south glazed opening increases the average temperature in 1.31 °C, during the exposed period. In both cases, the thermal lag between interior and exterior maximum temperature is about 8 h (from 3 p.m. to 11 p.m.), due to the effect of thermal inertia of the Earthbag walls. This effect is more visible in the simulated case of the building without glazed openings. If the glazed openings were covered, the thermal lag would be 7 h (from 3 p.m. to 10 p.m.) and the thermal amplitude 1.3 °C.

#2.2. Equinox free floating temperature.

- *Comparison of simulation and monitoring:*

During the equinox period simulation and monitoring data follow a similar trend (Figure 7). In both cases, the thermal amplitude ranges between 1.8–2.2 °C. While the outer maximum temperature is at 3 p.m., inside the Earthbag building the maximum is at 2 p.m., one hour after the maximum solar radiation. This peak of temperature is produced by the direct solar gain through the south glazed door. In this period of the year, the solar radiation incident in the east window is slightly noticed from 8–9 a.m. with a small increase of 0.5 °C in the indoor temperature, clearly visible in the simulation.

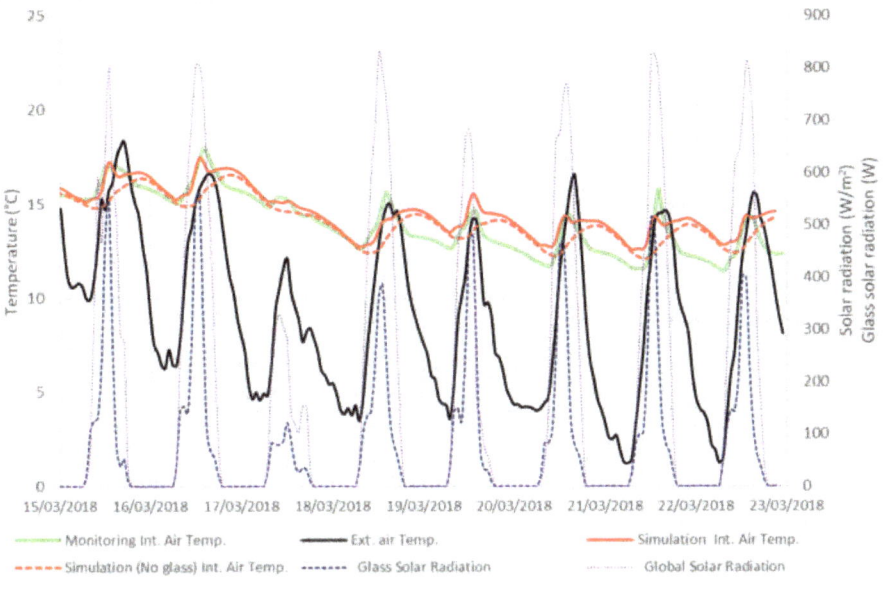

Figure 7. Interior simulated and monitored temperature, exterior temperature, solar radiation and radiation through the glazed surface, during the equinox.

- *Comparison between the two simulation cases:*

Compared to the simulated model with no windows, the effect of the direct solar gains increases the temperature in 1.37 °C. This increment of temperature is mainly due to the south glazed opening. In this case, the thermal lag of the simulated analysis between interior and exterior maximum temperature is about 7 h (from 3 p.m. to 10 p.m.), due to the effect of thermal inertia of the Earthbag walls. In the case of the Earthbag building with no glazed openings, the thermal lag is 8 h (from 3 p.m. to 11 p.m.) and an average of the thermal amplitude about 1.8 °C.

#2.3. Summer solstice free floating temperature.

- *Comparison of simulation and monitoring:*

In the summer solstice, the interior temperatures in the monitored and the simulated Earthbag building have a similar tendency (Figure 8). The thermal amplitude for both, simulation and monitoring is a maximum of 2.3 °C. The solar radiation entering through the glazed openings is visible for the three glazed surfaces in the monitoring, and barely visible in the simulation. Due to the relative position of the sun respect to the south facade, the radiation entering in the south glazed opening is inferior than in the other periods, with 200 W around midday corresponding to the maximum solar radiation (12 p.m.–1 p.m.). The shadow produced by the design of the awnings over the windows and the thickness of the Earthbag walls caused enough solar protection to minimize the solar heat gains. In this case, the maximum exterior temperature is around 4 p.m.

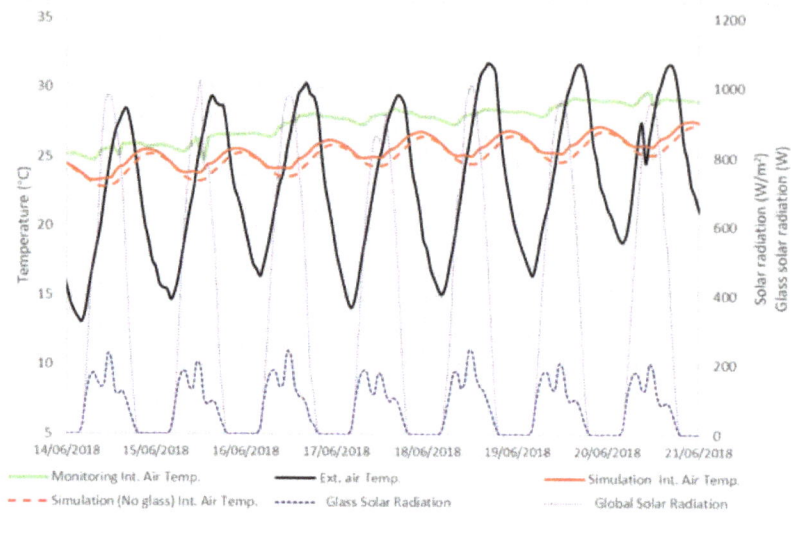

Figure 8. Interior simulated and monitored temperature, exterior temperature, solar radiation and radiation through the glazed surface, during the summer solstice.

- *Comparison between the two simulation cases:*

Compared to the simulated model with no windows, the effect of the direct solar gains increases the temperature in 0.52 °C, the lowest increase for the three analyzed periods. The thermal lag for the glazed Earthbag building simulation is about 6 h while for the simulation with no glasses is 7 h. The internal temperature thermal amplitude is about 2 °C for the building simulated with glazed openings and about 2.1 °C for the simulated building without glazed openings, when the outer temperature's amplitude is about 15 °C. When calculating the thermal amplitude for the monitored building, the value is about 1.2 °C.

#3.1. Summer 24 h cross ventilation.

The behavior of the Earthbag building under natural cross ventilation conditions (24 h per day) during summertime is presented in Figure 9. Both results of interior temperature for simulation and monitoring, present a similar trend, which validates the simulation. The thermal lag between exterior and interior temperatures is inferior to 1 h. Despite the ventilation, a decrement factor can still be observed. The mean exterior thermal amplitude is 10 °C, while the interior is 4 °C. During midday, interior temperature is almost reaching exterior temperatures (from 1 °C to 4 °C below the maximum peak temperatures). During the night, the effect of the thermal inertia of the Earthbag building is visible. Despite the ventilation, when the exterior temperature decreases drastically, the Earthbag building keeps the thermal energy and makes the interior temperature be higher than the exterior (from 3 °C to 5 °C over the minimum peak temperatures). The thermal inertia effect is also visible on 6 June (a cooler day than the previous) when the interior maximum temperatures are 2 °C over the exterior temperatures. The natural cross ventilation during all the day was not effective to cool down the Earthbag building because the exterior day temperatures were high enough to keep the daily average interior temperatures in a high range (20–24 °C). In this case, the solar radiation does not have much influence because the main temperature changes are due to the constant hot air circulation from the outside.

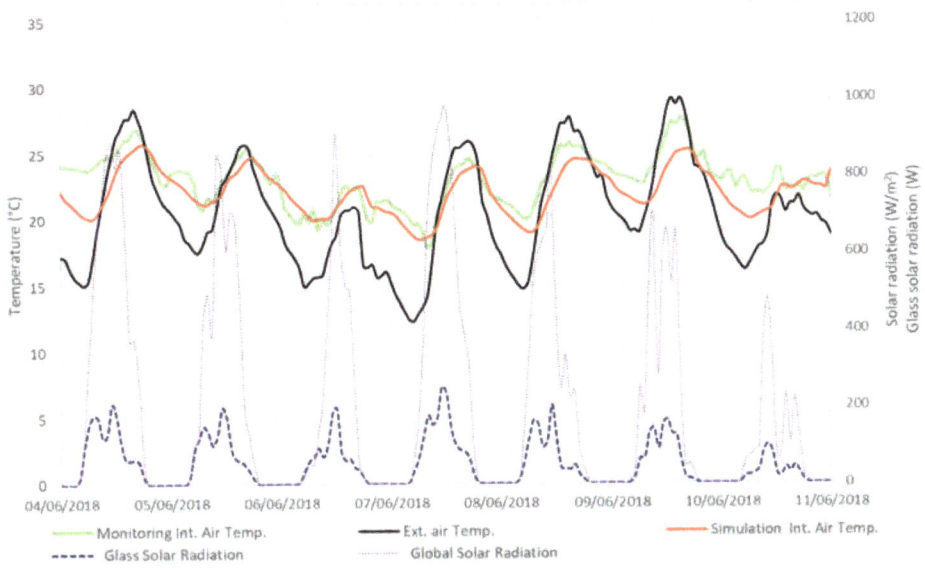

Figure 9. Indoor simulated and monitored temperature, exterior temperature and solar radiation, during 24 h natural cross ventilation scenario.

3.2. Summer night cross ventilation.

Similar to the previous scenario, the prototype was studied under night natural cross ventilation conditions, where the windows were opened from 8 p.m. to 8 a.m. every day. Figure 10 shows the behaviour of the indoor simulated and monitored temperature, the outdoor temperature and the solar radiation as well as the hours when natural cross ventilation is active. Both monitored and simulated indoor temperatures follow the same trend. In both cases, the maximum interior temperature is reached after the exterior maximum temperature, with a thermal lag about 1 h. The decrement factor is 0.4, higher than previous results, caused by the night ventilation. During the day time, the effect of solar gains through the glazed openings is visible in the small temperature peaks in the morning, midday and afternoon. The nigh ventilation produces also a 5.5 °C decrease from the maximum to the minimum temperatures. The effect of opening the windows can be clearly observed on 26, 27 and 28 July, with a sudden matching of interior and exterior temperatures.

#4. Winter controlled temperature.

Figure 11 shows the outdoor temperature and the power consumption per square meter to maintain the indoor temperature at 22 °C. Monitored and simulated consumptions behave quite similar. The oscillation presented by the monitored power consumption of the heater is due to its control system. However, the trend line of the monitored consumption is similar to the simulation, which does not present these oscillations. The average power consumption for heating was 56.56 W/m^2, for the analyzed period (Figure 6), having the maximum consumption around 90 W/m^2 during the coldest days, for outdoor temperatures under 0 °C.

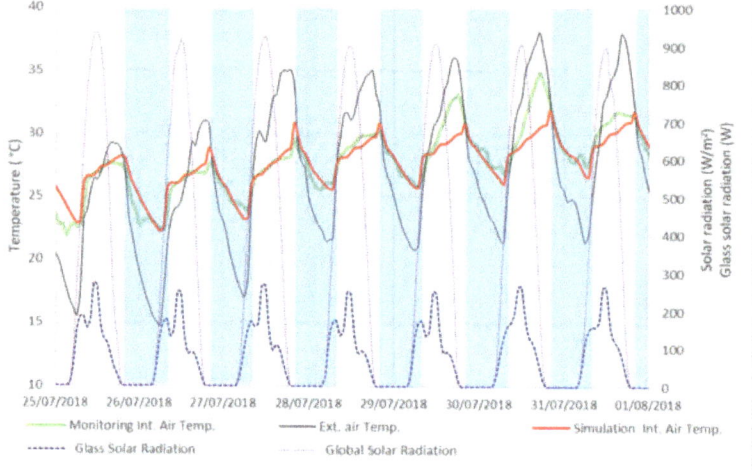

Figure 10. Interior monitored and simulated air temperature, solar radiation and exterior temperature during night cross ventilation (ventilation time represented in blue area).

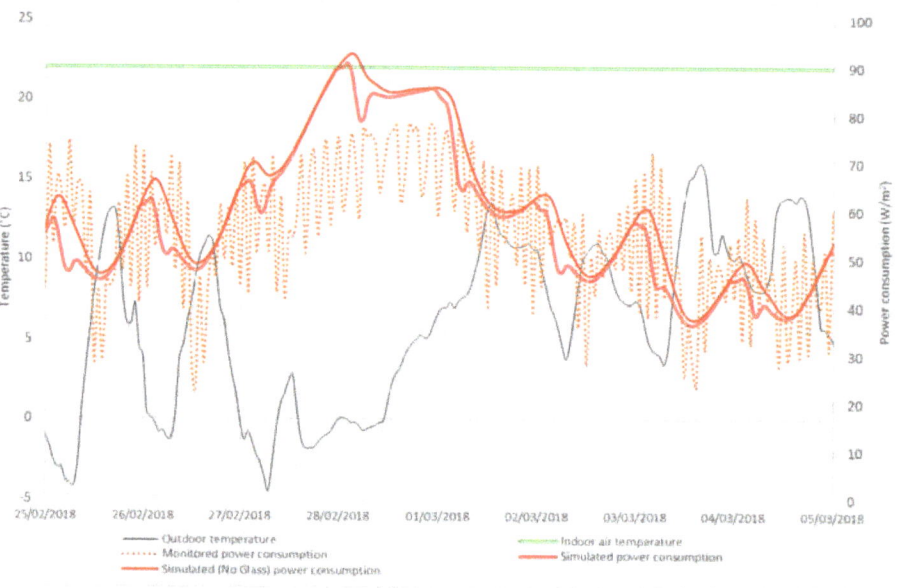

Figure 11. Power consumption of the heater, both simulated and monitored, during the indoor controlled temperature scenario.

4. Discussion

4.1. Characterization of the Thermal Properties of the Earthbag Walls

- The thermal transmittance (U-value) of the Earthbag wall evaluated in this study is 2.7 W/m²·K. It is interesting to highlight that no references about Earthbag transmittance were found in the literature for comparison. When comparing the Earthbag transmittance with transmittance values found in literature for rammed earth, it is seen that the value provided in this study is higher than the one found for a rammed earth wall of 30 cm, with U-values ranging from 1.3 [30] to 1.9 [1]. It is important to highlight that in many Mediterranean countries with low income and traditional buildings, no limits for thermal transmittance are still established, and the Earthbag buildings can definitely contribute to better comfort and less energy consumption than their traditional ones.

- The thermal conductivity (λ) of the Earthbag (35 cm thick) evaluated in this study is 2.18 W/m·K. This value is in the high range found in the literature for rammed earth, from 1.1 W/m·K with a density about 1900 kg/m^3 [25] to 1.2 W/m·K with a density of 2000 kg/m^3 [31]. However, it is within the range of other materials with similar density, such as limestone with 2200 kg/m^3 [25].

- When observing the thermal properties of an Earthbag building from a qualitative point of view, it is important to focus on air temperature stratification inside the Earthbag dome: A difference of 1.4 °C in the summer solstice and 2.8 °C in the equinox from the ground to the ceiling in the dome has been observed in the monitoring. The simulation does not take into account the air stratification in a zone, since it calculates a mean zone temperature. Differences in the use of the rooms or cultural differences play an important role and must be taken into consideration, when designing an Earthbag building. For example, in those cultures where the living room is used at ground level, the thermal comfort of the users would be reached at different temperatures than in a culture where the users are usually seated on a chair or in a table level.

- One of the characteristics of an Earthbag wall is its high thermal inertia. This is why thermal lag is a key factor when studying these buildings. In this study, the thermal lag in the three analyzed periods, in both monitoring and simulation, ranged from 6 to 9 h, similar to the theoretical calculated value (Table 4). The decrement factor ranged from 0.1 to 0.19, a range within the calculated theoretical value. The mean thermal amplitude ranged from 1.2 °C to 2.5 °C, depending on the period of the year. These values confirm the high thermal inertia of the Earthbag building, which contributes to an improvement of the thermal comfort in continental climates.

Table 4. Thermal lag and decrement factor of the Earthbag building in theoretical, simulated and monitoring cases for winter solstice, equinox and summer solstice.

Parameter	Data Taken From:	Theoretical	Test		
			Winter	Equinox	Summer
Thermal lag (h), Φ	Monitoring	8.1	8 (3 p.m.–11 p.m.)	9 (3 p.m.–12 a.m.)	9 (4 p.m.–1 a.m.)
	Simulation		8 (3 p.m.–11 p.m.)	7 (3 p.m.–10 p.m.)	6 (4 p.m.–10 p.m.)
	Simulation without glazed surface		7 (3 p.m.–10 p.m.)	8 (3 p.m.–11 p.m.)	7 (4 p.m.–11 p.m.)
Decrement Factor, μ	Monitoring	0.12	0.12	0.19	0.10
	Simulation		0.17	0.16	0.14
	Simulation without glazed surface		0.13	0.14	0.16

4.2. Passive Design Strategies in the Earthbag Building

- Cooling: natural cross ventilation

One of the most used and effective bioclimatic strategy in summer conditions is the natural cross ventilation. This is the reason why the Earthbag building, object of this study, was designed locating the two windows in the exact opposite side, in order to increase the effects of the natural cross ventilation.

In the case of night ventilation, from 8 p.m. to 8 a.m., natural cross ventilation takes advantage of the cooler temperatures at night to decrease the average temperature of the Earthbag building. This is an appropriate passive strategy for cooling. Night ventilation emphasized the positive effect of thermal inertia of the Earthbag walls, as the energy stored in the walls during the day is discharged at night to the outside air and the interior temperature is reduced.

As expected, the 24 h ventilation in summer conditions is not a good strategy for cooling during midday and afternoon hours. This type of ventilation makes interior temperatures almost reach the exterior high temperatures during the day, what makes night ventilation inefficient. Therefore, 24 h ventilation cannot be used as a passive strategy for cooling.

- Heating: direct solar gain

Direct solar gain through a glazed opening in the south facade is a commonly used passive design strategy for heating in middle latitudes. Theoretically, due to the position of the sun in the winter solstice, the south facade is the surface of the building that receives the highest amount of solar radiation and, therefore, any glazed opening is expected to act as a heater of the building, having more heat gains than heat losses in a sunny day. In this latitude, the elevation angle of the sun during wintertime reaches a maximum of 25° during the midday, so the solar radiation heats naturally the Earthbag building without any extra source of energy, due to the greenhouse effect.

Solar gains through the glazed openings are presented in Table 5. For comparison purpose with glazed openings, solar radiation is presented per glass area. Each glazed opening received the solar radiation during different periods of time, depending on the moment of the year and, therefore, on the position of the sun and the amount of solar radiation. The south glazed opening received the highest solar radiation in winter and the lowest in summer. East and west glazed openings received the maximum hours and solar radiation in summer solstice. To avoid overheating in summer, these glazed openings should be protected from solar radiation.

Table 5. Average solar radiation (W/m^2) and period of time (h) that receives solar radiation each glazed opening.

Glazed Opening	Concept	Test		
		Winter	Equinox	Summer
East	Period of time with solar radiation (h) Average solar radiation (W/m^2)	1 h (9 a.m.–10 a.m.) 862.50	3 h (9 a.m.–12 p.m.) 348.17	4 h (6 a.m.–10 a.m.) 576.43
South	Period of time with solar radiation (h) Average solar radiation (W/m^2)	3 h (11 a.m.–2 p.m.) 299.31	2 h (12 p.m.–2 p.m.) 159.04	1 h (12 p.m.–1 p.m.) 145.20
West	Period of time with solar radiation (h) Average solar radiation (W/m^2)	- -	2 h (5 p.m.–7 p.m.) 1296.90	4 h (4 p.m.–8 p.m.) 1959.29

The indoor temperature in the winter free floating test is below the thermal comfort limits. Therefore, a combination of an increase of the internal heat gains and an increase of the glazed surface in the south facade are possible strategies to achieve the thermal comfort. The option with active systems was heating. The effect of the passive strategy of direct solar gain meant a daily reduction of the energy consumption and an increment of the interior temperature, as shown in Table 6.

Table 6. Average reduction of heating energy consumption (Wh) and average increase of the interior air temperature (°C) when it is used the passive strategy of direct solar gains through the glazed openings.

Comparison of Glazed-No Glazed Simulation	Test	Period		
		Winter Solstice	Equinox	Summer Solstice
Decrease in heating energy consumption (Wh)	Controlled temperature	−22.67 (−2.3%)	−35.70 (−8.9%)	-
Increase in peak interior temperature (°C)	Free floating	+1.31	+1.37	+0.52

4.3. Thermal Comfort Analysis

The application of the ASHRAE Standard 55 Adaptive Comfort model [32] to all the hours of the monitored summer solstice week is clearly shown in Figure 12, plotting the internal operative temperature for the Earthbag building as a function of outdoor temperature. Simulated values are included as well for comparison. Note that most of the points are either within the comfort adaptive range or only a bit above the upper limit (0.5–3 °C) or slightly below the lower one (0.5 °C). These results of satisfactory thermal comfort conditions are in good agreement with the found for an Earthbag dwelling in Ougadougou (Burkina Faso) by energy simulations [6]. Note that in this paper the Earthbag dwelling improved a 99% the discomfort degree-days with respect to the traditional Burkinable dwelling. Thus, these experimental results confirm the good performance of the Earthbag building to achieve thermal comfort in hot climates.

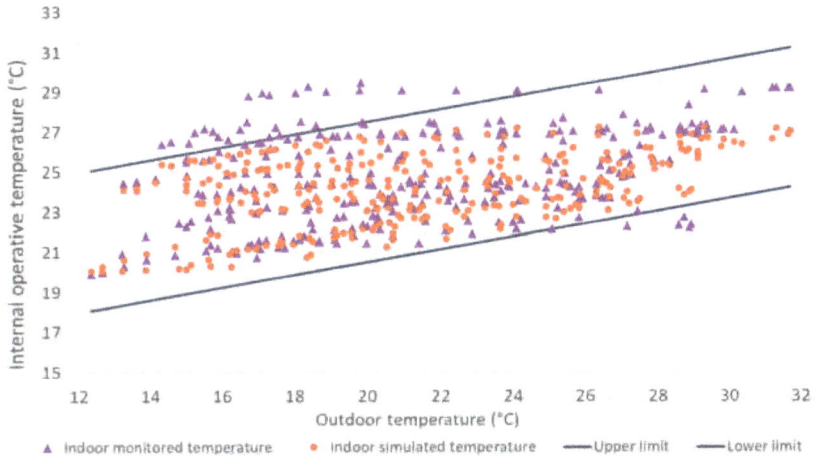

Figure 12. Thermal comfort of the prototype during the summer period according to the ASHRAE 55 Adaptive method [32].

5. Conclusions

The thermal performance of an Earthbag dwelling in Mediterranean continental climate is studied both experimentally and numerically. The experimental campaign was carried out along the different seasons of the year and included both free floating and temperature controlled tests. The building energy model was developed with EnergyPlus. Simulation and experimental results have a similar trend in free floating experiments, with and without ventilation, and in controlled temperature tests. Monitoring and simulation comparisons showed a good match in temperature behavior for the different scenarios studied.

The effect of stratification of the air temperatures inside the dome is not considered in EnergyPlus. However, qualitative results presented in this study show that stratification is significant and must be taken into account in the design phase of the Earthbag spaces. Experimental results showed a difference up to 1.4 °C to 2.8 °C between the floor and the ceiling.

Singularities in the construction process or material availability outcomes in differences in the Earthbag wall. Variations in density or humidity level result in different thermal conductivity and specific heat. The Earthbag of 35 cm, with an earth mixture of 0.80% fine gravel, 92.21% sand, 3.42% slime and 3.57% clay, and a density of 2190 kg/m^3, resulted in a conductivity of 2.18 W/m·K. The facade made of one Earthbag of 35 cm plus an exterior lime coating of 4 cm resulted in a thermal transmittance of 2.7 W/m^2·K. Using these experimental values in the simulation, instead of the generic data present in literature or building codes, provides more accurate simulation results.

Passive design strategies are a key point when designing an Earthbag building, contributing to achieving thermal comfort during summer and winter, despite the high U-value of the Earthbag walls. Those strategies were, for the winter period, the use of high thermal mass to store thermal energy and the heat gains by direct solar radiation, and, in the summer period, the use of night natural cross ventilation together with the use of high thermal mass to reduce the midday peak temperatures.

Theoretical calculations and experimental results matched in the thermal lag and decrement factor and verified the high thermal inertia of the Earthbag enclosure. The thermal lag was determined in a range of 6–9 h and the decrement factor in 0.1–0.2. Moreover, the thermal mass produced reductions of 90% in summer and 88% in winter, in the interior thermal amplitude with respect to exterior temperatures. The effect of the building floor in contact with the ground would also cause this high thermal inertia and it should be analyzed in future research. The low shape factor could contribute to a lower energy flux that also should be analyzed in future research.

The position of the glazed openings was fundamental in the direct collection of solar radiation in winter period, whose effect is increasing the interior temperature in free floating mode (reducing the heating energy consumption in controlled temperature mode). As expected, the greatest effect of temperature increase due to solar collection in winter time and the equinox, was caused by the south facade glass, with an increment of 1.31 °C and 1.37 °C, respectively.

Night ventilation in summer period in combination with high thermal inertia has a good performance as a passive system for semi-hot climates because it reduces the indoor temperatures to a comfort range without any active system. However, all day ventilation cannot be recommended because the high exterior day temperatures increase the average temperatures inside the building.

Despite the passive design strategies, the Earthbag building requires a heating system in winter period to achieve thermal comfort levels. In the case of the analyzed building, to achieve a comfort temperature of 22 °C, energy consumption in the range 1–1.7 kWh/m^2 per day was required.

Author Contributions: L.R., C.S., I.M. and M.M. conceived and designed the analysis. L.R., A.C. (Ariadna Carrobé), and I.M. collected the data. L.R., A.C. (Ariadna Carrobé), and M.M. contributed in the simulation process. L.R., A.C. (Ariadna Carrobé), M.M., and A.C.(Albert Castell) performed the analysis. L.R., A.C. (Ariadna Carrobé) and I.M. wrote the paper. A.C. (Albert Castell), M.M. and C.S. reviewed and edited the paper. All authors have read and agreed to the published version of the manuscript.

Funding: This research was funded by the Oficina de Desenvolupament i Cooperació (ODEC), the "Escola Politècnica Superior" of UdL, the "Consell Social" of UdL and the Catalan Government (AGAUR- Generalitat de Catalunya).

Acknowledgments: The authors would like to thank the "Oficina de Desenvolupament i Cooperació (ODEC)" of the University of Lleida (UdL), Spain, for the grant projects of 2014, 2015, 2016 and 2017, the "Escola Politècnica Superior" of UdL, the "Consell Social" of UdL for their donatives, and Domoterra association. The authors would also thank all the students that participated in the construction of the Earthbag building. Ariadna Carrobé would like to thank the Catalan Government (AGAUR- Generalitat de Catalunya) for her collaboration grant 2017–2018 in the Computer Sciences and Industrial Engineering Department at University of Lleida. The authors would like to thank the Catalan Government for the quality accreditation given to their research group (2017 SGR 659).

Conflicts of Interest: The authors declare no conflict of interest.

References

1. Minke, G. *Manual De Construcción En Tierra*; Fin De Siglo: Cuba, Havana, 2012; ISBN 9974-49-347-1.
2. Nader Khalili, P.V. *Earth Architecture and Ceramics*; Calearth: Hesperia, CA, USA, 1998.
3. Canadell, S.; Blanco, A.; Cavalaro, S.H.P. Comprehensive design method for earthbag and superadobe structures. *Mater. Des.* **2016**, *96*, 270–282. [CrossRef]
4. Sargentis, G.F.; Kapsalis, V.C.; Symeonidis, N. Earth building. Models, technical aspects, tests and environmental evaluation. In Proceedings of the 11th International Conference on Environmental Science and Technology Chania, Crete, Greece, 3–5 September 2009.
5. Adegun, O.B.; Adedeji, Y.M.D. Review of economic and environmental benefits of earthen materials for housing in Africa. *Front. Arch. Res.* **2017**, *6*, 519–528. [CrossRef]
6. Rincón, L.; Carrobé, A.; Martorell, I.; Medrano, M. Improving thermal comfort of earthen dwellings in sub-Saharan Africa with passive design. *J. Build. Eng.* **2019**, *24*, 100732. [CrossRef]
7. Merina, S.; Javier, F. *EURAU18 Alicante: Retroactive Research: Congress Proceedings*; Universidad de Alicante: Alicante, Spain, 2018; ISBN 978-84-13-02003-7.

8. Cal-Earth, The California Institute of Earth Art and Architecture, 2014. Cal-Earth Inc./Geltaftan Foundation. Available online: http://calearth.org/ (accessed on 1 February 2019).
9. Training Medical Center—Emsimision, (n.d.). Available online: http://www.emsimision.org/proyectos/training-medical-center/ (accessed on 6 April 2018).
10. Builders without Borders, Earthbag Housing in Port au Prince, Haiti. Available online: www.haitistrawbale.wordpress.com (accessed on 1 March 2019).
11. Rasin Fundation. Medical Center in Petite-Riviere, Haiti. 2010. Available online: www.rasinfoundation.org/programs/clinic (accessed on 1 March 2019).
12. Heats of Haiti. School in Lounglan, Haiti. 2010. Available online: www.howsitgoinginhaiti.blogspot.com (accessed on 1 March 2019).
13. Konbit Shelter. Community Center in Barriere Jeudy, Haiti. 2010. Available online: www.earthbagbuilding.com/projects/konbit2.htm (accessed on 1st March 2019).
14. Barrels of Hope. Impact Test to Evaluate Hurricane Effects on Earthbag Wall. Haiti. 2010. Available online: www.barrelsofhope.org (accessed on 1st March 2019).
15. Small-Earth. The Pegasus Children's Projectin Kathmandu Valley, Nepal. EarthbagBuilding Web Page. 2006. Available online: http://www.earthbagbuilding.com/projects/pegasus.htm (accessed on March 2019).
16. Palme, M.; Guerra, J.; Alfaro, S. Thermal Performance of Traditional and New Concept Houses in the Ancient Village of San Pedro De Atacama and Surroundings. *Sustainability* **2014**, *6*, 3321–3337. [CrossRef]
17. Desogus, G.; Di Benedetto, S.; Grassi, W.; Testi, D. Environmental monitoring of a Sardinian earthen dwelling during the summer season. *J. Phys. Conf. Ser.* **2014**, *547*, 012009. [CrossRef]
18. Martín, S.; Mazarrón, F.R.; Cañas, I. Study of thermal environment inside rural houses of Navapalos (Spain): The advantages of reuse buildings of high thermal inertia. *Constr. Build. Mater.* **2010**, *24*, 666–676. [CrossRef]
19. Catálogo Informático De Elementos Constructivos. Available online: https://www.codigotecnico.org/index.php/menu-catalogo-informatico-elementos-constructivos (accessed on 6 April 2018).
20. Martí, R. Anàlisi I Caracterització De La Terra Del Domo De La Universitat De Lleida (Analysis and Characterization of the Earth of the Dome of the University of Lleida). Bachelor's Thesis, University of Lleida, Catalunha, Espanha, 2016.
21. Rubel, F.; Kottek, M. Observed and projected climate shifts 1901–2100 depicted by world maps of the Köppen-Geiger climate classification. *Meteorol. Z.* **2010**, *19*, 135–141. [CrossRef]
22. Bienvenido-Huertas, D.; Moyano, J.; Marín, D.; Fresco-Contreras, R. Review of in situ methods for assessing the thermal transmittance of walls. *Renew. Sustain. Energy Rev.* **2019**, *102*, 356–371. [CrossRef]
23. Teni, M.; Krstić, H.; Kosiński, P. Review and comparison of current experimental approaches for in-situ measurements of building walls thermal transmittance. *Energy Build.* **2019**, *203*, 109417. [CrossRef]
24. Cabeza, L.F.; Castell, A.; Medrano, M.; Martorell, I.; Pérez, G.; Fernández, I. Experimental study on the performance of insulation materials in Mediterranean construction. *Energy Build.* **2010**, *42*, 630–636. [CrossRef]
25. ISO 6946:2017. Available online: http://www.iso.org/cms/render/live/en/sites/isoorg/contents/data/standard/06/57/65708.html (accessed on 14 December 2019).
26. Yáñez Paradera, G. *Arquitectura Solar E Iluminación Natural: Conceptos, Métodos Y Ejemplos*; Munilla-Lería: Madrid, Spain, 2008; ISBN 978-84-89150-81-2.
27. Mazzarella, L.; Pasini, M. CTF vs FD Based Numerical Methods: Accuracy, Stability and Computational Time's Comparison. *Energy Procedia* **2015**, *78*, 2620–2625. [CrossRef]
28. Liddament, M.W. *International Energy Agency Energy Conservation in Buildings and Community Systems Programme: IEA-ECB & CS. Annex 5 [...]: Air Infiltration and Ventilation Centre Air Infiltration Calculation Techniques: An Applications Guide*; Document AIC-AG; Air Infiltration and Ventilation Centre: Bracknell, Berkshire, 1986; ISBN 978-0-946075-25-6.
29. Taylor, B.N.; Kuyatt, C.E. *Guidelines for Evaluating and Expressing the Uncertainty of NIST Measurement Results*; Physics Laboratory National Institute of Standards and Technology: Gaithersburg, MD, USA, 1994.
30. Birkhäuser, G.M. *Building with Earth: Design and Technology of a Sustainable Architecture*; Walter de Gruyter: Berlim, Alemanha, 2012; ISBN 978-3-0346-0872-5.
31. Heathcote, K. El comportamiento térmico de los edificios de tierra. *Inf. Constr.* **2011**, *63*, 117–126. [CrossRef]
32. ANSI/ASHRAE Standard 55-2013. *Thermal Environmental Conditions for Human Occupancy*; American Society of Heating, Refrigeration and Air-Conditioning Engineers, Inc.: Atlanta, GA, USA, 2013.

 © 2019 by the authors. Licensee MDPI, Basel, Switzerland. This article is an open access article distributed under the terms and conditions of the Creative Commons Attribution (CC BY) license (http://creativecommons.org/licenses/by/4.0/).

Article

Thermal Performance and Comfort Condition Analysis in a Vernacular Building with a Glazed Balcony

Jorge Fernandes [1], Raphaele Malheiro [1], Maria de Fátima Castro [1], Helena Gervásio [2], Sandra Monteiro Silva [1] and Ricardo Mateus [1,*]

1. Department of Civil Engineering, University of Minho, Campus de Azurém, 4800-058 Guimarães, Portugal; jepfernandes@me.com (J.F.); raphamalheiro@gmail.com (R.M.); info@mfcastro.com (M.d.F.C.); sms@civil.uminho.pt (S.M.S.)
2. Institute for Sustainability and Innovation in Structural Engineering (ISISE), University of Coimbra, University of Coimbra, Faculdade de Ciências e Tecnologia, Rua Luís Reis Santos, Pólo II, 3030-788 Coimbra, Portugal; hger@dec.uc.pt
* Correspondence: ricardomateus@civil.uminho.pt

Received: 10 November 2019; Accepted: 22 January 2020; Published: 1 February 2020

Abstract: The increase in global environmental problems requires more environmentally efficient construction. Vernacular passive strategies can play an important role in helping reducing energy use and CO_2 emissions related to buildings. This paper studies the use of glazed balconies in the North of Portugal as a strategy to capture solar gains and reduce heat losses. The purpose is understanding thermal performance and comfort conditions provided by this passive heating strategy. The methodology includes objective (short and long-term monitoring), to evaluate the different parameters affecting thermal comfort and air quality, and subjective assessments to assess occupants' perception regarding thermal sensation. The results show that the use of glazed balconies as a passive heating strategy in a climate with cold winters is viable. During the mid-seasons, the rooms with balcony have adequate comfort conditions. In the heating season, it is possible to achieve comfort conditions in sunny days while in the cooling season there is a risk of overheating. Regarding indoor air quality, carbon dioxide concentrations were low, but the average radon concentration measured was high when the building was unoccupied, rapidly decreasing to acceptable values, during occupation periods when a minimum ventilation rate was promoted. Occupants' actions were essential to improving building behavior.

Keywords: glazed balcony; indoor comfort; passive strategies; thermal performance; vernacular architecture

1. Introduction

1.1. Context

The construction industry is one of the largest and most active sectors of the world economy. Regarding the importance of this sector and its influence on sustainable development issues, several organizations set different goals to achieve more efficient construction. For example, the European Union (EU) is committed to developing a sustainable, competitive, secure, and decarbonized energy system setting a goal for reducing carbon dioxide (CO_2) emissions by at least 40% by 2030 and by 80–95% until 2050, compared to 1990 values [1,2]. In parallel, it is intended to increase the proportion of renewable energy consumed and to improve Europe's energy security, competitiveness, and sustainability [3].

According to Directive (EU) 2018/844, it is essential to ensure that measures to improve the energy performance of buildings do not focus only on the building envelope. It should also include all relevant elements and technical systems in a building, such as passive elements that can contribute to reducing energy needs for heating or cooling, as well as energy use for lighting and ventilation, and hence improve thermal and visual comfort [3].

Therefore, one of the ways to improve the sustainability of buildings is to reduce the importance of active systems and give higher priority to architectural form and passive systems [4–6]. Passer et al. [6] demonstrated that technical equipment has a significant influence on the life cycle environmental impacts of buildings. These authors also concluded that, on a life cycle assessment approach, passive buildings have the lowest impacts associated with mechanical equipment, mainly because they have reduced needs for mechanical ventilation and air conditioning systems [6]. The introduction of passive strategies in buildings from the design stage reduces the amount and the need for these types of systems [5].

In the context of passive techniques aiming to reduce energy needs, it is important to analyze vernacular architecture to understand the way vernacular buildings were shaped to suit local climate constraints. Additionally, the strategies that are now the basis of sustainable construction derive from aspects and characteristics of this type of architecture [7]. In these construction projects, strategies used to mitigate the effects of climate and ensure thermal comfort conditions are usually passive in operation, low in technology, and do not depend on fossil energy to operate, making them particularly suitable for contemporary building applications, mainly in the design of passive buildings. For this reason, vernacular architecture continues to be the subject of several studies whose findings seek to contribute to the development of a more sustainable built environment. Although these studies have been taking part around the world, they adopted similar methodologies and reported similar conclusions and limitations. These conclusions highlight that the use of vernacular techniques and local materials in the design of buildings, developed on the basis of the need for adaptation to a specific territory and climate, will contribute to the reduction of waste, energy use, and consequently carbon emissions, among other environmental impacts [8–12].

Additionally, the study and valorization of the vernacular buildings and the inherent knowledge will contribute not only to its preservation but also to the dynamization of local economies [13].

1.2. Vernacular Strategies and the Built Environment

In the past, due to the lack of active systems, buildings were built using passive strategies to reduce thermal discomfort. These strategies were based on available endogenous resources and design principles arising from local geographical characteristics [10,14]: insolation; orientation; topography; shape; and materials, among others.

The relationship between the built and the natural environments, well described by the mythological concept of *Genius Loci*, is of prime importance in the design of buildings and their thermal performance. As an example, it was not random how the North African houses and the North European houses were designed, or, in the Portuguese context, the differences between northern and southern interior residential buildings.

Regarding the thermal performance of vernacular buildings, several quantitative studies conducted in different parts of the world have shown that these buildings achieve acceptable levels of thermal comfort throughout most of the year using only passive strategies, in some cases with indoor temperature remaining stable [7,15–19]. In some of these studies, vernacular buildings performed better than contemporary buildings, although several of the building solutions adopted do not meet current thermal regulation requirements. These results support the idea that passive strategies are, in many cases, feasible for application in contemporary buildings and can contribute to the reduction of energy requirements for air-conditioning. The adequate response of vernacular solutions to climate constraints reveals the importance of local specificities for contemporary construction, in terms of sustainability and energy efficiency [11,12,20,21].

In this context, some authors, like Ascione, et al. [22] defend that the building orientation and its passive design can positively affect the energy and environmental performance of buildings. For instance, glazed balconies act as a sunspace (Figure 1) and are a vernacular design solution that can contribute to the thermal performance of buildings during the heating season, since the sunspace heats the adjacent rooms, not compromising the thermal behavior of the building during summer [23]. This technique also has great potential to be used in contemporary buildings to improve energy efficiency, as shown by the case of the rehabilitation of the residential complex of Dornbirn, in Austria. The option of introducing this type of solution in the south-facing facades has increased the floor area of the dwellings and significantly reduced the heating energy bill [24]. The operation of these balconies as a buffer space allows the simultaneous capture and trapping of solar gains and reduction of heat losses. By being physically separated from the interior spaces of the dwellings, in situations where heat gain is undesirable, the balcony space can act as a shading device and promotes natural ventilation [24].

Figure 1. Balconies in the vernacular architecture of Beira Alta, Portugal.

The vernacular architecture strategies can contribute to improving the energy efficiency of buildings, whereas the local specificities should assume particular relevance. At a time of achieving high-performance buildings, defining the future of architecture and construction should seek to integrate tradition with modernity, at a crossroad that unites the best of today's technological potentialities with traditional materials and techniques [4].

1.3. Aim of This Research

The study of Portuguese vernacular architecture based on in situ measurements that allow a comprehensive demonstration of the effects of vernacular passive strategies on thermal performance is still lacking. In Portugal, there are only a few quantitative studies [25–28] focusing on passive strategies and their contribution to the thermal performance of buildings.

Analyzing the state-of-art for the specific context of vernacular buildings with glazed balconies, it is possible to verify that there are no quantitative studies developed so far on their thermal performance. Thus, this study aims to contribute to the development of this field of research by analyzing the thermal performance of a vernacular building with a glazed balcony, located in Northern Portugal (region of Beira Alta), considering the thermal comfort standards, and analyzing how the glazed balcony technique suit the local conditions. To fulfil this goal, the study consists of assessing the hygrothermal parameters that characterise the indoor thermal environment and that affect the occupants' thermal comfort conditions.

The number of existing vernacular buildings with glazed balconies identified in the Survey on Portuguese Popular Architecture [29] is decreasing, and it is becoming increasingly difficult to find this type of building in good condition. Therefore, this study intends to demonstrate the potential of this passive technique on improving the indoor environmental quality and reducing the energy needs of buildings. Presenting quantitative data about the thermal performance will contribute to a better understanding of this type of buildings and about the contribution of glazed balconies in maintaining indoor temperatures within the comfort range. This research will also contribute to the preservation of this type of building and their related knowledge.

2. Materials and Methods

To assess indoor thermal performance, in situ assessments were divided into short and long-term monitoring. In these assessments, the hygrothermal parameters that characterize the indoor thermal environment and that affect the body/environment heat exchange (air temperature, relative humidity, mean radiant temperature, and air velocity) were measured. The measurements were carried out from the autumn of 2014 to the summer of 2015.

2.1. Short-Term Monitoring

Short-term monitoring was carried out at least one day per season and consisted of objective measurements and subjective evaluation:

- Objective measurements had the purpose of quantitatively assess the thermal comfort conditions in a room using a thermal microclimate station (model Delta OHM 32.1) that measures air temperature, relative humidity, mean radiant temperature, and air velocity (Table 1), in compliance with standards ISO 7726 [30], ISO 7730 [31], and ASHRAE 55 [32]. The location of the equipment is chosen according to occupants' distribution in the room and in the rooms where occupants stay for more extended periods. The measurements were performed considering that the occupants were seated, as recommended in ASHRAE 55 [32]. The data recorded in these measurements was used to determine the operative temperature (the analysis procedure is explained below in this section).
- Subjective evaluation was carried out to assess the occupants' perceived indoor environment quality, using surveys. The case study building is occupied by two persons, which comfort level was surveyed. The survey was based in the "Thermal Environment Survey" from ASHRAE 55 [32] and was used to determine occupants' satisfaction according to ASHRAE thermal sensation scale.

Table 1. Location and characteristics of measurement equipment used.

Equipment	Specifications, Measurement Range and Accuracy	Location
Thermal microclimate station (model Delta OHM 32.1)	Probes installed: 1. Globe temperature probe Ø150 mm (range from −10 to 100 °C); 2. Omnidirectional hot-wire probe for wind speed measurement (range from 0 to 5 m/s); 3. Combined temperature and relative humidity probe (range from −10 to 80 °C and 5–98% RH); 4. Two-sensor probe for measuring natural wet bulb temperature and dry bulb temperature (range from 4 to 80 °C).	Living room/kitchen and bedroom with balcony

Table 1. Cont.

Equipment	Specifications, Measurement Range and Accuracy	Location
Thermo-hygrometer and datalogger (Klimalogg Pro, TFA 30.3039.IT) + Wireless thermo-hygrometer transmitters (model TFA 30.3180.IT) connected to the datalogger	Datalogger: • Temperature accuracy of ±1 °C and a measuring range between 0 and 50 °C with 0.1 °C resolution; • Relative humidity accuracy of± 3% and measuring range between 1 and 99% with 1% resolution. Transmitters: • Temperature accuracy of±1 °C and measuring range between 39.6 °C and +59.9 °C with 0.1 °C resolution; • Relative humidity accuracy of± 3% and measuring range of 1–99% with 1% resolution.	Datalogger: Living room/Kitchen Transmitters: Bedrooms, Bathroom
Thermo-hygrometers (Testo AG, model Testostor 175-2)	• Temperature accuracy of ±0.9 °C and a temperature measuring range between −10 °C and +50 °C, with 1 °C resolution. • Relative humidity measuring ranges from 0 to 100%, with a resolution of 1%.	Outdoor
Multifunction climate measuring instrument with the IAQ probe for CO_2 and absolute pressure (Testo AG, Testo 435)	Probe for ambient CO_2: • Measuring range from 0 to 10,000 ppm. • Accuracy ± (75 ppm ± 3% of mv) (0 to +5000 ppm) ± (150 ppm ± 5% of mv) (+5001 to 10,000 ppm). Absolute pressure: • Measuring range from +600 to +1150 hPa. • Accuracy of ±10 hPa.	All rooms
Determination of radon content using a portable ATMOS 12 PDX sensor	Instrument: • Measurement operation (Temperature range from 0 to 50 °C; Humidity range from 0 to 90%). • Pulse counting ionisation chamber. • 10% standard deviation at 800 Bq/m^3 and 10 min measurement time. • Upper limit for radon gas content detection is 100,000 Bq/m^3; • Air pump for continuous flow of 1.4 l/min. Airflow through the chamber 1.0 l/min. • Memory with capacity for 28 days of time distribution and 20 energy spectra; • 10 min interval measurements (it allows 1, 5, 10, 30 min and 1, 8, 24 h).	Living room/kitchen

2.2. Long-Term Monitoring

Long-term monitoring was carried out to measure the indoor and outdoor air temperatures and relative humidity throughout the measurement period. For this, thermo-hygrometer sensors were installed in the most representative rooms and outdoors (Table 1). The measurements were carried out during different monitoring campaigns for all seasons, in compliance with specified procedures and standards (ISO 7726 [30], ISO 7730 [31], and ASHRAE 55 [32]). The monitoring campaigns were carried out for periods of at least 25 days and with the sensors recording data in periods of 30 min. Results on indoor environmental parameters were correlated with the outdoor parameters. During the measurement period, occupants filled an occupancy table where they recorded how they used the building, i.e., if they used the heating or cooling systems and promoted ventilation, among other effect. These occupancy records were useful to understand, for example, sudden changes in air temperature and relative humidity profiles. Local weather data was collected from the nearest weather stations.

3. Model of Thermal Comfort

An adaptive model of thermal comfort was used in the analysis of thermal comfort conditions since this is the adequate model for naturally conditioned buildings. The chosen model was the Portuguese adaptive model of thermal comfort, to be more representative of the Portuguese reality [33]. This model is an adaptation to the Portuguese context of the models specified in standards ASHRAE 55 [32,34] and EN 15251 [35]. It considers the typical climate and ways of living and how buildings are conventionally designed and used. According to this model [33]: (i) occupants may tolerate broader temperature ranges than those indicated for mechanically heated and/or cooled buildings; and ii) the outdoor temperature has a strong influence on occupants' thermal perception/sensation.

In the application of the proposed model to the case study, the following conditions were assumed: (i) the occupants have activity levels that result in metabolic rates (met) ranging from 1.0 to 1.3 met (sedentary activity levels); (ii) occupants are free to adapt their clothing for thermal insulation; (iii) air velocity below 0.6 m/s; (iv) indoor operative temperature between 10 °C and 35 °C; and (v) outdoor running mean temperature between 5 °C and 30 °C. The building has no air-conditioning system, or its use is sporadic, and, therefore, in the analysis of the case study, the adaptive model for building without mechanical systems was applied.

Considering that an individual takes approximately one week to be fully adjusted to the changes in outdoor climate, the thermal comfort temperature (indoor operative temperature, Θ_o) is obtained from the exponentially weighted running mean of the outdoor temperature during the last seven days (outdoor running mean temperature, Θ_{rm}). The calculation of the exponentially weighted running mean of the outdoor temperature in the previous seven days is done using Equation (1).

$$\Theta_{rm} = (T_{n-1} + 0.8T_{n-2} + 0.6T_{n-3} + 0.5T_{n-4} + 0.4T_{n-5} + 0.3T_{n-6} + 0.2T_{n-7})/3.8 \quad (1)$$

where:

Θ_{rm} (°C)—exponentially weighted running mean of the outdoor air temperature;

T_{n-i} (°C)—outdoor mean air temperature of the previous day (i).

In this model, two comfort temperatures ranges are defined, one to be applied in spaces with active air-conditioning systems and the other in non-air-conditioned spaces (which do not have air-conditioning systems or systems which are turned off). The operative temperature limits defined in this model are for 90% of acceptability, these limits are up to 3 °C above or below the estimated comfort temperature both for non-air-conditioned spaces ($\Theta_o = 0.43\Theta_{rm} + 15.6$) and air-conditioned spaces ($\Theta_o = 0.30\Theta_{rm} + 17.9$).

The operative temperature was calculated based on the results obtained in the measurements from the thermal microclimate station. With the operative temperature (Θ_o) and the outdoor running mean temperature (Θ_{rm}) is possible to represent in the adaptive chart, the point that characterises the thermal environment condition in the moment of measurement.

4. Description of the Case Study

4.1. Site and Climate

The case study is located in the old village of Granja do Tedo, in the municipality of Tabuaço, district of Viseu, Northern Portugal (Figure 2). The Granja do Tedo territory has an ancient history, with a rich medieval past and some archaeological remains dating back to the Romans (as the bridge over the river Tedo) [36]. The village is strategically implanted in the lower part of a valley, next to the confluence between the river Tedo (that flows to the river Douro) and of other two streams. The village is divided by the river in lower and upper parts (Figure 3). The implantation favours a good solar exposure from south (particularly in the upper part of the village located on a south-facing slope), and the surrounding mountains offer protection against the wind (Figure 4). The implantation in the valley also provides a more favourable microclimate, warmer than the one of the higher areas of the territory. Nearby, the available soils are good for agriculture [37]. At a geological level, the area is dominated by granitoids of different types and ages (Figure 5), confirming the abundancy of this resource and its use in the village as the primary building material [38].

Figure 2. Case study's location. (**a**) country context; (**b**) Granja do Tedo's urban layout.

Figure 3. Granja do Tedo. (**a**) Upper part; (**b**) lower part.

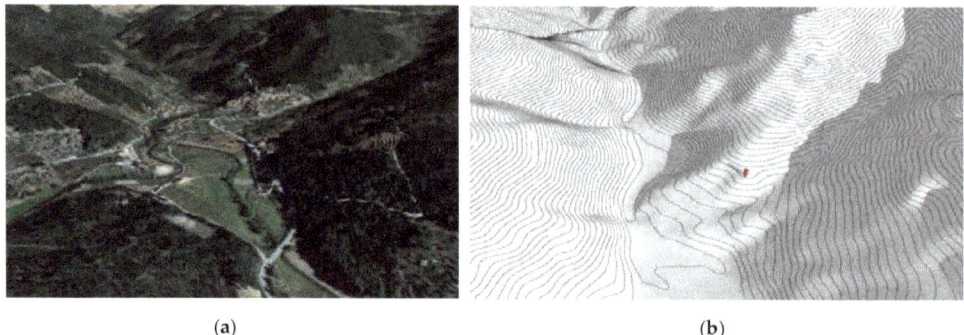

Figure 4. Granja do Tedo's context. (**a**) Aerial view with terrain relief (Google Earth); (**b**) Tridimensional model of the terrain showing the solar exposure at 9:30 a.m. on the winter solstice (case study location marked in red).

Figure 5. Geological map of Granja do Tedo area (adapted from [38]).

The village has a compact urban layout with narrow and winding streets, and most of the built area is implanted on a rocky massif (Figures 2 and 3), sparing the fertile agriculture land near the watercourses.

The village is mostly composed of two- and three-storey buildings, where the ground floor is commonly used to store goods and/or livestock, and the upper floors are for human occupancy. The wooden balconies (open or glazed) are frequent in the village. Due to sun exposure these were spaces used to dry grains and fruits and also for sewing. Additionally, like other constructions in regions with cold winters, buildings have very few and small openings to avoid heat losses. The compact layout and form also allow for reducing heat losses through the building envelope.

The Douro Valley region has a temperate climate—Type C, according to Köppen–Geiger Climate Classification, co-existing the sub-types Csa (temperate with hot and dry summer) and Csb (temperate with dry or temperate summer) (Figure 6a) [39]. Granja do Tedo is located in a narrow valley connected to the river Douro valley, and in the transition between the two climate subtypes [39]—The Csa in the valley and the Csb in the higher altitude areas. The annual average mean temperature is of 17.5 °C. The average mean temperature in winter is of 10.0 °C, while in summer is between 22.5 and 25.0 °C (Figure 6b,c) [39]. Winter is the harshest season in this area. Excluding the valley, the mean temperature in winter is of 7.5 °C. The average maximum air temperature in winter varies between 12.5 and 15.0 °C, while the average minimum air temperature is of 5.0 °C [39]. In winter, there are 10 to 20 days with a minimum temperature below or equal to 0 °C (Figure 6d), whereas the surrounding area has around 40 days [39].

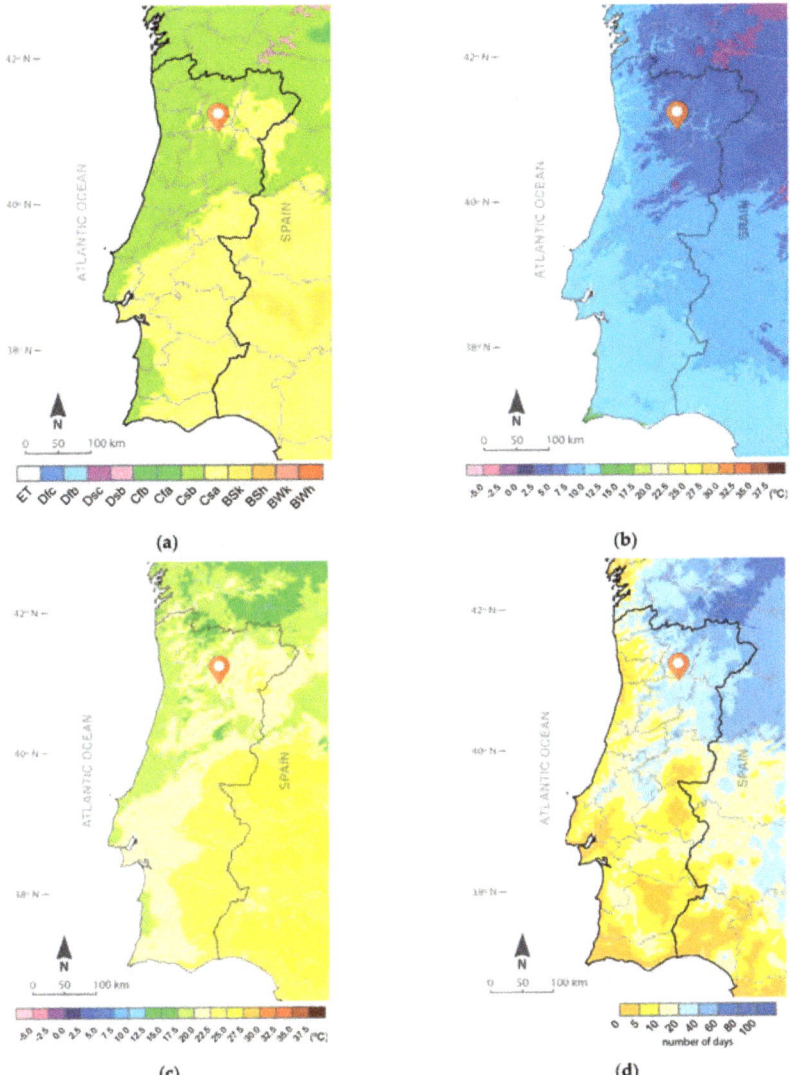

Figure 6. (**a**) Köppen-Geiger Climate Classification for Portugal; (**b**,**c**) Average mean temperature in winter and summer; (**d**) Average number of days with minimum temperature ≤0 °C in winter (adapted from [39]).

4.2. Building

The selected case study is a representative glazed-balcony building of Northern Portugal vernacular architecture [29], presenting a set of strategies to promote heat gains and reduce heat losses. The construction date is unknown, but considering the ages of neighbour buildings, and according to the owners, the case study is probably from the 18th century.

The building is a semi-detached single-family house, integrated into the urban mesh (Figure 7). It has an irregular floor plan and the main façade with the balcony is facing southwest, while the others are facing northeast, southeast, and west (Figure 8). As other constructions in regions with cold winters, and apart from the balcony that has the purpose of harvesting solar gains, the building has only two windows to avoid heat losses (one at the west, facing the street, a small one facing southeast and none at the north quadrant). The gross floor area is of approximately 50 m² divided into two floors.

Figure 7. External views. (**a**) southwest and southeast façades; (**b**) northeast and west facades.

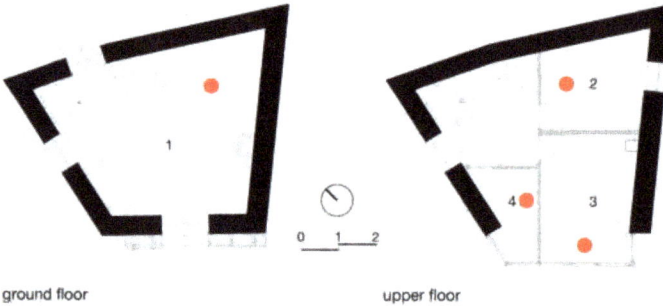

Figure 8. Floor plans showing the location of measuring instruments (1—living room/kitchen; 2—bedroom; 3—bedroom with balcony; 4—bathroom).

The building was renovated in 2005. During this intervention, some changes were introduced in the layout and use of some rooms. Some improvements were also implemented, such as the installation of a bathroom, renovation of windows and doors, the ground floor was paved, renovation of the timber balcony structure, and fitting thermal insulation to the ceiling. In the renovation, the ground floor was converted into a kitchen and living room (Figure 9a), and the upper floor layout was reorganized to accommodate two bedrooms and a bathroom (Figure 9b). In this modification of the floorplan, the partition wall of the balcony and other walls were removed to increase the floor area of the bedrooms and bathroom (Figure 9c).

Figure 9. (**a**) Kitchen view; (**b**) bathroom view; (**c**) bedroom with balcony; (**d**) closed wood-burning fireplace; (**e**) removable ventilation net; (**f**) smoke exhaust by the roof.

The building envelope consists of granite walls (50–55 cm thick) with a pitched roof, wooden doors, and wooden framed single glazed windows. Indoors, the partitions walls in *tabique* (earth-filled timber frame walls) were replaced by plasterboard walls. The ground floor is now paved with ceramic tiles, and the upper floor has a wooden floor with timber structure. Table 2 lists the thermal transmittance coefficient (*U*-value) of the building envelope. The building has no cooling system, and the heating system is a closed wood-burning fireplace (Figure 9d).

Table 2. Characteristics of the building envelope.

Envelope Element	Materials	*U*-Value (W/(m$^2 \cdot$°C)
External walls	Granite (50–55 cm)	2.87 [40]
Ceiling (in contact with ventilated roof)	Ceiling with timber structure with 4 cm of extruded polystyrene (XPS)	0.84 [41]
Doors	Wood	2.15 [41]
Windows	Wooden single glazed windows, indoor wooden shutters	3.40* [41]
Windows (balcony)	Wooden single glazed sash windows, indoor opaque curtains	4.30* [41]
Balcony (lower part)	Timber frame (double wooden panel) (10 cm)	1.70 [41]

* U_{wdn}—day–night thermal transmittance coefficient, including the contribution of the shading systems.

4.3. Passive Strategies

In the inland northern part of Portugal, to respond to a climate of harsher winter conditions and milder summers, vernacular architecture developed specific mitigation strategies. These had, in general, the purpose of increasing solar gains and reducing heat losses during winter, like the ones found in this case study:

- Balconies are an architectonic feature and identity of Northern Portugal vernacular architecture. It has to be taken into consideration that most of these buildings had low daylight levels and comfort conditions. Therefore, balconies were spaces used to enjoy the sun, work with daylight, and to heat the adjacent spaces, particularly on sunny winter days. The glazed balcony is an improved version of a balcony, that acts as a sunspace, allowing to harvest solar gains and reduce heat losses (Figure 9c). In the case study, the larger area of the balcony is facing southwest, with parts facing southeast and west. Therefore, in winter, the balcony is exposed to a higher solar radiation level during a larger number of sunshine hours. Although this strategy is aimed for the heating season, the cantilevered volume of the balcony and the possibility to keep windows open without compromising security also allows proper operation during the cooling season (Figure 9e), by shading the walls and promoting natural ventilation (Figure 10);

- To reduce heat losses, only a few windows (upper floor) face directly outdoors. In the original configuration of building, the balcony acted as buffer space and only some indoor rooms connected directly to the outdoors (Figure 9c); additionally, and although it was not possible to verify if it was the case of this building, sometimes to reduce heat losses by ventilation, buildings did not have chimneys and the exhaust of smoke was done through the roof, as it is still visible in a neighbouring building (Figure 9f);
- The use of high thermal inertia building elements, namely the massive granite walls and the massif rock where the building is laying, gives the building the capacity to stabilize indoor temperature;
- The functional arrangement of the indoor spaces in this type of buildings (as it was the case of this building before the renovation), can also reduce the heating needs. In this type of architecture, bedrooms rarely had exterior windows and were located next to the kitchen, taking advantage of the heat generated by the fireplace;
- The storage of the livestock on the ground floor was also a heating strategy. After the renovation, this strategy is mimicked by the closed wood-burning fireplace;
- The organic and compact urban layout, suited to the topography, can also be considered a passive strategy since the compactness of constructions allows to minimize the area of the envelope exposed to outdoor conditions and therefore reduce heat losses. The narrow and winding streets allow reducing wind speed, and in some places, the streets form small 'public-patios' sheltered from the prevailing winds (Figures 2 and 3).

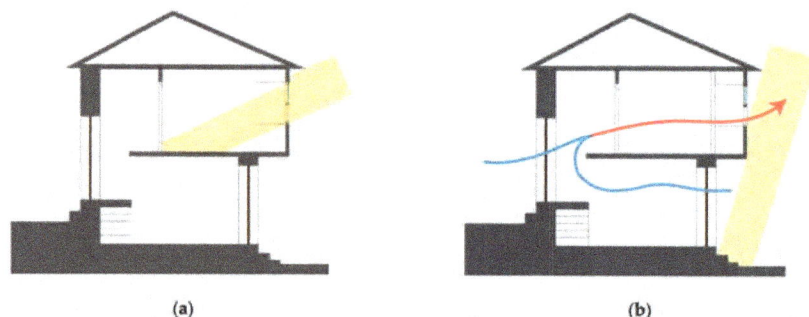

Figure 10. Schematic section of the glazed balcony operation. (**a**) Winter solstice; (**b**) Summer solstice.

The combination of all these passive strategies has the main purpose of achieving the better possible indoor thermal comfort conditions. The range of strategies highlights the poor living conditions and the need to understand and use the available resources the best way possible.

The dissemination of the abovementioned strategies in the region highlights their usefulness in mitigating the effects of the cold climate, as shown in previous studies [26,27]. Therefore, the quantitative study of the effectiveness of these passive strategies, particularly of the glazed balcony, and their effect on the thermal performance is useful to the discussion about the energy efficiency of buildings in this region. This is described and discussed in the following sections.

4.4. Occupancy Profile

It is essential to know the building occupancy profile since the daily occupants' habits have a direct influence on the thermal performance of the building [32]. The studied building is a holiday house, mainly used for weekends and holidays. During the summer period (vacations), it is occupied continuously during one or two months. The building is only used sporadically during the remainder of the year. Table 3 summarizes the main activities reported by the occupants (during the occupancy period) that may influence the thermal performance of the building. It is important to note that the building was unoccupied during most of the winter monitoring period.

Table 3. Building occupancy profile.

Season		Use and Description
Autumn	Heating/Cooling	The closed wood-burning fireplace was in operation.
	Ventilation	The windows remained closed.
	Shading	The curtains were usually opened in the morning (around 9:30 a.m.) and closed at night.
Winter	Heating/Cooling	The closed wood-burning fireplace was in operation from 6:00 p.m. until 12 p.m.
	Ventilation	Sporadic opening of windows for ventilation.
	Shading	The curtains were usually opened during the day and closed during the night.
Spring	Heating/cooling	No cooling system was used.
	Ventilation	Daily opening of the window for ventilation (8:30 a.m. to 6:30 p.m.).
	Shading	The curtains were usually opened during the day and closed during the night.
Summer	Heating/cooling	No cooling system was used.
	Ventilation	The windows were open day and night. Mosquito nets were placed in the windows to allow for ventilation during night time.
	Shading	The bedroom/balcony curtains remained open in the morning only until the direct sun passes through the window (around 1:00 p.m.).

5. Results and Discussion

5.1. Thermal Monitoring and Indoor Comfort Evaluation

The thermal performance monitoring included the assessment of the air temperature and relative humidity. Additionally, the indoor comfort conditions in the main rooms of the case study were characterized. These parameters were evaluated over one year, and data here presented are for 30 representative days of each season.

5.1.1. Autumn

During Autumn monitoring (from 8th November to 8th December 2014), the outdoor mean air temperature was of about 10.6 °C (Table 4). The daily maximum and minimum outdoor air temperatures had some variations during the monitoring period. In the second half of the monitoring period, starting from 23 November (Figure 11), these variations were more frequent and significative.

Table 4. Comparison between outdoor and indoor air temperatures and relative humidity values during autumn.

Place/Room	Outdoor	Kitchen/Living Room	Bedroom/Balcony	Bedroom	Bathroom
		Autumn			
		Temperature (°C)			
Mean	10.1	12.1	12.6	11.5	11.5
Maximum	24.6	14.3	18.9	15.2	16.4
Minimum	−0.3	9.2	6.5	8.5	6.6
		Relative Humidity (%)			
Mean	84.1	75.7	72.5	78.9	77.9
Maximum	96.8	79.0	79.0	82.0	85.0
Minimum	32.3	67.0	60.0	69.0	70.0

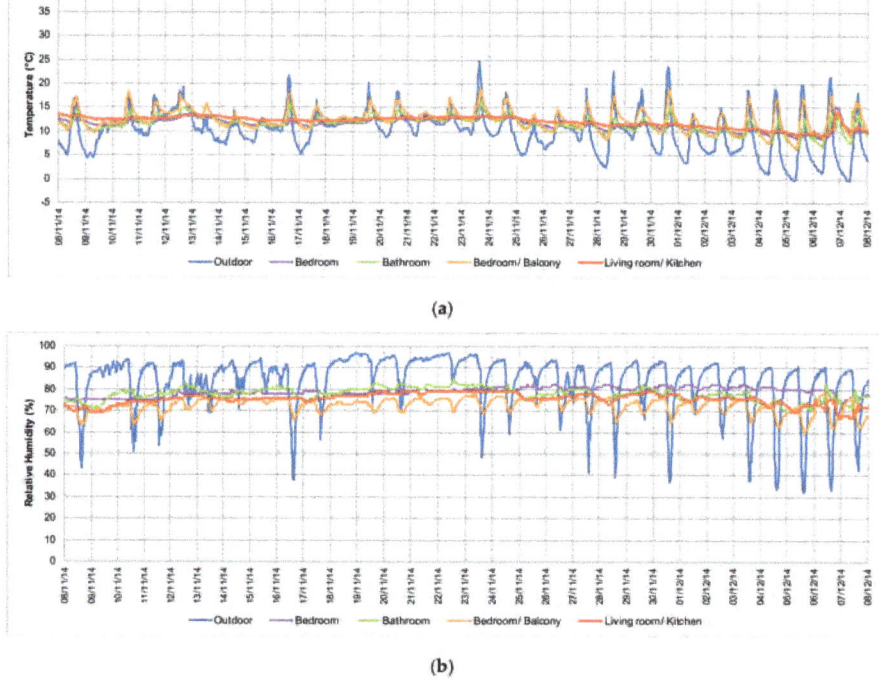

Figure 11. Autumn monitoring: (**a**) Indoor and outdoor air temperature profiles; (**b**) Indoor and outdoor air relative humidity profiles.

Figure 11a shows that indoor temperature remained stable in the rooms with a smaller glazing area, with a mean temperature of 12.1 °C in the living room/kitchen and 11.5 °C in the bedroom (Table 4). The reduced glazed area and the high thermal inertia of the building envelope allow stabilization of the indoor temperature in these rooms. On 7 December, when the outdoor temperature reaches a minimum value of 1.2 °C, it is possible to observe how building occupants can take correcting measures to improve the indoor thermal comfort conditions. The increase of the indoor temperatures in the living room/kitchen and bedroom (Figure 11a) is due to the use of the heating system (closed wood-burning fireplace). According to Table 4 and Figure 11a), in these rooms, the maximum temperatures were always below the comfort temperature range.

In the rooms where the glazing area is predominant, bedroom/balcony and bathroom, it was observed that the indoor temperature was not stable as it is strongly dependent on the outdoor climate conditions. The maximum temperature recorded in the bedroom/balcony was of 18.9 °C while in the bathroom was of 16.4 °C (Table 4). In these rooms, during the day, the indoor temperature followed the trend of the outdoor temperature (Figure 11a). The temperature profiles in both rooms were quite similar, but since the bedroom/balcony has a larger glazed area than the bathroom, it presented higher temperatures. The bedroom/balcony had the highest indoor temperature throughout the monitoring period, reaching temperatures close to the comfort threshold temperature. These results highlight the effect of the glazed balcony as a strategy to capture solar gains.

Concerning the outdoor relative humidity, it was found that there was a high daily variation, reaching values of around 90% during the night and lower figures of 32.3% during the day (Table 4). The average outdoor relative humidity value was also high, being 84.1% during the monitoring period (Table 4). In contrast, almost all indoor rooms had stable relative humidity profiles with small daily variations. The exception was the bedroom/balcony, where the fluctuations were slightly higher than the other rooms, due to higher solar radiation, but much lower than the variations outdoors. The indoor relative humidity values were high (about 70–80%), higher than those recorded outdoors during the day, but smaller than those verified outdoors during the night. The reduced ventilation rate of the rooms, due to the lack of occupancy, might be the main reason for the high indoor relative humidity levels. During the occupancy period (from 7 to 8 December, 2014) there was a slight decrease in the relative humidity level in the living room and bedroom (Figure 11b) due to the use of the heating systems. However, due to the low outdoor temperatures, the ventilation was minimized to reduce heat losses.

Regarding the assessment of the thermal comfort, the measurements in the living room/kitchen and bedroom/balcony were carried out when the heating system was not used. The influence of the curtains on the thermal comfort in the bedroom/balcony was also evaluated. In autumn and without the use of the heating system, the results showed that the thermal comfort conditions in the living room/kitchen were below the lower comfort limit (Figure 12a). In the survey, the two inhabitants answered as being "slightly cool" (1.0 met; 0.91 clo) and one as being "cool" (1.0 met; 0.95 clo), confirming the objective measurements. In what concerns the assessment of the thermal comfort conditions in the bedroom/balcony, it was possible to verify the influence of the glazing area. In this room, when the curtains were closed, the comfort conditions were within the thermal comfort limits, but close to the bottom threshold (blue dot in Figure 12b). In the survey, the two occupants answered as being 'neutral' (comfortable) (1.0 met; 0.91–0.95 clo), i.e., results were in line with the objective assessment. When the curtains were open, the solar gains increased the operative temperature, and thermal conditions were above the upper thermal comfort threshold, showing an overheating period (red dot in Figure 12b).

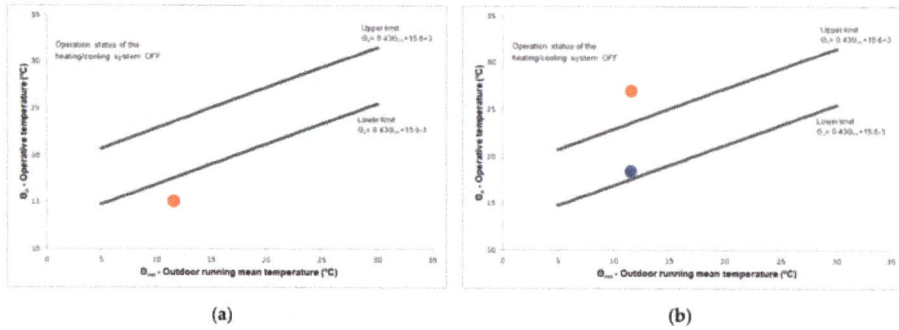

Figure 12. Adaptive comfort chart during a representative autumn day: (**a**) Thermal comfort temperature (operative temperature) in the living room/kitchen; (**b**) Thermal comfort temperature (operative temperature) in the bedroom/balcony for open curtains (red dot) and closed curtains (blue dot).

5.1.2. Winter

The winter monitoring was carried out between 27 December 2014 and 27 January 2015. In this period, the minimum outdoor temperature was very low, reaching a minimum value of −4.0 °C (Table 5), being around 0 °C most of the days. The maximum outdoor temperature reached 20.9 °C (at the end of the monitoring period), and mean temperature did not exceed 4.6 °C.

Table 5. Comparison between outdoor and indoor air temperatures and relative humidity values during the winter.

	Winter				
Place/Room	Outdoor	Kitchen/Living Room	Bedroom/Balcony	Bedroom	Bathroom
	Temperature (°C)				
Mean	4.6	6.4	7.4	6.0	6.1
Maximum	20.9	8.0	15.7	8.2	12.8
Minimum	−4.0	5.2	3.0	4.2	3.1
	Relative Humidity (%)				
Mean	77.8	75.5	68.8	79.4	74.5
Maximum	95.2	80.0	76.0	83.0	85.0
Minimum	14.7	68.0	58.0	77.0	63.0

From the analysis of Figure 13a, it is possible to conclude that the living room/kitchen and the bedroom (the rooms with smaller glazed area and not in contact with the glazed balcony), showed a stable profile with low daily thermal variation and a mean temperature of 6.0 °C and 6.4 °C, respectively (Table 5). Beyond the reduced glazed area, the thermal inertia of the envelope is the main reason for this steady behavior. The fact the building was not occupied during this period of the monitoring campaign explains the lower temperature values and their uniformity during the period, since there was no human action to achieve thermal comfort conditions (i.e., active heating to increase the indoor temperatures). Although considerably below the comfort limits, even in a free-running mode, it has to be highlighted that indoor mean temperature was always higher than outdoors.

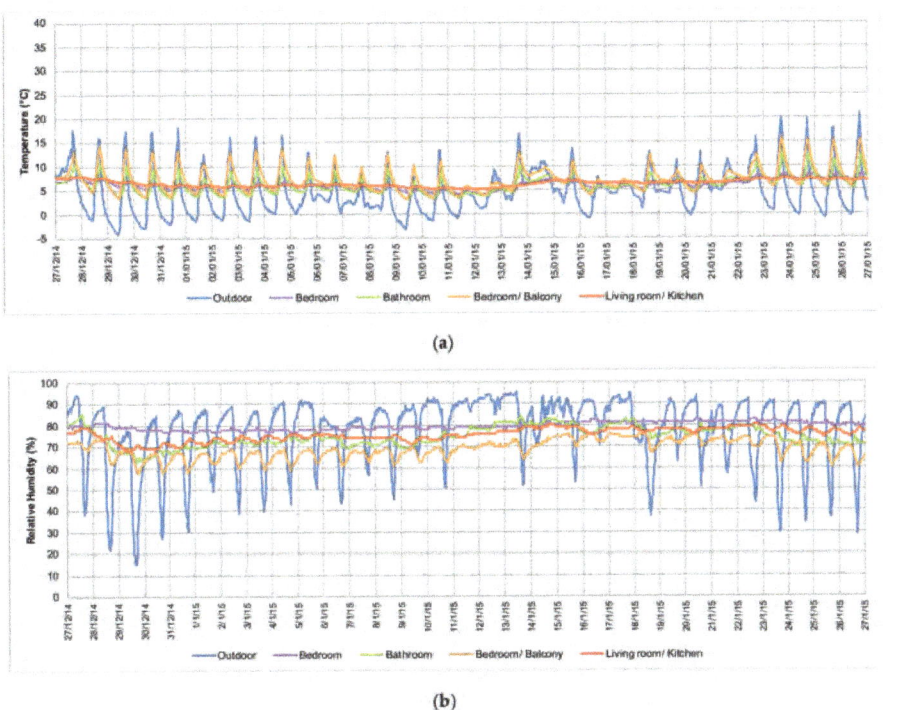

Figure 13. Winter monitoring: (**a**) Indoor and outdoor air temperature profiles; (**b**) Indoor and outdoor air relative humidity profiles.

Both in the bedroom/balcony and the bathroom, it was observed that temperature profiles were not stable and followed the outdoor temperature variation during the day (Figure 13a). The maximum indoor temperature recorded was of 15.7 °C in the bedroom/balcony and 12.8 °C in the bathroom, both in days with higher outdoor temperatures. Due to the large glazed area of the balcony, the effect of sunny days is visible in temperature peaks close to the thermal comfort boundary, even with this strategy not being used with full potential, since the opaque curtains were closed and therefore part of solar radiation was reflected. Consequently, in days with more incident radiation and if the curtains were open, it was expected that temperature would reach or be much closer to the comfort boundaries (similar to the condition measured during autumn and shown in Figure 12a, where active heating was only necessary as a backup). Nevertheless, there is also a drawback resulting from the greater glazed area, since these rooms also have more heat losses and therefore the minimum temperature recorded is lower than in the ground floor (Table 5). Moreover, the lack of thermal mass to store the heat gained during the day is a disadvantage, since the rooms have lightweight wooden floor and walls (as the original earth-filled timber frame walls—*tabique*—were replaced by plasterboard walls).

Considering that the glazed area is an important strategy to harvest solar gains, it was expected that these rooms had temperatures close to the comfort conditions, but mean temperature during the monitoring period was very low (7.4 °C) (Table 5). The non-occupation of the building and the use of the internal shading curtains during all monitoring period are the aspects that explain this behaviour.

Additionally, the temperature differences between ground and upper floors show how well the functional distribution of the rooms was before the building renovation. Originally, the ground floor was designed for storage and not for human occupancy, and in that case the stable and warm temperatures were an advantage.

The outdoor relative humidity had significative daily variation, reaching values close to 96% during the night and a minimum of 14.7% during the day (Table 5). The average relative humidity of 77.8% is also high (Table 5). In general, the rooms have stable relative humidity profiles with little daily fluctuations. Rooms with smaller glazing area are the ones with the most stable temperatures. The bedroom/balcony showed the highest daily variation among the studied rooms, of about 8.0%, being most of the monitoring period between 60 to 70%.

Regarding the assessment of the thermal comfort (Figure 14), the measurement of the thermal environment conditions was performed during a typical winter day, in the bedroom/balcony and in the living room/kitchen. In the living room/kitchen, the analysis was carried out for two situations: (i) when the heating system was not in operation (Figure 14a); and (ii) when the heating system was in operation (Figure 14b). Results showed that when the heating system was not in operation the thermal environment was very uncomfortable (Figure 14a). The influence on the thermal comfort of using the closed wood-burning fireplace is quite evident, since when the heating system was in operation, the living room/kitchen had a comfortable thermal environment (Figure 14b). In the survey, the occupants also expressed their thermal sensation for the same two situations. When the heating system was not in operation, one occupant (1.0 met; 1.48 clo) answered as being "cool" and the other (1.0 met; 0.92 clo) as being "cold". When the heating system was in operation, one occupant (1.0 met; 1.48 clo) answered as being "neutral" and the other (1.0 met; 0.92 clo) as being "slightly cool". These results confirm the ones from the objective measurements. The differences between the answers of the two occupants are related to the different clothing insulation levels, which influenced their thermal sensation.

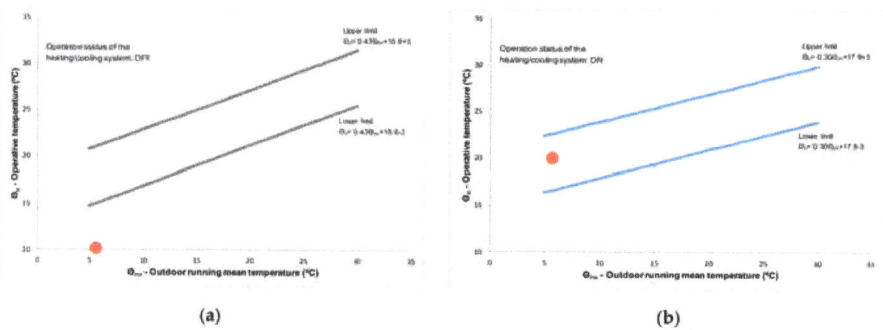

Figure 14. Adaptive comfort chart during a representative winter day: (**a**) Heating system OFF—Thermal comfort temperature (operative temperature) in the living room/kitchen; (**b**) Heating system ON—thermal comfort temperature (operative temperature) in the living room/kitchen.

In the bedroom/balcony, the measurements were carried out only when the heating system was not in operation. The thermal comfort conditions in this room were outside the comfort boundaries (Figure 15). Although the operative temperature was outside the comfort limits, it was very close to the lower comfort threshold. It is likely that the regular building occupation and, consequently, the appropriated use of the glazed balcony, would lead to an operative temperature within the comfort limits. It must be stressed that during the measurements, the sky was cloudy and thus solar gains were very low. In the survey, the two occupants answered as being "slightly cool" (1.0 met; 0.92–1.48 clo), which confirms the objective measurements.

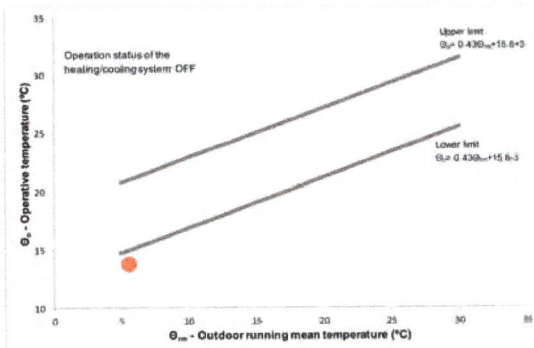

Figure 15. Adaptive comfort chart during a representative winter day. Thermal comfort temperature (operative temperature) in the bedroom/balcony.

5.1.3. Spring

During this monitoring campaign (carried out from 14 April to 14 May 2015), the outdoor mean air temperature was of about 16.0 °C, the maximum temperature was often below 20.0 °C, and the minimum values varied between 5.0 °C and 10.0 °C (Table 6 and Figure 16). The outdoor air temperatures had significative daily variations during the period, with maximum and minimum values having a slight increment in the last days of the period (Figure 16). The maximum temperature recorded was 34.2 °C, while the minimum was below 4 °C (Table 6).

Table 6. Comparison between outdoor and indoor air temperatures and relative humidity values during the spring.

	Spring				
Place/Room	Outdoor	Kitchen/Living Room	Bedroom/Balcony	Bedroom	Bathroom
	Temperature (°C)				
Mean	16.0	15.2	18.1	17.2	17.9
Maximum	34.2	19.2	28.9	24.0	28.7
Minimum	3.8	13.2	11.0	13.5	12.4
	Relative Humidity (%)				
Mean	65.9	70.3	59.6	67.4	60.4
Maximum	92.8	78.0	72.0	77.0	74.0
Minimum	11.3	62.0	46.0	47.0	43.0

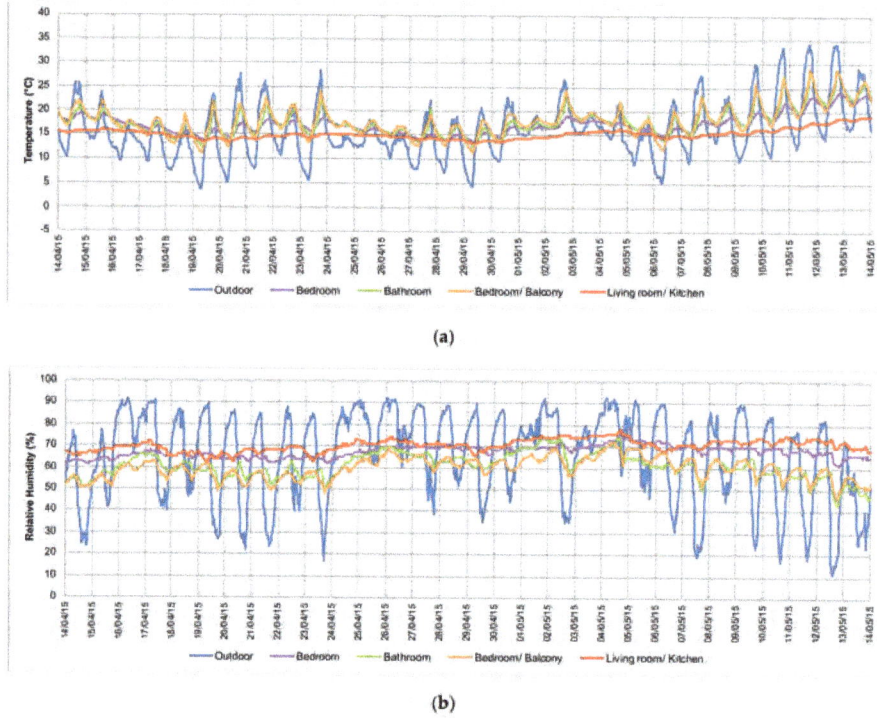

Figure 16. Spring monitoring: (**a**) Indoor and outdoor air temperature profiles; (**b**) Indoor and outdoor air relative humidity profiles.

In the spring, it was observed a relevant difference between the indoor air temperatures of the rooms located on the ground floor and those on the upper floor. Concerning the ground floor, the living room/kitchen had a very stable indoor temperature during the monitoring period, with a mean air temperature of 15.2 °C (Table 6). In the upper floor, it was observed that the indoor temperature was less stable, particularly in the rooms in the glazed balcony. The increase in the outdoor temperature and the number of hours of solar radiation had a strong influence on the temperature of these rooms. The bedroom had a more stable temperature profile, since it has fewer solar gains through the windows and higher thermal inertia due to the granite walls. The bedroom/balcony had the highest indoor temperature in the building during the monitoring period. The maximum temperature in the bedroom/balcony always remained below the outdoor temperature, since during the monitoring period, the curtains were closed. Nevertheless, from May onwards, when the outdoor temperature begins to rise, closing the curtains is the right decision to reduce solar gains. However, since the glazed area is protected by an inside shading device (opaque curtains), it is difficult to avoid overheating both in the bedroom/balcony and in the bathroom, as shown in Figure 16a).

Regarding the outdoor relative humidity, it was found that there is a high daily fluctuation, reaching values near 93% during the night and minimum values of 11.3% during the day (Table 6). Indoors, the values were stable, with daily variations around 10%. The bedroom/balcony and the bathroom showed higher daily variation, and the mean relative humidity was of around 60% (Table 6). The relative humidity is within the recommended levels for human health and comfort [35]. The living room/kitchen and the bedroom also had a very stable relative humidity profile, with mean values around 70% (Table 6).

The thermal comfort assessment was carried out both in the bedroom/balcony and the living room/kitchen, without the heating system in operation. From the analysis of the adaptive comfort charts, it is possible to conclude that the thermal comfort conditions in the living room/kitchen are below the lower comfort limit (Figure 17a), even with an operative temperature of 18.9 °C and an outdoor running mean temperature above 20 °C. The low heat gains and mainly the high thermal inertia of the envelope are the main factors affecting these results. In the survey, the two occupants (1.0 met; 0.44–0.58 clo) answered as being "slightly cool".

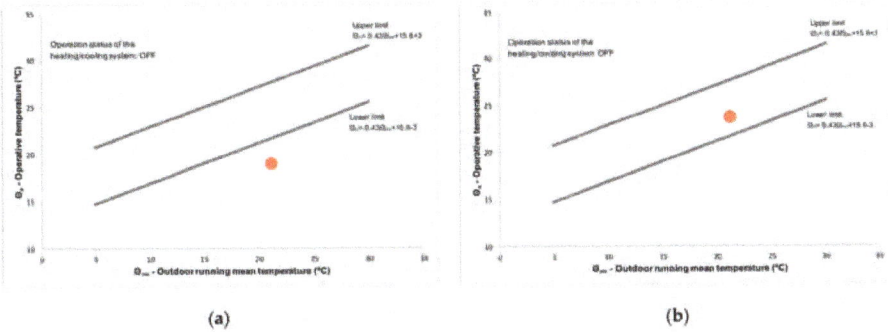

Figure 17. Adaptive comfort chart during a representative spring day: (**a**) Thermal comfort temperature (operative temperature) in the living room/kitchen; (**b**) Thermal comfort temperature (operative temperature) in the bedroom/balcony.

In contrast, the bedroom/balcony had a thermal condition within the comfort range. The operative temperature was higher than in the ground floor due to the heat gains provided by the glazed balcony. In the comfort survey, the two occupants (1.0 met; 0.44 and 0.58 clo) answered as being "neutral", which confirms the measurements.

5.1.4. Summer

The summer monitoring was carried out from 18 July to 18 August 2015. In this period, the mean outdoor temperature was of 24 °C, there was a high daily thermal amplitude, with several days reaching most of the time maximum values around 35.0 °C (and a peak of 39.1 °C), and minimum values around 15.0 °C (Table 7).

Table 7. Comparison between outdoor and indoor air temperatures and relative humidity values during the summer.

		Summer			
Place/Room	Outdoor	Kitchen/Living Room	Bedroom/Balcony	Bedroom	Bathroom
		Temperature (°C)			
Mean	23.7	24.1	26.8	26.8	27.1
Maximum	39.1	26.2	35.0	31.0	35.2
Minimum	12.4	21.4	19.6	22.7	21.5
		Relative Humidity (%)			
Mean	54.1	51.8	46.0	48.1	46.5
Maximum	89.4	63.0	64.0	60.0	65.0
Minimum	13.8	35.0	27.0	30.0	28.0

From the analysis of Figure 18, it is possible to conclude that the living room/kitchen had the most stable temperature profile, with a mean temperature of 24.1 °C (Figure 18 and Table 7). This is due to the higher thermal inertia and lower direct solar gains of the room. As mentioned before, this room was initially for storage and thus, during summer, it had the advantage of keeping the temperature stable. In its current use, during summer it is the room with the best thermal comfort conditions.

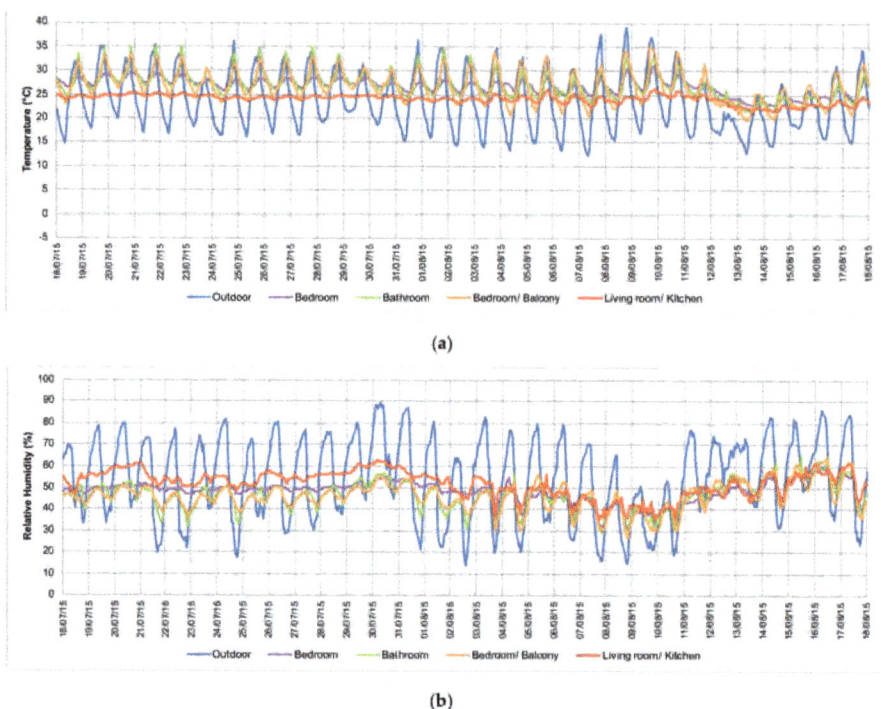

Figure 18. Summer monitoring: (**a**) Indoor and outdoor air temperature profiles; (**b**) Indoor and outdoor air relative humidity profiles.

In the upper floor, the bedroom is the room with the most stable air temperature profile with slight day to night temperature variations (usually around 3 °C). The reason for the small differences in this room can be related to the higher thermal inertia than the other rooms in the upper floor. Nevertheless, when it was unoccupied, and therefore without ventilation, the maximum air temperature in this room was around 30 °C.

Regarding the rooms in the balcony, as in the seasons previously presented, it was observed that the indoor temperature had significant daily variations. In these rooms, indoor temperature follows the outdoor temperature profile during the day (due to both solar gains and heat losses through the glazing area). The minimum mean indoor temperature stabilizes around 25 °C, while the minimum outdoor temperature was usually 10 °C lower (Table 7 and Figure 18). The larger glazed area of these spaces, facing southwest, is the reason for these rooms having higher temperatures due to the solar gains.

In this season, the building was occupied during the entire month of August. From the moment the building began to be occupied, it was expected that the promotion of natural ventilation would change the indoor temperature profile, but this is not noticeable in the graphs (Figure 18a). The maximum temperature in the rooms remained similar (Figure 18a) since the flow of warmer air from the outdoors into the building during the day does not favor its cooling. On the other hand, during the night, the minimum temperature slightly dropped due to the ventilation since the outdoor air temperature was lower during the night. During the occupation period, the inhabitants closed the curtains during the morning to avoid solar gains (usually until 2 p.m.). However, since there are no external shading devices in addition to the fact that windows were kept open for ventilation, it is difficult to control the solar gains through the glazed area of the balcony.

Nonetheless, the airflow in the building can improve occupants' thermal sensation by increasing convective heat losses from their bodies. The most-recommended solutions to avoid solar gains in the cooling season are to use an external shading device and to use night ventilation to remove diurnal thermal loads. At this point, it is worth mentioning that if the balcony had its original configuration (i.e., if it was a space separated from the indoor rooms by a wall) it would influence in a very positive way the thermal behavior of the building during this season. The reasoning for this is that it would act as a buffer space between outdoor and indoor rooms and would work as a shading device of the openings that existed in the demolished wall.

Regarding the outdoor relative humidity, it showed significant daily variations, with maximum values around 70–80% and sometimes near to 90% during the night, and minimum values varying from near 40% to minimum values of 14% during the day (Table 7). The mean value is of around 55% (Table 7). The indoor relative humidity has lower daily variations, being relatively stable (Figure 18b). The rooms with the most stable relative humidity profiles are the living room/kitchen and the bedroom. In general, the relative humidity decreases during the day due to the warmer dry air and increases during the night due to the cooler outdoor humid air that flows into the building. This is particularly visible in the rooms with the balcony, where daily variations are higher.

The period of occupation (starting on 4 August) influenced indoor relative humidity profiles, increasing the daily humidity variation, even in rooms with stable profiles. This reduction in relative humidity values is related to the ventilation and circulation of hot air from outdoors. The relative humidity slightly raised during some rainy days and then decreased again.

In the thermal comfort assessment, the living room/kitchen and the bedroom/balcony showed a thermal environment within the comfort range (Figure 19a). The operative temperature in the living room/kitchen is more stable due to the higher thermal inertia, and therefore this room had a better thermal condition during the summer. The results of the survey confirmed the measurements, since the two occupants (1.0 met; 0.27–0.43 clo) answered as being "neutral".

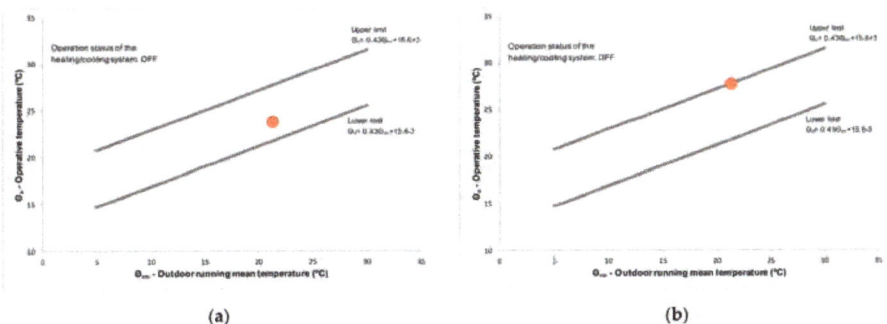

Figure 19. Adaptive comfort chart during a representative summer day: (**a**) Thermal comfort temperature (operative temperature) in the kitchen/living room; (**b**) Thermal comfort temperature (operative temperature) in the bedroom/balcony.

In the bedroom/balcony, the operative temperature was close to the upper comfort limit, mainly due to the solar gains through the glazed envelope (Figure 19b). In the survey, one occupant (1.0 met; 0.43 clo) answered as being "slightly warm" and the other (1.0 met; 0.27 clo) as being "neutral". The difference between the answers is mainly related to different clothing insulation levels.

5.2. Indoor Air Quality Monitoring

5.2.1. Carbon Dioxide Concentration

In this section, the carbon dioxide (CO_2) concentration in the case study is evaluated and classified according to the categories defined by EN 15251 [35]. The CO_2 concentrations were measured in different rooms during a representative day of each season. During the winter, the measurements were carried out in two situations, i.e., with and without the closed wood-burning fireplace in operation, to verify the influence of the fireplace use in the CO_2 concentrations. From the measurements, it was verified that the use of the fireplace slightly increased the CO_2 concentrations, but they did not exceed the design values for category I (high level of expectation) (Table 8). The small differences between outdoor and indoor carbon dioxide concentrations are due to the low occupation density of the building, to the natural ventilation and infiltration rate, and to the efficiency of the closed fireplace exhaust system. In the records, two values correspond to category III. A possible explanation for this situation is that those two rooms were closed until the beginning of the measurements, and therefore the CO_2 concentrations were higher. Although the case study is an old building, the results showed that the CO_2 concentrations are, most of the time, within the boundaries of the most demanding category.

Table 8. Classification of indoor air quality in representative rooms.

Season		Place/Room	Concentration (ppm)	Difference above Outdoor	Category *	Pressure (hPa)
Autumn		Outdoor	496	-	-	975.3
		Kitchen/Living room	797	301	I	
		Bedroom/Balcony	725	229	I	
		Bedroom	1210	714	III	
		Bathroom	686	190	I	
Winter	Heating OFF	Outdoor	450	-	-	974.7
		Kitchen/Living room	589	139	I	
		Bedroom/Balcony	915	465	II	
		Bedroom	596	146	I	
		Bathroom	641	191	I	
	Heating ON	Kitchen/Living room	725	275	I	-
		Bedroom/Balcony	642	192	I	
		Bedroom	730	280	I	
		Bathroom	720	270	I	
Spring		Outdoor	483	-	-	982.8
		Kitchen/Living room	620	137	I	
		Bedroom/Balcony	492	9	I	
		Bedroom	555	72	I	
		Bathroom	560	77	I	
Summer		Outdoor	405	-	-	977.4
		Kitchen/Living room	680	275	I	
		Bedroom/Balcony	610	205	I	
		Bedroom	520	115	I	
		Bathroom	480	75	I	

* classification according to EN 15251 standard.

5.2.2. Radon Gas Concentration

The concentration of carbon dioxide is a good indicator of air quality in buildings where occupants are the main source of pollution. However, since the building is located on a granitic area, it is also necessary to measure the radon gas concentration [42]. Radon—without color, odor, or taste—results from the decay of the radium and is found in rocks and soils, as in the granitic massif where the building is located. Its infiltration in buildings generally takes place through the foundations. The high concentration of radon in the environment has health risks, since the element is lodged in the lungs by inhalation and its main effect is the lung cancer (risk potential increases in about 16% for each 100 Bq/m^3 in long-term average radon concentration) [43]. According to the World Health Organization

(WHO), radon is the second leading cause of lung cancer, after smoking in smokers, and the first among those who have never smoked [43]. Directive 2013/59/EURATOM [44] states that the reference level for the annual average concentration of activity in the air should not exceed 300 Bq/m^3 per year in new construction homes and workplaces, whose approximate equivalence is 10 mSV annual, according to recent calculations by the International Radiological Protection Community [44]. In the Portuguese context, and according to national legislation [42], it is mandatory to study and measure the concentrations of radon in granitic sites, as the one where the case study is located.

In the case study, the concentration of radon was measured during the heating season, when the ventilation rate was lower. The living room/kitchen was the room chosen for the measurements since it has the lower ventilation rates, has granite walls, is located in the ground floor, and it sits on a granitic massif. The measurements took place during 28 days, with integration periods of 10 min, started after a period of stabilisation of the radon sensor (about 60 days). Figure 20 shows the results of the measurements. It is possible to see an irregular distribution of values with several peaks. The peaks in the radon concentration are considerably above the maximum defined by the Portuguese law (400 Bq/m^3) [42], with a maximum of 2660 Bq/m^3, and an average concentration of 1432 Bq/m^3. Although the concentration of radon was high, it has to be taken into consideration that the building was unoccupied most of the time and thus had low ventilation rates.

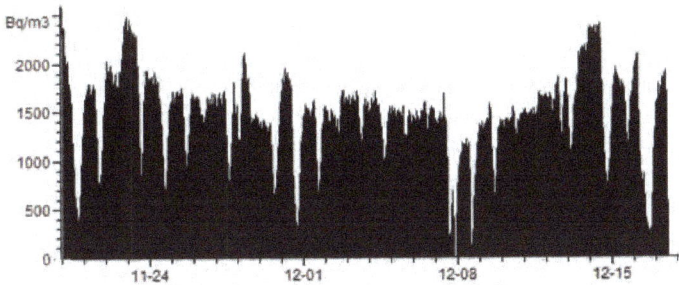

Figure 20. Concentrations of radon in the living room/kitchen during the winter period.

During a short period of occupation, 7 and 8 December, even with low ventilation rates (windows and doors were only open sporadically), as it was winter, the concentration of radon sharply fell to values below 300 Bq/m^3, as recommend by the Directive 2013/59/EURATOM [44]. Although the air change rate of the case study building was not measured, results show that the way the occupants use the building is sufficient to maintain the radon concentrations within the mandatory values. Therefore, ventilation must not be neglected in this type of buildings, particularly after renovations when the airtightness of the envelope increases due to the replacement of windows and doors, and no other measures are implemented to mitigate the ingress of radon into the building. In buildings located in granitic areas, it is necessary to maintain a minimum hourly air change rate to remove radon or to renovate the ground floor, by introducing, for example, a waterproofing membrane that does not allow the flux of radon gas from the ground to the indoor environment. In the renovation of this case study, measures to prevent the ingress of radon gas into the building were not introduced and therefore ventilation is the only way to control the radon gas concentration.

5.3. Conditions and Limitations of the Study

The outcomes of this study are based on the analysis of the annual thermal behavior of just one case study, since in this region of Portugal it was not possible to identify other case studies in good conservation conditions or that are still occupied. Nevertheless, this building is representative of the typical glazed balcony vernacular buildings of Northern Portugal and Spain, and has the typical functional organization of this type of house, with the ground floor used to store goods and/or livestock (coldest part of the building) and the occupied area on the first floor (part of the house with higher solar gains and comfort levels). In the case study building, the glazed balcony is the focus of the study, and it is representative of this type of architecture, due to its size and orientation. Additionally, the case study was refurbished and therefore presented good conditions to carry out the research.

Since it was the only case study available to evaluate, it was not possible to statistically analyze and compare the behavior of this building with similar ones in the same region.

The results presented are specific to this zone due to the particular type of climate. Nonetheless, the benefits of the glazed balconies can be extrapolated to other areas with similar climates and not only to buildings with similar characteristics.

Another limitation of this study is the fact that the building is a vacation house that is only used during weekends and holidays. Since this building is mainly used for short periods, especially during winter, it is possible that some of the inhabitants' potential actions to improve the indoor environmental quality were not fully addressed.

6. Conclusions

The results of the research work presented showed the viability of using glazed balconies as a passive heating strategy in a climate with cold winters. This type of building is common in the North of Portugal, in the North of Spain, and in other regions where passive principles (as the glazed balcony) are implemented in buildings to increase solar gains during the heating season.

The glazed balconies act as a sunspace, increasing the contribution of solar gains in the maintenance of the thermal comfort conditions during the heating season. In Portuguese vernacular architecture, these elements are normally well oriented, and there is a proportional relationship between their dimensions and the ones of the adjacent rooms. Glazed balconies are always on the upper floors, for better sun exposure, and are adjacent to living spaces (usually living rooms and bedrooms).

During occupation period of the mid-seasons, the rooms in the balcony had adequate comfort conditions, since the occupants can easily control the solar gains using the shading system (opaque curtains). Not controlling the solar radiation increases the risk of overheating periods, as seen during autumn (when the building was not occupied).

In winter, the results showed that it is difficult to achieve adequate thermal comfort conditions without an active heating system. Nevertheless, during the thermal comfort assessment, performed on a cloudy day, the operative temperature was close to the lower limit of the thermal comfort range. Even when the building was in free-running mode, that was the case for most of the monitoring period, during sunny days and even with the solar shading active, it was possible to verify that the indoor air temperature increases considerably.

During summer, the results showed that the thermal comfort conditions are within the comfort limits, but with some risk of overheating. The use of an external solar shading device will be more effective to reduce the risk of excessive solar gains and overheating during summer than the existing curtains.

From the results presented, it was possible to compare periods with and without occupation, which highlights the importance of the occupants' actions in optimising the solar gains through the glazed balcony and, therefore, regulate their comfort conditions by activating/deactivating solar shading and promoting natural ventilation (useful to remove air pollutants and heat loads—particularly during night-time).

Since the glazed balcony is the main passive strategy in this building, it is important to note that by removing the partition wall between the glazed balcony and the other rooms, the original buffer zone was eliminated. The removed *tabique* wall thermal inertia was also useful, both in winter and summer, to keep the indoor temperature more stable. The balcony would also act as a sunspace in winter, increasing the solar heat gains, and as a buffer space, reducing heat losses. In summer, with the windows open, the glazed balcony will work as a shading device for the building walls.

The floor area of this kind of building is small for current living standards and therefore, during refurbishment operations, the partition walls between the rooms and the glazed balcony were removed to increase the net floor area. Additionally, the traditional materials were replaced by modern industrial materials (e.g., aluminium, steel, and plasterboard). The lack of knowledge on the advantages of using this passive strategy is resulting in the destruction of this vernacular technique that is one of the architectonic identities of Northern Portugal vernacular architecture. Hence, during the renovation of this type of building, it is necessary to take into account the balance between the functional needs of the spaces and the effectiveness of existing passive strategies in order to harmonize them.

Additionally, further studies are needed to complement and corroborate the research presented, to understand better the effectiveness of this strategy, and to disseminate its advantages on improving thermal comfort conditions and reducing the energy needs for heating. Moreover, it is necessary to promote its use in new buildings, since the benefits have also already been discussed in other studies.

Regarding indoor air quality, even after a renovation where the airtightness of the envelope was improved, the concentrations of carbon dioxide in the building did not exceed the most demanding design values for new buildings, according to EN 15251, even when the closed wood-burning fireplace was in operation. The measurements of the radon gas concentrations conducted during a long period without occupation showed average values above the maximum defined by national legislation. During the occupation period and even with low ventilation rates, the radon gas concentration rapidly decreased to acceptable values, thus not harming the occupants' health. Nevertheless, the need to maintain a minimum hourly air change rate to remove air pollutants and assure a healthy indoor environment must be emphasized.

Author Contributions: J.F. undertook the main part of the research that was the base of this article. He developed the research method and analyzed the results with the contribution of R.M. (Raphaele Malheiro) and M.d.F.C., who wrote the document with the input of S.M.S. and R.M. (Ricardo Mateus). S.M.S. and R.M. (Ricardo Mateus) helped to develop the discussion sections of the paper and provided critical judgment on the undertaken research. Additionally, they supervised all the works and revised the document. H.G. helped in the preparation of the article. All authors have read and agreed to the published version of the manuscript.

Funding: The authors would like to acknowledge the support granted by the FEDER funds through the Competitively and Internationalization Operational Programme (POCI) and by national funds through FCT (Foundation for Science and Technology) within the scope of the project with the reference POCI-01-0145-FEDER-029328, and of the Ph.D. grant with the reference PD/BD/113641/2015, that were fundamental for the development of this study.

Acknowledgments: The authors also wish to thank José Pombo and his family, and Tabuaço's Municipality for helping this research work.

Conflicts of Interest: The authors declare no conflict of interest. The funders had no role in the design of the study; in the collection, analyses, or interpretation of data; in the writing of the manuscript, or in the decision to publish the results.

References

1. EEA Directive 2012/27/EU of the European Parliament and of the Council of 25 October 2012 on Energy Efficiency, Amending Directives 2009/125/EC and 2010/30/EU and Repealing Directives 2004/8/EC and 2006/32/EC. 2012. Available online: https://eur-lex.europa.eu/LexUriServ/LexUriServ.do?uri=OJ:L:2012:315:0001:0056:en:PDF (accessed on 27 January 2020).
2. European Council. *EUCO 169/14—2030 Climate and Energy Policy Framework and Economic Issues*; European Council: Brussels, Belgium, 2014.
3. EEA Directive (EU) 2018/844 of the European Parliament and of the Council of 30 May 2018 Amending Directive 2010/31/EU on the Energy Performance of Buildings and Directive 2012/27/EU on Energy Efficiency. 2018. Available online: https://eur-lex.europa.eu/legal-content/EN/TXT/PDF/?uri=CELEX:32018L0844&from=EN (accessed on 27 January 2020).
4. Abalos, I. *Harvard Design Magazine 30: (Sustainability) + Pleasure, Vol. 1.*; Harvard University: Cambridge, MA, USA, 2009; pp. 14–17.

5. Li, J.; Colombier, M. Managing carbon emissions in China through building energy efficiency. *J. Environ. Manag.* **2009**, *90*, 2436–2447. [CrossRef] [PubMed]
6. Passer, A.; Kreiner, H.; Maydl, P. Assessment of the environmental performance of buildings: A critical evaluation of the influence of technical building equipment on residential buildings. *Int. J. Life Cycle Assess.* **2012**, *17*, 1116–1130. [CrossRef]
7. Cardinale, N.; Rospi, G.; Stefanizzi, P. Energy and microclimatic performance of Mediterranean vernacular buildings: The Sassi district of Matera and the Trulli district of Alberobello. *Build. Environ.* **2013**, *59*, 590–598. [CrossRef]
8. Kimura, K. Vernacular technologies applied to modern architecture. *Renew. Energy* **1994**, *5*, 900–907. [CrossRef]
9. Gallo, C. Bioclimatic architecture. *Renew. Energy* **1994**, *5*, 1021–1027. [CrossRef]
10. Coch, H. Chapter 4—Bioclimatism in vernacular architecture. *Renew. Sustain. Energy Rev.* **1998**, *2*, 67–87. [CrossRef]
11. Cañas, I.; Martín, S. Recovery of Spanish vernacular construction as a model of bioclimatic architecture. *Build. Environ.* **2004**, *39*, 1477–1495. [CrossRef]
12. Singh, M.K.; Mahapatra, S.; Atreya, S.K. Solar passive features in vernacular architecture of North-East India. *Sol. Energy* **2011**, *85*, 2011–2022. [CrossRef]
13. ICOMOS. *Charter on the Built Vernacular Heritage*. Mexico. 1999. Available online: https://www.icomos.org/charters/vernacular_e.pdf (accessed on 27 January 2020).
14. Oliveira, E.V.; Galhano, F. *Arquitectura Tradicional Portuguesa*; Publicações Dom Quixote: Lisboa, Portugal, 1992.
15. Sayigh, A.; Marafia, A.H. Chapter 2—Vernacular and contemporary buildings in Qatar. *Renew. Sustain. Energy Rev.* **1998**, *2*, 25–37. [CrossRef]
16. Martín, S.; Mazarrón, F.R.; Cañas, I. Study of thermal environment inside rural houses of Navapalos (Spain): The advantages of reuse buildings of high thermal inertia. *Constr. Build. Mater.* **2010**, *24*, 666–676. [CrossRef]
17. Singh, M.K.; Mahapatra, S.; Atreya, S.K. Thermal performance study and evaluation of comfort temperatures in vernacular buildings of North-East India. *Build. Environ.* **2010**, *45*, 320–329. [CrossRef]
18. Dili, A.S.; Naseer, M.A.; Varghese, T.Z. Passive environment control system of Kerala vernacular residential architecture for a comfortable indoor environment: A qualitative and quantitative analyses. *Energy Build.* **2010**, *42*, 917–927. [CrossRef]
19. Shanthi Priya, R.; Sundarraja, M.C.; Radhakrishnan, S.; Vijayalakshmi, L. Solar passive techniques in the vernacular buildings of coastal regions in Nagapattinam, TamilNadu-India—A qualitative and quantitative analysis. *Energy Build.* **2012**, *49*, 50–61. [CrossRef]
20. Singh, M.K.; Mahapatra, S.; Atreya, S.K. Bioclimatism and vernacular architecture of north-east India. *Build. Environ.* **2009**, *44*, 878–888. [CrossRef]
21. Barbosa, J.A.; Bragança, L.; Mateus, R. Assessment of land use efficiency using BSA tools: Development of a new index. *J. Urban Plan. Dev.* **2015**, *141*, 04014020. [CrossRef]
22. Ascione, F.; Bianco, N.; De Rossi, F.; De Masi, R.F.; Vanoli, G.P. Concept, Design and Energy Performance of a Net Zero-Energy Building in Mediterranean Climate. *Procedia Eng.* **2016**, *169*, 26–37. [CrossRef]
23. Asdrubali, F.; Cotana, F.; Messineo, A. On the Evaluation of Solar Greenhouse Efficiency in Building Simulation during the Heating Period. *Energies* **2012**, *5*, 1864–1880. [CrossRef]
24. Küess, H.; Koller, M.; Hammerer, T. *Detail Green—English Edition*; Institut für Internationale Architektur-Dokumentation GmbH & Co.: Munich, Germany, 2011; pp. 44–49.
25. Costa Carrapiço, I.; Neila-González, J. Study for the rehabilitation of vernacular architecture with sustainable criteria. In Proceedings of the Vernacular Heritage and Earthen Architecture: Contributions for Sustainable Development; Correia, M., Carlos, G., Rocha, S., Eds.; CRC Press/Balkema: Vila Nova da Cerveira, Portugal, 2013; pp. 581–586.
26. Fernandes, J.; Mateus, R.; Bragança, L.; Correia da Silva, J.J. Portuguese vernacular architecture: The contribution of vernacular materials and design approaches for sustainable construction. *Archit. Sci. Rev.* **2015**, *58*, 324–336. [CrossRef]
27. Fernandes, J.; Pimenta, C.; Mateus, R.; Silva, S.M.; Bragança, L. Contribution of Portuguese Vernacular Building Strategies to Indoor Thermal Comfort and Occupants' Perception. *Buildings* **2015**, *5*, 1242–1264. [CrossRef]

28. Fernandes, J.; Mateus, R.; Gervásio, H.; Silva, S.M.; Bragança, L. Passive strategies used in Southern Portugal vernacular rammed earth buildings and their influence in thermal performance. *Renew. Energy* **2019**, *142*, 345–363. [CrossRef]
29. AAVV. *Arquitectura Popular em Portugal*, 3rd ed.; Associação dos Arquitectos Portugueses: Lisboa, Portugal, 1988.
30. International Organization for Standardization. ISO7726. In *Ergonomics of the Thermal Environment E Instruments for Measuring Physical Quantities*; ISO: Geneva, Switzerland, 2002.
31. International Organization for Standardization. ISO7730. In *Ergonomics of the Thermal Environment: Analytical Determination and Interpretation of Thermal Comfort Using Calculation of the Pmv and Ppd Indices and Local Thermal Comfort Criteria*; ISO: Geneva, Switzerland, 2005.
32. ASHRAE. ASHRAE 55—Thermal Environmental Conditions for Human Occupancy. In *ANSI/ASHRAE Stand. 55*; ASHRAE: Atlanta, GA, USA, 2013.
33. Matias, L. *TPI65—Desenvolvimento de um Modelo Adaptativo Para Definição das Condições de Conforto Térmico em Portugal*; Laboratório Nacional de Engenharia Civil/National Laboratory of Civil Engineering: Lisboa, Portugal, 2010; ISBN 978-972-49-2207-2.
34. ASHRAE. *ANSI/ASHRAE Standard 55—Thermal Environmental Conditions for Human Occupancy*; ASHRAE: Atlanta, GA, USA, 2004.
35. EN15251. *Indoor Environmental Input Parameters for Design and Assessment of Energy Performance of Buildings-Addressing Indoor Air Quality, Thermal Environment, Lighting and Acoustics*; BSI: London, UK, 2007.
36. Monteiro, J.G. *Tabuaço: Esboços e Subsídios Para Uma Monografia*; Câmara Municipal de Tabuaço: Tabuaço, Portugal, 1991.
37. SNIAmb. Soil Map of—Environment Atlas—SNIAmb—Agência Portuguesa do Ambiente. Available online: https://sniamb.apambiente.pt/content/geo-visualizador?language=pt-pt (accessed on 14 June 2018).
38. LNEG. Geological Map of Portugal at 1:1 000 000—Geoportal do Laboratório Nacional de Energia e Geologia. Available online: http://geoportal.lneg.pt/geoportal/mapas/index.html (accessed on 14 June 2018).
39. Agencia Estatal de Meteorología. *Atlas Climático Ibérico: Temperatura do Ar e Precipitação (1971–2000)/Iberian Climate Atlas: Air Temperature and Precipitation (1971/2000)*; Agencia Estatal de Meteorología, Ministerio de Medio Ambiente y Medio Rural y Marino, Instituto de Meteorologia de Portugal: Madrid, Spain, 2011; ISBN 978-84-7837-079-5.
40. Pina dos Santos, C.A.; Rodrigues, R. *ITE54—Coeficientes de Transmissão Térmica de Elementos Opacos da Envolvente dos Edifícios*; Laboratório Nacional de Engenharia Civil/National Laboratory of Civil Engineering: Lisboa, Portugal, 2009; ISBN 978-972-49-2180-8.
41. Pina dos Santos, C.A.; Matias, L. *ITE50—Coeficientes de Transmissão Térmica de Elementos da Envolvente dos Edifícios*; Laboratório Nacional de Engenharia Civil/National Laboratory of Civil Engineering: Lisboa, Portugal, 2006.
42. Diário da República. Regulamento de Desempenho Energético dos Edifícios de Comércio e Serviços (RECS)—Requisitos de Ventilação e Qualidade do Ar Interior. 2013. Available online: https://dre.pt/application/conteudo/331868 (accessed on 27 January 2020).
43. Ting, D.S. WHO Handbook on Indoor Radon: A Public Health Perspective. *Int. J. Environ. Stud.* **2010**, *67*, 100–102. [CrossRef]
44. EEA Council Directive 2013/59/Euratom of 5 December 2013 Laying Down Basic Safety Standards for Protection against the Dangers Arising from Exposure to Ionising Radiation, and Repealing Directives 89/618/Euratom, 90/641/Euratom, 96/29/Euratom, 97/43/Euratom a. 2013. Available online: https://eur-lex.europa.eu/LexUriServ/LexUriServ.do?uri=OJ:L:2014:013:0001:0073:EN:PDF (accessed on 27 January 2020).

© 2020 by the authors. Licensee MDPI, Basel, Switzerland. This article is an open access article distributed under the terms and conditions of the Creative Commons Attribution (CC BY) license (http://creativecommons.org/licenses/by/4.0/).

Article

Evaluation on Overheating Risk of a Typical Norwegian Residential Building under Future Extreme Weather Conditions

Zhiyong Tian [1],*, Shicong Zhang [2], Jie Deng [3] and Bozena Dorota Hrynyszyn [1]

1. Department of Civil and Environmental Engineering, Norwegian University of Science and Technology, 7491 Trondheim, Norway; bozena.d.hrynyszyn@ntnu.no
2. Institute of Building Environment and Energy, China Academy of Building Research, Beijing 100013, China; zhangshicong01@126.com
3. School of the Built Environment, University of Reading, Reading RG6 6DF, UK; deng-jie2@163.com
* Correspondence: tianzy0913@163.com

Received: 28 November 2019; Accepted: 27 January 2020; Published: 4 February 2020

Abstract: As the temperature in the summer period in Norway has been always moderate, little study on the indoor comfort of typical Norwegian residential buildings in summer seasons can be found. Heat waves have attacked Norway in recent years, including in 2018 and 2019. Zero energy buildings, even neighborhoods, have been a hot research topic in Norway. There is overheating risk in typical Norwegian residential buildings without cooling devices installed under these uncommon weather conditions, like the hot summers in 2018 and 2019. Three weather scenarios consisting of present-day weather data, 2050 weather data, and 2080 weather data are investigated in this study. The overheating risk of a typical Norwegian residential building is evaluated under these three weather scenarios. 72 scenarios are simulated in this study, including different orientations, window-to-wall ratios, and infiltration rates. Two different overheating evaluation criteria and guidelines, the Passive House Planning Package (PHPP) and the CIBSE TM 59, are compared in this study.

Keywords: overheating risk; evaluation; indoor comfort; cold climates

1. Introduction

Since the Energy Performance of Buildings Directive (EPBD) was published in 2010, low energy building has become a hot topic in Europe. In June 2018, the EPBD was revised. Health and well-being of building users is promoted under the new revised EPBD (2018/844/EU).

Unusually hot weather occurred in many European countries in the summers of 2018 and 2019. In northern Europe, from Ireland to the Baltic countries through southern Scandinavia, the outdoor temperatures have risen by 3–6 °C above average. All previous temperature records were broken in many weather stations of northern Europe in May, 2018. In Norway, the outdoor temperature in 2018 was 4 °C hotter than that in previous years. The year 2018 was the third hottest on record, which underlined "the clear warming trend" in the last four decades [1]. Many Norwegian cities have recorded temperatures in excess of 30 °C, up to 35.6 °C in the summer of 2019. Hot summer such as that in 2018 is predicted to become common by 2050 [2]. High temperatures linked to climate changes and heat waves are already causing premature deaths in northern Europe. The heat-related mortality will be more serious if high temperature weather is more and more common in the near future. For instance, more than 1500 people have died from heat waves in all of Sweden in recent decades [3].

Normally there are few installations of cooling devices, even electrical fans, in the household in the Nordic area, including Norway. Many residents felt very uncomfortable indoors in the hot summers of 2018 and 2019 because of the lack of cooling devices. In general, there are no efficient shading facilities in Norwegian residential buildings. Most of the current venetian blinds in the household are just for glare prevention. As far as we know, there are no publications which evaluate the overheating risk of residential buildings in Norway built today, while this problem is taking place more and more frequently.

Many researchers have carried out investigations of overheating risk in apartments. Pathan et al. [4] found that there was a significant risk of overheating based on the measurements from 122 residential buildings in London in the summers of 2009 and 2010. Jenkins et al. [5] simulated the domestic overheating of a dwelling under different climate scenarios for the different locations in UK. Bertug Ozarisoy assessed the overheating risk issues of a typical house during the heatwave period in the England [6]. The monitoring data in the summer of 2018 showed that there was a heavy overheating risk and discomfort in many occupied spaces. Masoud et al. [7] investigated the overheating risk of social housing flats built to passive house standards in the UK. It was found that more than two-thirds of flats had the overheating exceeding the benchmark. Ji et al. [8] created a simulation of overheating risk of a typical house in Manchester under future weather scenarios. Gupta et al. [9] suggested that attention should be paid to the overheating risk in the south-east of England in the future. They found that the most effective (passive) solutions for reducing future overheating were to improve envelope and decrease internal heat gains. Peacock et al. [10] investigated the possible overheating risk in UK dwellings for the future climate change. It was predicted that 18% of the dwellings in the south of England had to install domestic air conditioners by 2030.

Psomas et al. [11] determined the overheating risk of retrofitting of single-family buildings and found that ventilation and shading systems were useful for reducing overheating. Most of the studies mentioned focus on the overheating risk in the south-east of England. Furthermore, Ibrahim et al. [12] highlighted the overheating risk of a retrofitted residential building in Sheffield, in the north of England. It was suggested that solar shading systems and night ventilation systems were the most effective passive overheating strategies. Sehizadeh et al. [13] investigated the influence of possible climate changes on the overheating of a house retrofitted to the international EnerPHit standard in Canada. It was found that the overheating risk of a typical house retrofitted to the international EnerPHit standard would significantly increase in the near future. Grussa et al. [14] evaluated the use of solar shading and night ventilation in a residential retrofit case study located in London in order to reduce the overheating risk. They concluded that night ventilation and shading systems during the daytime could decrease the overheating risk significantly. Salem et al. [15] investigated the impacts of changing weather conditions on the overheating risk and energy performance for a village adapted to the nearly zero energy building standards in the UK. It was shown that night ventilation, double glazing (low-e) windows, and shading devices were not enough to reduce the overheating risk.

Mitchell et al. [16] analyzed the overheating risk of UK passive residential buildings by collecting high-resolution indoor temperature data from 82 homes across the UK. It was suggested that the overheating should be identified in individual rooms, not at the whole-building level. Petrou et al. [17] investigated the indoor temperatures of English buildings. 26% of the residential buildings monitored had overheating. Roberts et al. [18] analyzed the overheating risk with the dynamic thermal models.

Figure 1 shows the land surface temperature difference compared to the average temperature for the period of 2000–2015. It can be seen that the temperature in most of southern Norway in the summer season of 2018 was 5 °C higher than that in normal years. Temperatures of 35 °C outside will result in an uncomfortable indoor environment. There are several publications on the overheating risk in southern Europe. Mlakar et al. [19] identified different energy gains and the impact factors on the overheating risk in a passive building in Slovenia. The results showed that night ventilation in the summer seasons, shading, and reduction of the inter heat gains were enough to decrease the overheating risk. Overheating discomfort also may be one of the unintended consequences in the building sector even in Norway, while the extreme heat waves in the summer will happen more frequently in the future. No publications on this topic in Norway have been published, as far as we know.

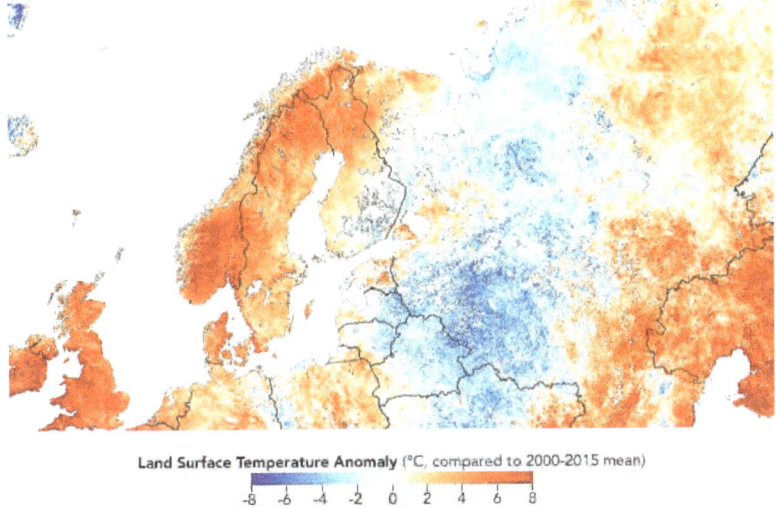

Figure 1. Temperature anomaly in Northern Europe in July (Source: NASA).

A building simulation model based on a typical residential building built in Norway, according to the Norwegian building code (Pbl/TEK17), was set up in the computer-aided design (CAD) application software Rhinoceros®(Robert McNeel & Associates, Seattle, WA, USA). The Energy Plus platform was used to do energy calculations. The overheating risk of a typical Norwegian residential building in the summer under different weather scenarios was determined in this paper. Three weather scenarios consisting of present-day weather data, 2050 s weather data and 2080 s weather data were investigated. The overheating risk of a typical Norwegian residential building under three weather scenarios was evaluated. Different orientations, window-to-wall ratios, and infiltration rates were simulated. Two different overheating evaluation criterial guidelines, including the Passive House Planning Package (PHPP) and the CIBSE TM 59, were compared in this study. The results can provide some design basis for architects and real estate developers in Norway.

2. Method

2.1. Simulation Tool

Geometry of the simulated building was drawn in the Rhinoceros 5.0. Building performance simulation was carried out in Energy Plus engine via Ladybug and Honeybee plugins.

CCWorldWeatherGen tool was used to generate the future weather parameters in the 2050 and 2080 scenarios. The output data of the HadCM3 was combined with the Intergovernmental Panel on Climate Change (IPCC) A2 emission scenario through the morphing method to generate future EnergyPlus weather files.

2.2. Evaluation Criteria and Guideline

2.2.1. Passive House Planning Package (PHPP)

The Passive House Planning Package has been developed by the Passive House Institute in Darmstadt. The PHPP methodology defines the risk of overheating of a building by the percentage of the hours when the indoor temperature is higher than a limit value during one whole year. The default limit value is 25 °C. The comfort range is 10%.

2.2.2. CIBSE TM 59 (Adaptive Thermal Comfort)

The Predicted Mean Vote (PMV), developed by Povl Ole Fanger (1970), has been widely used in many standards to describe thermal comfort of mechanically heated/cooled spaces. Many parameters, including indoor environment parameters, metabolic rate and clothing insulation were considered in the PMV.

Developed for commercial buildings, Technical Memorandum 52 (TM52) is based on BS EN 15251:2007. The method of overheating evaluation in CIBSE TM59 Design Methodology was amended from the CIBSE TM52. The bedrooms should meet two requirements. The first criterion for the bedrooms is that the number of hours when temperature difference is bigger than or equal to one degree (K) from May to September shall not be more than 3% of the occupied hours.

$$\Delta T = T_{op} - T_{max}, \qquad (1)$$

$$T_{max} = 0.33 T_{rm} + 21.8, \qquad (2)$$

$$T_{rm} = (T_{od-1} + 0.8 T_{od-2} + 0.6 T_{od-3} + 0.5 T_{od-4} + 0.4 T_{od-5} + 0.3 T_{od-6} + 0.2 T_{od-7})/3.8, \qquad (3)$$

where, T_{op} is the operative temperature, °C; T_{max} is the maximum permissible temperature, °C; T_{rm} is the exponentially weighted running average ambient temperature, °C; T_{od-1} is the daily average ambient temperature for the day before, °C; T_{od-2} is the daily average ambient temperature for the day before the previous day, °C.

The second criterion is that the hours when the operative temperature in the bedrooms from 10 pm to 7 a.m. is bigger than 26 °C shall not be more than 1% of annual hours (33 h).

3. Typical Residential Building in Norway

Figure 2 shows the typical newly-built Norwegian residential building studied in this paper. Facades are shown in the Figure 3. Figure 4 presents the layout of the first floor. Kitchen and dining room are located on the first floor. Figure 5 shows the layout of the second floor. Three bedrooms are located on the second floor. One typical bedroom (highlighted in Figure 1) is selected to simulate indoor comfort in this study. In the different orientations (south and north), this bedroom can be used to show the different situations. The total floor area is 130 m². U-values of the building envelope components (minimum requirements in TEK 17) are listed in Table 1. Internal shading with roller blinds was assumed for windows. The window consists of an insulated frame and two-layer glass with argon in the cavity with 1.2 W/(m² K). The external wall is insulated by 20 cm mineral wool. The roof is insulated by 30 cm mineral wool. The floor is insulated by 30 cm extruded polystyrene (XPS). As there are no cooling devices in the building studied, the design parameters of infiltration rates are very important for the indoor comfort. Three scenarios (0.0001 m³/s per m² facade-tight building, 0.0003 m³/s per m² facade-average building, and 0.0006 m³/s per m² facade-general building) were investigated in this study. The window-to-wall ratio (WWR) is the definition of the fraction on dividing the window area by the external wall area. Four WWR scenarios (0.35, 0.5, 0.75 and 0.9) were simulated. In addition, two orientations (south and north) and three weather conditions (present-day, 2050 and 2080) were the inputs for the simulation. There were 72 scenarios in total based on the parameters mentioned in this study.

Figure 2. Typical residential building in Norway (Source: Norgeshus).

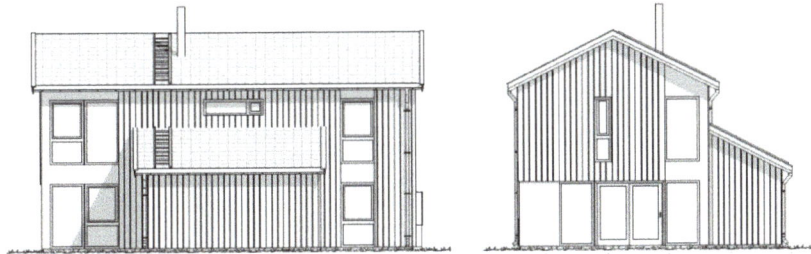

Figure 3. Typical residential building in Norway, facades (Source: Norgeshus).

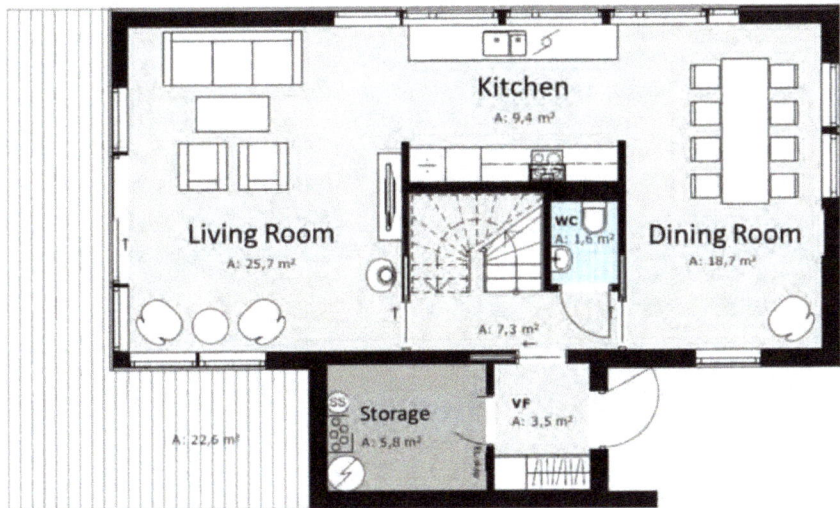

Figure 4. Layout of the first floor (Source: Norgeshus).

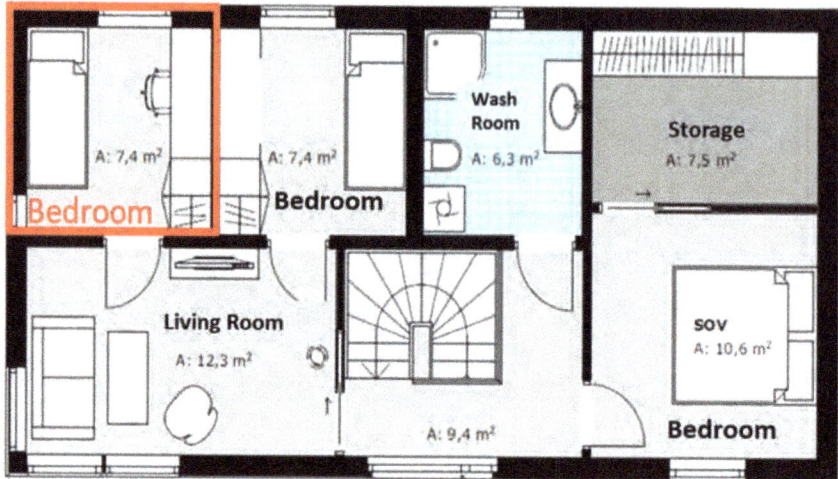

Figure 5. Layout of the second floor (Source: Norgeshus).

Table 1. Building envelope components.

	External Wall	Ground Floor	Roof	Window
U-Value (W/(m² K))	0.22	0.18	0.18	1.2

4. Weather Scenarios

Figure 6a shows the present-day hourly dry bulb temperature in Oslo, Norway. It can be found that the temperatures of few days from 1 May to 30 September can be higher than 30 °C Currently only few residential buildings have cooling devices installed. Figure 6b,c shows the hourly dry bulb temperature in Oslo in the future, in 2050 and 2080 respectively. In Figure 6c, it can be seen that the temperatures in the hottest days are expected to rise to close to 35 °C, which are similar to the temperatures in the hot summers of 2018 and 2019 in Norway. Heat waves experienced during the summer of 2018 and 2019 may become very commonplace by 2080. The simulated indoor comfort in 2080 may provide some references to design strategies for the extreme hot summer conditions.

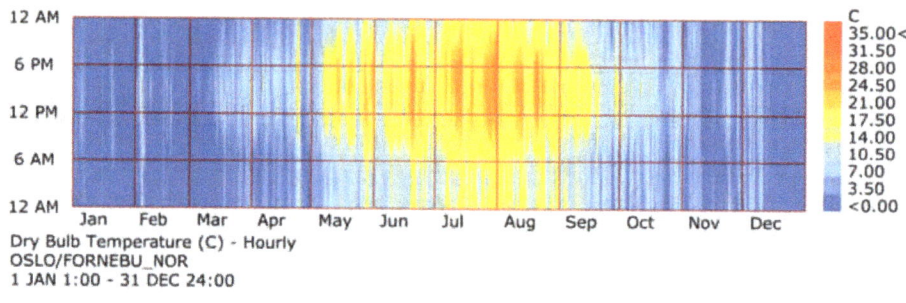

(a)

Figure 6. *Cont.*

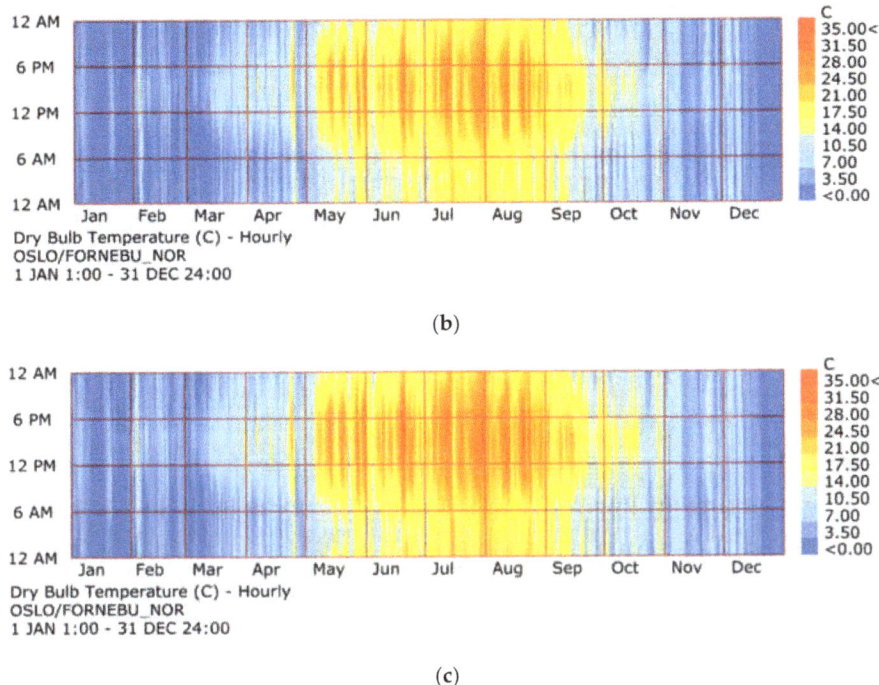

Figure 6. Hourly dry bulb temperature during one year in Oslo, Norway (**a**) present-day weather file, (**b**) 2050 weather file, (**c**) 2080 weather file.

5. Results and Discussion

5.1. PHPP Method

Figure 7 shows the annual percentage beyond the acceptable temperature 25 °C of the tight building model (0.0001 m^3/m^2). By comparison, the annual overheating percentage of the bedroom with a southern orientation is 1.5 times more than that of the bedroom with a northern orientation. Modern buildings trend to be designed with a higher window-to wall ratio (WWR). It can be seen that the overheating risk increases with the growth of the window-to-wall ratio, based on the PHPP method. Under the present-day weather conditions, the scenarios with WWR 0.35 and WWR 0.5 have no overheating risk, while the scenarios with WWR 0.75 and WWR 0.9 have obvious overheating risk based on the PHPP method. The scenarios with higher WWR also have higher heating loss in the Norwegian winter. The solar radiation resource is low and the solar gains in winter are low as well. Thus, it is not recommended to use a WWR that is too big in Norwegian buildings, based on the PHPP method.

Figure 8 shows the annual percentage beyond the acceptable temperature of 25 °C of the average building model (0.0003 m^3/m^2). In the Figure 8, the room with a southern orientation (WWR 0.75 and WWR 0.9) has higher than 10% percentage beyond the acceptable temperature, except in the scenario with a WWR of 0.75 under the present-day weather condition.

Figure 9 shows the annual hours and percentage beyond the acceptable temperature of 25 °C for the general building model (0.0006 m^3/m^2). As shown in Figure 9, there are less overheating hours in the less tight building model based on the PHPP method. When the WWR is 0.9, the overheating percentages of the three building models are similar. If the bedroom with a southern orientation has the smallest WWR of 0.35, the higher infiltration can reduce the overheating risk for the bedroom.

When the bedroom with a southern orientation has a bigger WWR, the high infiltration only reduces the overheating risk slightly.

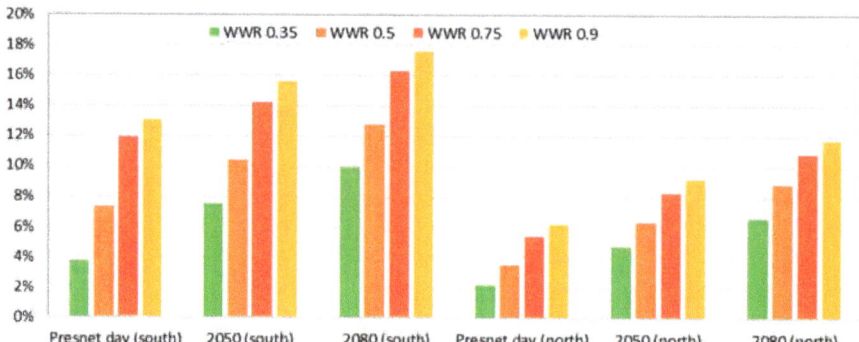

Figure 7. Annual percentage beyond the acceptable temperature 25 °C (tight building model) based on the Passive House Planning Package (PHPP).

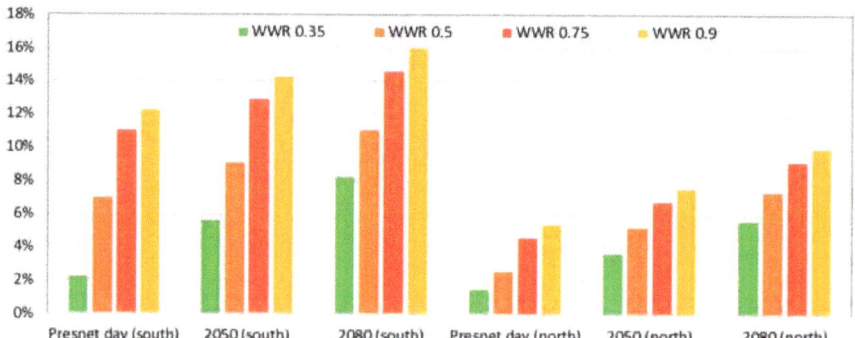

Figure 8. Annual percentage beyond the acceptable temperature 25 °C based on PHPP (average building).

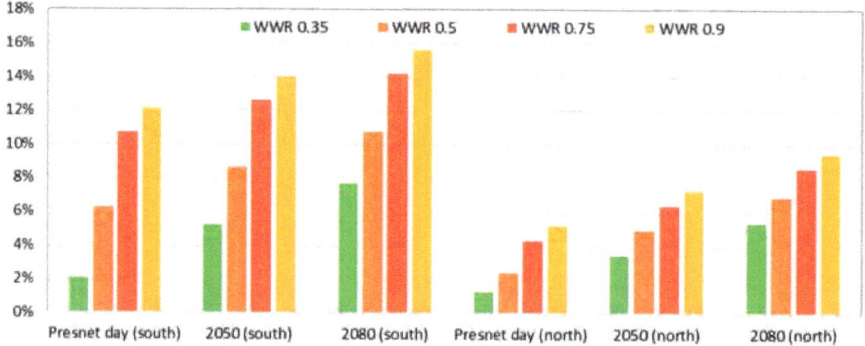

Figure 9. Annual percentage beyond the acceptable temperature 25 °C based on PHPP (general building).

5.2. CIBSE TM 59 (Adaptive Thermal Comfort)

The operative temperatures during the period of 1 May to 30 September are not more than 1K higher than the maximum permissible temperature. The indoor climate conditions of the studied bedroom meet the criterion 1 of the CIBSE TM 59 (adaptive thermal comfort). Figures 10–12 show the accumulative hours with temperatures exceeding 26 °C during the occupied period (criterion 2 of CIBSE TM 59). In contrast to the results shown in Figures 7–9, the room with a northern orientation tends to have more severe overheating risk than the room with a southern orientation. That is because all hourly indoors temperatures based on the annual basis are considered in the PHPP method. There are many hours with high temperatures during the daytime in the summer seasons. Only the occupied time of the bedroom (from 10 p.m. to 7 a.m.) is taken into consideration in the CIBSE TM 59 method (adaptive thermal comfort). The indoor temperature decreases to below 26 °C during the night time. For the scenarios with bedrooms with a northern orientation, the bedroom keeps warm from 10 p.m. to 7 a.m. due to the late sunset in the most of Norway in the summer season. In addition, the bedroom has a western wall with window. The bedroom with a southern orientation does not tend to have overheating risk, except in the scenario of the tight building with WWR 0.9, under the 2080 weather conditions. Under the present-day weather conditions, the bedroom with a northern orientation does not tend to have overheating. However, under the future weather conditions (2050 and 2080), the big WWR ratios increase the overheating risk of the bedroom with a northern orientation.

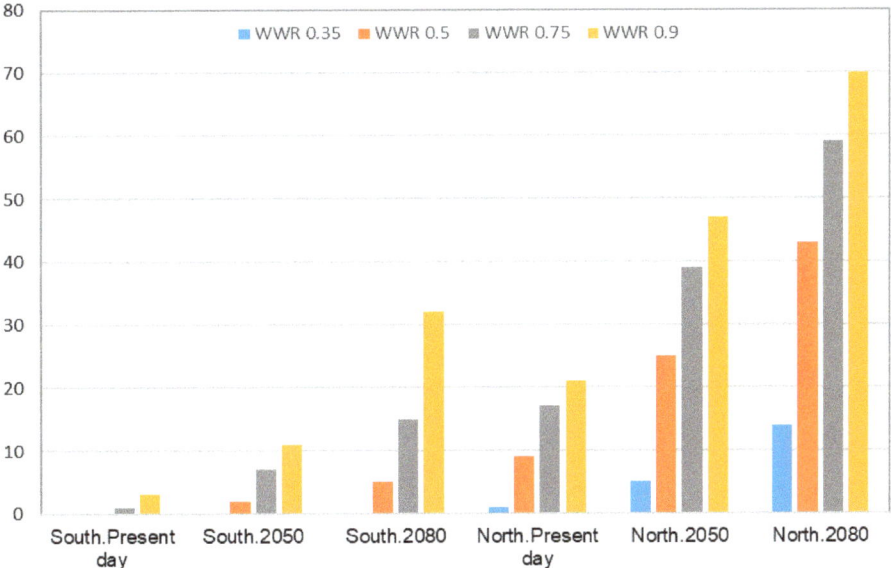

Figure 10. Hours of temperature exceeding 26 °C during the occupied period based on CIBSE TM 59 (Tight building).

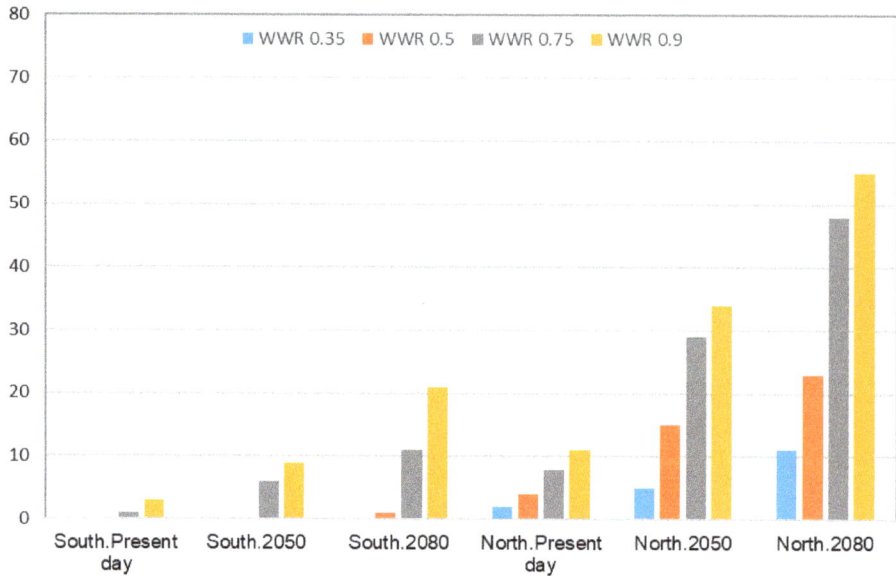

Figure 11. Hours of exceedance 26 °C during the occupied period based on CIBSE TM 59 (Average building).

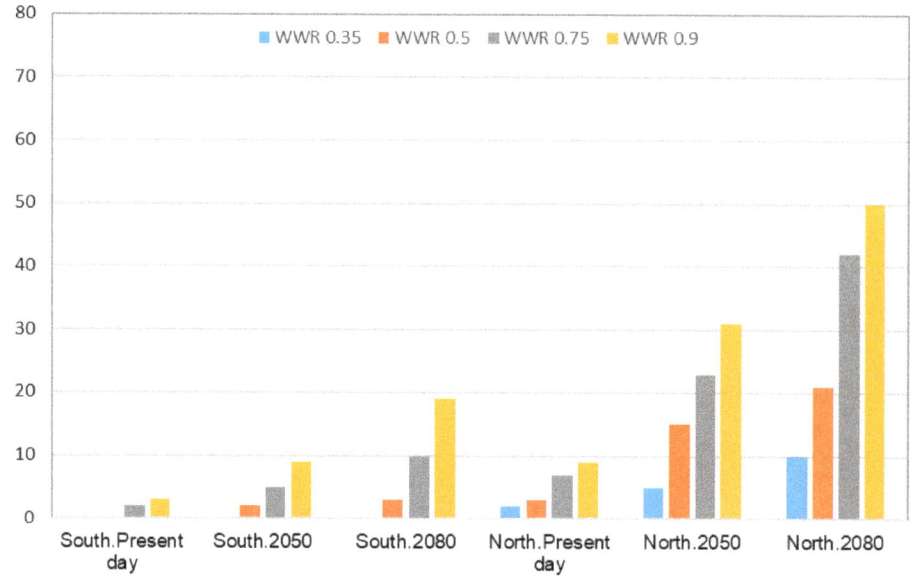

Figure 12. Hours of exceedance 26 °C during the occupied period based on CIBSE TM 59 (General building).

6. Conclusions and Outlook

The overheating risk of a typical Norwegian residential building under present-day, 2050 and 2080 weather conditions was evaluated in this study. Two different overheating evaluation criterial guidelines (the Passive House Planning Package and CIBSE TM 59) were compared. The following conclusions could be drawn:

1. The evaluation method recommended in the PHPP is not very precise to evaluate the specific overheating risk for bedrooms without considering occupied time. The adaptive thermal comfort method is recommended to evaluate the overheating risk for single rooms in residential buildings.
2. Large window-to-wall ratios (WWR) are not recommended for Norwegian residential buildings. Too large WWR will result in overheating risk in the summer, particularly in the future extreme weather conditions. In the north-western oriented bedrooms with windows faced north, the use of a large WWR is not recommended.
3. In very airtight residential buildings, overheating risk can take place in the future climate scenarios analyzed.

Overheating risks in Norway should be paid attention to, as more and more extreme heatwaves have taken places in recent years. There are mainly two methods, namely passive and active, to tackle the overheating risks. The passive methods are mainly natural ventilation and shading systems. For example, a green roof can be one possible solution to reduce the indoor temperature in the summer seasons due to the shading effect of the plants [20]. The active methods are mainly mechanical cooling devices, such as air-conditioners, and heat dissipation panels [21]. If the passive measures are not enough to reduce the overheating hours, the mechanical methods are required to be installed to keep the indoor environment comfortable.

In future research, the indoor comfort of a typical Norwegian existing residential building retrofitted to the international EnerPHit standard will be investigated. A sensitivity study on other weather parameters, such as diffuse solar radiation and direct normal irradiation (DNI), and wind speed, will be done in future work.

Author Contributions: Z.T. proposed the structure of the paper. Z.T. wrote the paper. S.Z. and J.D. contributed to the model and proofread of the paper. B.D.H. contributed to formulating and developing of the research question and proofread of the paper. All authors have read and agree to the published version of the manuscript.

Funding: The authors would like to acknowledge the Norwegian University of Science and Technology (NTNU), the Faculty of Engineering and the Department of Civil and Environmental Engineering for financial support for the research.

Acknowledgments: Norgeshus provides the information of the typical building model analyzed. Thanks are expressed to Tobias Skov Pedersen from the Technical University of Denmark for the hints on the software.

Conflicts of Interest: The authors declare no conflict of interest.

References

1. Copernicus Climate Change Service. Available online: https://cds.climate.copernicus.eu/#!/home (accessed on 1 April 2019).
2. Hughes, C.; Natarajan, S. Summer thermal comfort and overheating in the elderly. *Build. Serv. Eng. Res. Technol.* **2019**, *40*, 426–445. [CrossRef]
3. Åström, D.O.; Forsberg, B.; Ebi, K.L.; Rocklöv, J. Attributing mortality from extreme temperatures to climate change in Stockholm, Sweden. *Nat. Clim. Chang.* **2013**, *3*, 1050–1054. [CrossRef]
4. Pathan, A.; Mavrogianni, A.; Summerfield, A.; Oreszczyn, T.; Davies, M. Monitoring summer indoor overheating in the London housing stock. *Energy Build.* **2017**, *141*, 361–378. [CrossRef]
5. Jenkins, D.; Ingram, V.; Simpson, S.; Patidar, S. Methods for assessing domestic overheating for future building regulation compliance. *Energy Policy* **2013**, *56*, 684–692. [CrossRef]
6. Ozarisoy, B.; Elsharkawy, H. Assessing overheating risk and thermal comfort in state-of-the-art prototype houses that combat exacerbated climate change in UK. *Energy Build.* **2019**, *187*, 201–217. [CrossRef]
7. Sameni, S.M.T.; Gaterell, M.; Montazami, A.; Ahmed, A. Overheating investigation in UK social housing flats built to the Passivhaus standard. *Build. Environ.* **2015**, *92*, 222–235. [CrossRef]
8. Ji, Y.; Fitton, R.; Swan, W.; Webster, P.; Swan, W. Assessing overheating of the UK existing dwellings—A case study of replica Victorian end terrace house. *Build. Environ.* **2014**, *77*, 1–11. [CrossRef]
9. Gupta, R.; Gregg, M.; Gregg, M. Preventing the overheating of English suburban homes in a warming climate. *Build. Res. Inf.* **2013**, *41*, 281–300. [CrossRef]

10. Peacock, A.; Jenkins, D.; Kane, D. Investigating the potential of overheating in UK dwellings as a consequence of extant climate change. *Energy Policy* **2010**, *38*, 3277–3288. [CrossRef]
11. Psomas, T.; Heiselberg, P.; Duer, K.; Bjørn, E. Overheating risk barriers to energy renovations of single family houses: Multicriteria analysis and assessment. *Energy Build.* **2016**, *117*, 138–148. [CrossRef]
12. Ibrahim, A.; Pelsmakers, S.L. Low-energy housing retrofit in North England: Overheating risks and possible mitigation strategies. *Build. Serv. Eng. Res. Technol.* **2018**, *39*, 161–172. [CrossRef]
13. Sehizadeh, A.; Ge, H. Impact of future climate change on the overheating of Canadian housing retrofitted to the passivehaus standard. In Proceedings of the 2009 IBPSA Conference, Glasgow, Scotland, 27–30 July 2009.
14. De Grussa, Z.; Andrews, D.; Lowry, G.; Newton, E.J.; Yiakoumetti, K.; Chalk, A.; Bush, D. A London residential retrofit case study: Evaluating passive mitigation methods of reducing risk to overheating through the use of solar shading combined with night-time ventilation. *Build. Serv. Eng. Res. Technol.* **2019**, *40*, 389–408. [CrossRef]
15. Salem, R.; Bahadori-Jahromi, A.; Mylona, A. Investigating the impacts of a changing climate on the risk of overheating and energy performance for a UK retirement village adapted to the nZEB standards. *Build. Serv. Eng. Res. Technol.* **2019**, *40*, 470–491. [CrossRef]
16. Mitchell, R.; Natarajan, S. Overheating risk in Passivhaus dwellings. *Build. Serv. Eng. Res. Technol.* **2019**, *40*, 446–469. [CrossRef]
17. Petrou, G.; Symonds, P.; Mavrogianni, A.; Mylona, A.; Davies, M. The summer indoor temperatures of the English housing stock: Exploring the influence of dwelling and household characteristics. *Build. Serv. Eng. Res. Technol.* **2019**, *40*, 492–511. [CrossRef]
18. Roberts, B.M.; Allinson, D.; Diamond, S.; Abel, B.; Das Bhaumik, C.; Khatami, N.; Lomas, K.J. Predictions of summertime overheating: Comparison of dynamic thermal models and measurements in synthetically occupied test houses. *Build. Serv. Eng. Res. Technol.* **2019**, *40*, 512–552. [CrossRef]
19. Mlakar, J.; Štrancar, J. Overheating in residential passive house: Solution strategies revealed and confirmed through data analysis and simulations. *Energy Build.* **2011**, *43*, 1443–1451. [CrossRef]
20. Schweitzer, O.; Erell, E. Evaluation of the energy performance and irrigation requirements of extensive green roofs in a water-scarce Mediterranean climate. *Energy Build.* **2014**, *68*, 25–32. [CrossRef]
21. Zuazua-Ros, A.; Martín-Gómez, C.; Ramos, J.C.; Bermejo-Busto, J. Towards cooling systems integration in buildings: Experimental analysis of a heat dissipation panel. *Renew. Sustain. Energy Rev.* **2017**, *72*, 73–82. [CrossRef]

© 2020 by the authors. Licensee MDPI, Basel, Switzerland. This article is an open access article distributed under the terms and conditions of the Creative Commons Attribution (CC BY) license (http://creativecommons.org/licenses/by/4.0/).

Article

Analytical Methods to Estimate the Thermal Transmittance of LSF Walls: Calculation Procedures Review and Accuracy Comparison

Paulo Santos *, Gabriela Lemes and Diogo Mateus

ISISE, Department of Civil Engineering, University of Coimbra, Pólo II, Rua Luís Reis Santos, 3030-788 Coimbra, Portugal; gabriela.lemes@uc.pt (G.L.); diogo@dec.uc.pt (D.M.)
* Correspondence: pfsantos@dec.uc.pt; Tel.: +351-239-797-199

Received: 20 January 2020; Accepted: 11 February 2020; Published: 14 February 2020

Abstract: An accurate evaluation of the thermal transmittance (U-value) of building envelope elements is fundamental for a reliable assessment of their thermal behaviour and energy efficiency. Simplified analytical methods to estimate the U-value of building elements could be very useful to designers. However, the analytical methods applied to lightweight steel framed (LSF) elements have some specific features, being more challenging to use and to obtain a reliable accurate U-value with. In this work, the main analytical methods available in the literature were identified, the calculation procedures were reviewed and their accuracy was evaluated and compared. With this goal, six analytical methods were used to estimate the U-values of 80 different LSF wall models. The obtained analytical U-values were compared with those provided by numerical simulations, which were used as reference U-values. The numerical simulations were performed using a 2D steady-state finite element method (FEM)-based software, THERM. The reliability of these numerical models was ensured by comparison with benchmark values and by an experimental validation. All the evaluated analytical methods showed a quite good accuracy performance, the worst accuracy being found in cold frame walls. The best and worst precisions were found in the Modified Zone Method and in the Gorgolewski Method 2, respectively. Very surprisingly, the ISO 6946 Combined Method showed a better average precision than other two methods, which were specifically developed for LSF elements.

Keywords: lightweight steel frame; LSF walls; thermal transmittance; U-value; analytical methods; calculation procedures; accuracy

1. Introduction

The use of lightweight steel frame (LSF) systems has emerged as a viable alternative to traditional construction and its usage are increasing every year, mostly because of its great advantages, such as: cost efficiency, reduced weight, mechanical resistance, fast assemblage and others [1,2]. However, the high thermal conductivity of the steel could lead to thermal bridges effects resulting in a poor thermal performance of the building if those issues are not properly addressed (e.g., at design stage) [3].

A usual LSF wall is mainly composed of three parts: (1) steel frame internal structure (cold formed profiles); (2) sheathing panels (internal and external, e.g., gypsum plasterboard and OSB); (3) the insulation layers (cavity/batt insulation, such as mineral wool, and/or ETICS-exterior thermal insulation composite system) [3]. Notice that the batt insulation, besides the thermal insulation function, can also perform an important acoustic insulation function [4]. Moreover, the effectiveness of thermal insulation depends on its position in the LSF element [5], as well as on the type of LSF construction [6].

In fact, the existence of an insulation layer and its position on the wall determines the type of LSF construction. According to Santos et al. [1], a LSF construction element can be classified into three wall frame typologies: (1) cold, (2) hybrid, and (3) warm. On cold frame constructions, all the thermal insulation is placed inside the wall air cavity, between the vertical studs and limited to the stud depth. The opposite happens with the warm frame construction, where all thermal insulation is continuous and located outside of the steel frame (ETICS). Given its advantages, the hybrid construction type is used more often [4], being an intermediate solution between cold and warm construction and has both types of insulation applied: continuous exterior (ETICS) and between the steel studs (cavity insulation).

An accurate evaluation of the thermal transmittance (U-value) of building envelope elements is fundamental for a reliable assessment of their thermal behaviour and energy efficiency [7]. LSF elements are even more challenging given the very reduced thickness of the cold formed steel profiles and the strong contrast between its thermal properties (e.g., thermal conductivity) and the thermal insulation materials (e.g., mineral wool) [8]. In buildings, despite thermal bridges originated by the high thermal conductivity of the steel frame [9], it is needed to account for flanking thermal losses around and in the intersection of building LSF components [10].

There are several approaches to obtain the thermal resistance/transmittance of building elements: (i) analytical, (ii) numerical, and (iii) measurements [11]. Regarding thermal performance measurements, they could be accomplished in-situ or in laboratory settings, being crucial for the validation of numerical and analytical methods [7]. In-situ non-destructive thermal transmittance measurements in existing buildings are very important for energy audit and retrofitting actions [7], which is very challenging to perform since the properties of materials are often unknown, components frequently degrade over time, and the experiments should be fast, simple and non-destructive [12]. Measurements under laboratory conditions have the advantages of well-known controlled environmental conditions, geometries, configurations and materials [7], but could be very time-consuming and expensive. There are various measurement methods, the most used ones being [7]: the heat flow meter (HFM) [13]; the guarded hot plate (GHP) [14]; the hot box (HB) [15], which could be calibrated (CHB) [16] or guarded (GHB) [17]; infrared thermography (IRT) [18].

Numerical simulations could be performed with more simpler two-dimensional (2D) models [5,6] or more complex/detailed three-dimensional (3D) models [9,10]. They have the advantage of allowing a quick comparison between several building component solutions/configurations. However, they need a specific software tool, skills to use it and to ensure the reliability of the obtained results the used models should be validated with measurements or at least verified by comparison with benchmark results.

The use of analytical formulas could be the simplest approach of all these three methods, being very useful and easy to use by designers [8]. However, this analytical approach is usually only available for simpler configurations; its applicability being, most often, very limited. Moreover, these analytical calculation formulations frequently consider a simplified steady-state one-dimensional (1D) heat transfer and do not take into account the heat storage inside the material, nor the thermal properties variation (e.g., with temperature or humidity) [19].

The use of analytical methods to calculate the thermal resistance (R-value) and transmittance (U-value) of a building element could be a complicated subject, especially when the element has inhomogeneous layers with very dissimilar thermal properties. For LSF constructions, those analytical calculations can be harder than in other forms of construction, as the methodology must include the effects of the non-homogeneous layers, the thermal bridges and the large difference between materials thermal conductivities [8].

One of the most used simplified analytical methods to calculate the thermal resistance and transmittance of a building component containing homogeneous and inhomogeneous layers is prescribed by standard ISO 6946 [20]. The total thermal resistance of a component is computed by combining its upper and lower limits, and thus this methodology is often designated as the ISO 6946 Combined Method. These R-value limits are computed making use of the parallel path method (upper limit) and the isothermal planes method (lower limit). The total R-value of the building element is calculated as the average of the upper and lower limits, as previously mentioned. However, this simplified analytical R-value calculation methodology should not be applicable to building elements where the insulation is bridged by metal [20], as happens in cold and hybrid LSF elements.

This applicability limitation of the ISO 6946 Combined Method has motivated several researchers to seek for a specific analytical methodology suitable to calculate the R- and U-values for LSF building elements. The ASHRAE Zone Method [19] was one of the first analytical simplified methods to be developed to calculate the R-value of a LSF element. The ASHRAE zone method is a modification of the parallel path method [21], where instead of considering only the thickness of the steel stud web, it considers a larger zone of influence of the metal thermal bridge within the LSF wall. The width of the area affected by the steel thermal bridge depends on the length of the steel stud flange and on the distance from this metal flange to the wall surface, i.e., the sheathing layers thickness [21].

Given the reported unsatisfactory accuracy of the ASHRAE Zone Method, Kosny et al. [21,22] developed a new improved methodology, often designated as Modified Zone Method [19]. This enhanced method to estimate the R-value of metal frame walls was developed based on computer-simulation results and experimental measurements of different LSF wall configurations, taking into account several wall parameters, such as stud spacing, stud (depth) and flange sizes, stud metal thickness, the thermal resistance of the cavity insulation and thermal resistance of exterior sheathing [22]. It was concluded that the differences in the thermal calculations are caused by the metal stud zone area estimation. Thus, a more precise estimation technique to define the thermally affected zones caused by steel studs of LSF walls was developed and implemented.

More recently, Gorgolewski [8] adapted the ISO 6946 Combined Method to a more accurate analytical U-value calculation methodology for LSF building components, including cold and hybrid frames. In this new suggested analytical methodology, the upper and lower R-values limits are still being used, but instead of an average between these limits Gorgolewski found an "algorithm" for estimating the adequate weighting between them [8]. It was proposed and compared the accuracy of three analytical methods, being taken into account some parameters of the steel frame elements, such as flange width, stud spacing and stud depth. The third method developed by Gorgolewski was found to exhibit the best accuracy performance and thus it was adopted in the United Kingdom for LSF buildings code of practice [23].

As reviewed before, there are several analytical simplified methods available in the literature to compute the thermal resistance or transmittance of LSF building elements. However, it was not found in the bibliography any research work with the evaluation and comparison of the accuracy performance of these different analytical methodologies. Moreover, ISO 6946 [20] states that the prescribed Combined Method is not suitable to estimate the U-value of cold and hybrid LSF elements, but it is not known how large is this methodology calculation error.

In this context, the main aim of this work—besides to perform a review of the analytical methods calculation procedures—is to evaluate and compare the accuracy of the above mentioned simplified analytical methods. With this goal, six analytical methods were used to estimate the thermal transmittance values of eighty different LSF walls. The obtained analytical U-values were compared with those provided by numerical simulations, which were used as reference U-values. The numerical simulations were performed using a 2D steady-state finite element method (FEM) based software, THERM [24]. The reliability of these numerical models was ensured via comparison with benchmark values. Additionally, an experimental validation of some LSF numerical models was also accomplished.

This paper is structured as follows. After this introduction, the six evaluated analytical methods are described, namely the ISO 6946 Combined Method, the three Gorgolewski methods and the two ASHRAE methods (zone and modified zone). Next, the numerical reference FEM models are described, including the benchmark verification and experimental validation, the boundary conditions used and the air-layer modelling. Subsequently, all the assessed 80 LSF walls are described, including the parameters evaluated, the variables changed and the values considered in the assessment; then we present the dimensions and thermal properties of the materials used in this study. Afterwards, the obtained results are presented and discussed. Finally, the main concluding remarks of this work are described.

2. Methods and Materials

2.1. Analytical Simplified Methods

As mentioned before, a building component could have homogeneous and/or inhomogeneous layers. When the building element is constituted by n homogeneous plane layers (j), which are perpendicular to the heat flow, the originated heat flow transfer is one-dimensional and the total thermal resistance (environment to environment) could be computed as prescribed by ISO 6946 [20],

$$R_{\text{tot}} = R_{\text{si}} + \sum_{j=1}^{n} R_j + R_{\text{se}} \qquad (1)$$

where R_{si} and R_{se} are the internal and external surface resistances [m^2·K/W], R_j is the thermal resistance of each homogeneous layer j. Notice that the results presented in Section 3 are thermal transmittances (U-values), which were computed by the reciprocal of the total thermal resistance (R_{tot}), including the internal (0.13 m^2·K/W) and external (0.04 m^2·K/W) surface resistances, as prescribed by ISO 6946 [20] for horizontal heat flow.

When there are inhomogeneous layers in the building component, the heat flow starts being two-dimensional, instead of one-dimensional, given the different thermal conductivities and consequent different thermal resistances. These two-dimensional heat flow features get stronger when the discrepancies between the thermal properties of the materials (e.g., conductivity) within the same layer are more significant.

There are several analytical simplified methods available to compute the thermal resistance or transmittance of building elements containing inhomogeneous layers (e.g., LSF elements). In the next sections several analytical methods will be briefly described, namely: (1) ISO 6946 Combined Method [20]; (2) three methods proposed by Gorgolewski [8]; (3) ASHRAE Zone Method [19]; (4) ASHRAE Modified Zone Method [21].

2.1.1. ISO 6946 Combined Method

One of the most commonly used analytical simplified method to compute the thermal resistance of building elements consisting of homogeneous and inhomogeneous layers, which may contain air layers up to 0.30 m thick, is described in the international standard ISO 6946 [20] and therefore is often identified as ISO 6946 Combined Method, since the total thermal resistance (R_{tot}) is computed by combining two different methods: (1) Parallel Path Method; (2) Isothermal Planes Method, as will be explained next.

According to ISO 6946 [20], this simplified analytical approach is only valid for the cases where the ratio of the upper limit to the lower limit of the thermal resistance does not exceeds 1.5. Furthermore, this method is not applicable to building elements where thermal insulation is bridged by metal (e.g., steel studs), i.e., when there is a significant difference between the thermal conductivity of the materials in the layer providing the most important thermal resistance of the building element. Thus, the ISO 6946 Combined Method (theoretically) is not valid for cold and hybrid LSF construction elements.

Moreover, ISO 6946 [20] prescribes (in Annex F) simplified corrections to the thermal transmittance values for: (1) air voids, (2) mechanical fasteners, and (3) inverted roofs, whenever the total correction exceeds 3%.

Upper Limit of the Total Thermal Resistance: Parallel Path Method

The upper limit of the total thermal resistance ($R_{\text{tot;upper}}$) is determined making use of the parallel path method, i.e., assuming one-dimensional heat transfer perpendicular to the surfaces of the building element.

This assumption is suitable when the materials on the same layer have close (i.e., same order of magnitude) thermal conductivity values, as for example on wood frame walls [19]. As illustrated in Figure 1, usually two main paths are considered in stud cavity walls: Path A, where the heat flux is typically higher given the higher thermal conductivity of the stud material and Path B, with habitually lower heat flux given the lower thermal conductivity of the cavity insulation. Figure 1b displays these equivalent parallel path circuits for both paths.

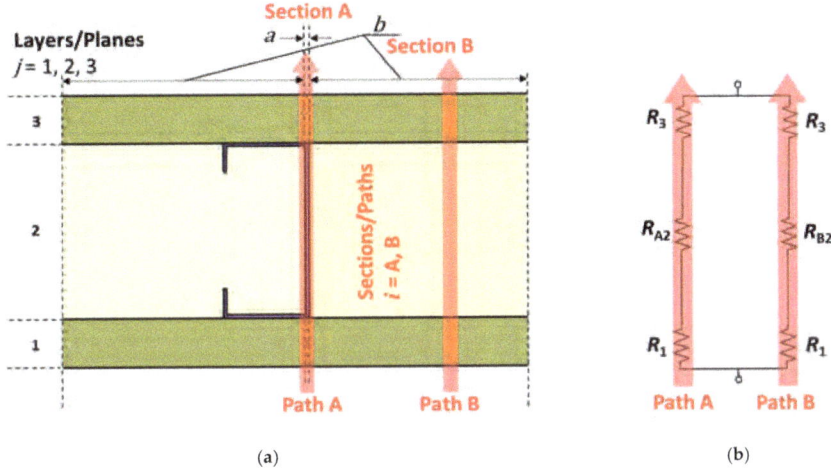

Figure 1. Parallel path method schematic illustration: (**a**) lightweight steel frame (LSF) wall cross-section; (**b**) Equivalent parallel path circuit.

This methodology does not take into account the steel stud horizontal parts (flanges) and stud returns, only considering the web of the stud, being the width of Section A delimited by the web stud thickness.

Assuming these calculation principles, the upper limit of the total thermal resistance is given by Equation (2),

$$\frac{1}{R_{tot;upper}} = \frac{f_A}{R_{tot;A}} + \frac{f_B}{R_{tot;B}} \qquad (2)$$

where f_A and f_B are the fractional areas of sections A and B, respectively, $R_{tot;A}$ and $R_{tot;B}$ are the total thermal resistances of each section/path [m²·K/W]. These total thermal resistances are computed as the summation of the thermal resistances in series for each path (Figure 1b), i.e., assuming homogeneous layers are perpendicular to the heat flow [20], including the internal and external surface thermal resistances.

Lower Limit of the Total Thermal Resistance: Isothermal Planes Method

The lower limit of the total thermal resistance ($R_{tot;lower}$) is determined by making use of the isothermal planes method, i.e., by assuming that all planes parallel to the building element surface are isothermal surfaces. In this method, it is assumed that the heat can flow laterally in any component and the thermal resistances of adjacent components are combined in parallel, resulting on a path with series-parallel resistance combined [21]. This assumption is appropriate when adjacent materials of the same layer/plane have conductivity values moderately different, as with masonry walls [19]. Figure 2a illustrates the three layers (1, 2 and 3) and two sections (A and B) considered in a cold formed stud cavity wall. As mentioned before, given the batt insulation placed in the cavity, the thermal resistance of section B (cavity insulation) is much greater than section A (stud). As in the previously method (parallel path), only the web of the steel stud is considered for heat transfer calculation purposes.

Figure 2b displays the equivalent series-parallel circuit assuming isothermal planes. Since layer 2 is inhomogeneous, both thermal resistances (R_{A2} and R_{B2}) are represented in parallel.

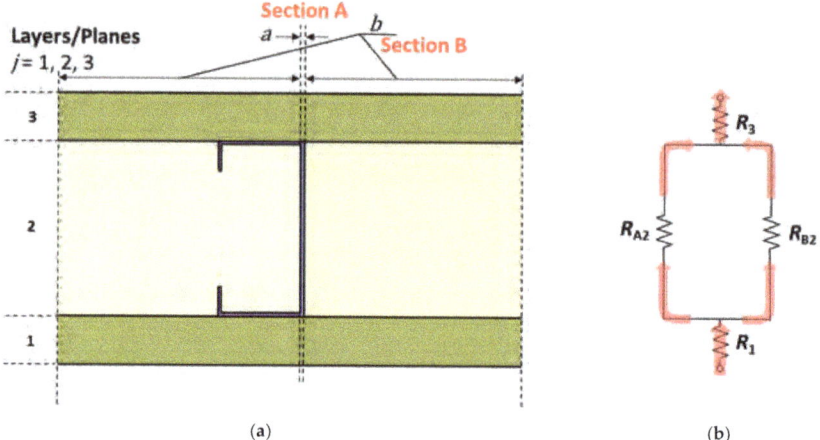

Figure 2. Isothermal planes method schematic illustration: (**a**) LSF wall cross-section; (**b**) equivalent series-parallel circuit.

The calculation of the lower limit of the total thermal resistance is divided into two stages [20]. First, the equivalent thermal resistance (R_j) of each thermally inhomogeneous layer (j) is calculated (layer 2 for the wall in Figure 2) making use of the parallel path method according to Equation (3) [20], which could be simplified for the LSF wall illustrated in Figure 2, resulting in Equation (4).

$$\frac{1}{R_j} = \frac{f_A}{R_{Aj}} + \frac{f_B}{R_{Bj}} + \ldots + \frac{f_Q}{R_{Qj}} \quad (3)$$

$$\frac{1}{R_2} = \frac{f_A}{R_{A2}} + \frac{f_B}{R_{B2}} \quad (4)$$

Second, the lower limit of the total thermal resistance ($R_{tot;lower}$) is calculated as a summation of the series resistances,

$$R_{tot;lower} = R_{si} + R_1 + R_2 + R_3 + R_{se} \quad (5)$$

including the equivalent thermal resistance of the inhomogeneous layer (R_2) previously obtained in Equation (4), as well as the internal and external surface thermal resistances.

Total Thermal Resistance: Combined Method

According with the ISO 6946 Combined Method, the total thermal resistance ($R_{tot;ISO}$) is computed as an arithmetic average of the total upper ($R_{tot;upper}$) and lower ($R_{tot;lower}$) thermal resistances,

$$R_{tot;ISO} = \frac{R_{tot;upper} + R_{tot;lower}}{2} \quad (6)$$

which means that the two *R*-values (upper and lower limits) have the same weight (0.5) on the total resistance calculation [20].

2.1.2. Gorgolewski Methods

As the ISO 6946 Combined Method *U*-value calculation excludes wall configurations—in which insulating layers are bridged by linear metal elements, like on lightweight steel frame (LSF) construction—from its scope, Gorgolewski [8] proposed three new methods based on similar principles used in that standard [20], adapting it to increase the accuracy for this type of construction. Using the same calculation methodology proposed on ISO 6946 to reach upper ($R_{tot;upper}$) and lower ($R_{tot;lower}$) limits of the thermal resistances, Gorgolewski's method differs on the total resistance calculation by applying different weights for the upper and lower resistance values and considering a factor p, between 0 and 1, such that the total thermal resistance (R_{tot}) is given by Equation (7),

$$R_{tot;gorg} = p\, R_{tot;upper} + (1-p)\, R_{tot;lower} \qquad (7)$$

Thus, the total thermal resistances provided by the Gorgolewski methods ranges in the interval $[R_{tot;lower};\, R_{tot;upper}]$, as illustrated in the following expression,

$$R_{tot;gorg} = \begin{cases} R_{tot;upper} & \text{if } p = 1.0 \\ R_{tot;ISO} & \text{if } p = 0.5 \\ R_{tot;lower} & \text{if } p = 0.0 \end{cases} \qquad (8)$$

being equal to the ISO 6946 total resistance when the p-value is 0.5.

Notice that for warm LSF elements, i.e., when there is only external insulation, it was assumed a p-value equal to 0.5 [23]. Thus, the obtained total thermal resistance for any of the Gorgolewski methods is equal to the one provided by ISO 6946 Combined Method [20].

The accuracy of the several methods proposed by Gorgolewski was verified by comparison to the results provided by 2D numerical FEM models for 52 different LSF walls and roof slabs [8].

Gorgolewski Method 1

The p-value for the first (refined) method proposed by Gorgolewski [8] is expressed in Equation (9),

$$p = 0.8\left(\frac{R_{tot;lower}}{R_{tot;upper}}\right) + 0.1 \qquad (9)$$

This p-value depends directly on the ratio between the lower and upper limits of the total thermal resistance.

Gorgolewski Method 2

The p-values for the second method proposed by Gorgolewski [8] are displayed in Table 1. These values take into account the stud spacing, having as reference 500 mm, and whether the LSF element is a hybrid or cold frame type.

Table 1. Tabulated p-Values for Gorgolewski Method 2 [8].

p-Values	Frame Type	
	Hybrid	Cold
Stud spacing ≥ 500 mm	0.50	0.30
Stud spacing < 500 mm	0.40	0.25

Analysing the proposed p-values, being all of them lower or equal to 0.5, and looking to Equation (7), it can be concluded that the total thermal resistance predicted by this method will be closer to the lower limit, $R_{tot;lower}$, for cold frame construction or whenever the stud spacing is lower than 500 mm. This is to be expected, given the higher amount of thermal insulation bridged by steel webs and the higher amount of steel, respectively, thus reducing the overall thermal resistance of the LSF element.

Gorgolewski Method 3

The *p*-value for the third method developed by Gorgolewski [8] is derived from the previous ones and it is expressed in Equation (10),

$$p = 0.8\left(\frac{R_{\text{tot;lower}}}{R_{\text{tot;upper}}}\right) + 0.44 - 0.1\left(\frac{fl}{0.04}\right) - 0.2\left(\frac{0.6}{ss}\right) - 0.04\left(\frac{sd}{0.1}\right) \tag{10}$$

As with Method 1 (Equation (9)), the *p*-value directly depends on the ratio between the lower and upper limits of the total thermal resistance. Additionally, besides the constant 0.44 parcel, there are more three variables, namely: the flange length (*fl*), the stud spacing (*ss*) and stud depth (*sd*), all dimensions of which are given in metres [m].

2.1.3. ASHRAE Methods

Some of the previously described methods (e.g., parallel path method) assume that the heat flow is perpendicular to the wall. Although when the wall structure contains steel framing members next to materials with low thermal conductivity (e.g., thermal insulation), the two-dimensional effects caused by thermal bridges become more relevant [21]. The ASHRAE methods were developed for structures with widely spaced metal members of substantial cross-sectional areas and when the adjacent materials have very high different conductivities (two order or more of magnitude), as what happens on typical LSF constructions [19].

The ASHRAE methods are an adjustment of the parallel path method, where an area "weighting factor" is applied to the wall section influenced by the steel stud thermal bridge [21]. This section is defined by the width of the steel thermal bridge influence zone (Figure 3) and, thus, it is named section W. The remaining portion of the wall cavity without the thermal bridge influence it is called section CAV.

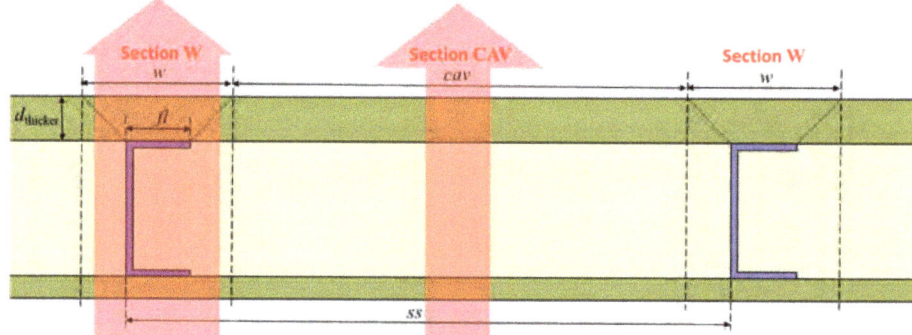

Figure 3. LSF wall cross-section illustration for the ASHRAE methods: Sections W and CAV.

The section W represents the area where the metal stud has influence on the heat path, being centred on the metal part of the wall cross-section and its length, *w*, is determined by,

$$w = fl + zf\, d_{\text{thicker}} \tag{11}$$

where *fl* is the flange length [m], *zf* is the zone factor (which will distinguish both ASHRAE methods as explained in the following sections) and d_{thicker} is the thickness [m] of the thicker sheathing side (internal or external).

For both sections paths, the thermal resistances values are computed and them combined using the parallel path method and the average thermal transmittance per unit overall area is calculated by reversing the total thermal resistance [19], as detailed in the next subsection.

ASHRAE Zone Method

The first method proposed by ASHRAE, the Zone Method, uses Equation (11) to calculate the length of section W and the zone factor, zf, is equal to 2.0. The remaining calculations for total thermal resistance and transmittance are the same for both ASHRAE methods and will be presented next. The detailed dimensions of Section W, which were already presented in Figure 3, are illustrated in Figure 4a. Moreover, the equivalent series-parallel circuit used in the simplified heat transfer calculations is displayed in Figure 4b. Notice that the steel frame is taken into account in the web and both flanges, though is neglected at the steel lip/return.

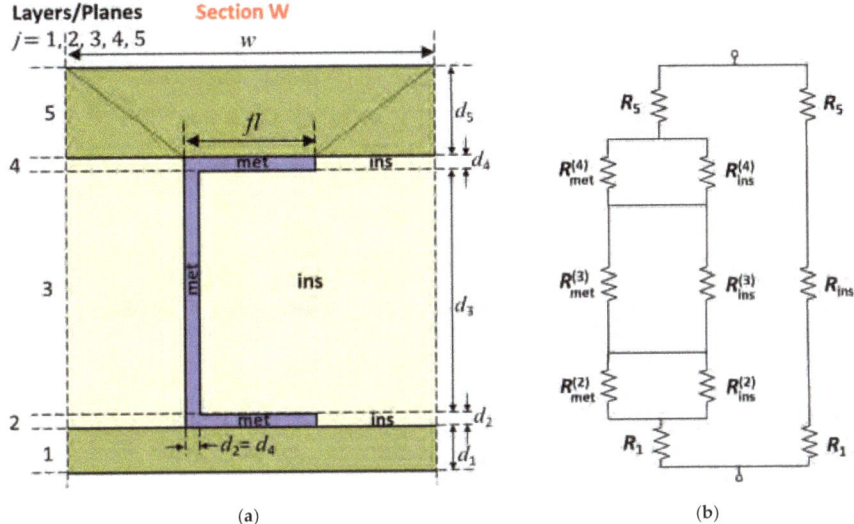

Figure 4. ASHRAE methods schematic illustration: (**a**) Section W of the LSF wall cross-section; (**b**) Equivalent series-parallel circuit.

The total thermal resistance, R_{tot}, of a generic LSF wall displayed in Figure 3 is computed by applying the parallel path method to both considered sections (W and CAV),

$$\frac{1}{R_{tot;ASHRAE}} = \sum_{i=1}^{2} \frac{f_i}{R_i} = \frac{w/ss}{R_{tot;w}} + \frac{cav/ss}{R_{tot;cav}} \quad (12)$$

where $R_{tot;w}$ and $R_{tot;cav}$ are the total thermal resistances [m²·K/W] of sections W and CAV, respectively, w and cav are the lengths [m] of sections W and CAV, respectively, and ss is the studs spacing [m].

The total thermal resistance of the homogeneous layers of the LSF wall cavity, $R_{tot;cav}$, is calculated as the summation of the thermal resistances of all the layers in series, including the internal and external surface thermal resistances,

$$R_{tot;cav} = R_{si} + R_1 + R_{ins} + R_5 + R_{se} \quad (13)$$

where R_{ins} is the thermal resistance of the insulation layer [m²·K/W].

The total thermal resistance of the Section W, $R_{tot;w}$, is computed making use of the isothermal planes method. First, the equivalent thermal resistance (R_j) of each thermally inhomogeneous layer

(j = 2, 3, 4) is calculated making use of the parallel path method to both metal (*met*) and insulation (*ins*) materials,

$$\frac{1}{R_2} = \sum_{i=1}^{2} \frac{f_i^{(2)}}{R_i^{(2)}} = \frac{fl/w}{R_{met}^{(2)}} + \frac{(w-fl)/w}{R_{ins}^{(2)}} \tag{14}$$

$$\frac{1}{R_3} = \sum_{i=1}^{2} \frac{f_i^{(3)}}{R_i^{(3)}} = \frac{d_2/w}{R_{met}^{(3)}} + \frac{(w-d_2)/w}{R_{ins}^{(3)}} \tag{15}$$

$$\frac{1}{R_4} = \frac{1}{R_2} \tag{16}$$

Next, these three equivalent thermal resistances are used to compute the total thermal resistance of Section W, having taken into account all the five layers considered, including the ones for the sheathing homogeneous layers (R_1 and R_5), as well as the surface thermal resistances,

$$R_{tot;w} = R_{si} + \sum_{j=1}^{5} R_j + R_{se} \tag{17}$$

Modified Zone Method

The Modified Zone Method is very similar to the Zone Method, making use of the same equations (Equations (12)–(17)). However, it uses a modified zone factor (zf) value, which is not a constant, nor necessarily equal to 2. In the Modified Zone Method, the width (w) of the steel stud influence zone (Section W in Figure 3), besides the flange length, fl, depends on three parameters [19]: (1) the ratio between thermal resistivities of sheathing material and cavity insulation material; (2) the size (depth) of the stud; (3) thickness of the sheathing material.

The modified zone factor, zf, is usually obtained from a chart [19] (when the thickness of the sheathing materials is higher than 16 mm) and depends on the ratio between the average resistivity of the external sheathing material (r_{sheat}) and cavity insulation material (r_{ins}) for the first 25 mm, combined with the stud size (usually one curve for each stud type). Notice that the thermal resistivity r of a material is the reciprocal of its thermal conductivity λ, i.e., $r = 1/\lambda$.

In this work, the authors adjusted two power trend-lines to the points obtained from reference [19] for C90 and C150 steel studs, as illustrated in Figure 5. The R-squared determination coefficient was very good, i.e., equal to 0.999 in both curves. These power functions were used in the computations for both steel profile sizes (C90 and C150). The authors were not able to find the modified zone factor curves/points for C170 and C200 studs. Thus, it was assumed by approximation that the C170 zf factors were similar to the ones provided by the C150 curve. Regarding the LSF walls with C200 steel studs, they were not computed by this method in this work (only five LSF walls).

The condition for using the chart presented on Figure 5 is that—for at least one of the sides of the wall—the total thickness of the sheathing layers must be thicker than 16 mm. If both interior and exterior sheathings have a total thickness smaller than 16 mm, the zf values should be obtained according to the following conditions [19]:

$$zf = \begin{cases} -0.5 & \text{if } r_{sheat} \leq 10.4 \text{ m·K/W} \\ +0.5 & \text{if } r_{sheat} > 10.4 \text{ m·K/W} \end{cases} \tag{18}$$

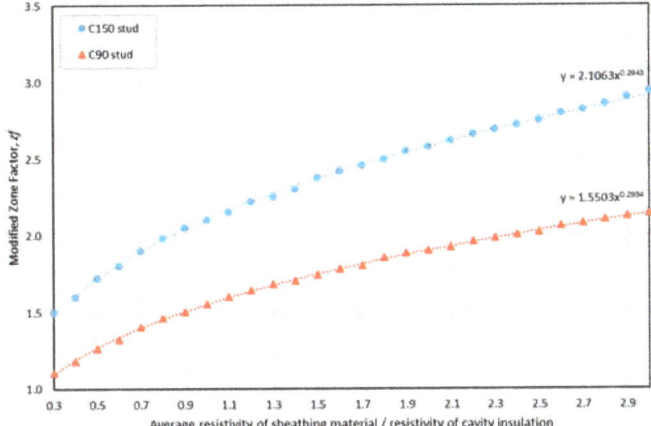

Figure 5. Modified zone factor curves for LSF walls with cavity insulation whenever the total thickness of sheathing materials is higher than 16 mm.

2.2. Numerical Reference 2D FEM Computations

There are several numerical computational methods that are able to reproduce highly detailed models of building components and accurately calculate and predicted their thermal behaviour under pre-established conditions. The numerical computational tool used in this work was the 2D FEM software, THERM [24]. For all LSF wall models the maximum error admitted on the FEM computations was 2%.

2.2.1. Accuracy Verification and Models Validation

The accuracy of THERM models used for the LSF wall thermal performance evaluation was checked under two different verifications: (1) benchmark values for two test cases presented on ISO 10,211 [25], and (2) comparison with the analytic U-value provided for a wall assuming homogeneous layers. Despite the fulfilled verifications, a validation with laboratorial measurements was also performed.

ISO 10,211 Test Cases Verification

To verify the accuracy of 2D calculation algorithms, the standard ISO 10,211 [25] provides, in Annex C, two test reference cases (Case 1 and 2). According with these 2D standard test cases, the FEM THERM software [24] is classified as a steady-state high precision algorithm. The authors also implemented two test cases to ensure and demonstrate their ability to accurately make use of this software to model heat transfer problems. In both test cases, the difference between the standard solution given for each point and the temperature computed by the algorithm should not exceed 0.1 °C [24]. In test case 1, we provided a sketch of a half square column with 28 grid points placed equidistantly, for which the corresponding temperatures for each point are known (Figure 6a). Figure 6b displays the temperature values calculated by THERM in these reference grid points, all of them being equal to the ones provided by ISO 10,211 when using one decimal place temperature values.

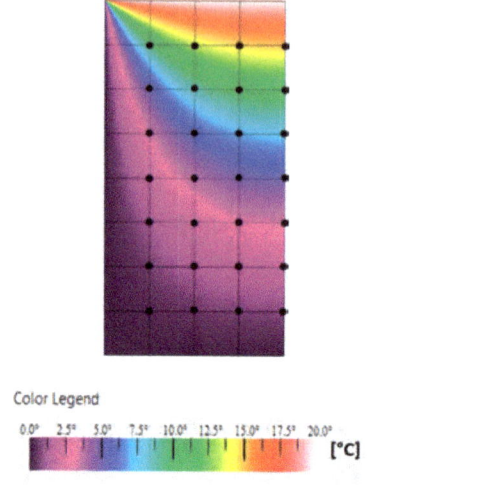

Figure 6. ISO 10,211 obtained results for test case 1: (**a**) Temperature distribution and reference grid points; (**b**) Computed temperatures at reference grid points.

In the second test case, the difference between the heat flow calculated and the reference value shall not exceed 0.1 W/m. Figure 7a illustrates the computed temperature distribution, as well as the points where the reference temperatures are provided (points A to I). Figure 7b displays the previously mentioned computed temperatures and the heat flow calculated by THERM for this model. For all reference points, the temperatures obtained were exactly the same as prescribed by ISO 10211. The calculated heat flow rate was only 0.01 W/m lower the reference value (9.5 W/m), but still far below the difference limit of 0.1 W/m.

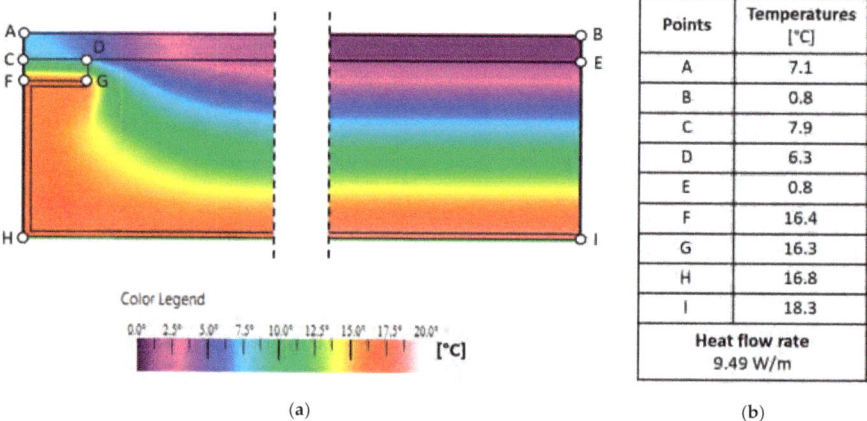

Figure 7. ISO 10,211 test case 2 obtained results: (**a**) Temperature distribution and reference points; (**b**) Computed values.

Homogeneous Wall Layers Verification

An additional simple verification that was made was to compare the analytical and the numerical results for a simplified model of the LSF wall, i.e., the same wall composed only for homogeneous layers (without the steel studs). For those homogeneous walls, the analytical solution is known, being the total thermal resistance calculated as a sum of the layer's resistances as described in Equation (1).

For this verification, the LSF wall (defined in Section 2.3) was used as a reference, but without steel studs. The numerical simulation result provided by THERM was $U = 0.227$ W/(m^2·K). As expected, a simple analytical approach for the same homogeneous wall brings up exactly the same thermal transmittance result.

Experimental Lab Measurements Validation

To validate the numerical simulation results provided by THERM software, used as reference values to evaluate the accuracy of the analytical methods, the thermal transmittance (U-value) of some LSF walls were measured on a laboratory facility. In these experiments it was used the heat flow meter method, prescribed in standard ISO 9869-1 [13]. To ensure a controlled temperature gradient between the two surfaces of the LSF wall test-specimen two small thermally insulated boxes were used (Figure 8): (1) a hot box, heated by an electrical resistance (70 watts), and (2) a cold box, cooled by a refrigerator attached to it (Figure 8b). The steady-state set-point temperatures considered in the hot and cold boxes were 40 °C and 5 °C, respectively.

Figure 8. Mini hot box apparatus: (a) Cold and hot boxes with the wall sample; (b) Refrigerator attached to the cold box.

The wall test samples used in these measurements have the following height and width dimensions: 1030 × 1060 mm, respectively, and the stud spacing was equal to 400 mm, as illustrated in Figure 9a. The cold formed steel profiles used in these experiments have a type C cross-sectional shape and have the following dimensions: C90 × 43 × 15 × 1.5 mm.

Four heat flux meters (Hukseflux HFP01, precision: ±3%) were used in these LSF wall experiments, being the measurements performed at four different locations: two at the hot surface and the remaining two at the cold wall surface. In both test-specimen wall surfaces a measurement location was chosen in the vicinity of the vertical steel stud (HFM1) and another one in the middle of the insulation cavity (HFM2), as illustrated in Figure 9b.

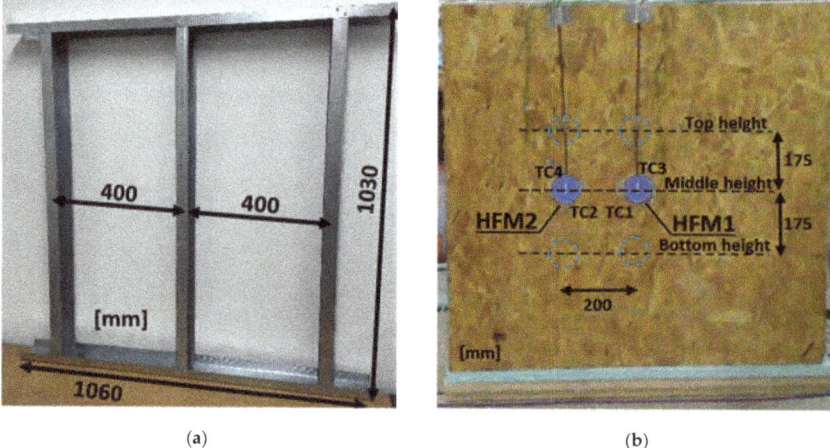

Figure 9. LSF small-scale test sample wall: (**a**) Steel frame; (**b**) Sensors locations on exterior wall surface (HFM–Heat flux meter; TC-Thermocouples).

The temperatures were measured making use of 12 thermocouples (TC), certified with class 1 precision, with half of them being used in each side of the wall (hot and cold). From these six TC, two measured the environment air temperature inside each box (TC5 and TC6), another two measured the air temperature between the radiation shield and the wall surface (TC3 and TC4), while the remaining two measured the wall surface temperatures (TC1 and TC2), as illustrated in Figure 9b.

In order to ensure the repeatability of the experimental measurements, one test was performed for each wall at three height locations (Figure 9b): (1) top, (2) middle, and (3) bottom, the average of these three tests being the considered measured U-value of the LSF wall. Each test had a duration of 24 h.

The data recorded during the experiments (temperatures and heat fluxes) were recorded in two PICO TC-08 data-loggers (precision: ±0.5 °C); one for each side of the LSF wall test-specimen (hot and cold). Making use of the data recorded (heat fluxes and temperatures) and applying the HFM method, prescribed in standard ISO 9869-1 [13], two distinct U-values were obtained: (1) a higher value for location 1, i.e., in the vicinity of the steel studs, and (2) a lower value between the steel studs, i.e., in the middle of the insulation cavity. The overall U-value of the wall was obtained by computing an area weighted of both U-values. The steel stud influence zone area was defined as prescribed by ASHRAE zone method.

Two LSF walls were tested to validate the numerical simulations: (1) an air cavity wall, and (2) a Mineral Wool (MW) insulation filled cavity wall. All the other exterior and interior sheathing materials were the same, i.e., an outer and an inner OSB layer (12 mm), attached to the C90 steel studs, as well as a Gypsum Plaster Board (GPB), this being the innermost layer (12.5 mm).

Table 2 display the obtained overall U-values measured under controlled laboratory conditions (three tests for each wall) and the predicted values by 2D FEM numerical simulations.

The U-value predicted by the numerical simulations for the MW LSF wall exactly matches the measured one (0.621 W/(m^2·K)), while for the air cavity LSF wall it was found to have an error of about 2%—the predicted U-value (1.931 W/(m^2·K)) being slightly lower than the measured one (1.969 W/(m^2·K)). Given the uncertainties related to the measurements (e.g., sensor precision and material properties) and the maximum error estimated for the FEM numerical simulations (under 2%), these results allowed to reiterate and ensure the reliability of the numerical simulations used in this work as reference values.

Table 2. Thermal Transmittance Values Measured in Lab and Computed by 2D FEM Numerical Simulations (THERM).

	Test	Sensors Location	U-Value [W/(m^2·K)]
Air Cavity LSF Wall	1	Top	1.984
	2	Middle	2.001
	3	Bottom	1.922
	-	Average Measured	1.969
	-	Computed by THERM	1.931
	-	Percentage Error	−2%
MW LSF Wall	1	Top	0.602
	2	Middle	0.614
	3	Bottom	0.648
	-	Average Measured	0.621
	-	Computed by THERM	0.621
	-	Percentage Error	0%

MW—Mineral Wool; LSF—Lightweight Steel Frame.

2.2.2. Boundary Conditions

In this section are briefly presented the boundary conditions used in the numerical simulations of the LSF wall cross-sections. The temperatures for the inside warm and outside cold environments were set to 20 °C and 0 °C, respectively. The surface thermal resistance values used in the simulations were obtained from ISO 6946 [20] for horizontal heat flow (walls): 0.13 (m^2·K)/W for internal thermal resistance (R_{si}) and 0.04 (m^2·K)/W for external thermal resistance (R_{se}). Additionally, two adiabatic surfaces were defined in both extremities of the LSF wall model cross-section.

2.2.3. Air Layers Modelling

Some LSF walls evaluated do not present a full-filled insulation cavity or have an empty air cavity, being necessary to model air gaps inside the LSF wall. The thermal resistances of those unventilated air layers were modelled with a solid-equivalent thermal conductivity, using the thermal resistance values prescribed by ISO 6946 [20] for horizontal heat flow.

2.3. Walls Description and Material Characterization

In this work, all the evaluated walls were derived from a typical reference exterior LSF wall (hybrid frame construction), as illustrated in Figure 10 and described in Table 3. The vertical steel studs (C90 × 43 × 15 × 1.5 mm) were spaced 600 mm apart. The exterior sheathing was constituted of an oriented strand board (OSB) panel (12 mm thick), while the external thermal insulation composite system (ETICS) was made of EPS (Expanded Polystyrene), 50 mm thick. The interior sheathing was made of an OSB panel (12 mm) and a gypsum plaster board (GPB), 12.5 mm thick, while the air cavity was filled with mineral wool (MW) batt insulation.

Modifying some parameters and variables (listed on Table 4) on the reference LSF wall (Figure 10), eighty different LSF walls models were obtained. The evaluated parameters were the cold formed steel studs, the cavity insulation, the exterior continuous insulation and the studs facing sheathing materials.

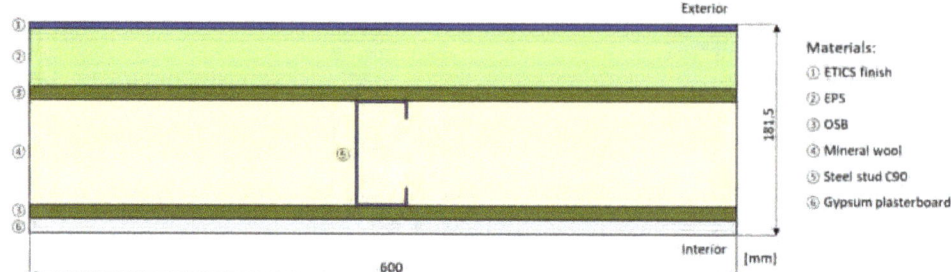

Figure 10. Reference LSF wall cross-section materials and dimensions.

Table 3. Reference LSF Wall Material Thickness (*d*) and Thermal Conductivities (λ).

Material (from Outer to Innermost Layer)	*d* [mm]	λ [W/(m·K)]	Ref.
ETICS [1] finish	5	0.450	[26]
EPS [2]	50	0.036	[27]
OSB [3]	12	0.100	[28]
MW [4]	90	0.035	[29]
Steel studs (C90 × 43 × 15 × 1.5 mm)	90	50.000	[30]
OSB [3]	12	0.100	[28]
GPB [5]	12.5	0.175	[31]
Total Thickness	181.5	-	-

[1] ETICS-External Thermal Insulation Composite System; [2] EPS-Expanded Polystyrene; [3] OSB-Oriented Strand Board; [4] MW-Mineral Wool; [5] GPB-Gypsum Plaster Board.

Regarding the steel studs, four different values were modelled for the spacing (300–800 mm range) and depth (90–200 mm range) of the studs. Five different values for the steel studs thickness were evaluated, ranging from 0.6 mm up to 2.0 mm. Two different studs flange lengths were modelled: 43 and 70 mm, as obtained from the Pertecno cold-formed steel profiles manufacturer catalogue [32].

Concerning the cavity insulation thickness, three different levels of batt insulation were evaluated for each one of the four assessed steel studs (C90, C150, C170 and C200): (1) no cavity insulation; (2) half of the cavity filled with batt insulation; (3) cavity full filled with batt insulation. Moreover, two different batt insulation materials were considered: (1) the reference one, i.e., mineral wool (MW) with thermal conductivity equal to 0.035 W/(m·K), and (2) a better performance insulation material (aerogel insulation blanket-AIB) with 0.018 W/(m·K).

Regarding exterior continuous insulation, eight different thicknesses were evaluated, ranging from 0 mm up to 80 mm. Furthermore, two different materials were considered: (1) the reference one, i.e., EPS with thermal conductivity equal to 0.036 W/(m·K), and (2) a worst performance insulation material (insulation cork board-ICB) with 0.045 W/(m·K).

Finally, concerning the sheathing parameter, besides the OSB and GPB panels, three different materials were evaluated, namely cement wood board (CWB), fibre cement board FCB and glass-fibre reinforced board (GRB). The thermal conductivities of these sheathing materials ranges from 0.100 W/(m·K) for OSB up to 0.500 W/(m·K) for GRB.

Usually LSF walls are grouped into warm, hybrid and cold frame construction depending on the thermal insulation type/location, i.e., cavity batt insulation and/or exterior continuous thermal insulation [1]. Table 5 displays the total number of LSF walls evaluated (80) as well as the number of LSF walls by frame type: Warm (22 walls), Hybrid (43 walls), and Cold (15 walls). Additionally, this table also shows the range of *U*-values evaluated, being the minimum thermal transmittance equal to 0.153 W/(m^2·K) for a hybrid frame construction, while the maximum value is 0.983 W/(m^2·K) for warm frame construction.

Table 4. Evaluated Parameters, Variables and Values Used in the Simulations, and Range of Obtained Thermal Transmittances (U-Values).

Parameter	Variable		Evaluated Values	U-Value [W/(m²·K)] (Min–Max.)
Steel Studs	Spacing [mm]		300, 400, 600 *, 800	0.260–0.319
	Depth [mm]		90 *, 150, 170, 200 [32]	0.199–0.272
	Thickness [mm]		0.6, 1.0, 1.2, 1.5 *, 2.0	0.264–0.274
	Flange [mm]		43 *, 70 [32]	0.272–0.223
Cavity Insulation	Thickness [mm]	C90 *	0, 45, 90 *	0.272–0.489
		C150	0, 75, 150	0.224–0.489
		C170	0, 85, 170	0.223–0.489
		C200	0, 100, 200	0.199–0.489
	Thermal Conductivity [W/(m·K)]	AIB [1]	0.018 [33]	0.153–0.287
		MW *[2]	0.035 * [29]	0.199–0.381
Exterior Insulation (ETICS [10])	Thickness [mm]		0, 5, 10, 15, 20, 30, 50 *, 80	0.221–0.869
	Thermal Conductivity [W/(m·K)]	EPS *[3]	0.036 * [27]	0.221–0.869
		ICB [4]	0.045 [30]	0.246–0.346
Sheathing	Thermal Conductivity [W/(m·K)]	OSB *[5]	0.100 * [28]	0.221–0.983
		GPB *[6]	**0.175 * [31]**	
		CWB [7]	0.220 [34]	
		FCB [8]	0.390 [35]	
		GRB [9]	0.500 [36]	

* Reference value; [1] AIB-Aerogel Insulation Blanket; [2] MW-Mineral Wool; [3] EPS-Expanded Polystyrene; [4] ICB-Insulation Cork Board; [5] OSB-Oriented Strand Board; [6] GPB-Gypsum Plaster Board; [7] CWB-Cement Wood Board; [8] FCB-Fibre Cement Board; [9] GRB-Glass-fibre Reinforced Board; [10] ETICS-Exterior Thermal Insulation Composite System.

Table 5. Number of Evaluated LSF Walls by Frame Type and Range of Obtained Thermal Transmittances (U-values).

Frame Type	Number of Evaluated LSF Walls	U-Value [W/(m²·K)]	
		Min.	Max.
Warm	22	0.348	0.983
Hybrid	43	0.153	0.608
Cold	15	0.384	0.869
Total	80	0.153	0.983

3. Results and Discussion

3.1. All LSF Walls

The U-values obtained by the six analytical methods for all the evaluated LSF walls are plotted on Figure 11. Each point in these graphics represents a different LSF wall, being the value on the horizontal axis the reference U-value provided by the numerical 2D FEM simulations, while the value on the vertical axis is the analytical U-value estimated by the respective method: (a) ISO 6946 Combined Method; (b) Gorgolewski Method 1; (c) Gorgolewski Method 2; (d) Gorgolewski Method 3; (e) ASHRAE Zone Method; (f) Modified Zone Method.

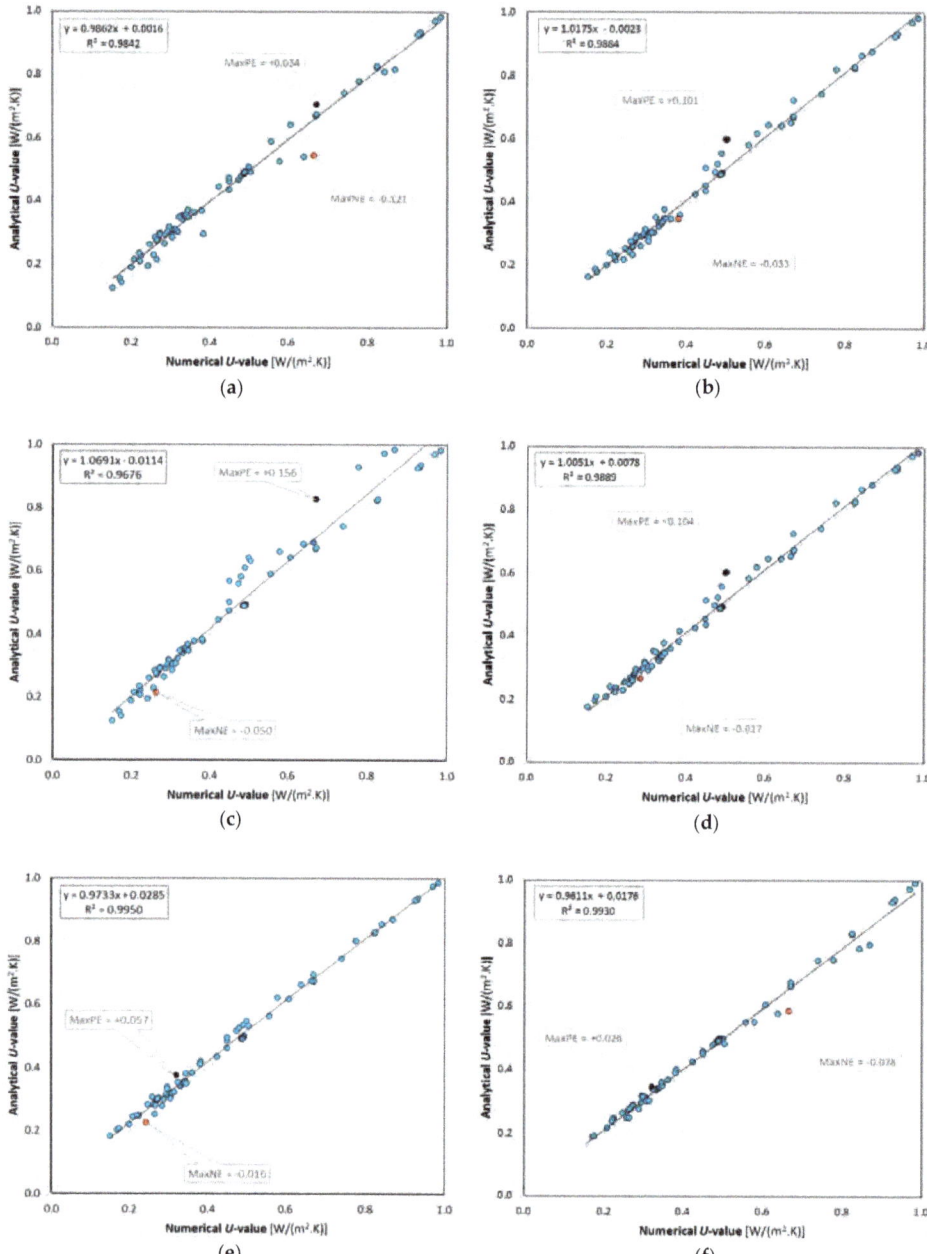

Figure 11. Thermal transmittances (*U*-values) comparison between the evaluated analytical methods and the numerical 2D FEM results used as reference: (**a**) ISO 6946 Combined Method; (**b**) Gorgolewski Method 1; (**c**) Gorgolewski Method 2; (**d**) Gorgolewski Method 3; (**e**) ASHRAE Zone Method; (**f**) Modified Zone Method.

Moreover, these plots also display a linear trend-line, the R-squared coefficient of determination, the maximum positive error (MaxPE), as well as the maximum negative error (MaxNE) for each analytical method, as well as a 45 degrees' inclination line, that corresponds to the plots position for a virtual perfect match between the analytical and the numerical methods.

First of all, it could be concluded that there is a quite good agreement between the U-values provided by the analytical methods evaluated and the numerical reference ones, evidencing a pretty good accuracy of these analytical methods.

Gorgolewski Method 2 (Figure 11c) exhibits a larger dispersion of values mainly for higher U-values (greater than 0.4 W/(m^2·K)). This feature is ensured by the linear trend line, which has the biggest slope (1.0691), being above the 45° diagonal line and—even so—exhibiting the smallest determination coefficient (0.9676). Additionally, a major positive error was also found in this analytical method (+0.156 W/(m^2·K)). This LSF wall is cold framed (without ETICS), having the air cavity 50% filled with mineral wool.

Moreover, the major negative error (−0.121 W/(m^2·K)) was found in the ISO 6946 Combined Method (Figure 11a). This LSF wall is cold-framed (without ETICS), having GRB sheathing panels. Quite surprisingly, this analytical method provides pretty good accuracy, since according with standard ISO 6946 [20] it should not be applicable to building elements where the insulation is bridged by metal, as happens in these cold and hybrid LSF walls.

Looking to both ASHRAE methods (Figure 11e,f), though they have different trends, they both have a very good determination factor (0.995 and 0.993). The ASHRAE Zone Method (Figure 11e) has a very good precision for higher U-values (e.g., >0.6 W/(m^2·K)), whereas for lower values exhibits a conservative trend, i.e., giving U-values bigger than the real ones. On the other hand, the Modified Zone Method (Figure 11f) has a good precision for lower U-values (e.g., <0.6 W/(m^2·K)), but for lower values exhibits an overoptimistic trend, i.e., U-values smaller than the real ones.

Notice that the linear trend-lines presented before and the corresponding determination factors are not the most adequate features to accurately quantify the precision of each analytical method, since they do not correlate the analytical U-values with the numerical reference ones, but instead they correlate the analytical values with the corresponding trend-line, which could be very different from the 45° diagonal line.

Thus, it was decided to also compute the root mean square error (RMSE), as an absolute value and as a percentage, which is graphically displayed in Figure 12. These plots also contain the maximum positive errors (MaxPE) and the maximum negative errors (MaxNE).

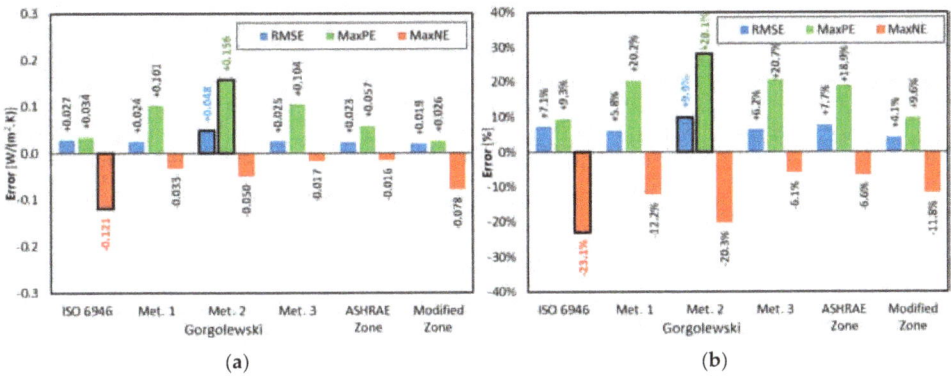

Figure 12. U-values errors for all LSF walls: (**a**) Absolute errors; (**b**) Percentage errors.

The Gorgolewski Method 2 exhibits the major RMS error (+0.048 W/(m^2·K); +9.9%), as well as the higher maximum positive error (+0.156 W/(m^2·K); +28.1%), confirming the relatively bad accuracy performance of this method. As mentioned before, the maximum negative error was found in the ISO 6946 Combined Method (−0.121 W/(m²·K); −23.1%).

Looking now to the smaller RMS error, the lowest value was found for the Modified Zone Method (+0.019 W/(m²·K); +4.1%). This analytical method also exhibits the lowest absolute MaxPE (+0.026 W/(m²·K)) and the second lowest percentage MaxPE (+9.6%), confirming the relatively good accuracy performance of this method. The lowest absolute negative error was found in the ASHRAE Zone Method (−0.016 W/(m²·K)), while the lowest percentage value was found in the Gorgolewski Method 3 (−6.1%).

Taking into account only the RMSE percentage values, the accuracy performance of these analytical methods could be ranked, from the better to the worst, as follows: (1) Modified Zone Method (+4.1%); (2) Gorgolewski Method 1 (+5.8%); (3) Gorgolewski Method 3 (+6.2%); (4) ISO 6946 Combined Method (+7.1%); (5) ASHRAE Zone Method (+7.7%); (6) Gorgolewski Method 2 (+9.9%).

In the following sections, a similar analysis will be performed, but separating the LSF walls into groups, depending on the frame type: (1) warm; (2) hybrid; (3) cold.

3.2. Warm Frame Walls

Figure 13 shows the absolute and percentage thermal transmittance error values obtained for the warm frame walls. Comparing these values with the previous ones (Figure 12), the first remarkable feature is that these error values now appear too small. This is justifiable by the existence of only continuous external thermal insulation (ETICS), existing no insulation in the air cavity between the steel studs and therefore without any thermal bridge effect.

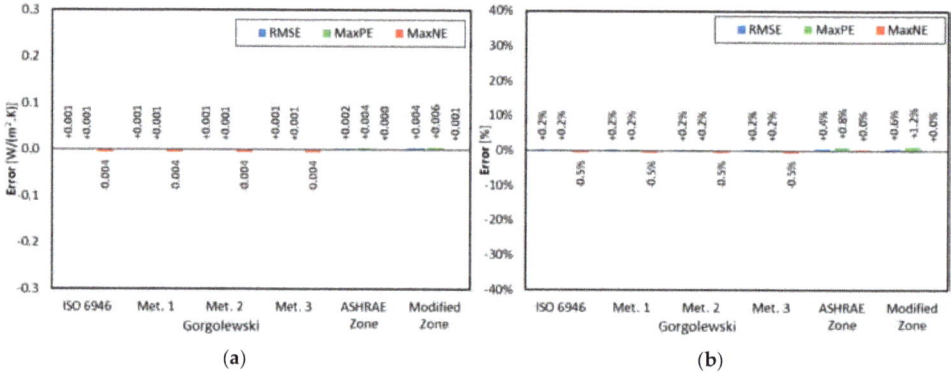

Figure 13. *U*-values errors for warm frame walls: (**a**) Absolute errors; (**b**) Percentage errors.

The errors are so small in absolute values, ranging between +0.006 W/(m²·K) and −0.004 W/(m²·K), as well as in percentage (+1.2%; −0.5%), that it does not worth it to make a more detailed analysis.

3.3. Hybrid Frame Walls

Figure 14 illustrates the error values obtained for the hybrid frame walls. These error values are considerably higher when compared to the previous warm frame ones (Figure 13). This is to be expected, as besides the continuous external thermal insulation, there is also batt insulation which is bridged by the steel studs.

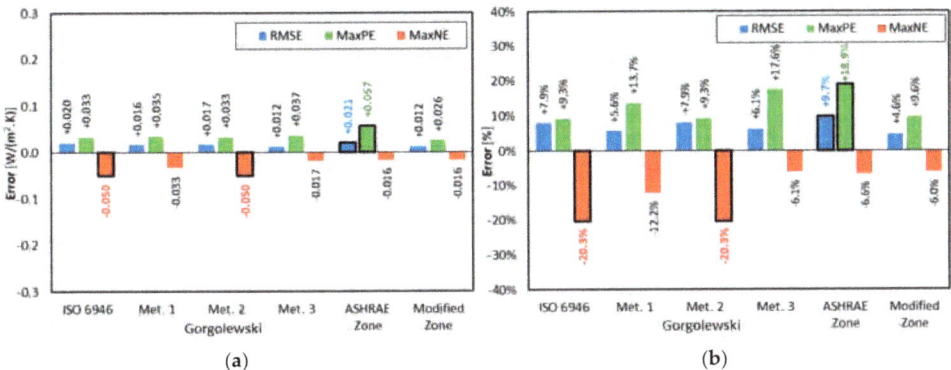

Figure 14. U-values errors for hybrid frame walls: (**a**) Absolute errors; (**b**) Percentage errors.

In these hybrid frame walls, the major U-value RMS error was obtained by the ASHRAE Zone Method (+0.021 W/(m²·K); +9.7%), as well as the higher maximum positive error (+0.057 W/(m²·K); +18.9%), showing a relatively bad accuracy performance of this method. The maximum negative error was achieved by both ISO 6946 Combined Method and Gorgolewski Method 2 (−0.050 W/(m²·K); −20.3%).

The minor U-value RMS error was obtained by the Modified Zone Method (+0.012 W/(m²·K); +4.6%), exhibiting also the lowest absolute positive error (+0.026 W/(m²·K)) and the smaller percentage negative error (−6.0%), demonstrating the relatively good accuracy of this analytical method for hybrid frame walls.

Having taken into account only the RMSE percentage values, the accuracy performance of these analytical methods, regarding the computation of hybrid frame walls U-values, could be ranked, from the better to the worst, as enumerated next: (1) Modified Zone Method (+4.6%); (2) Gorgolewski Method 1 (+5.6%); (3) Gorgolewski Method 3 (+6.1%); (4) ISO 6946 Combined Method and Gorgolewski Method 2 (+7.9%); (5) ASHRAE Zone Method (+9.7%).

3.4. Cold Frame Walls

The thermal transmittance error values for cold frame walls are displayed in Figure 15. In general, these error values are even higher than the ones obtained for hybrid fame walls (Figure 14). These could be explained by the fact that all thermal insulation of the LSF wall is now bridged by the steel studs, exhibiting no continuous external thermal insulation (ETICS).

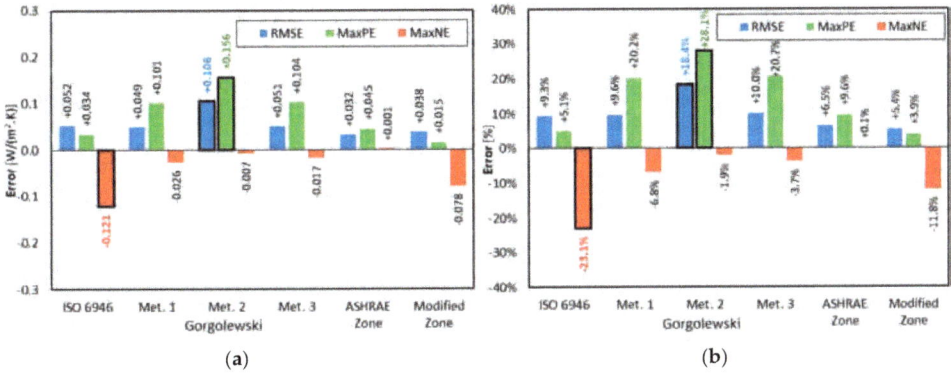

Figure 15. U-values errors for cold frame walls: (**a**) Absolute errors; (**b**) Percentage errors.

The major U-value RMS error was obtained by the Gorgolewski Method 2 (+0.106 W/(m²·K); +18.4%), as well as the higher maximum positive error (+0.156 W/(m²·K); +28.1%), showing a relatively bad accuracy performance of this method. However, the maximum negative error is the smaller one (−0.007 W/(m²·K); −1.9%). These two facts suggest that this analytical method tends to provide much too conservative U-values—i.e., higher than the real ones—originating from a relatively lower precision performance.

The maximum negative error was found in the ISO 6946 Combined Method (−0.121 W/(m²·K); −23.1%), evidencing that this method could provide significant overoptimistic U-values, i.e., lower than the real ones, being this trend already seen also for hybrid frame walls (Figure 14).

Observing now the lowest U-value RMS errors, the smaller error was obtained by the ASHRAE Zone Method in absolute value (+0.032 W/(m²·K) and by the Modified Zone Method in percentage value (+5.4%). The Modified Zone Method also exhibits the lowest maximum positive error (+0.015 W/(m²·K); +3.9%), revealing also in this parameter a relatively good accuracy performance. Another interesting feature is that all the errors provided by the ASHRAE Zone Method are positive, the minimum value being equal to +0.001 W/(m²·K) and +0.1%, showing a conservative trend.

Taking into account only the RMSE percentage values obtained for the cold frame walls, the accuracy performance of the evaluated analytical methods could be ranked, from best to the worst, as listed next: (1) Modified Zone Method (+5.4%); (2) ASHRAE Zone Method (+6.5%); (3) ISO 6946 Combined Method (+9.3%); (4) Gorgolewski Method 1 (+9.6%); (5) Gorgolewski Method 3 (+10.0%); (6) Gorgolewski Method 2 (+18.4%).

3.5. Overview

In order to provide a better perception and an easier comparison between the average accuracy performance of the six analytical methods evaluated, Figure 16 displays the percentage RMS errors for all LSF walls and also grouped by frame types.

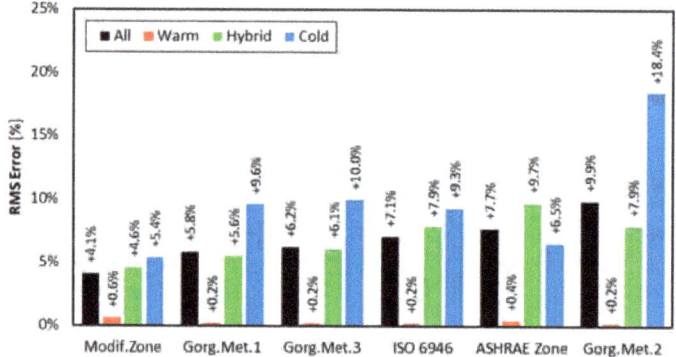

Figure 16. Root mean square U-values errors obtained for the evaluated analytical methods.

Looking to the four grouped RMSE values and comparing their relative values, the highest RMS percentage errors occur in the cold frame walls, followed by the hybrid frame walls (which exhibits values relatively closer to all LSF walls) and by the warm frame walls. There is only one exception: the ASHRAE Zone Method, where the higher error is in the hybrid frame (9.7%) instead of in the cold frame (6.5%). As mentioned before, this trend is related to the amount of thermal insulation that is bridged by the steel frames, which increases the error of the analytical methods in cold frame walls, the error being significantly reduced when there is only continuous external insulation (warm frame walls).

The major RMS U-value error occurred in cold frame walls evaluated by the Gorgolewski Method 2 (18.4%), while for the same frame type the smaller error occurs in the Modified Zone Method (5.4%).

Regarding the most common LSF wall type (hybrid frame), the major error occurred in the ASHRAE Zone Method (9.7%), while the smaller error happened in the Modified Zone Method (4.6%), confirming the relatively good accuracy performance of this method.

Observing all the LSF walls evaluated, the RMS U-value errors ranges between 4.1% (Modified Zone Method) and 9.9% (Gorgolewski Method 2). According to these RMS U-value errors, the accuracy performance of the evaluated methods could be ranked as displayed in Figure 16 horizontal axis from the left (better) to the right (worst), being this ranking previously presented in Section 3.1.

In order to assess the statistical significance of the previously presented percentage U-values errors, Table 6 displays the standard deviations and the confidence intervals (upper and lower limits, and amplitude) for each evaluated analytical method, computed for a level of significance equal to 5%, i.e., for a 95% confidence level. The standard deviation ranges from 4.0% (Modified Zone Method) up to 9.4% (Gorgolewski Method 2), while the amplitude of the confidence intervals ranges from 1.9% up to 4.1%, respectively. Thus, it can be concluded that these error measures are statistically significant.

Table 6. Standard Deviations and Confidence Intervals for the Analytical U-Values Errors.

		Modified Zone	Gorgol. Met.1	Gorgol. Met.3	ISO 6946	ASHRAE Zone	Gorgol. Met.2
	Stand. Dev.	4.0%	5.8%	5.7%	7.0%	5.7%	9.4%
Confid. Interval	Upper Lim.	1.8%	2.3%	3.8%	0.4%	6.4%	5.2%
	Lower Lim.	−0.1%	−0.2%	1.3%	−2.7%	3.9%	1.1%
	Amplitude	1.9%	2.5%	2.5%	3.1%	2.5%	4.1%

To better visualize the statistical distribution of the obtained analytical U-values errors, a box and whisker graph is displayed in Figure 17. This plot allows us to verify the higher statistical reliability of the Modified Zone Method given the lowest interquartile interval amplitude, being these values very close to zero, including its average (0.9%). These issues ensure a good accuracy performance of this method. Looking to the outliers, Gorgolewski Method 2 exhibits the greatest variability, having the biggest outlier range, ranging between −20.3% and +28.1%. Considering now the other intermediate methods, they have a similar statistical behaviour, standing out the absence of outlier values for the ASHRAE Zone Method and the existence of only negative outlier values for the ISO 6946 Combined Method, down to −23.1%. Moreover, this method is the only which exhibits a negative average (−1.2%), confirming its trend to be over optimistic, predicting a better thermal performance (i.e., a lower U-value) than the real one.

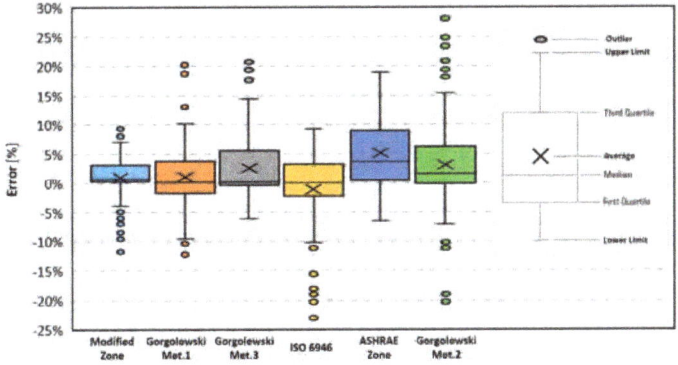

Figure 17. Statistical box and whisker graph for the evaluated analytical methods.

4. Conclusions

In this work, the accuracy performance of six analytical methods to compute the thermal transmittance (U-value) of LSF walls was evaluated. These methods were described and applied to estimate the U-values of 80 LSF walls—their precision being evaluated by comparison with the results provided by 2D FEM numerical simulations which were experimentally validated. Trend-lines and respective determination coefficients were obtained for each method. Moreover, the root mean square error (RMSE), the maximum positive error (MaxPE) in absolute and percentage values, as well as the maximum negative error (MaxNE) for each method for all LSF walls were also computed, as well as for each frame type: (1) warm; (2) hybrid; (3) cold. To assess the statistical significance of the obtained analytical percentage U-values errors, standard deviations and confidence intervals were computed, and a statistical box and whisker graphs were plotted.

All the evaluated analytical methods showed a quite good accuracy performance, with the RMSE ranging from 0.019 W/(m^2·K) up to 0.048 W/(m^2·K), or in percentage from 4.1% up to 9.9%. The maximum positive and negative U-values errors were +0.156 W/(m^2·K) and −0.121 W/(m^2·K), respectively. In percentages these error values were +28.1% and −23.1%, respectively.

As expected, given the different LSF walls thermal insulation configuration, the precision of the analytical methods for warm frame walls (RMS errors up to 0.6%, in the ISO 6946 Modified Zone Method) was considerably higher than the one observed on hybrid (RMS errors up to 9.7%, in the ASHRAE Zone Method) and cold frame walls (RMS errors up to 18.4%, in the Gorgolewski Method 2). Moreover, the worst accuracy of the evaluated analytical methods was found in cold frame walls, where all the batt thermal insulation is bridged by the steel studs.

Having taken into account all eighty LSF walls studied alongside the obtained RMS U-values errors expressed in percentage, the best accuracy performance was found in the Modified Zone Method (4.1%), while the worst was found in the Gorgolewski Method 2 (9.9%), with the latter also having the higher maximum positive error (+0.156 W/(m^2·K); +28.1%), evidencing a tendency to provide conservative U-values. The other two Gorgolewski methods (1 and 3) were ranked second (5.8%) and third (6.2%), respectively.

Very surprisingly, since the ISO 6946 standard [20] states that elements where insulation is bridged by metal (e.g., cold and hybrid LSF walls) are out of the scope of this method, the ISO 6946 Combined Method was ranked as the fourth most accurate methodology, exhibiting better performance (7.1%) than other two analytical methods (ASHRAE Zone Method and Gorgolewski Method 2), which were specifically developed for LSF elements. Nevertheless, the use of this analytical method should be performed with some caution, since it was the one that exhibit the larger negative error (−0.121 W/(m^2·K); −23.1%), evidencing some trend to provide over-optimistic U-values.

Author Contributions: All the authors participated equally to this work. All authors have read and agreed to the published version of the manuscript.

Funding: This research was funded by FEDER funds through the Competitivity Factors Operational Programme—COMPETE and by national funds through FCT—Foundation for Science and Technology within the scope of the project POCI-01-0145-FEDER-032061.

Acknowledgments: The authors also want to thank the support provided by the following companies: Pertecno, Gyptec Ibéria, Volcalis, Sotinco, Kronospan, Hulkseflux, Hilti and Metabo.

Conflicts of Interest: The authors declare no conflict of interest.

Nomenclature

Symbols

R	thermal resistance [m^2·K/W]
U	thermal transmittance [W/(m^2·K)]
a	width of section A (thickness of the steel stud web) [m]
b	width of section B (wall insulation cavity) [m]
cav	width of section CAV (the remaining wall cavity zone) [m]
d	layer sheathing thickness [m]
f	fractional area [—]
fl	flange length [m]
p	weight factor for the Gorgolewski method [—]
r	thermal resistivity [m·K/W]
sd	stud depth [m]
ss	stud spacing [m]
w	width of section W (steel stud influence zone) [m]
zf	zone factor [—]
λ	thermal conductivity [W/(m·K)]

Subscripts

ins	insulation
lower	lower limit
met	metal
n	number of layers or planes
q	number of sections or paths
se	external surface
sheat	sheathing
si	internal surface
thicker	thicker sheathing side (interior or exterior)
tot	total
upper	upper limit
i	sections, paths (A, B, C, ...)
j	layers, planes (1, 2, 3, ...)

Acronyms

1D	One-Dimensional
2D	Two-Dimensional
3D	Three-Dimensional
AIB	Aerogel Insulation Blanket
ASHRAE	American Society of Heating, Refrigerating and Air-conditioning Engineers
CBW	Cement Wood Board
CHB	Calibrated Hot Box
EPS	Expanded Polystyrene
ETICS	External Thermal Insulation Composite System
FCB	Fibre Cement Board
FEM	Finite Element Method
GHB	Guarded Hot Box
GHP	Guarded Hot Plate
GPB	Gypsum Plasterboard
GRB	Glassfibre Reinforced Board
HB	Hot Box
HFM	Heat Flow Meter
ICB	Insulation Cork Board
IRT	Infrared Thermography
ISO	International Standards Organization

LSF	Lightweight Steel Frame
MaxNE	Maximum Negative Error
MaxPE	Maximum Positive Error
MW	Mineral Wool
OSB	Oriented Strand Board
RMSE	Root Mean Square Error
TC	Thermocouple

References

1. Santos, P.; da Silva, L.S. *Energy Efficiency of Light-Weight Steel-Framed Buildings*, 1st ed.; Technical Committee 14—Sustainability & Eco-Efficiency of Steel Construction; European Convention for Constructional Steelwork (ECCS): Mem Martins, Portugal, 2017; ISBN 978-92-9147-105-8.
2. Soares, N.; Santos, P.; Gervásio, H.; Costa, J.J.; Da Silva, L.S. Energy efficiency and thermal performance of lightweight steel-framed (LSF) construction: A review. *Renew. Sustain. Energy Rev.* **2017**, *78*, 194–209. [CrossRef]
3. Santos, P. Chapter 3—Energy Efficiency of Lightweight Steel-Framed Buildings. In *Energy Efficient Buildings*; Eng Hwa Yap, Ed.; InTech: London, UK, 2017; pp. 35–60.
4. Roque, E.; Santos, P.; Pereira, A.C. Thermal and sound insulation of lightweight steel-framed façade walls. *Sci. Technol. Built Environ.* **2019**, *25*, 156–176. [CrossRef]
5. Roque, E.; Santos, P. The Effectiveness of Thermal Insulation in Lightweight Steel-Framed Walls with Respect to Its Position. *Buildings* **2017**, *7*, 13. [CrossRef]
6. Santos, P.; Lemes, G.; Mateus, D. Thermal Transmittance of Internal Partition and External Facade LSF Walls: A Parametric Study. *Energies* **2019**, *12*, 2671. [CrossRef]
7. Soares, N.; Martins, C.; Gonçalves, M.; Santos, P.; da Silva, L.S.; Costa, J.J. Laboratory and in-situ non-destructive methods to evaluate the thermal transmittance and behaviour of walls, windows, and construction elements with innovative materials: A review. *Energy Build.* **2019**, *182*, 88–110. [CrossRef]
8. Gorgolewski, M. Developing a simplified method of calculating U-values in light steel framing. *Build. Environ.* **2007**, *42*, 230–236. [CrossRef]
9. Martins, C.; Santos, P.; da Silva, L.S. Lightweight steel-framed thermal bridges mitigation strategies: A parametric study. *J. Build. Phys.* **2016**, *39*, 342–372. [CrossRef]
10. Santos, P.; Martins, C.; da Silva, L.S.; Bragança, L. Thermal performance of lightweight steel framed wall: The importance of flanking thermal losses. *J. Build. Phys.* **2014**, *38*, 81–98. [CrossRef]
11. Santos, P.; Gonçalves, M.; Martins, C.; Soares, N.; Costa, J.J. Thermal Transmittance of Lightweight Steel Framed Walls: Experimental Versus Numerical and Analytical Approaches. *J. Build. Eng.* **2019**, *25*, 100776. [CrossRef]
12. Sassine, E. A practical method for in-situ thermal characterization of walls. *Case Stud. Therm. Eng.* **2016**, *8*, 84–93. [CrossRef]
13. *ISO 9869, Thermal Insulation—Buildins Elements—In-Situ Measurement of Thermal Resistance and Thermal Transmittance—Part 1: Heat Flow Meter Method*; ISO—International Organization for Standardization: Geneva, Switzerland, 2014.
14. Salmon, D. Thermal conductivity of insulations using guarded hot plates including recent developments and sources of reference materials. *Meas. Sci. Technol.* **2001**, *12*, R89. [CrossRef]
15. *ISO 8990, Thermal Insulation—Determination of Steady-State Thermal Transmission Properties—Calibrated and Guarded Hot Box*; ISO—International Organization for Standardization: Geneva, Switzerland, 1994.
16. Klems, J.H. A calibrated hotbox for testing window systems—Construction, calibration, and measurements on prototype high-performance windows. In Proceedings of the ASHRAE/DOE Conference on the Thermal Performance of the Exterior Envelopes of Buildings, ASHRAE, Orlando, FL, USA, 3–5 December 1979.
17. Basak, C.K.; Sarkar, G.; Neogi, S. Performance evaluation of material and comparison of different temperature control strategies of a Guarded Hot Box U-value Test Facility. *Energy Build.* **2015**, *105*, 258–262. [CrossRef]
18. Lucchi, E. Applications of the infrared thermography in the energy audit of buildings: A review. *Renew. Sustain. Energy Rev.* **2018**, *82*, 3077–3090. [CrossRef]

19. ASHRAE. *Handbook of Fundamentals (SI Edition)*; ASHRAE—American Society of Heating, Refrigerating and Air-conditioning Engineers: Atlanta, GA, USA, 2017.
20. *ISO 6946, Building Components and Building Elements—Thermal Resistance and Thermal Transmittance—Calculation Methods*; ISO—International Organization for Standardization: Geneva, Switzerland, 2017.
21. Kosny, J.; Christian, J.E.; Barbour, E.; Goodrow, J. *Thermal Performance of Steel-Framed Walls*; ORNL report; Oak Ridge National Laboratory: Oak Ridge, TN, USA, 1994.
22. Kosny, J.; Christian, J.E. Reducing the uncertainties associated with using the ASHRAE zone method for R-value calculations of metal frame walls. In *ASHRAE Transactions, Technical and Symposium Papers*; Refrigerating and Air-Conditioning Engineers, Inc.: Atlanta, GA, USA, 1995; Volume 101, pp. 779–788.
23. Doran, S.M.; Gorgolewski, M.T. *BRE Digest 465—U-Values for Light Steel-Frame Construction*; BRE—Building Research Institute: London, UK, 2002.
24. THERM. *Software Version 7.6.1*; Lawrence Berkeley National Laboratory, United States Department of Energy: Berkeley, CA, USA, 2017. Available online: https://windows.lbl.gov/software/therm (accessed on 14 February 2019).
25. *ISO 10211, Thermal Bridges in Building Construction—Heat Flows and Surface Temperatures—Detailed Calculations*; ISO—International Organization for Standardization: Geneva, Switzerland, 2017.
26. WeberTherm Uno, Technical Specifications: Weber Saint-Gobain ETICS Finish Mortar. 2018. Available online: https://www.pt.weber/files/pt/2019-04/FichaTecnica_weberthermuno.pdf (accessed on 14 March 2019). (In Portuguese)
27. TincoTerm, Technical Sheet: EPS 100. 2015. Available online: http://www.lnec.pt/fotos/editor2/tincoterm-eps-sistema-co-1.pdf (accessed on 14 March 2019). (In Portuguese)
28. KronoSpan, Technical Sheet: KronoBuild OSB 3. 2019. Available online: https://de.kronospan-express.com/public/files/downloads/kronobuild/kronobuild-en.pdf (accessed on 14 March 2019).
29. Volcalis, Technical Sheet: Alpha Mineral Wool. 2019. Available online: https://www.volcalis.pt/categoria_file_docs/fichatecnica_volcalis_alpharolo-253.pdf (accessed on 14 March 2019). (In Portuguese)
30. Santos, C.; Matias, L. *ITE50—Coeficientes de Transmissão Térmica de Elementos da Envolvente dos Edifícios*; LNEC—Laboratório Nacional de Engenharia Civil: Lisboa, Portugal, 2006. (In Portuguese)
31. Gyptec Ibérica, Technical Sheet: Standard Gypsum Plasterboard. 2019. Available online: https://www.gyptec.eu/documentos/Ficha_Tecnica_Gyptec_A.pdf (accessed on 14 March 2019). (In Portuguese)
32. Pertecno, Catálogo Light Steel Frame. 2015. Available online: http://www.pertecno.pt/pdf/Catálogo-LightSteelFraming.pdf (accessed on 14 March 2019).
33. Thermablok, Thermablok® Aerogel Insulation Blanked. 2011. Available online: www.thermablok.co.uk (accessed on 14 March 2019).
34. Viroc, Cement Wood Board. 2019. Available online: http://www.viroc.pt/ResourcesUser/Documentos_Viroc/Dossiers_Tecnicos/Viroc_Dossier_Tecnico_PT.pdf (accessed on 14 March 2019).
35. Equitone, Fibre Cement Board. 2012. Available online: https://www.equitone.com/pt-pt/materiais/natura/ (accessed on 14 March 2019).
36. GRCA. *Practical Design Guide for Reinforced Concrete (GRC)*; GRCA—International Glassfibre Reinforced Concrete Association: Northampton, UK, 2018.

© 2020 by the authors. Licensee MDPI, Basel, Switzerland. This article is an open access article distributed under the terms and conditions of the Creative Commons Attribution (CC BY) license (http://creativecommons.org/licenses/by/4.0/).

Article

Model Simplification on Energy and Comfort Simulation Analysis for Residential Building Design in Hot and Arid Climate

Sara Elhadad [1,2,3,*], Chro Hama Radha [4], István Kistelegdi [1,5], Bálint Baranyai [1,5] and János Gyergyák [6]

1. Energia Design Building Technology Research Group, Szentágothai Research Centre, Ifjúság útja 20, H-7624 Pécs, Hungary; kistelegdisoma@mik.pte.hu (I.K.); balint.baranyai@mik.pte.hu (B.B.)
2. Department of Architecture, Faculty of Engineering, Minia University, Minia 61111, Egypt
3. Marcel Breuer Doctoral School, Faculty of Engineering and Information Technology, University of Pécs, Boszorkány u. 2, H-7624 Pécs, Hungary
4. Technical College of Engineering, City Planning Department, Sulaimani Polytechnic University, Sulaimani Polytechnic University, Sulaymaniyah 46001, Iraq; chro.radha@spu.edu.iq
5. Department of Building Constructions and Energy Design, Faculty of Engineering and Information Technology, University of Pécs, Boszorkány út 2, H-7624 Pécs, Hungary
6. Department of Architecture and Urban Planning, Faculty of Engineering and Information, Technology, University of Pécs, Boszorkány u. 2, H-7624 Pécs, Hungary; gyergyak.janos@mik.pte.hu
* Correspondence: sarareda@mu.edu.eg; Tel.: +0036-705590080

Received: 14 March 2020; Accepted: 10 April 2020; Published: 12 April 2020

Abstract: Accurate building physics performance analysis requires time-consuming, detailed modeling, and calculation time requirement. This paper evaluates the impact of model simplifications on thermal and visual comfort as well as energy performance. In the framework of dynamic zonal thermal simulation, a case study of a residential building in hot climate is investigated. A detailed model is created and simplified through four scenarios, by incrementally reducing the number of thermal zones from modeling every space as a separate zone to modeling the building as a single zone. The differences of total energy and comfort performance in the detailed and simplified models are analyzed to evaluate the grade of the simplifications' accuracy. The results indicate that all simplification scenarios present a marginal average deviation in total energy demand and thermal comfort by less than 20%. Combining rooms with similar thermal features into a zone presents the optimal scenario, while the worst scenario is the single-zone model. Results showed that thermal zone merging as a simulation simplification method has its limitations as well, whereas a too intensive simplification can lead to undesired error rates. The method is well applicable in further early-stage design and development tasks, specifically in large-scale projects.

Keywords: Model simplifications; Thermal and visual comfort; Energy performance; IDA ICE; Residential building

1. Introduction

High consumption of energy is unavoidable at a global scale [1–6]. It measures the economic success of a given country. The operation of residential and commercial buildings attributes one third of the world's energy consumption [7]. Thus, there is great potential for decreasing global energy consumption through improving the building design [8]. All advanced countries concerned on building-energy problem in various ways to preserve the energy sources and to use energy in a rational way [9]. Based on the U.S. Department of Energy report, buildings are attributed to the majority of total annual energy consumptions and greenhouse gas emissions by the range of 40% to 50% [10,11], and similar results are shown in Europe [12]. Thus, different supranational and national initiatives, regulations, and different programs of private sectors such as CASBEE, LEED, BEEAM, DGNB, and others identify the parameters and standards to assess buildings' sustainability level and to minimize energy use. The role of appliances and residents' behaviors of users should be taken into account in sustainable building design as this role is strictly connected to energy savings and indoor comfort [13,14]. Becherini et al. [15] suggested and modeled several scenarios through which occupant behavior and thermal coating can contribute to the thermal performance of the building. Proper implementation of the framework, materials, knowledge, and system from design stage to construction and operation stages is required to obtain efficient buildings. The "Integrated Building Design" approach [16] is one of the possible solutions to integrate all these elements in the building sector.

Building-energy simulation is an essential support tool to design and commission green buildings. Many available, validated building-energy simulation tools, as Energy Plus, IDA ICE, TRNSYS, BLAST, ESP-r, Radiance, DOE-2 and eQUEST promise high accuracy level and effectivity for comprehensive simulation of building designs [17,18], but they require detailed input for model analysis, composing of zero thickness partitions or walls between thermal zones [19]. The operation and input of building-energy simulation parameters are quite complex [20], including geometric modeling, division of thermal zones, software selection, and selection of meteorological data. Geometric modeling represents the first stage of simulation and often consumes about half of the time of the simulation procedure [21]. Thus, simplification of geometric modeling is considered to be one of the most crucial way to enhance the simulation process. Converting a detailed model back to the spatial model is a complex task for the user and represents some of unfortunate challenges [19]. Despite the proliferation of several building-energy analysis tools in recent years, architects still face difficulties to use the basic tools of energy analysis [22]. The outputs confirmed that the majority of energy simulation tools are not appropriate for the working needs and methods of architects [23–25]. Usually, simplifications occur during translating real building geometry into an energy simulation model due to the lack of modeler software, or model simplifications serve the reduction of computational effort and calculation time. Though some previous studies such as Liu and Henze [26], Westphal and Lamberts [27] and Capozzoli et al. [28] investigated the effects of simplifications on the energy analysis of buildings, it is often underestimated or neglected. Therefore, it is essential to develop a simplification methodology of building physics modeling tools to reduce time and costs of thermal and lighting building simulations, without adverse impact on the quality of results. Complex building geometries are often simplified to perform energy performance simulation [29]. Zhao et al [20] identified three common types of geometric model simplifications as follows:

(A) Calculating the load for one floor and multiplying it based on the number of floors,
(B) Simplifying the fenestration of modeling (e.g., merging windows in one space's façade),
(C) Reducing the number of internal thermal mass and thermal zones of the building.

Several studies have examined the effect of model simplification on the result accuracy. Amitrano et al. [30] investigated the effect of the level of detail on the accuracy of the energy simulation in office buildings. Their study concluded that more detailed geometry can enhance the reliability of simulation by 5 to 15%. Picco and Marengo [16] assessed the effect of different simplifications in building construction types, thermal zoning, and building obstructions, for instance. The findings showed that strong simplifications on the building geometries do not make significant change on the outputs, compared to the detailed model. Bosscha [31] applied a sensitivity analysis by varying the material properties, geometry, and heating, ventilations and air conditioning (HVAC) settings to compare the accuracy of the calculations with the detailed model. The results concluded that the increase in accuracy obtained by more detailed zoning and geometry is highly relying on the HVAC simulation type. Korolija and Zhang [32] compared the predicted annual energy use of the detailed model in which every room was modeled as a separate zone with a simplified model, in which each floor is was modeled as a single zone. The output results showed that thermal zoning simplifications decreased the simulation time by 30% and the mean absolute error of annual heating demand was 10.6%. Klimczak et al. [33] explored the effect of model simplifications on the quality of energy simulation results of a residential building case. The simplifications consisted of the reduction of thermal zones and internal walls, removal of shading elements, and calculations were carried out in different iterations. The findings showed that the exclusion of the shading devices on the south façade had a considerable effect, thus, in future studies this simplification should not be applied. Heo [34] estimated the impacts of internal load, scheduling, and thermal zoning simplifications for domestic buildings in the United Kingdom. They concluded that the differences in annual heating demand are 26% and 17% in the simplification with one single zone for the entire dwelling and one thermal zone per floor, respectively. Dipasquale et al. [35] studied the impact of defining the physical and geometric characteristics of buildings, such as the presence of internal walls, thermal capacity, thermal bridges, the gross or net surfaces, and the number of zones during the modeling stage for heat load assessment. The findings of these results concluded that the reduction of the number of zones has the highest effect on the loads, almost 22% in the cooling demand and 12.5% in heating demand. Chatzivasileiadi et al. [19] explored the impact of simplifying the complex geometries through a systematic analysis of different test cases on the accuracy of energy performance simulation results. The results concluded that orthogonal prisms as simplified surrogates for buildings should be avoided where it is possible, as it showed the worst-case scenario. Akkurt et al. [36] concluded that the simplification of geometry is often unavoidable for use in building-energy performance simulation, but inaccuracies resulted from oversimplification in some geometrical characteristics must be avoided. Zhao et al. [21] investigated the appropriate level of geometric modeling simplification through thermal zone, typical floor and fenestration in energy analysis for office buildings and they found that the more accurate case is modeling the exterior wall in regarding to internal edge. Samuelson et al [37] assessed the accuracy of 18 design-phase building-energy models to enhance the simulation predictions compared to measured energy data.

Despite the valuable results of the aforementioned studies, they just evaluate the impact of model simplification on energy simulation in residential buildings or in office types. The impacts of modeling simplification on the thermal comfort analysis are usually not investigated properly. A study of Korolija and Zehan [32] analyzed the effect of modeling simplification on thermal comfort analysis, but with a different method and focus as they considered one simplification scenario of treating each floor by a single zone and they assessed the thermal comfort performance through annual operation of carbon emission and overheating risk. Consequently, it can be stated that there is no study about the effect of model simplifications on the thermal and visual comfort published yet.

Accurate energy and thermal comfort analysis of buildings requires a lot of time, especially in complex cases it may require up to several weeks. Minimizing the required time of analysis is necessary to be compatible with design duration. Therefore, the main aim of this paper is to assess the impact of model simplifications through different scenarios considering the simulation time, modeling time, and accuracy level of the derived results in both energy demand and thermal comfort in residential houses. The paper evaluates the impact of simplifications by comparing the simulation outputs of the detailed reference model and the simplified models with incremental reduction in the number of thermal zones, until the whole house is modeled as a single zone. Moreover, the investigations explore what level of simplified thermal zoning is required to support energy and thermal comfort analysis of residential buildings. The study is carried out in the simulation framework of IDA ICE, and it also identifies the optimal scenario of the proposed simplification scenarios.

2. Model Simplification Methodology

This study examines the impact of reducing the number of thermal zones on the prediction accuracy of energy and comfort of residential buildings. A thermal zone represents the division of a dwelling for the convenient calculation of the energy and thermal comfort simulation of the building. The thermal properties and parameters are relatively consistent in the same thermal zone. Obviously, to get more accurate results of energy and thermal comfort, the simulation model should be more accurate regarding the number of modeled thermal zones of the building, but at the same time it would need more calculation time and, as a result, modeling work expenses. Many countries have provided relevant regulations for the division of thermal zones of the buildings. American National Standards Institute / American Society of Heating, Refrigerating and Air-Conditioning Engineers (ANSI/ASHRAE) 90.1 [38] reported that multiple spaces can be represented as one thermal zone with the following requirements: the usage of the spaces, the air conditioning and heating systems applied in the spaces and the orientation of the exterior walls and windows should be the same. The Building Research Establishment Ltd. [39] stated that a thermal zone is an area that has the same set points for cooling and heating, identic operating times of the plant, the same ventilation provisions and set-back conditions. In addition, they should be served by the same primary plant and terminal device type. The Canadian standard EE4 [40] stipulates that a thermal zone must have the following features: (1) same air conditioning system and heating with similar operations and functions, and similar heating and cooling loads; (2) the surrounding and the internal space should be distributed into different thermal zones; (3) rooms for laundry, equipment, power distribution, corridors, cloakrooms, and stairs cannot be modeled as a single partition.

For the purpose of the model simplifications, a multifamily house as a reference is proposed, representing a generic, typical residential building type in the largest building sector of the world. This reference building model is derived form an existing, common residential house, built in 2005 in New Minia, Egypt at 30.73 E longitude, 28.08 N latitude (Figure 1). The building consists of nine apartments. The ground floor is represented by one apartment and consists of a lounge, dining room, bathroom, and kitchen, with the total floor area of 180 m^2. Each floor of the repeated floors consists of two identical apartments, with 220 m^2 net floor area. Every apartment includes reception, master bedroom, two children rooms, bathroom, and kitchen as shown in Figure 1 and occupied by a couple with two children based on the real evaluation from the field. The composition of building elements was used on the basis of the Egyptian standards, as shown in Table 1. IDA ICE has been used to simulate thermal and visual comfort as well as energy performance in a detailed model about the reference building and in several simplification scenarios, whereas the reference model is modified according to the simplification concepts. Table 1 presents an overview of the major parameters and input data.

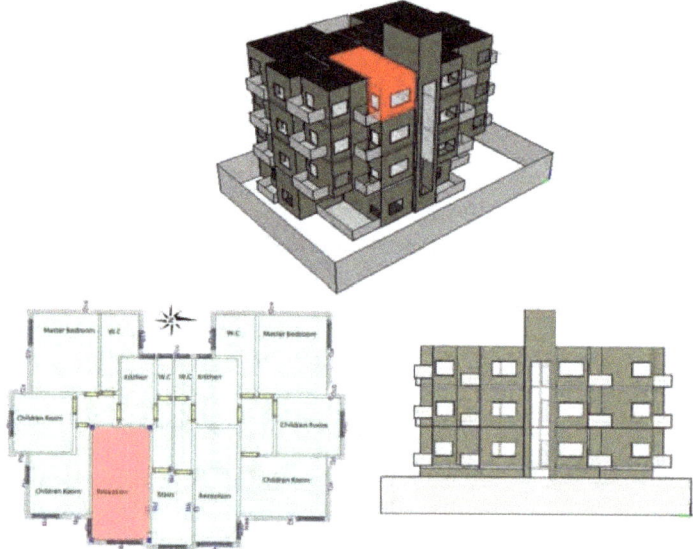

Figure 1. Generic residential building as a reference for model simplification tests.

Table 1. Boundary conditions for the simulation.

Boundary Conditions	Model Characteristics
Location	Minya
Simulation Weather File	EGY_MINYA_623870_IW2.PRN (ASHRAE 2013)
Modeling Software	IDA Indoor Climate and Energy
House Type	Family house
Plot Area	300 m^2
Glazing Type	20 mm single glazed glass, U-value = 5.9 W/(m^{2}K)
External Walls	5 mm Plaster + 25 mm Egyptian Portland cement mortar + 250 mm Double red brick + 25 mm Egyptian Portland cement mortar + 5 mm Plaster. U-value = 1.546 W/(m^{2}K)
Internal Walls	5 mm plaster + 25 mm Egyptian Portland cement mortar + 125 mm single red brick + 25 mm Egyptian Portland cement mortar + 5 mm plaster U-value = 2.281 W/(m^{2}K)
Internal Floors	10 mm concrete tiles + 20 mm Egyptian Portland cement mortar + 50 mm sand + 200 mm plain concrete. U-value = 1.824 W/(m^{2}K)
Roof	10 mm concrete tiles + 20 mm Egyptian Portland cement mortar+ 50 mm sand + 20 mm betomine damp insulation + 150 mm rein force concrete. U-value = 1.707 W/(m^{2}K)
External Floor	10 mm Concrete tiles + 50 mm sand + 20 mm Egyptian Portland cement mortar + 200 mm plain concrete+ 250 mm soil. U-value = 1.172 W/(m^{2}K)
Basement Wall Towards Ground	5 mm Plaster + 25 mm Egyptian Portland cement mortar + 250 mm double red brick + 25 mm Egyptian Portland cement mortar + 5 mm plaster. U-value = 1.546 W/(m^{2}K)
Infiltration	7 ACH
Internal Gains	- Occupant: Activity level 1.0 MET Constant clothing 0.85 ± 0.25 CLO (clothing is automatically adapted between limits to obtain comfort) Occupancy time: 1- Living room: fully present (1) [7:00–8:00, 17:00–22:00], half present (0.5) [15:00–17:00], 0 otherwise, 2- Bedroom 0 [7:00–22:00], 1 otherwise (remaining) Emitted heat per person 75 W - Equipment usage time: 1- Living room: full intensity 1 [7:00–8:00, 17:00–22:00], half intensity 0.5 [15:00–17:00], 2- Bedroom: 0 [7:00–22:00], 1 otherwise Luminous efficiency 12 lm/W - Artificial lighting use: 1- living room From 1 Jan to 14 Apr all days:1 [7:00–8:00, 17:00–22:00], 0.5[15:00–17:00], 0 otherwise From 16 Oct to 31 Dec all days:1[7:00–8:00, 17:00–22:00], 0.5[15:00–17:00], 0 otherwise From 15 Apr to 15 Oct all days:1 [19:00–22:00], 0 otherwise All days: 0 2- Bedroom From 1 Jan to 14 Apr all days:1 [6:00–7:00, 22:00–23:00], 0 otherwise From 16 Oct to 31Dec all days:1 [6:00–7:00,22:00–23:00], 0 otherwise All days: 0
Schedules	Independ in different spaces
Daylight	Meteonorm database diffuse and direct radiation (W/m^2)
HVAC	No mechanical ventilation. Generic heating and cooling in the zones to compensate heat losses and loads.

In the following, four different simplification scenarios of the thermal zones are proposed as shown in Figure 2. Summary of the simulated scenarios is presented in Table 2. First, in the base scenario (BS) model, each space is modeled as a single independent zone (Figure 1). Then, scenario S1 combines spaces with similar characteristics (e.g., orientation, operation schedules, same use, etc.) into one thermal zone (Figure 2). Then, scenario S2 combines the same oriented spaces for all of the 4 floors into one thermal zone (Figure 2). In scenario S3, all spaces on the same floor are merged into one single zone, and scenario S4 models the entire building as one single thermal zone (Figure 2).

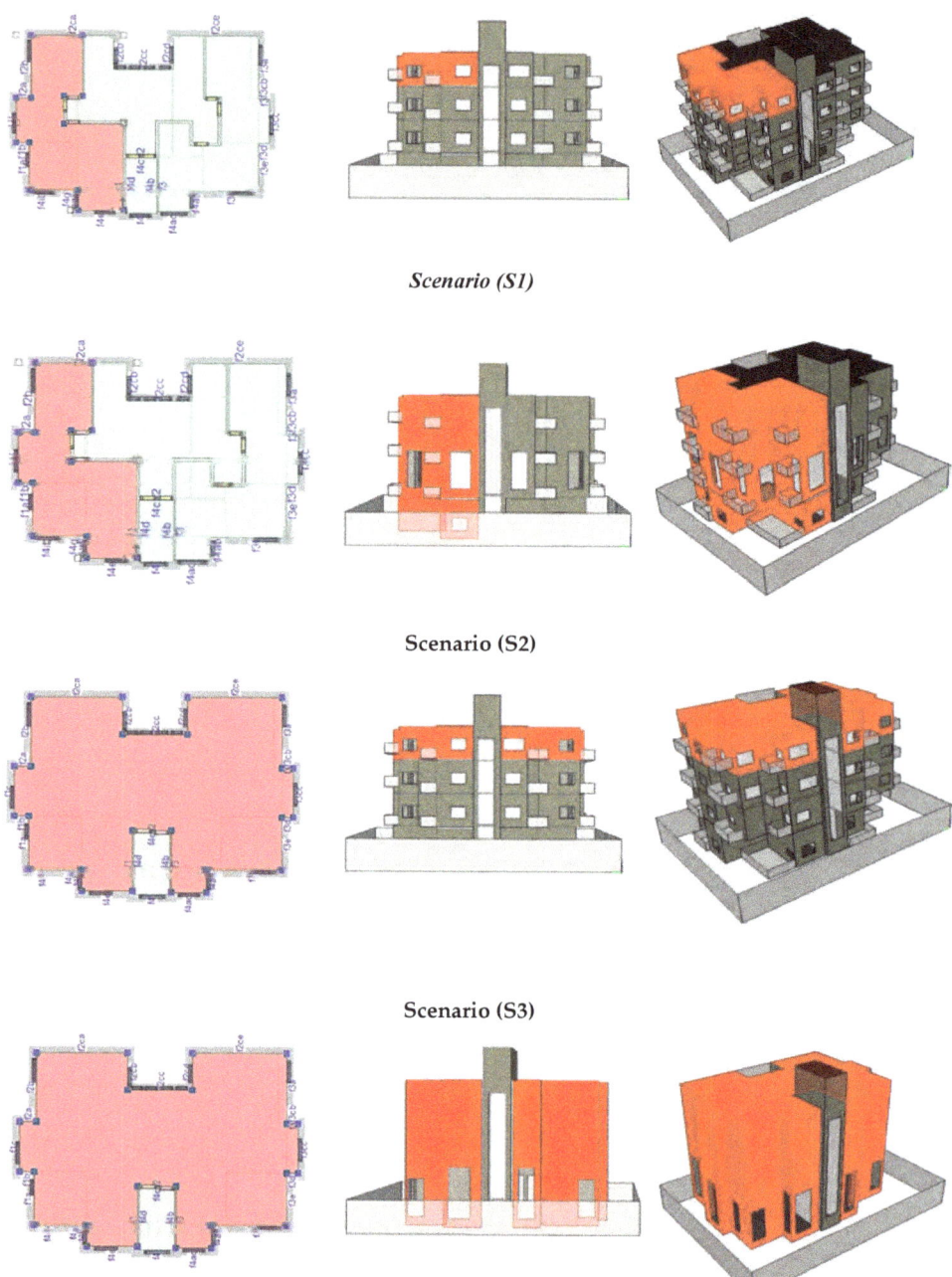

Figure 2. Simplification scenarios—simulation models (plan, side, and 3D view).

Table 2. Simulated simplification scenarios.

Scenario	Description of Investigated Thermal Zones	Number of Thermal Zones
BS	Base model: Each building space is modeled as a single zone.	64
S1	Floor by floor, all identically oriented spaces with the same function are merged into one zone with the same operation schedules, use, etc.	14
S2	The same oriented spaces with the same use for all of the 4 floors are combined into one thermal zone, i.e., bedrooms on ground floor, 1st floor, 2nd floor, and 3rd floor are merged with circulation areas into one thermal zone.	8
S3	All rooms on the same floor are merged into one thermal zone, thus in this scenario the whole building has 4 zones.	4
S4	The entire building is modeled as one single thermal zone.	1

3. Results and Discussion

3.1. Building-Energy Assessment

IDA ICE has been used to simulate energy consumption and indoor comfort performance of the studied building for the BS model and all the simplification scenarios. Figure 3 summarizes the energy results for the BS model and the simplification scenarios in comparison to BS model. The simplification scenarios have minor effect on the lighting, facility, equipment, tenant, and DHW results due to their similar input parameter and cumulated settings. On the other hand, electric cooling and heating show larger differences. In BS scenario, the cooling demand accounts to 67% of the total energy consumption, while the heating demand attributes to 18%, as the case study located in a hot and dry climate. Lighting, facility, equipment, tenant and DHW accounted to 15% of the total energy consumption. In S1, the cooling demand increased by 9.6% and the heating demand decreased by 3.1% with respect to BS (Table A1). S2 and S3 scenarios performed an increased cooling demand by 15.1% and 10.6% respectively, while the heating demand decreased by 3.5% and 0.3%, respectively, compared to BS (Table A1). In scenario S4, the heating demand decreased by 23.6% in respect to BS model, while the cooling demand increased by 12.2% compared to BS (Table A1). Similar reports of the simplification on the energy performance are available in the literature, e.g., Heo et al. [34] and Ren et al. [41] have reported that merging rooms with similar characteristics into one zone (scenario S1) and modeling a single zone for the entire building (scenario S4) underestimated the annual heating demand by 7% and 24%, respectively, in comparison to modeling every room as a separate zone (detail model) for domestic buildings in UK. Picco et al [17] have also reported that cooling and heating loads was underestimated by 9.29% and 8.12%, respectively for scenario S3 (Every floor was represented by one individual zone) compared to the detailed model in an office building built, located in Bolzano, Italy. Picco and Marengo [16] have reported similar finding of simplification on cooling and heating demands. They reported that when the number of thermal zones are reduced to one thermal zone per floor (scenario S3), the annual heating and cooling demand are underestimated by 0.86% and 6.25%, respectively. Consistent with the present result, Dipasquale et al. [35] have also reported that reducing the whole floor to one thermal zone underestimated the annual heating and cooling demand by 12.5% and 22%, respectively with respect to the detailed model. Korolija, and Zhang [32] have also reported that treating each floor of a house as a single thermal zone underestimated the annual heating demand by 10.6%. The change in the total energy consumption evolved in the first, second, third, and the fourth scenarios as follows, +5.8%, +9.5%, +7.1%, +4.0% in respect to the BS model (Table A1). Although the fourth scenario represented the worst scenario considering only the cooling and heating demand individually, it had the smallest change in total energy consumption compared to BS model, because the heating and cooling deviations equaled each other out, resulting in the least difference in total. The thermal envelope is the same in all of the models, hence the fundamental differences can be derived from the complexity level of the actual modeled thermal mass, (walls, slabs) that affect mostly the cooling and heating demand, although the geometrically "missing" thermal mass was added to the model variations as individual mass elements respectively. Case S4's lowest heating demand is caused by the least floor space to be heated.

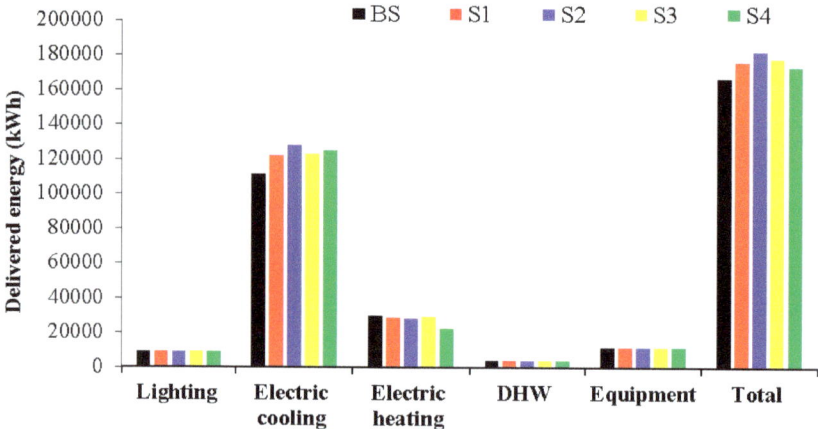

Figure 3. Delivered energy for the detailed and simplified models.

3.2. Simulation Time and Modeling Time

The total modeling time of the BS model was 215 min, while it decreased to 45, 35, 22, and 11 minutes in the simplification scenarios, as shown in Table 3. The most decisive difference in modeling expenditure of time takes place in modeling of one story of a building, since multifamily houses possess a great diversity of apartment sizes, room arrangements and room geometries. After completion of a floor, the typically identic domestic levels can be copied above each other to complete the building model; therefore, this modeling work duration is insignificant. As the number of thermal zones are reduced in a story, the simulation time decreases decisively. Considering the geometry and structure creation as well as the editing and parametrization working time, the required modeling time is approx. proportional—with a rate of 1:1—to the number of zones. At the same time, the total simulation time of the BS model was 86 minutes, and it decreased to 32, 14, 23, and 5 minutes in the scenarios. With a decreasing number of thermal zones, the simulation time decreases significantly. The scenarios saved 79 to 95% of the modeling time and 63 to 94% calculation duration compared to BS, demonstrating a huge potential in model simplification and workflow conservation.

Table 3. Modeling and calculation duration of the detailed and simplified models and respective differences.

	BS	S1	S2	S3	S4
Modeling time (Minutes)	215	45	35	22	11
Modeling time difference (%)	0	−79	−84	−90	−95
Calculation time (Minutes)	86	32	14	23	5
Calculation time difference (%)	0	−63	−84	−73	−94

3.3. Assessment of Building Thermal Comfort

3.3.1. Evaluation of Predicted Mean Vote (PMV)

In this study, PMV was evaluated as one of the main indices to assess the thermal comfort in an occupied zone [42,43]. PMV refers to thermal 7-stage sensation scale [44] through seven points range from −3 to +3 as follow −3 = cold, −2 = cool, −1 = slightly cool, 0 = neutral, 1 = slightly warm, 2 = warm, and 3 = hot [45]. Three categories A, B, and C were proposed in ISO 7730, PMV is ranged in the interval of [−0.2, +0.2]; for Category A, in the interval [−0.5, +0.5] for Category B and, in the interval [−0.7, +0.7] for Category C [46]. Category B represents the normal level of applicability based on ISO 7730. Figure 4 shows the average number of annual hours of PMV, category B in the detailed and simplified models' separated as well as merged thermal zones. In the simplified models, the average annual hours of PMV, category B is calculated by an area weighted averaging of the annual hours of PMV, category B for each thermal zone, as presented in Equation (1)

$$N_{PMV} = \frac{\sum_{i=1}^{i=n} N_i \cdot A_i}{\sum_{i=1}^{i=n} A_i} \quad (1)$$

where N_{PMV}. means the average annual hours of PMV, category B for the whole model, N_i represents the number of annual hours of PMV, category B for thermal zone i, A_i the total area of each thermal zone [m^2], "n" is the total number of thermal zones of the model. For the complete building in BS, the annual hours of PMV, category B were 7781 h, while 6642 were accounted for S1. The annual hours of PMV, category B increased by 6 hours for S2 and, while the annual hours decreased by 875 and 64 hours for S3 models and one-zone model (S4), respectively compared to the BS model, as shown in Table A2. In S2, the difference in the annual hours of PMV, category B increased by 3.2% in the south side and decreased by 29.3% in the north side, related to the BS model (Table A2). Reason for that: in the south oriented zone, solar gains enabled higher level of PMV, while in the north zone, the contrary effect evolved, because the high thermal zones (3-storey high) are more difficult to heat. In S3, the PMV decreased by 4.7% and 1.7% on the 2rd floor and the 3nd floor respectively, with respect to BS. Reason for that: in the 3rd floor the highest zone is the warmest in summer because of thermal gradient and less thermal mass. However, this greatest deviation is more still at a marginal scale, hence, in general, a consistent calculated thermal comfort sensation was observed in each model.

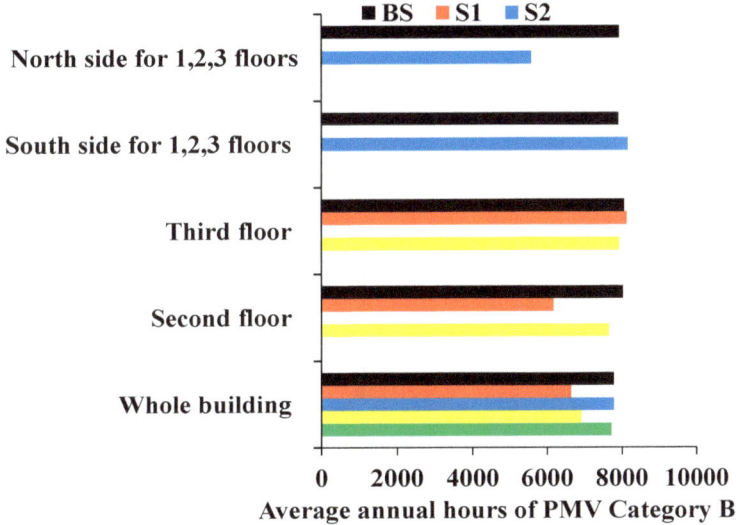

Figure 4. Average number of annual hours of PMV, Category B for whole and some parts of the building in the detailed.

3.3.2. Carbon Dioxide Level Assessment

Concentration of Carbon Dioxide (CO_2) was applied as an indicator of indoor air quality [47]. The connection between indoor air quality and indoor CO_2 concentration originates from the fact that at the same time people are generating odor-causing bio effluents and producing CO_2 [47]. In European Standard CEN-EN 13779:2007 [48], CO_2 concentration is also applied to classify indoor air quality, and the maximum value of CO_2 concentration level is 1500 ppm, while they recommend keeping CO_2 concentration level below 1000 ppm. In this particular study, the number of annual hours is estimated, when the CO_2 concentration level is above 1000 ppm in the models. The results are compared at three scales (i.e., whole building, 2nd and 3rd floors, south and north sides of the building in all floors). Figure 5 presents the annual hours with CO_2 concentration level above 1000 ppm in the detailed and simplification models. Additionally, an area weighting such as Equation (1) was used to calculate average annual hours of CO_2 concentration level. Regarding to the complete building, the number of annual hours of CO_2 level above 1000 ppm in BS scenario was 2248 h, while S1 was accounted to 2130 h. In the scenarios S2, S3, and S4 this value decreased by 7.2%, 8.4%, and 5.9%, respectively, compared to the BS scenario Table A3. Consistent with the present result, Korolija and Zhang [32] have also reported that treating each floor of a house as a single thermal zone (scenario S3) underestimated the carbon emission by 8%. In scenarios S1 and S3, the differences in the air quality were 0.1% and 21.7% respectively in the second floor, while 48.9% and 8.7%, differences occurred in third floor. In the south side of the whole building (S2 – building high thermal zone), the air hygiene decreased by 17.4% at the north side of the building with respect to the BS scenario while and the south side accounted to the same hours of BS scenario (Table A3). The merged, simplified zones have more space to be window-ventilated, since they include the corridors and secondary spaces (elevator/stairs) as well. That is why they perform higher CO_2 level. Generally, the distribution of CO_2 concentration shows great inhomogeneity in the different sized thermal zones.

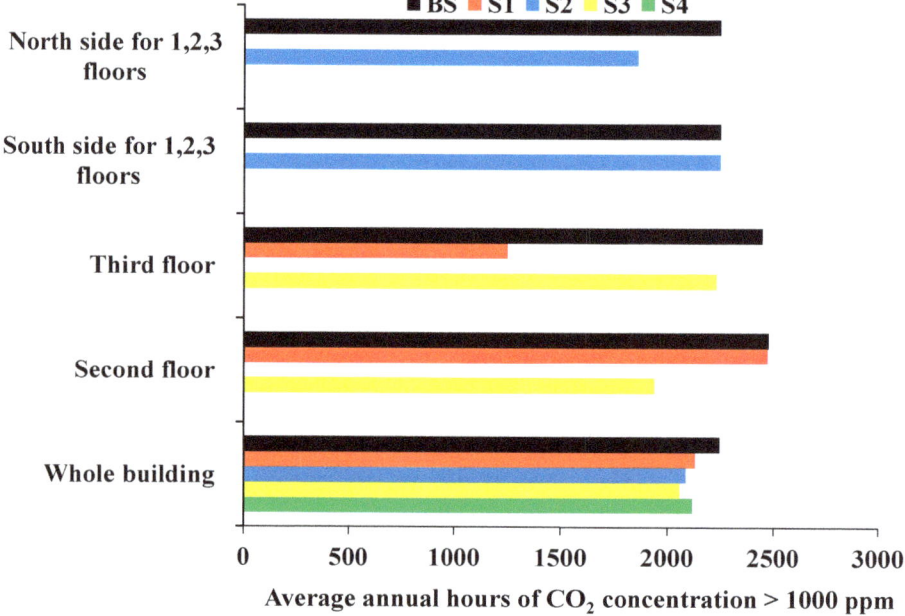

Figure 5. Average number of annual hours with CO_2 concentration > 1000 ppm for the whole and some parts of the building in the detailed and simplified models.

3.3.3. Daylight Factor Assessment

Daylighting as visual comfort is an effective parameter in sustainable and energy efficient building design [49] and it is becoming an essential part of the environmentally friendly building design [50]. Adequate level of daylight is not only important to illuminate all year long and secondarily to heat in wintertime the interior, but it is also an essential source of the occupant's emotional and physiological well-being. Besides ensuring low level of odor and noise, daylight provision is an essential parameter in indoor environment investigations for maintaining the enjoyment of a property. Daylighting performance strongly relies on the illuminance under direct, respectively diffuse sky conditions.

Since the daylight provision under direct illuminance (clear sky conditions) in Minia region possesses high level of daylight autonomy in interior spaces, in this study, the visual comfort assessment focused on the Daylight Factor (DF), representing the illuminance performance of the spaces under mixed sky circumstances, as a kind of 'worst-case scenario'. Satisfying the minimum required DF limit means a whole year long secured daylighting quality. The DF value is a ratio that represents the amount of illuminance available indoors relative to the illuminance level present outdoors at the same time, under overcast sky [51]. DF at a point of the room is the ratio of the indoor illuminance Ei to the outdoor horizontal illuminance, Eo, [52], expressed as percentage in the following Equation (2):

$$DF = \frac{E_i}{E_o} \times 100 \: [\%] \qquad (2)$$

Calculating Equation (2), the required value of DF for Minia city is 2.1, by applying the required Ei as 300 lx and Eo (median external diffuse illuminance) as 14012 lx according to EN17037 Daylight in Buildings and ASHRAE database. The DF was assessed in all models. Illuminances were computed using meteorological data taken from Meteonorm 7 database [53]. Figure 6 presents the ratio of floor area performing a DF above (corresponding to adequate daylight space partition) and below (equals to inadequate daylight space partition) the DF (2.1) threshold value. In case of BS, 21.3% the floor area is adequately daylight. In S1 and S3 the appropriately daylight floor area increased by 6.1% and 21.6% with respect to BS, while in S2 and S4 delivered significant, 19.8% and 60.3% differences compared to the reference. In S1 the abandonment of all internal walls caused the weaker DF performance and in S3 the additionally merged, deep spaces of the whole story thermal zones indicated the lower level of DF. The reason of the anomalies in S2 and S4 were the different height of the zones in the S2 and S4 models.

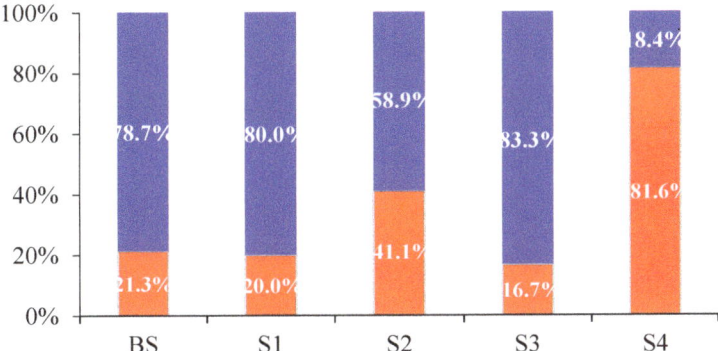

Figure 6. Floor area ratio with daylight factor above (red color) and below (blue color) the minimum DF (2.1%) value.

4. Optimal Scenario of the Proposed Model Simplifications

To determine the optimal scenario of the proposed simplifications, two crucial criteria should be taken into account: the required simulation time and the accuracy of the energy and comfort results. In respect to calculation duration, obviously the single-zone model (S4) represents the fastest model as shown in Table 3, followed by S2 model, S3 model, and S1 model. For the accuracy criteria, Table 4 presents the absolute differences of energy demand (heating and cooling) and indoor comfort (PMV, CO_2 level and DF) in respect to the BS model. More simplification leads to more inaccurate results, as in S4 model the high differences in energy demand and DF distribution demonstrate. In comparison to BS, (S1) presented the optimal accuracy case of the proposed simplification scenarios, resulting in 6.8% average difference of all parameters in energy demand and comfort performance. At the same time, S1 saves over 63% of simulation time. S1 is followed by scenarios 3, 2, and 4. Consequently, the model simplification can be accomplished until the anomalies appear due to the simplified geometry.

Table 4. Absolute % differences of heating and cooling demand, PMV, CO_2 concentration, and DF between the simplification scenarios and BS.

Parameter	Simplification Scenarios			
	Absolute % Differences with Respect to the BS			
	S1	S2	S3	S4
Heating Demand	3.1	3.5	0.3	23.6
Cooling Demand	9.6	15.1	10.6	12.2
PMV	14.6	0.1	11.3	0.8
CO_2 Concentration	5.3	7.2	8.4	5.9
DF	1.5	20.1	4.5	56.6
Average Differences	6.8	9.2	7.0	19.8
Order	1	3	2	4
% Save in Simulation Time	63	84	73	94

5. Conclusions

Buildings are attributed to a tremendous amount of energy consumption due to their continuous operation and extensive lifetime. Performing Building-energy simulations is an essential part of a decision-making process as it helps designers to assess the energy and comfort effect of different building design options. Since the impacts of building physics simulation model simplifications on the accuracy of the results are not well studied and reported, the proposed simplification scenarios seek to overcome the obstacle of long calculation time and according design costs by providing a simpler and faster way to carry out building-energy and comfort simulations. The main aspect of the methodology is to achieve an adequate level of accuracy that can promote the simulation results of energy demand and thermal comfort analysis by simultaneously minimizing calculation time. The detailed reference building physics simulation model contained all separate rooms modeled as individual thermal zones. The model was then simplified in scenario S1, whereas all spaces with similar use and orientation were merged into one-zone floor by floor. The same oriented spaces for all of the 4 floors were combined into one thermal zone in scenario S2. Every floor was represented by one individual zone in scenario S3, and the whole building was treated as one single zone for scenario S4. Multiple effects of the model simplification methods on energy consumption, CO_2 level, PMV, and DF performance were evaluated in a common residential building in New Minia, Egypt.

A model simplification method that merges all spaces with similar use and orientation into one-zone floor by floor (scenario S1), enables the shortening of the required modeling time of 79% and the acceleration of the required solver calculation duration by 63%. At the same time, the comfort performance values possess 21.4% deviations, while the energy performance results are underestimated by 12.7% in comparison to the detailed model. Combining the same oriented spaces with the same use for all of the 4 floors into one thermal zone (scenarios S2) reduces the simulation time by 84%, while the deviation in total energy demand and thermal comfort are 18.6% and 27.4%, respectively, compared to the detailed model. When the number of thermal zones is further reduced to one thermal zone per floor (scenarios S3), the simulation time is saved by 73%, while the energy and thermal comfort are underestimated by 10.9% and 24.2%. However, modeling the entire building by a single zone (scenarios S4) saves 95% and 94% of the required modeling time and the simulation time, respectively, the energy and thermal comfort are underestimated by 35.8% and 63.3%, respectively. The interdependency of result accuracy and calculation time proved that the optimal simplification method merges all spaces with similar use and orientation into one-zone floor by floor (scenario S1). It is obvious that besides the advantages the geometrical simplifications might carry some limitations as well. Results showed that thermal zone merging as a simulation simplification method has its limitations as well, whereas a too intensive simplification can lead to undesired error rates. Furthermore, the essentially geometry related daylight distribution interpretation can be affected due to the different depth of the merged zones. In addition, the orientation should be considered with consciousness, since the different oriented zones should not be combined to avoid different solar heat load (summer) or heat gain (winter) effect to be mixed in one greater unified zone to confuse both energy and comfort behavior.

Important to mention that taking only the energy results into consideration during the simplification process is not sufficient to get a truly comparable model version to the original detailed building model, rather it is inevitable to consider all determinate indoor comfort indices as well. Analysis of both comfort and energy results is the only way to identify the optimum on model simplification level. The gained thermal zoning simplification method can imply a high design feedback acceleration effect, offering a great potential for building design optimization. An until now unreached quality level of design optimization evolves, since testing of significantly higher number of design cases in the same amount of available planning time is getting to be possible. The thermal zone geometry simplification's result inaccuracy level should be further reduced by compensation solutions for thermal mass and the central, deeper settled zone sections, which distort to a certain measure the simulation results. The described methodology can help to reduce the duration requirements for a dynamic simulation and it can be seen as a 1st step in a multi-level model simplification strategy, consisting of next stages in simplifications techniques for fenestration, shading, thermal mass, HVAC systems, as well as controlling automation strategies. It can be concluded that the analysis results will be useful for modelers to determine the optimal level of model simplification in the modeling process depending on the achievable accuracy level of energy performance and thermal comfort. The method provided promising results for further applications and it is intended to be further tested in next multifamily projects and office buildings to prove its reliability in building industry standard practice.

Author Contributions: Conceptualization, S.E.; Data curation, S.E., B.B. and C.H.R.; Formal analysis, S.E.; Methodology, S.E. and B.B.; Project administration, I.K. and B.B.; Software, S.E. and B.B.; Supervision, I.K., B.B. and J.G.; Validation, S.E. and C.H.R.; Visualization, I.K., B.B. and C.H.R.; Writing – original draft, S.E.; Writing – review & editing, I.K., B.B., J.G. and C.H.R. All authors have read and agreed to the published version of the manuscript.

Funding: This research received no external funding

Acknowledgments: The present scientific contribution is dedicated to the 650th anniversary of the foundation of the University of Pécs, Hungary. The first author would like to thank the Egyptian Ministry of Higher Education (MoHE) and Tempus Public Foundation for providing him the Stipendium Hungaricum Scholarship. Further we appreciate the support of the János Szentágothai Research Centre, University of Pécs, Energia Design Research Group, led by Prof Dr István Kistelegdi. The research project is conducted at the University of Pécs, Hungary, within the framework of the Biomedical Engineering Project of the Thematic Excellence Programme 2019 (TUDFO/51757-1/2019-ITM). The research project is conducted at the University of Pécs, Hungary, within the framework of the Biomedical Engineering Project of the Thematic Excellence Programme 2019 (TUDFO/51757-1/2019-ITM).

Conflicts of Interest: "The authors declare no conflict of interest."

Appendix A

Table A1. Delivered energy for detailed and simplified models and their differences with respect to detailed model.

Delivered Energy	BS	S1	S2	S3	S4
Lighting, Facility (kWh)	9199	9209	9213	9205	9205
Difference %	0	0.1	0.1	0.1	0.1
Electric Cooling (kWh)	111,501	122,223	128,313	123,326	125,104
Difference %	0	9.6	15.1	10.6	12.2
Electric Heating (kWh)	29,755	28,824	28,716	29,653	22,723
Difference %	0	−3.1	−3.5	−0.3	−23.6
DHW (kWh)	4246	4246	4246	4246	4246
Difference %	0.0	0.0	0.0	0.0	0.0
Equipment, Tenant (kWh)	11,746	11,764	11,750	11,753	11,761
Difference %	0	0.2	0.0	0.1	0.1
Total (kWh)	166,447	176,166	182,237	178,182	173,038
Difference %	0	5.8	9.5	7.1	4.0

Table A2. Average number of annual hours of PMV, Category B for whole and some parts of the building in the detailed and simplified models and differences of simplified models with respect to detailed model.

	BS	S1	S2	S3	S4
	Average Annual Hours of PMV Category B				
Whole Building (hours)	7781	6642	7787	6906	7717
Difference %	0.0	−14.6	0.1	−11.3	−0.8
Second Floor (hours)	8026	6176	-	7646	-
Difference %	0.0	−23.1	-	−4.7	-
Third Floor (hours)	8063	8128	-	7928	-
Difference %	0.0	0.8	-	−1.7	-
South Side for 1,2,3 Floors (hours)	7900	-	8153	-	-
Difference %	0.0	-	3.2	-	-
North Side for 1,2,3 Floors (hours)	7921	-	5585	-	-
Difference %	0.0	-	−29.3	-	-

- = Zero (no value), as there was no simulation in this zone for the given scenario.

Table A3. Average number of annual hours with CO_2 concentration > 1000 ppm for whole and some parts of the building in the detailed and simplified models and differences of simplified models with respect to detailed model.

	BS	S1	S2	S3	S4
	Average annual hours of CO_2 concentration > 1000 ppm				
Whole Building (hours)	2248	2130	2086	2058	2116
Difference %	0.0	−5.3	−7.2	−8.4	−5.9
Second Floor (hours)	2476	2473	-	1940	-
Difference %	0.0	−0.1	-	−21.7	-
Third Floor (hours)	2445	1249	-	2232	-
Difference %	0.0	−48.9	-	−8.7	-
South Side for 1,2,3 Floors (hours)	2249	-	2248	-	-
Difference %	0.0	-	0.0	-	-
North Side for 1,2,3 Floors (hours)	2248	-	1858	-	-
Difference %	0.0	-	−17.4	-	-

- = Zero (no value), as there was no simulation in this zone for the given scenario.

References

1. Solangi, K.H.; Islam, M.R.; Saidur, R.; Rahim, N.A.; Fayaz, H. A review on global solar energy policy. *Renew. Sustain. Energy Rev.* **2011**, *15*, 2149–2163. [CrossRef]
2. International Energy Agency (IEA). *World Energy Outlook 2012*; Organisation for Economic Cooperation and Development (OECD): Paris, France, 2012; ISBN 978-92-64-18134-2.
3. Wolfram, C.; Shelef, O.; Gertler, P. How Will Energy Demand Develop in the Developing World? *J. Econ. Perspect.* **2012**, *26*, 119–138. [CrossRef]
4. Pérez-Lombard, L.; Ortiz, J.; Pout, C. A review on buildings energy consumption information. *Energy Build.* **2008**, *40*, 394–398. [CrossRef]
5. Malko, J. Energia dla wszystkich. Globalne wyzwanie dla sektora energii. *Polityka Energetyczna* **2015**, *18*, 5–13.
6. Keho, Y. What drives energy consumption in developing countries? The experience of selected African countries. *Energy Policy* **2016**, *91*, 233–246. [CrossRef]
7. Ürge-Vorsatz, D.; Eyre, N.; Graham, P.; Harvey, D.; Hertwich, E.; Jiang, Y.; Kornevall, C.; Majumdar, M.; McMahon, J.E.; Mirasgedis, S.; et al. Energy End-Use: Buildings. In *Global Energy Assessment (GEA)*; Johansson, T.B., Nakicenovic, N., Patwardhan, A., Gomez-Echeverri, L., Eds.; Cambridge University Press: Cambridge, UK, 2012; pp. 649–760. ISBN 978-0-511-79367-7.

8. Urban, B.; Glicksman, L. The mit design advisor—A fast, simple tool for energy efficient building design. In Proceedings of the Simbuild 2006 Second National IBPSA-USA Conference, Cambridge, MA, USA, 2–4 August 2006; pp. 270–276.
9. Elhadad, S.; Baranyai, B.; Gyergyák, J.; Kistelegdi, I.; Salem, A. Passive design strategies for residential buildings in a hot desert climate in upper Egypt. In Proceedings of the Multidisciplinary Scientific GeoConference SGEM 2019, Albena, Bulgaria, 28 June–7 July 2019; Volume 19, pp. 495–502.
10. Chen, A. Working Toward the Very Low Energy Consumption Building of the Future. Available online: https://newscenter.lbl.gov/2009/06/02/working-toward-the-very-low-energy-consumption-building-of-the-future/ (accessed on 19 March 2019).
11. Elhadad, S.; Rais, M.; Boumerzoug, A.; Baranyai, B. Assessing the impact of local climate on the building energy design: Case study Algeria-Egypt in hot and dry regions. In Proceedings of the International Conference, Istanbul, Turkey, 20–21 November 2019; pp. 21–24.
12. Economidou, M.; Atanasiu, B.; Despret, C.; Maio, J.; Nolte, I.; Rapf, O.; Laustsen, J.; Ruyssevelt, P.; Staniaszek, D.; Strong, D.; et al. *Europe's Buildings under the Microscope: A Country by Country Review of the Energy Performance of Buildings*; BPIE: Buildings Performance Institute of Europe: Brussels, Belgium, 2011; pp. 98–122.
13. Roberti, F.; Oberegger, U.F.; Lucchi, E.; Gasparella, A. Energy retrofit and conservation of built heritage using multi-objective optimization: Demonstration on a medieval building. In Proceedings of the Building Simulation Applications, Bolzano, Italy, 4–6 February 2015; pp. 189–197.
14. Elhadad, S.; Baranyai, B.; Gyergyák, J. Energy consumption indicators due to appliances used in residential building, A case study New Minia, Egypt. In Proceedings of the 6th International Academic Conference on Places and Technologies, Pécs, Hungary, 9–10 May 2019; pp. 188–193.
15. Becherini, F.; Lucchi, E.; Gandini, A.; Barrasa, M.C.; Troi, A.; Roberti, R.; Sachini, M.; Di Truccio, M.C.; Arrieta, L.G.; Pockelé, L.; et al. Characterization and thermal performance evaluation of infrared reflective coatings compatible with historic buildings. *Build. Environ.* **2018**, *134*, 35–46. [CrossRef]
16. Picco, M.; Marengo, M. On the Impact of Simplifications on Building Energy Simulation for Early Stage Building Design. *J. Eng. Archit.* **2015**, *3*, 66–78. [CrossRef]
17. Picco, M.; Lollini, R.; Marengo, M. Towards energy performance evaluation in early stage building design: A simplification methodology for commercial building models. *Energy Build.* **2014**, *76*, 497–505. [CrossRef]
18. Elhadad, S.; Baranyai, B.; Gyergyák, J. The impact of building orientation on energy performance: A case study in new Minia, Egypt. *Pollack Period.* **2018**, *13*, 31–40. [CrossRef]
19. Chatzivasileiadi, A.; Lannon, S.; Jabi, W.; Wardhana, N.M.; Aisha, R. Addressing Pathways to Energy Modelling Through Non-Manifold Topology. In Proceedings of the 2018 Symposium on Simulation for Architecture and Urban Design (SimAUD 2018), Delft, The Netherlands, 4–7 June 2018.
20. Fonseca, A.; Ortiz, J.; Garrido, N.; Fonseca, P.; Salom, J. Simulation model to find the best comfort, energy and cost scenarios for building refurbishment. *J. Build. Perform. Simul.* **2018**, *11*, 205–222. [CrossRef]
21. Zhao, J.; Wu, Y.; Shi, X.; Jin, X.; Zhou, X. Impact of Model Simplification at Geometric Modelling Stage on Energy for Office Building. In Proceedings of the 4th Building Simulation and Optimization Conference, (BSO 2018), Cambridge, UK, 11–12 September 2018; pp. 402–406.
22. Punjabi, S.; Miranda, V. Development of an integrated building design information interface. In Proceedings of the Ninth International IBPSA Conference 2005, Montréal, QC, Canada, 15–18 August 2005; p. 8.
23. Van Dijk, E.J.; Luscuere, P. An architect friendly interface for a dynamic building simulation program. In Proceedings of the Sustainable Building 2002 Conference, Oslo, Norway, 23–25 September 2002.
24. Gratia, E.; De Herde, A. A simple design tool for the thermal study of an office building. *Energy Build.* **2002**, *34*, 279–289. [CrossRef]
25. Attia, S.; Beltrán, L.; Herde, A.D.; Hensen, J. Architect friendly: A comparison of ten different building perfor-mance simulation tools. In Proceedings of the 11th International IBPSA Conference on Building Simulation, Glasgow, UK, 27–30 July 2009; pp. 204–211.
26. Liu, S.; Henze, G.P. Calibration of building models for supervisory control of commercial buildings. In Proceedings of the of the Ninth International IBPSA Conference 2005, Montreal, QC, Canada, 15–18 August 2005; p. 8.s.
27. Westphal, F.S.; Lamberts, R. Building simulation calibration using sensitivity analysis. In Proceedings of the Ninth International IBPSA Conference 2005, Montreal, QC, Canada, 15–18 August 2005; p. 8.

28. Capozzoli, A. Impacts of architectural design choices on building energy performance applications of uncertainty and sensitivity techniques. In Proceedings of the 11th International IBPSA Conference on Building Simulation, Glasgow, Scotland, 27–30 July 2009; p. 9.
29. Smith, L.; Bernhardt, K.; Jezyk, M. Automated Energy Model Creation for Conceptual Design. In Proceedings of the 2011 Symposium on Simulation for Architecture and Urban Design, Boston, MA, USA, 4–7 April 2011.
30. Amitrano, L.; Isaacs, N.; Saville-Smith, K.; Donn, M.; Camilleri, M.; Pollard, A.; Babylon, M.; Bishop, R.; Roberti, J.; Burrough, L.; et al. *Building Energy End-Use Study (BEES), Part 1: Final Report*; BRANZ Study Report No. SR 297/1; BRANZ Ltd.: Judgeford, New Zealand, 2014.
31. Bosscha, E. Sensitivity Analysis Comparing Level of Detail and the Accuracy of Building Energy Simulations. Bachelor's Thesis, University of Twente, Enschede, The Netherlands, 2013. Available online: https://purl.utwente.nl/essays/64511 (accessed on 12 April 2020).
32. Korolija, Y.D.Q.; Zhang, Y. Impact of model simplification on energy and comfort analysis for dwellings. In Proceedings of the 13th Conference of International Building Performance Simulation Association, Chambéry, France, 26–28 August 2013; pp. 1184–1192.
33. Klimczak, M.; Bojarski, J.; Ziembicki, P.; Kęskiewicz, P. Analysis of the impact of simulation model simplifications on the quality of low-energy buildings simulation results. *Energy Build.* **2018**, *169*, 141–147. [CrossRef]
34. Heo, Y.; Ren, G.; Sunikka-Blank, M. Investigating an Adequate Level of Modelling for Energy Analysis of Domestic Buildings. In Proceedings of the 3rd Asia conference of International Building Performance Simulation Association - ASim2016, Jeju(Cheju) Island, South Korea, 27–29 November 2016.
35. Dipasquale, C.; Antoni, M.D.; Fedrizzi, R. The effect of the modelling approach on the building's loads assessment. In Proceedings of the Energy Forum on Advanced Building Skins, Bressanone, Italy, 5–6 November 2013.
36. Akkurt, G.G.; Aste, N.; Borderon, J.; Buda, A.; Calzolari, M.; Chung, D.; Costanzo, V.; Del Pero, C.; Evola, G.; Huerto-Cardenas, H.E.; et al. Dynamic thermal and hygrometric simulation of historical buildings: Critical factors and possible solutions. *Renew. Sustain. Energy Rev.* **2020**, *118*, 109509. [CrossRef]
37. Samuelson, H.W.; Ghorayshi, A.; Reinhart, C.F. Analysis of a simplified calibration procedure for 18 design-phase building energy models. *J. Build. Perform. Simul.* **2016**, *9*, 17–29. [CrossRef]
38. ASHRAE. *ANSI/ASHRAE/IESNA Standard 90.1-2013 Energy Standard for Buildings Except Low Rise Residential Buildings*; American Society of Heating, Refrigerating and Air-Conditioning Engineers: Atlanta, GA, USA, 2013.
39. Building Research Establishment Ltd. *National Calculation Methodology (NCM) modelling guide (for buildings other than dwellings in England)*; Communities and Local Government: London, UK, 2013. Available online: https://www.uk-ncm.org.uk/filelibrary/NCM_Modelling_Guide_2013_Edition_20November2017.pdf (accessed on 12 April 2020).
40. Natural Resources Canada; CANMET Energy Technology Centre. *EE4 Software Version 1.7: Modelling Guide*; Natural Resources Canada: Ottawa, ON, Canada, 2008.
41. Ren, G.; Heo, Y.; Sunikka-Blank, M. Investigating an adequate level of modelling for retrofit decision-making: A case study of a British semi-detached house. *J. Build. Eng.* **2019**, *26*, 100837. [CrossRef]
42. Ho, S.H.; Rosario, L.; Rahman, M.M. Thermal comfort enhancement by using a ceiling fan. *Appl. Therm. Eng.* **2009**, *29*, 1648–1656. [CrossRef]
43. Ismail, A.R.; Jusoh, N.; Makhtar, N.K.; Zakaria, J.S.M.; Zainudin, M.K.; Omar, Z.C.; Ghani, R.A. Assessment of Thermal Comfort: A Study at Closed and Ventilated Call Centre. *Am. J. Appl. Sci.* **2010**, *7*, 402–407. [CrossRef]
44. Fanger, P.O. *Thermal Comfort*; Danish Technical Press: Copenhagen, Denmark, 1970.
45. Holmes, M.J.; Hacker, J.N. Climate change, thermal comfort and energy: Meeting the design challenges of the 21st century. *Energy Build.* **2007**, *39*, 802–814. [CrossRef]
46. Carlucci, S.; Pagliano, L. A review of indices for the long-term evaluation of the general thermal comfort conditions in buildings. *Energy Build.* **2012**, *53*, 194–205. [CrossRef]
47. Batog, P.; Badura, M. Dynamic of Changes in Carbon Dioxide Concentration in Bedrooms. *Procedia Eng.* **2013**, *57*, 175–182. [CrossRef]

48. European Committee for Standardization (CEN)-EN 13779:2007. *Ventilation for Non-Residential Buildings—Performance Requirements for Ventilation and Room-Conditioning Systems*; European Committee for Standardization: Brussels, Belgium, 2007.
49. Ihm, P.; Nemri, A.; Krarti, M. Estimation of lighting energy savings from daylighting. *Build. Environ.* **2009**, *44*, 509–514. [CrossRef]
50. Kim, G.; Lim, H.S.; Lim, T.S.; Schaefer, L.; Kim, J.T. Comparative advantage of an exterior shading device in thermal performance for residential buildings. *Energy Build.* **2012**, *46*, 105–111. [CrossRef]
51. Waldram, P.J. The Natural and Artificial Lighting of Buildings. *J. R. Inst. Br. Archit.* **1925**, *32*, 405–426, 441–446.
52. Commission Internationale de l'Eclairage (CIE). *Daylight*; Technical Report No. CIE 016-1970; CIE CIE Central Bureau: Vienna, Austria, 1970; ISBN 978 3 901906 66 4.
53. Meteotest, FabrikStrasse 14, CH-3012, Bern, Meteonorm 2000, Global Meteorological Database for Solar Energy and Applied Meteorology. Available online: https://www.thenbs.com/PublicationIndex/Documents/Details?DocId=306747 (accessed on 20 February 2019).

© 2020 by the authors. Licensee MDPI, Basel, Switzerland. This article is an open access article distributed under the terms and conditions of the Creative Commons Attribution (CC BY) license (http://creativecommons.org/licenses/by/4.0/).

Article

Trombe Wall Thermal Behavior and Energy Efficiency of a Light Steel Frame Compartment: Experimental and Numerical Assessments

Victor Lohmann and Paulo Santos *

ISISE, Department of Civil Engineering, University of Coimbra, Pólo II, Rua Luís Reis Santos, 3030-788 Coimbra, Portugal; valohmann@gmail.com
* Correspondence: pfsantos@dec.uc.pt; Tel.: +351-239-797-199

Received: 29 April 2020; Accepted: 27 May 2020; Published: 30 May 2020

Abstract: Buildings are seeking renewable energy sources (e.g., solar) and passive devices, such as Trombe walls. However, the thermal performance of Trombe walls depends on many factors. In this work, the thermal behavior and energy efficiency of a Trombe wall in a lightweight steel frame compartment were evaluated, making use of in situ measurements and numerical simulations. Measurements were performed inside two real scale experimental identical cubic modules, exposed to natural exterior weather conditions. Simulations were made using validated advanced dynamic models. The winter Trombe wall benefits were evaluated regarding indoor air temperature increase and heating energy reduction. Moreover, a thermal behavior parametric study was performed. Several comparisons were made: (1) Sunny and cloudy winter week thermal behavior; (2) Office and residential space use heating energy; (3) Two heating set-points (20 °C and 18 °C); (4) Thickness of the Trombe wall air cavity; (5) Thickness of the thermal storage wall; (6) Dimensions of the interior upper/lower vents; (7) Material of the thermal storage wall. It was found that a Trombe wall device could significantly improve the thermal behavior and reduce heating energy consumption. However, if not well designed and controlled (e.g., to mitigate nocturnal heat losses), the Trombe wall thermal and energy benefits could be insignificant and even disadvantageous.

Keywords: passive solar; Trombe wall; light steel frame; thermal behavior; energy efficiency; Mediterranean climate; office use; residential use; heating set-points

1. Introduction

Energy is one of the main concerns when addressing sustainable development, especially since the world's energy matrix is still very dependent on fossil fuels, as oil and coal. The building's sector plays an important role, as buildings consume approximately 40% of the total energy in Europe, being also responsible for about 36% of the CO_2 emissions [1]. Aiming to improve the energy efficiency of buildings, the European Union (EU) has established the energy performance of buildings directive (EPBD) [2], in which two key concepts are defined: (1) the cost-optimal energy, regarding cost-efficiency of strategies [3], and (2) the nearly zero-energy buildings (nZEB)—buildings with very high energy efficiency—that cover their energy needs with energy produced by renewable sources, on-site or nearby [4]. To meet the EPBD requirements, the optimization of construction systems and the development of strategies to decrease energy consumption by buildings are key [5].

A sustainable strategy to improve the thermal and energy performance of buildings is exploiting solar energy, which also meets the EPDB establishments. A Trombe wall (TW) is a passive solar device that can be present in a building's façade to accumulate solar heat, heating, and even cooling indoor spaces, fostering natural ventilation [6]. This passive solar device was patented in 1881 by the

American engineer Edward Morse and popularized in the 1960s by the French engineer Felix Trombe and architect Jacque Michel, as mentioned by Saadatian et al. [7]. The classical configuration of Trombe walls is an outer glazed area to allow solar radiation to reach a massive storage wall, promoting the greenhouse effect. The storage wall usually has two interior vents (ventilated TW), connecting the indoor space to an air cavity between the wall and the glass panel—one at a lower height and other at an upper height [8]. To reduce heat losses through the TW device during cold winter nights, it is often used as an external night shutter [6]. Additionally, in warmer climates, exterior shading devices or overhangs are often used to mitigate overheating risk, as well as external upper and lower vents, promoting natural air-ventilation cooling effect during the summer season [9].

The operation of a Trombe wall is based on heat transfer principles. It absorbs solar heat in its high thermal mass storage wall during daytime and transfers part of this heat to the interior space of the building through conduction, radiation, and convection. The wall stores heat during the day and releases it during evening and night times, when the occupants require it and outdoor temperature decreases. The TW system, when exposed to direct solar radiation, exploits the greenhouse effect that occurs in the glazed air cavity, absorbing and storing heat in a massive wall. When the air cavity is warmed up by the heated storage wall, the air will flow upward due to buoyancy or thermosiphon effect. This heated air goes to the interior of the adjacent compartment through an upper vent, while colder air comes from the same room through a lower vent, re-entering to the TW air cavity [6].

Trombe walls have attracted attention over the last years, with different types studied, incorporating modern materials and construction methods, such as the incorporation of phase change materials [10] and photovoltaic cells on the glazed area [11].

Recently, Zhou et al. [12] studied the thermal performance of a composite Trombe wall under steady-state conditions. They compared three types of Trombe walls: traditional (TTW), water (WTW), and glass-water (GWTW). They optimized the thermal performance of the composite Trombe walls by defining two operating modes: (1) heat-collecting mode during the daytime, and (2) heat-preservation mode during night-time. The WTW exhibited the best efficiency during daytime (3.3% higher than the TTW) and also during night-time, allowing a heat loss reduction of 31% compared to TTW.

Besides space heating, researchers are also trying to develop new application advantages for Trombe walls. Hu et al. [13] made some experimental and numerical studies of a novel water blind-Trombe wall system. This new TW system, besides space heating and natural ventilation, could also provide domestic hot water since it made use of orientated steel blinds filled with flowing water and a hot water tank. They performed a comparison with conventional (i.e., without a glazing panel) and traditional TWs. A significant annual overall thermal load reduction was found compared to conventional (−42.6%) and traditional (−13%) Trombe walls. They also concluded that the new water blind-Trombe wall system, besides achieving a favorable insulation performance during winter, was also able to take advantage of the undesired solar radiation during the summer season to heat the water for domestic uses.

As mentioned before, Trombe walls could be very useful during the winter season to reduce space heating energy, but during the cooling season, this may have a negative impact due to limited control capability. Hong et al. [14] analyzed the thermal performance of a Trombe wall with an integrated Venetian blind during the cooling season. They evaluated the TW cooling mode operational control to regulate shading (from orientable Venetian blind slats within the TW air cavity) and natural ventilation (outside and cross). Several building occupation schedules were compared, i.e., service, office, and domestic buildings. It was found that the studied Venetian blind integrated TW could effectively prevent overheating through shading and ventilation. Moreover, they also concluded that the outside circulation mode was a more effective ventilation strategy to reduce cooling energy (5.0% to 5.8%) in comparison with the cross ventilation mode (2.5% to 4.6%).

Obviously, the thermal behavior and energy efficiency of buildings also depend on the buildings' envelope and construction system. In Tunisia, Abbassi et al. [15] performed numerical simulations, for a small single zone building (4 m × 4 m), to evaluate the heating energy savings provided by a

Trombe wall for different heavyweight building envelope façade walls (e.g., brick and stone), having different thermal transmittances (U-values), ranging from 2.035 W/(m²·K) down to 0.388 W/(m²·K) for a higher insulated exterior wall. For a smaller TW (3 m² area), they predicted heating energy savings, ranging from 28% up to 69%, for lower and higher thermal insulation levels, respectively. For a larger TW (6 m² area), the analogous heating energy savings ranged from 66% up to 98%.

An interesting alternative to traditional reinforced concrete and ceramic masonry construction is the lightweight steel frame (LSF) system, which has been attracting attention worldwide, given its functional, economic, and environmental advantages [16,17]. This lightweight innovative system presents construction flexibility and adaptability due to its modularity [18], safety at work, and construction economy due to the industrialized nature of the components, which also facilitates series production, prefabrication, and transportation [19]. In fact, several previous research studies have addressed the LSF system-related benefits, including sustainability [20], life cycle energy balance [21], and operational energy [22]. Nevertheless, an effort has been made to mitigate eventual drawbacks related to the thermal behavior of LSF construction, aiming to mitigate thermal bridges originated by the high thermal conductivity of the steel elements [23,24] and to increase the thermal inertia of this type of construction [25].

As mentioned before, the thermal behavior and energy efficiency of a Trombe wall depend on many factors, such as geometric (e.g., area, height, thickness, and orientation of the TW; existence and dimension of overhangs), materials' properties (e.g., storage wall thermal properties; glazed pane optical and thermal properties; shutter thermal properties; thermal insulation), fluid dynamics (e.g., dimensions and control of inner/outer and upper/lower vents; thickness of the air channel; natural or forced airflow), location (e.g., latitude; north or south hemisphere), and weather (e.g., solar radiation level and incidence angle; nocturnal cloudy or clear sky; temperature; wind speed, and direction) [6]. Thus, it is not an easy task to adequately design and control a TW device to take full thermal, energy, and economic advantages [26,27].

As stated before, despite the LSF system advantages, there are also possible drawbacks, such as the reduced thermal inertia, due to its natural weightlessness, compared to traditional concrete structures [28]. Thus, it would be interesting to evaluate the effect of a solar passive Trombe wall device, which is characterized by having a massive storage wall, on an LSF construction system, having low thermal inertia and reduced mass. However, this kind of research has not been found in the literature. Moreover, research works on water Trombe walls are very scarce. Therefore, in this work, the influence of a passive solar water Trombe wall (TW) device on the thermal behavior and energy efficiency of a lightweight steel frame (LSF) compartment, located in Coimbra (Portugal), was studied, being this evaluation based in numerical simulations and in situ measurements. Measurements of indoor air temperature were performed inside two real scale experimental identical cubic modules, exposed to natural exterior weather conditions, while simulations were performed using advanced dynamic models, validated experimentally.

First, the experimental approach has been described, regarding the LSF experimental modules, the TW prototype, the weather stations, and temperature/humidity data-logger sensors. After, the numerical approach has been detailed, including the 2D thermal computations to obtain the U-values of the LSF components and the advanced numerical simulations. Next, the calibration and model validation has been reported for both reference and TW LSF models, and some computational fluid dynamics (CFD) results have also been reported. Afterward, the obtained results have been discussed and grouped in TW benefits and parametric study. The winter TW benefits were evaluated regarding indoor air temperature increase and heating energy reduction. The thermal behavior parametric study was performed for several TW key-factors, such as the thicknesses of the air cavity and storage wall and dimensions of the internal vents and the storage wall materials. Finally, some concluding remarks about this research work have been highlighted.

2. Materials and Methods

The materials and methods used in this research have been described in detail in this section, starting with the experimental and numerical approaches, followed by the calibration and validation of the advanced dynamic thermal simulation models of the LSF modules and water Trombe wall.

2.1. Experimental Approach

2.1.1. LSF Experimental Modules

The experimental measurements were performed on two similar lightweight steel frame (LSF) modules constructed near the Department of Civil Engineering (DEC) of the University of Coimbra (Portugal), as illustrated in Figure 1, having a GPS coordinates: 40.1855° N, 8.4167° W. Those experimental modules were two identical cubic compartments constructed in LSF, with inner dimensions: (L) 2.75 m × (W) 2.75 m × (H) 2.80 m. Module 1 was used as a reference (for results comparison), while module 2 had a water Trombe wall prototype on its south facade.

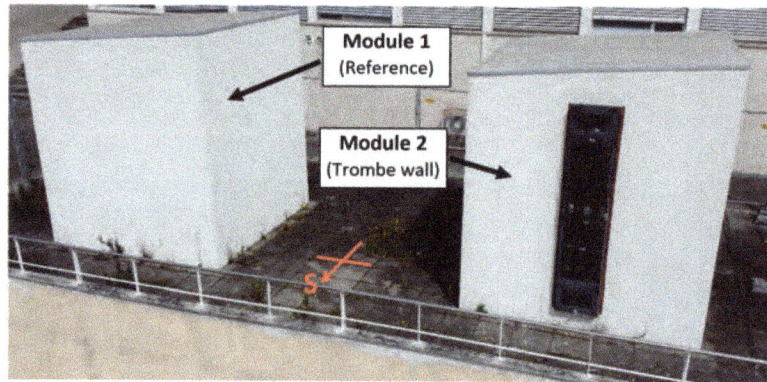

Figure 1. LSF (lightweight steel frame) experimental modules constructed at the University of Coimbra, Engineering Campus (GPS: 40.1855° N, 8.4167° W).

The external dimensions of the experimental modules, as well as the material specifications of the LSF construction elements, such as the number of layers, materials, and thicknesses, are schematically illustrated in Figure 2, while Table 1 displays the thermal conductivities of the materials. In these experimental modules, the LSF system B(A)[a] was adopted and manufactured by Urbimagem company [29], making use of steel profiles C100 × 45 × 1.5 mm. The structural sheathing was provided by 12 mm oriented strand board (OSB) panels [30] on both sides of the walls' steel frame. The ceiling was also inferiorly lined with OSB panels, as well as the upper side of the roof steel frame beams. To allow access to the interior, both modules had a similar wooden door (2.00 m high by 0.78 m wide), which was thermally insulated with the same expanded polystyrene (EPS) external thermal insulation composite system (ETICS) system of the walls. There were no windows in the experimental TW modules. This was justified by the intention to isolate the TW effect in the evaluated compartments. A glazed window (e.g., south orientated) would provide additional solar heat gains, which would be overlapped and more difficult to distinguish from the heat gains provided by the TW device.

Notice that, as illustrated in Figure 2, the experimental modules were designed to have gypsum plasterboard (GPB) as an inner sheathing layer of walls and ceiling, but later it was decided not to apply these GPB panels. The batt insulation was provided by 100 mm mineral wool (MW) [31], fulfilling the air-cavity between the steel frame. The exterior thermal insulation composite system (ETICS) was made with EPS thermal insulation [32] (50 mm thick) and finished by a reinforced plaster layer (5 mm). The exterior thermal insulation of the roof was made of extruded polystyrene (XPS) [33] with the same thickness. To avoid moisture direct contact from the ground, the floor was 300 mm elevated, creating a small crawl space below, as illustrated in Figure 2, having an 18 mm OSB panel [30] below and another above the continuous XPS [34] thermal insulation layer (60 mm thick). The inclined flat roof was waterproofed by a polyvinyl chloride (PVC) membrane [35] (1.5 mm thick), forming a plenum above the ceiling with variable thickness.

Table 1. Thermal conductivity (λ) of the materials used in the lightweight steel frame (LSF) modules.

Materials		λ ((m·K)/W)	Reference
Reinforced plaster (ETICS [1] finish)		0.720	[37]
EPS [2] (ETICS [1] thermal insulation)		0.036	[32]
OSB [3] (LSF sheathing)		0.130	[30]
Mineral wool (cavity insulation)		0.037	[31]
Steel (profiles C100 × 45 × 1.5 mm)		50.000	[38]
XPS [4]	(roof insulation)	0.036	[33]
	(floor insulation)	0.035	[34]
Vinyl floor cover		0.250	[39]
PVC [5] membrane (roof waterproofing)		0.170	[35]
Wooden door		0.144	[40]

[1] ETICS, external thermal insulation composite system; [2] EPS, expanded polystyrene; [3] OSB, oriented strand board; [4] XPS, extruded polystyrene; [5] PVC, polyvinyl chloride.

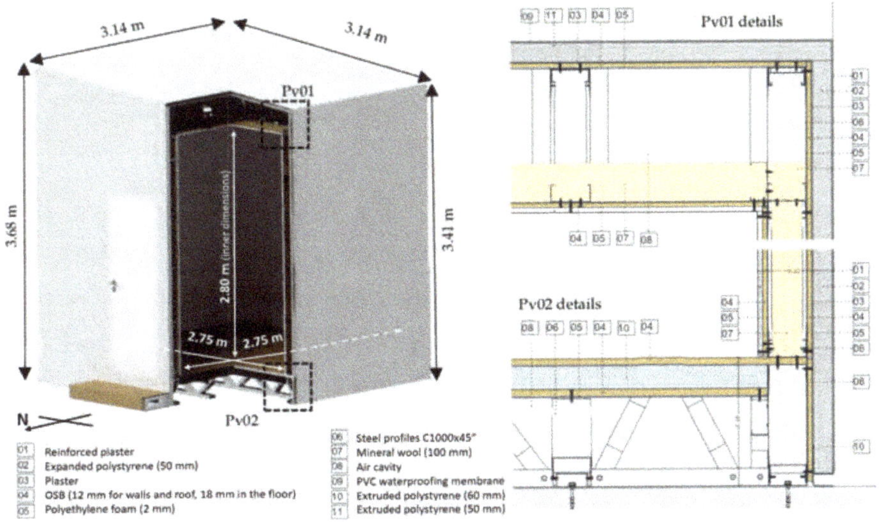

Figure 2. Schematic details of the LSF modules construction elements (adapted from [36]).

Table 2 displays, for each LSF element, the materials and thicknesses of the layers, as well as the computed thermal transmittance (U-value). Notice that two types of layers were assessed in these LSF elements: (1) homogeneous, where the steel frame was not included in the thermal computations, given its location outside the insulation and sheathing materials, and (2) inhomogeneous, where the steel frame crossed through the insulation materials (e.g., mineral wool). The U-value for the elements with homogeneous layers (floor, roof, and door) was computed following the analytical calculation procedures prescribed by standard ISO 6946 [41]. The U-values of the LSF elements containing inhomogeneous layers (walls and ceiling) were computed, making use of bi-dimensional (2D) finite element method (FEM) models built in the THERM software [42], as has been detailed next in Section 2.2.1. The obtained U-values (Table 2) ranged from 0.326 W/(m^2·K) in the walls up to 0.670 W/(m^2·K) in the ceiling.

Table 2. Materials, thicknesses (d), and thermal transmittances (U) of the LSF elements.

Element (Layers Type)	Materials (Layers from Outer to Inner Surfaces)	d (mm)	U-Value (W/(m^2·K))
Walls (inhomogeneous)	Reinforced plaster (ETICS [1] finish)	5	0.326
	EPS [2] (ETICS [1] thermal insulation)	50	
	OSB [3] (LSF sheathing)	12	
	Mineral wool (cavity insulation)	100	
	OSB [3] (LSF sheathing)	12	
	Total thickness =	179	
Floor (homogeneous)	OSB [3] (LSF sheathing)	18	0.426
	XPS [4] (floor slab insulation)	60	
	OSB [3] (LSF sheathing)	18	
	Vinyl floor cover	3.4	
	Total thickness =	99.4	
Ceiling (inhomogeneous)	Mineral wool (cavity insulation)	100	0.670
	OSB [3] (LSF sheathing)	12	
	Total thickness =	112	
Roof (homogeneous)	PVC [5] membrane (roof waterproofing)	1.5	0.613
	XPS [4] (exterior thermal insulation)	50	
	OSB [3] (LSF sheathing)	12	
	Total thickness =	63.5	
Door (homogeneous)	Reinforced plaster (ETICS [1] finish)	5	0.534
	EPS [2] (ETICS [1] thermal insulation)	50	
	Wooden door	44	
	Total thickness =	99	

[1] ETICS, external thermal insulation composite system; [2] EPS, expanded polystyrene; [3] OSB, oriented strand board; [4] XPS, extruded polystyrene; [5] PVC, polyvinyl chloride.

2.1.2. Trombe Wall Prototype

The Trombe wall prototype (2.80 m high and 0.55 m wide) was placed on the south-oriented wall of module 2 (Figure 1). Figure 3a schematically illustrates the geometry of this Trombe wall prototype, which was developed and executed during a Ph.D. research work [36]. Notice that the dimensions of this modular TW prototype were defined, taking into account the ceiling height (2.80 m) and the usual vertical steel stud spacing in LSF construction (0.60 m). The thermal storage wall was made with a black-painted steel sheet tank fulfilled with water, having 50 mm of thickness. On the outer side, there was an aluminum frame glazing system with double glass (4 mm + 16 mm of argon + planistar 6 mm), having an effective solar absorption area of 1.1 m^2. The glazing panel had a solar heat gain coefficient (SHGC) equal to 0.743, while the direct solar transmission was 0.667, and the thermal transmittance was 2.552 W/(m^2·K), as displayed in Figure 3b.

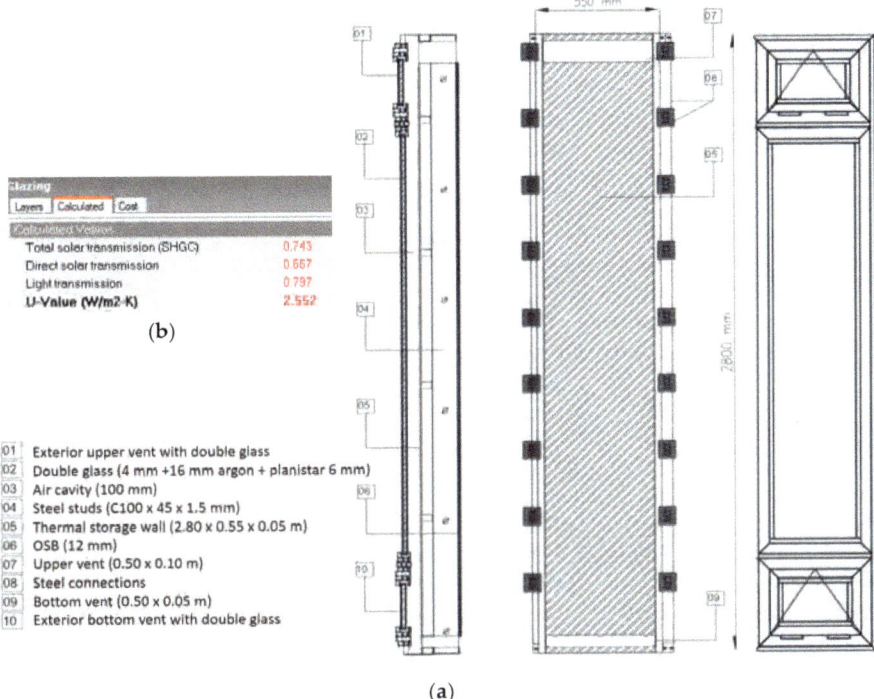

Figure 3. Trombe wall prototype: (**a**) Schematic geometry details (adapted from [36]); (**b**) Glazing optical and thermal properties.

This glazed aluminum frame had a top and lower exterior openings for exterior ventilation, which were not used during these experiments, being all the time closed. Between the storage wall and the outer glazing, there was a 100 mm thick air cavity. On the inner surface of the storage wall, there was a layer of 0.10 m of mineral wool, covered by an OSB panel (12 mm). To allow air circulation between the outer air cavity and the indoor environment, there were two rectangular air vents on the Trombe wall: (1) an upper air vent, 0.50 m wide by 0.10 m high, and (2) a bottom air vent with the same width but a smaller height (0.05 m).

2.1.3. Monitoring Equipment

To reproduce the thermal behavior of the experimental modules exposed to exterior weather conditions, it was needed to have access to hourly weather data recorded nearby. With this purpose, two weather stations were used: (1) Department of Mechanical Engineering (DEM) [43], also located in the Engineering campus of the University of Coimbra (GPS: 40.1849° N, 8.4132° W), and (2) CoolHaven company [44], located in Coimbra iParque, Antanhol (GPS: 40.1792° N, 8.4654° W).

The nearest weather data station (DEM) was used for most of the data needed to perform advanced dynamic simulations, including air temperature, dew-point temperature, relative humidity, wind direction, wind speed, atmospheric pressure, and precipitation. However, this weather station did not provide some additional relevant weather data, such as the parameters related to solar radiation, i.e., global horizontal radiation, diffuse horizontal radiation, and direct normal radiation. This essential detailed solar radiation information was obtained in the CoolHaven weather station, located about 7 km from the experimental modules.

Regarding the hardware, the DEM weather station is a wireless Davis Vantage Pro2 Plus [45], while the CoolHaven is constituted of several sensors, with the pyranometer being a sunshine sensor Delta-T BF5 [46].

Notice that according to the Köppen–Geiger climate classification [47], the city of Coimbra (Portugal) is located in a Csb climate region, which is characterized by a temperate climate with rainy winter and dry summer slightly hot, being a very frequent climate within the Mediterranean region [16].

The indoor air temperature and humidity were measured simultaneously, inside both LSF modules, to monitor their thermal behavior and verify the influence of the solar Trombe wall. With this purpose, one Tinytag Ultra 2—TGU-4500 [48] air temperature and humidity sensor was installed inside each module, being suspended in the middle ceiling, at mid-height. These sensors were factory calibrated, having a precision of ±0.45 °C for temperature and ±3% for relative humidity. The measured data was averaged and recorded every 10 minutes, having a sampling interval of 10 seconds. The in situ measurements took place from the 26th of July 2019 until the 19th of January 2020.

2.2. Numerical Approach

2.2.1. 2D FEM Thermal Computations

As mentioned before (see Section 2.1.1), the U-values of the inhomogeneous LSF elements (walls and ceiling) were computed, making use of bi-dimensional (2D) models implemented in a finite element method (FEM) software: THERM [42]. The FEM mesh was refined to have a maximum error of 2%.

LSF Ceiling Element

For the ceiling element, as the steel profiles are placed only in one direction (see the yellow region in Figure 4a), the U-value was directly obtained from the 2D FEM model, as illustrated in Figure 4b. The model had a width of 600 mm, i.e., equal to the distance between the steel studs within the ceiling. The steel C stud was positioned in the middle of the model, as shown in Figure 4b, and this is a representative part of the LSF ceiling slab. Moreover, the ceiling mineral wool (MW) insulation was considered only between steel sections since, in practice, it was not possible to put MW inside the corresponding steel lattice beam, where it was considered an air gap. Figure 4c displays the temperature distribution predicted in the ceiling cross-section, where the thermal bridged effect was clear due to the MW thermal insulation discontinuity. The global U-value computed from the THERM model was 0.670 W/(m²·K). Notice that assuming homogeneous layers, i.e., considering continuous MW insulation and neglecting the steel studs, the U-value obtained was 0.334 W/(m²·K), being 50% smaller.

LSF Wall Element

Since the LSF walls had steel studs in vertical, horizontal, and diagonal planes (see Figure 5a), the bi-dimensional U-value computation procedure was different from the ceiling element, where the U-value was directly obtained from the THERM model. It is well known that an insulated LSF element has two distinct thermal zones [49,50]: (1) an increased heat transfer zone (lower thermal resistance) in the vicinity of the steel studs, given the high thermal conductivity of steel, and (2) a more reduced heat transfer zone (higher thermal resistance) in the insulated cavity between the steel studs. Thus, the global thermal transmittance (U_{global}) of LSF elements with complex steel frame could be estimated, making an area-weighted summation of the U-values for each thermal zone mentioned before ("stud" and "cav"), as given in the following equation:

$$U_{global} = \frac{U_{stud} \cdot A_{stud} + U_{cav} \cdot A_{cav}}{A_{global}} \quad (1)$$

where A_{global} is the total area of the LSF element (internal dimensions), A_{stud} is the total area of influence of the steel stud on the LSF element, and A_{cav} is the remaining cavity area of the LSF wall. For this specific LSF wall, the areas considered in the computations are displayed in Figure 5a.

Both U-values (U_{stud} and U_{cav}) were obtained, making use of a THERM model, as illustrated in Figure 5b. This simplified LSF wall model had a length equal to the spacing between the vertical steel studs, i.e., 600 mm. To obtain the two representative U-values, two "measurement" zones were simulated in the LSF wall model: one right under the steel stud and another one in the edge of the wall cavity. These "measurement" zones were modeled having the same width as the steel stud flange, i.e., 45 mm, and is delimited in Figure 5 by two dashed white lines.

Figure 5c displays the obtained temperature (°C) color distribution along the cross-section of the LSF wall model and is well visible in the thermal bridge originated by the central steel stud and its correspondent temperature disturbance. Figure 5d shows the computed heat flux (W/m²) distribution within the cross-section of the LSF wall, as well as the two U-values computed in the steel stud vicinity and in the edge of the wall cavity. As expected, the U_{stud} (0.797 W/m²·K) was considerably higher (+260%) than the U_{cav} (0.221 W/m²·K), confirming the huge relevance of the steel stud (only 1.5 mm thick) in the thermal performance of the LSF wall.

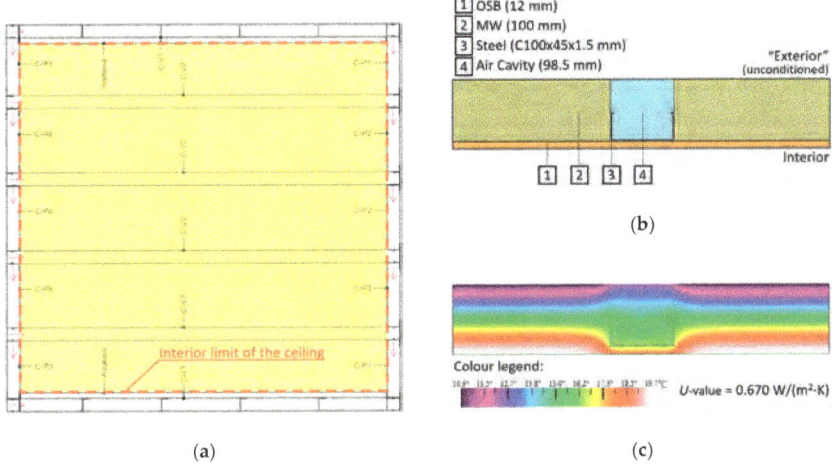

Figure 4. LSF ceiling element: (**a**) Plan view of the ceiling steel frame; (**b**) THERM model; (**c**) Temperature color distribution and obtained U-value.

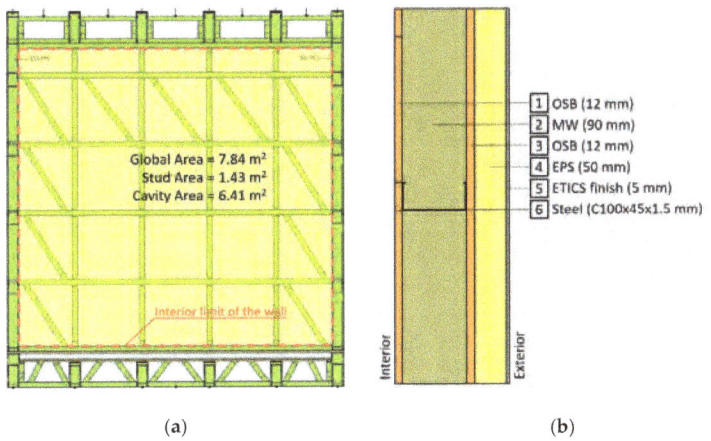

Figure 5. *Cont.*

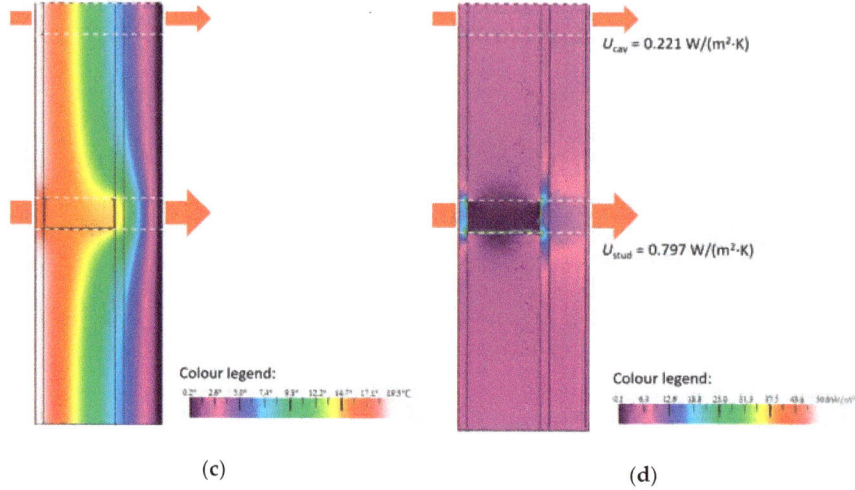

Figure 5. LSF wall element: (**a**) Frontal view of the wall steel frame; (**b**) THERM model; (**c**) Temperature color distribution; (**d**) Heat flux distribution and local *U*-values.

Finally, knowing the three areas (Figure 5a) and the two *U*-values (Figure 5d) and making use of Equation (1), a global *U*-value equal to 0.326 W/(m²·K) was obtained. Notice that when the steel studs were neglected and homogenous layers were assumed, the *U*-value reduced to 0.225 W/(m²·K) (31% smaller).

It is important to highlight that there are several strategies to mitigate the thermal bridges originated by steel studs within an LSF component, reducing their *U*-value, such as the use of thermal break (TB) strips within steel studs flange [51]. These TB strips could be made of different materials, such as recycled tire rubber [52]. Shortly, it was intended to use this type of TB strips to improve the thermal performance of these experimental LSF modules.

2.2.2. Advanced Dynamic Simulations

The advanced dynamic thermal simulations were performed in the software DesignBuilder version 5.5.0.012 (DesignBuilder Software Ltd, Stroud, Gloucester, UK) [37]. The computations were performed, making use of hourly interval data. A replica of the two LSF experimental modules photographed in Figure 1 was modeled, taking into account the location/climate, the geometry/dimensions, the construction elements composition (e.g., walls, floor, ceiling, roof, door, and Trombe wall), the material properties, the airtightness, the activity, and occupation parameters. Figure 6 exhibits a print-screen view of the two models: (1) module 1, used as reference (Figure 6a), and (2) module 2, containing the Trombe wall (Figure 6b).

The airtightness of these experimental modules was measured in-situ [36], and the obtained value (0.05 air changes per hour) was implemented in the DesignBuilder model as a constant value and without any natural ventilation since, during the measurements, the openings (back door and Trombe wall exterior vents) were always closed. Moreover, the modules were kept empty, i.e., without anyone inside. Thus, the occupancy was set as "null", and the activity tab as "none". Notice that the color of the materials was also reproduced, in particular, the black color of the Trombe wall (Figure 6b).

2.3. Calibration and Model Validation

To ensure good reliability of the DesignBuilder [37] advanced dynamic models (Figure 6) thermal behavior predictions, the obtained simulation results were compared with the air temperature in-situ measurements (see Section 2.1.3), performed inside the LSF modules (Figure 1), subjected to natural outdoor weather conditions (recorded nearby, as previously explained in Section 2.1.3), allowing to validate these models, as shown next.

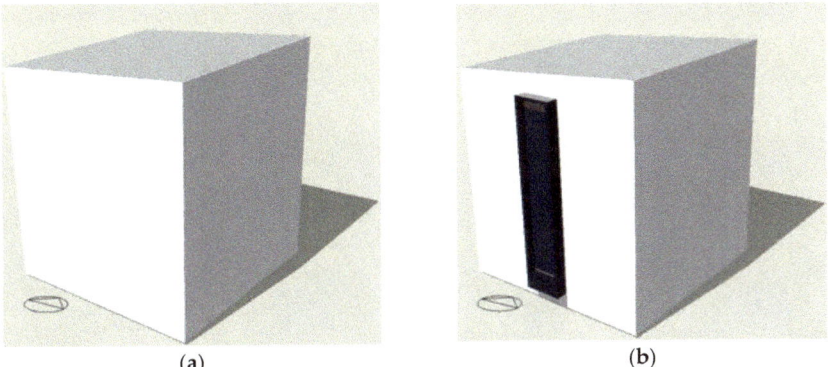

Figure 6. DesignBuilder models southeast views: (**a**) Module 1 (Reference); (**b**) Module 2 (Trombe wall, TW).

2.3.1. Reference LSF Model

Figure 7 presents a graph with a comparison among predicted and measured indoor air temperatures in the reference LSF module (module 1) during one week (2–8 September 2019). A good agreement between the DesignBuilder model predictions and the in-situ indoor air temperatures was observed. In fact, both average temperatures were very similar: 26.4 °C (recorded) and 26.3 °C (predicted). Moreover, the root mean square error (RMSE) was only 0.3 °C, allowing to conclude that this DesignBuilder advanced dynamic simulation reference LSF model was calibrated and experimentally validated.

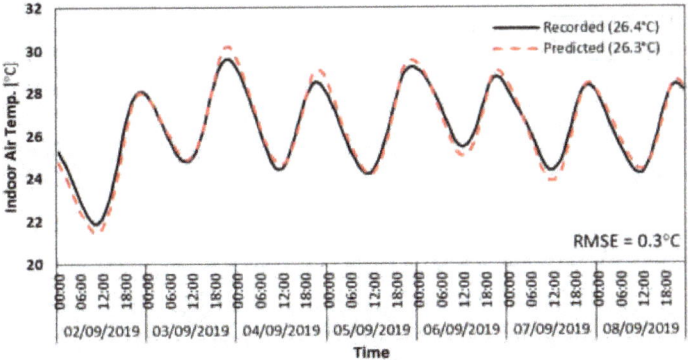

Figure 7. Predicted and measured indoor temperatures in module 1 (reference).

2.3.2. Trombe Wall LSF Model

The accuracy of the Trombe wall LSF model was also verified by comparison among predicted and measured indoor air temperatures. Figure 8 displays the obtained results plot, in which a good agreement between both curves was observed. The RMSE for this model was 0.5 °C, confirming also a good accuracy performance of this second model.

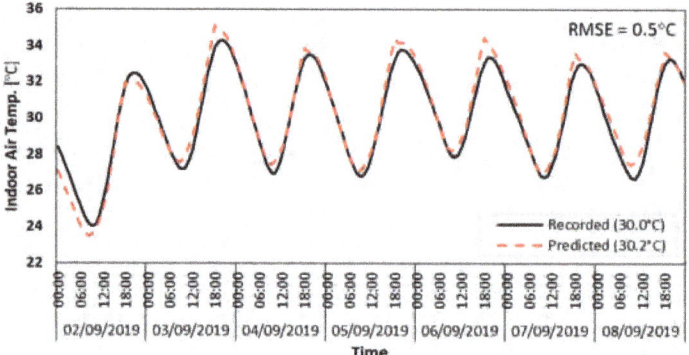

Figure 8. Predicted and measured indoor temperatures in module 2 (with a Trombe wall).

2.3.3. Trombe Wall CFD Assessment

To verify if the modeled Trombe wall is operating coherently, a computational fluid dynamics (CFD) analysis was also conducted on DesignBuilder, which has a built-in CFD tool. Figure 9 displays the results of the CFD analysis, carried for the 16:00 hours of the 4th of September, with both air velocity and temperature in a color scale being displayed, as well as velocity vectors.

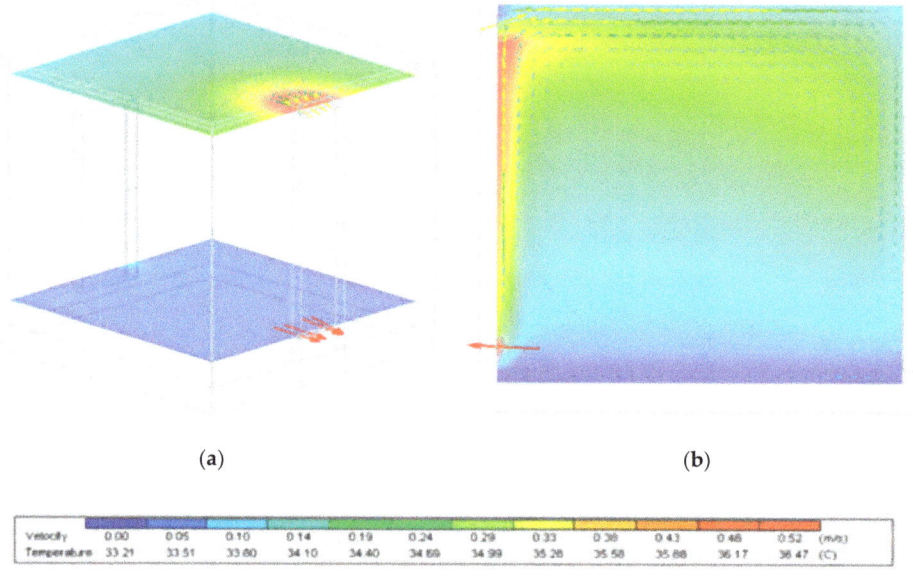

Figure 9. CFD (computational fluid dynamics) analysis (air velocity and temperature) of module 2 (with a Trombe wall): (**a**) Horizontal planes at vent levels; (**b**) Vertical plane in front of the Trombe wall.

Looking at the results of the horizontal plane plotted in Figure 9a was well visible the colder air entrance to the Trombe wall air cavity through its lower vent, as well as the warmer air flowing out of the upper vent near the ceiling. Moreover, in Figure 9b (the vertical plane in front of the Trombe wall), the air stratification in height and also the air being heated near the Trombe wall were again visible, which was exposed to direct solar radiation (4 pm) and, consequently, was flowing up to the ceiling. Therefore, these CFD simulation results made sense and were coherent with the expected ones for a compartment with a Trombe wall exposed to direct solar radiation, which ensured the reliability of the implemented models.

3. Results and Discussion

In this section, the obtained results have been presented and discussed, starting with the Trombe wall benefits, regarding the thermal behavior and heating energy savings. Thereafter, the results of the sensibility analysis, for several Trombe wall parameters, have been described and discussed.

3.1. Trombe Wall Benefits

In this section, the water Trombe wall benefits were assessed, making use of in situ indoor air temperature measurements (Section 3.1.1.) and advanced dynamic numerical simulations for the heating energy reduction predictions (Section 3.1.2.). These assessments were performed by comparison between module 1 (the reference one) and module 2 (the one with a Trombe wall) located in the city of Coimbra (Portugal), during winter.

3.1.1. Indoor Temperature Increase

The indoor air temperature comparisons were made using the data from measurements taken simultaneously with the temperature and humidity sensors [48], on both modules (with and without the Trombe wall) and are plotted in Figure 10, as well as the exterior environment air temperature. Two distinct winter weeks were chosen to demonstrate the behavior of the modules under different weather conditions. In Figure 10a, the records for a sunny week (from 28th of December to 3rd of January) are displayed, while in Figure 10b, the measurements for a cloudy week (from 16th to 22nd of December) are shown.

In the sunny winter week (Figure 10a), the indoor air temperature increase in module 2 due to the Trombe wall was well visible, having an average temperature of 16.2 °C, i.e., a temperature increase of 3.3 °C relative to module 1. Notice that even with a Trombe wall, the indoor comfort air temperature (e.g., 18 °C) was not reached. Another interesting feature was that the daily indoor air temperature amplitude (or fluctuation) was also greater in the experimental module with the Trombe wall (module 2), having a higher temperature increase rate during the day (due to the solar heat gains) and also a higher temperature decrease rate during the night (due to the higher heat losses through the Trombe wall, which did have any night shutter device).

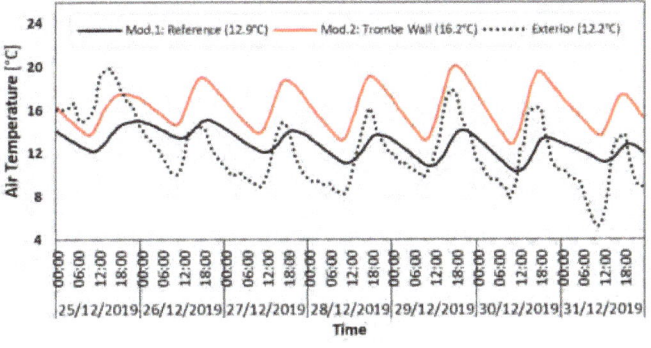

(a)

Figure 10. *Cont.*

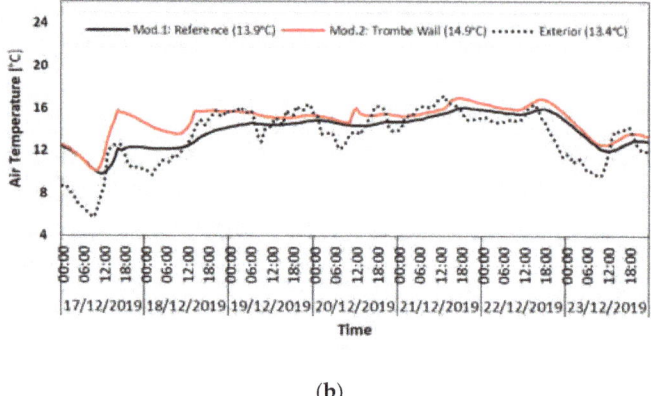

(**b**)

Figure 10. Recorded indoor air temperatures with and without a Trombe wall: (**a**) Winter sunny week; (**b**) Winter cloudy week.

When the sky was cloudy (Figure 10b), as expected, the daily temperature variation was very smothered, and the air temperature difference inside the modules became very reduced, which was only 1 °C higher for this week inside module 2. Comparing both weeks (sunny and cloudy), the average environment exterior air temperature was lower during the sunny week (Figure 10a) (12.2 °C) in comparison with the cloudy week (13.4 °C), which was 1.2 °C higher. This was due to the night cooling effect, which was much higher in a winter clear sky in comparison with a cloudy sky. Thus, this feature also demonstrated how important it was to control the night heat losses, mainly when the sky was clear, in order to optimize the thermal performance of the Trombe wall during the heating season.

3.1.2. Heating Energy Decrease

In this section, the heating energy decrease due to the existence of a Trombe Wall was predicted, making use of advanced numerical dynamic simulation models, as previously detailed in Section 2.2.2 and validated in Section 2.3. The hourly weather data was obtained from the EnergyPlus IWEC database [53] for Coimbra city (Portugal), and the computations were performed for all winter season (from 22nd of December until the 20th of March). The modeled air-conditioning heating system was a "split" type with no fresh air, having a coefficient of performance (COP) for heating mode equal to 2.35, with the adopted energy source the electricity from the grid.

To compare its relevance in the heating energy demand, two heating set-points were simulated, namely, 20 °C and 18 °C, respectively; the former and current thermal comfort temperatures considered for calculating residential heating energy needs in Portugal [54].

Moreover, two occupation schedules and use types were considered, namely, (1) an office space occupied from 08:00 to 18:00 during weekdays (Monday to Friday), and (2) a residential space occupied from 19:00 to 07:00 during all days. The predicted energy demand for heating (electricity) was displayed and analyzed as a total value (kWh) and as normalized values (kWh/m^2).

Residential Space Use (Heating during the Night)

The heating energy demand predicted for residential space use (night occupation) is displayed in Figure 11 for both LSF modules and two heating set-points. As expected, reducing the heating set-point (18 °C instead of 20 °C) allowed reducing also the heating energy consumption. This energy reduction was significant (Figure 11b), ranging from −33%, in the reference LSF module 1, to −40% in the Trombe wall LSF module 2.

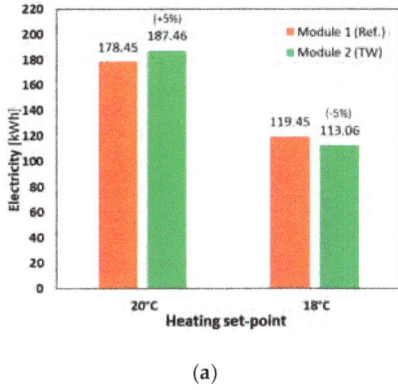 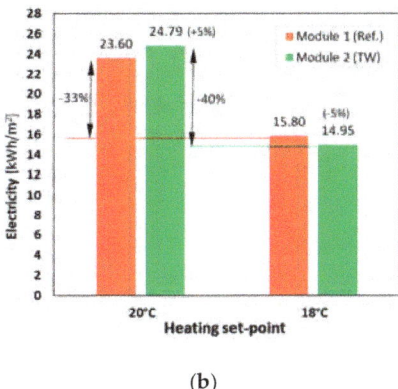

Figure 11. Predicted heating energy consumption (electricity) during the winter season in Coimbra (Portugal), assuming residential space use, for two different set-points: (**a**) Total values; (**b**) Normalized values.

The heating energy consumption in module 2 was 5% lower than in module 1 for an 18 °C heating set-point, confirming the energy efficiency advantage of the Trombe wall (TW) in the second LSF module. However, when the heating set-point was higher (20 °C), the computed results showed a 5% increase in the heating energy for the TW module 2 (24.79 kWh/m^2) in comparison with the reference module 1 (23.60 kWh/m^2). This surprising feature was related to the increased heat losses during the night due to the existence of the TW in module 2, which were not enough to balance the solar heat gains during the daytime, and this assumption has been explained in detail in the following paragraphs.

The space heating energy demand, besides the efficiency of the air-conditioning system (assumed to be 2.35 for the heating mode in this work), depended on the heat balance (gains versus losses) for each module. When this heat balance was positive (e.g., during a sunny day due to significant solar heat gains), the indoor temperature arose. When this heat balance was negative (e.g., during the night due to the exterior temperature drop and absence of solar radiation), the indoor temperature decreased.

As measured and previously plotted in Figure 10a, the indoor temperature increase rate during the day was bigger in module 2 (red line) due to the higher solar heat gains provided by the Trombe wall. However, as also displayed in the same figure, during the night, the indoor temperature decrease rate was also bigger in the TW module 2, compared to the reference module 1 (black line), due to higher heat losses through the Trombe wall.

In fact, the thermal transmittance (*U*-value) of the TW device, due to air circulation between the glazed air-cavity and the interior of the module, was increased to the *U*-value of the glazing panel (2.552 W/(m^2·K), see Figure 3b). Comparing this *U*-value with the one provided by the LSF wall (0.326 W/(m^2·K), see Table 2), for the same area and temperature difference, the heat losses through the glazing panel of the TW were almost 7 times higher (+683%).

Obviously, when the indoor air temperature set-point was elevated from 18 °C up to 20 °C, the temperature difference between indoor and outdoor conditions also increased, leading also to an increase in the heat losses, which originated a higher space heating energy consumption to maintain the defined set-point indoor temperature. Once again, this feature reinforced the importance of mitigating heat losses through the TW, mainly during winter season night-time, for example, making use of a controllable night shutter device.

Office Space Use (Heating during the Day)

The heating energy demand simulation results, assuming an office space use, i.e., during the daytime, in both LSF modules, are displayed in Figure 12. Now, the energy efficiency benefits of the TW use were significantly higher in comparison with the residential daytime use (Figure 11). The heating energy reduction ranged from −14%, for a 20 °C heating set-point, to −27% for an 18 °C set-point. This improved energy efficiency was because the heating schedule of the air-conditioning system matched the higher TW solar heating gains during the daytime. Consequently, the indoor temperature increased, and the heating energy use decreased for both heating set-points.

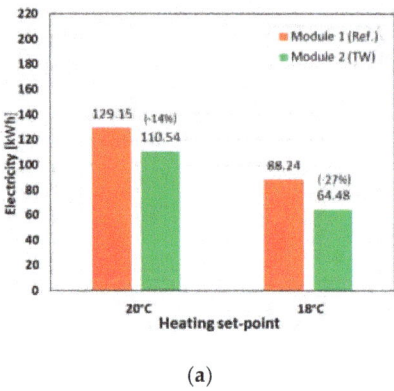 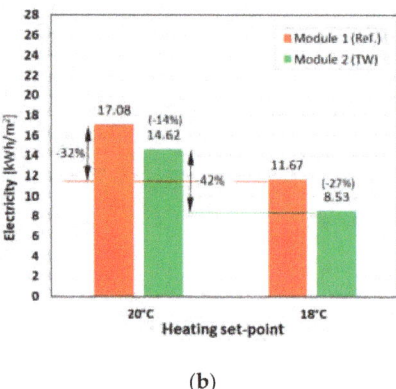

(a) (b)

Figure 12. Predicted heating energy consumption (electricity) during the winter season in Coimbra (Portugal), assuming office space use, for two different heating set-points: (**a**) Total values; (**b**) Normalized values.

Comparing the energy demand for both heating set-points, the energy reduction in percentages was similar to the previous ones, i.e., residential space use (Figure 11b), ranging from −32% up to −42% (Figure 12b), for reference LSF module 1 and TW module 2, respectively. However, in absolute values, this energy consumption reduction was smaller, i.e., −5.41 kWh/m^2 (office daytime use) instead of −7.80 kWh/m^2 (residential night-time use) for module 1, while for module 2, it was −6.09 kWh/m^2 instead of −9.84 kWh/m^2, for office and residential space use, respectively.

Jaber and Ajib [55] also performed hourly energy computer simulations to analyze the energy performance of a Trombe wall system for a typical Jordanian residential building (Mediterranean region). The studied house had a rectangular shape, having a floor area of about 154 m^2. The heavyweight façade walls had a very reduced thermal transmittance value, 0.133 W/(m^2·K), which corresponded to 41% of the LSF walls' U-value in the experimental modules, i.e., 0.326 W/(m^2·K) (see Table 2).

Their simulations were performed for a 20 °C heating set-point [55]. The predicted normalized heating energy consumption for the Jordanian building, without a Trombe wall, was 15.27 kWh/m^2, which was reduced to 12.09 kWh/m^2 (−21%), simulating a TW filling 18% of the south-oriented façade area (two bedrooms). They performed several simulations for different TW area ratios, ranging from 0% up to 50%, and based on the obtained results, they adjusted a polynomial curve (2nd order regression) to estimate the percentage of energy saving.

Making use of the previously mentioned estimation curve and applying the area ratio for the modular water TW evaluated in this paper, which was about 20%, the predicted energy saving would be around 22%. Not surprisingly, due to our reduced exterior walls insulation level, this energy-saving prediction was considerably higher than the ones obtained here for the 20 °C indoor set-point temperature.

3.2. Parametric Study

After analyzing the Trombe wall (TW) benefits in terms of indoor air temperature increase and heating energy decrease, in this section, a parametric study was conducted to assess the impact of the changes of some TW-related parameters on its thermal behavior. In this sensibility analysis, all the simulations were performed for the TW LSF module 2, having as reference for comparison the

DesignBuilder model, previously validated in Section 2.3.2, i.e., an unoccupied module. Notice that only one parameter was changed for each evaluated scenario, as displayed in Table 3. Four different parameters were evaluated: (1) Air cavity thickness; (2) Air vents dimensions; (3) Storage thickness; (4) Thermal storage material. For each parameter, two additional scenarios were assessed, besides the reference model scenario. Again, the hourly weather data for Coimbra (Portugal) was used [53], and a sunny winter week was chosen (23rd–29th January) for these simulations.

Table 3. Overview of evaluated parameters, models' identifications, and used values.

Parameter	Model	Value	
Air cavity thickness	Reference	10 cm	
	Scenario 1	20 cm	
	Scenario 2	30 cm	
Air vents dimensions	Reference	Lower	Upper
		50 × 5 cm	50 × 10 cm
	Scenario 3	50 × 8 cm	50 × 13 cm
	Scenario 4	50 × 11 cm	50 × 16 cm
Storage wall thickness	Reference	5 cm	
	Scenario 5	10 cm	
	Scenario 6	15 cm	
Thermal storage material	Reference	Water	
	Scenario 7	Concrete	
	Scenario 8	Basalt stone	

3.2.1. Air Cavity Thickness

The first TW parameter analyzed was the air cavity thickness between the storage wall and the glazed exterior frame. Three different air cavity thicknesses were evaluated: 10 cm (reference), 20 cm (scenario 1), and 30 cm (scenario 2), as illustrated in Figure 13. The increase in the air cavity thickness originated an indoor air temperature decrease. While the reference model had an average temperature of 18.2 °C, when the air cavity thickness was doubled (20 cm) and tripled (30 cm), the indoor temperature decreased to 0.9 °C and 1.2 °C, respectively. These results allowed to conclude that, for this TW configuration, the better thermal performance was achieved for the smaller air cavity (10 cm), which could be related to the lower air volume to be heated inside the air cavity and the higher buoyancy effect, promoting an increased upwards air convection and consequent higher heat flow through the upper vent to the interior of the module.

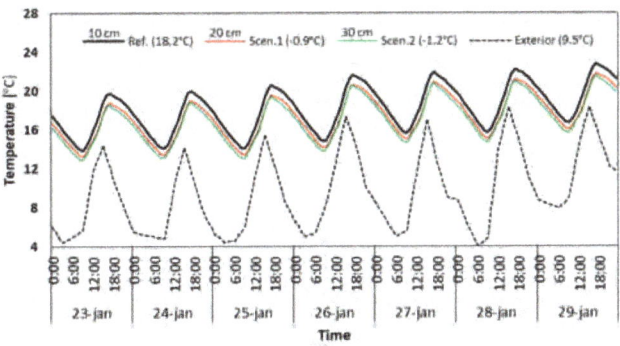

Figure 13. Influence of different air cavity thicknesses.

Hong et al. [56] performed a three-dimensional CFD thermal simulation of a Trombe wall with Venetian blind structure located in Hefei (China), assuming adiabatic surfaces for the air vents and internal wall. They compared several air cavity thicknesses, ranging from 8 cm up to 18 cm, with an increment of 2 cm. No significant thermal performance improvement was found for a thickness of the air cavity higher than 14 cm. Thus, they suggested a thickness equal to 14 cm.

3.2.2. Air Vents Dimensions

The second parameter analyzed was the dimension of the interior vents present on the storage wall to allow vertical air convection and airflow to/from the LSF module. The reference model had an upper vent with dimensions of 50 × 10 cm and a lower vent with 50 × 5 cm. Two additional scenarios were evaluated by modeling increased vents dimensions: 50 × 13 cm (upper) and 50 × 8 cm (lower) in scenario 3, and; 50 × 16 cm (upper) and 50 × 11 cm (lower) in scenario 4.

Figure 14 displays the obtained results, where a slightly indoor air temperature increase was visible with an increase in the dimensions of the air vents (+0.4 °C for scenario 3 and +0.5 °C for scenario 4). As expected, this indoor temperature increase was greater during the daytime, near noon, when the solar radiation was also higher. This better thermal performance could be justified by the increased natural air convection and airflow exchange between the TW air cavity and the interior of the module. Moreover, it could be deduced that forced air convection, making use of small fans, might improve, even more, the TW thermal performance.

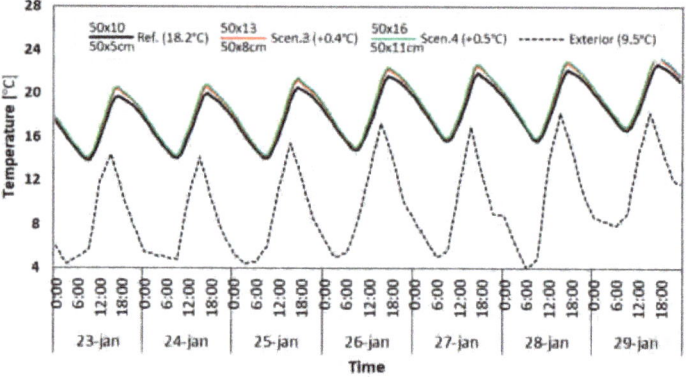

Figure 14. Influence of different dimensions of the air vents.

Hong et al. [56] also evaluated the influence of the inlet/outlet vent dimensions in the Trombe wall (2.00 m high × 1.00 m width) thermal performance. They assumed equal sized upper and lower vents and fixed their height to 10 cm. The vents width ranged from 20 cm up to 70 cm, with an increment of 10 cm. They found a slight decrease in the TW thermal performance for 70 cm width vents and suggested the use of vents with the following dimensions: 60 cm width × 10 cm height.

3.2.3. Storage Wall Thickness

The third parameter analyzed was the thickness of the water storage wall of the Trombe wall. The reference model had a 5 cm water storage wall composed of black painted steel, filled with water. Two additional scenarios with increased storage wall thickness were evaluated: 10 cm for scenario 5 and 15 cm for scenario 6.

Figure 15 exhibits the obtained results, where a decrease in indoor air temperature was visible in scenarios 5 (−0.7 °C) and 6 (−1.0 °C). This worst TW thermal performance could be justified by the larger volumes of water to be heated, inside the storage walls, by the same solar radiation and the consequent lower temperatures achieved.

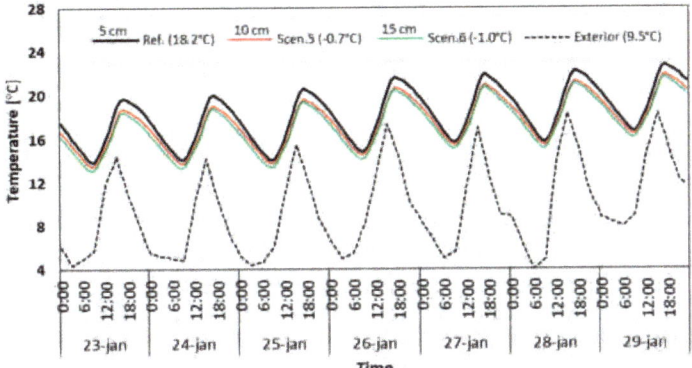

Figure 15. Influence of different storage wall thicknesses.

Briga-sá et al. [9] also evaluated the influence of the storage wall thickness (15 cm up to 40 cm), made of concrete, on ventilated and non-ventilated Trombe walls for the climate of Vila Real, a city located in the north of Portugal. Making use of a simplified calculation methodology prescribed by standard ISO13790:2008, they found that the heat gains were reduced when increasing the thickness for non-ventilated TWs, while for ventilated TWs, the heat gains increased.

3.2.4. Thermal Storage Material

The fourth and last parameter studied was the thermal storage material of the Trombe wall. As stated before, the reference TW thermal storage material was water. Two additional scenarios were simulated, making use of two other materials: concrete in scenario 7 and basalt stone in scenario 8. The thermal properties (thermal conductivity, specific heat, and density) of these three materials are displayed in Table 4. Regarding the optical properties, all these materials were modeled as being black painted, i.e., having solar and visible absorptances equal to 0.9.

Table 4. Thermal conductivity (λ), specific heat (c), and density (ρ) of thermal storage materials evaluated [37].

Material	λ ((m·K)/W)	c (J/(kg·K))	ρ (kg/m^3)
Water [1]	0.630	4190	990
Concrete	1.130	1000	2000
Basalt stone	3.490	840	2880

[1] For 40 °C temperature.

Figure 16 exhibits the obtained results, showing a slight decrease in the average indoor air temperature inside module 2 for the newly evaluated thermal storage materials: −0.4 °C for concrete (scenario 7) and −0.8 °C for basalt stone (scenario 8). Concrete storage material exhibited a higher temperature increase rate but also the higher temperature decrease rate during the cooling afternoon and night time, perhaps due to the significant lower specific heat (about four times smaller) and higher thermal conductivity (almost two times greater). The basalt stone temperature curve (scenario 8) exhibited a similar trend to the water temperature curve (Ref.), but with slightly lower indoor air temperature values (−0.8 °C).

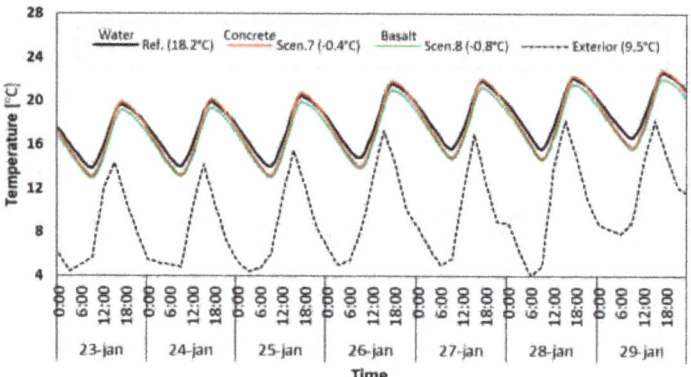

Figure 16. Influence of different thermal storage materials.

As stated by Saadatian et al. [7], "Because the specific heat of water (c) is higher than that of other types of building material, such as concrete, bricks, adobe, and stone, water stores more heat than the other materials. Similarly, because water convects, the transfer of heat to the interior space occurs faster than with classic Trombe walls.". Hu et al. pointed out another advantage of water as a thermal storage material: "Because the specific heat of water is higher than that of the building materials, the water's surface temperature does not rise as high as that of the masonry. Therefore, less heat is reflected back through the glazing." Nevertheless, Saadatian et al. [7], regarding water TWs, also stated that: "in harsh colder climates the glass layer should be insulated. Otherwise, the loss of heat from the warm wall to the outside would be significant.".

4. Conclusions

In this work, the influence of a passive modular water Trombe wall (TW) in the thermal behavior and energy efficiency of a lightweight steel frame (LSF) compartment was evaluated. Two real scale experimental identical LSF cubic modules, located in Coimbra (Portugal), exposed to natural exterior weather conditions, were used for in situ measurements. Module 1 was used as a reference, while the other one (module 2) was used to measure the influence of the TW, positioned in the south façade, on their thermal behavior by making a direct comparison between both modules. Additionally, these measurements allowed to calibrate and validate two numerical models (without and with a TW), with very good accuracy, i.e., having a root mean square error (RMSE) equal to 0.3 °C, for the reference model, and 0.5 °C for the TW model. These two validated models were used to perform advanced dynamic thermal simulations, making use of DesignBuilder software. Finally, these validated models allowed to predict the TW benefits in the heating energy consumption, as well as to perform a parametric study to evaluate the influence of four TW-related parameters on its thermal performance.

The first conclusion remark was that in this work, it was possible to evaluate the thermal behavior influence of a TW by in situ direct measurements and also performing advanced thermal dynamic simulations. The assessment was performed by quantifying the TW benefits (thermal and heating energy) and carrying out a thermal behavior parametric study. Several comparisons were performed, regarding (1) Sunny and cloudy winter week thermal behavior; (2) Office and residential space use heating energy; (3) Two heating set-points (20 °C and 18 °C); (4) Thickness of the TW air cavity; (5) Thickness of the thermal storage wall; (6) Dimensions of the interior upper/lower vents, and (7) Material of the thermal storage wall.

Regarding the obtained results for the TW benefits evaluation, the following main conclusions could be pointed out:

- In both sunny and cloudy winter weeks, the measured temperature was higher in module 2 (with a TW passive device). However, the warmer effect of the TW was much more effective during

the sunny week, increasing the average indoor air temperature significantly, i.e., +3.3 °C and +4.0 °C relative to the interior of module 1 (reference) and exterior environment temperatures, respectively.
- During the winter season, it was found that a TW was significantly more efficient for an office use schedule (during daytime), instead of a residential use schedule (during nigh-time). The heating energy consumption was reduced from 14.95 kWh/m^2, for residential space, down to 8.53 kWh/m^2 for office space (−43%), for an 18 °C indoor comfort temperature.
- A smaller heating set-point (18 °C instead of 20 °C) allowed to significantly reduce the heating energy consumption with and without a TW device, more than 40% and 32% reductions, respectively.
- A 27% reduction in heating energy due to TW device for an office 18 °C set-point was found, and this energy reduction was smaller (−14%) for the heating 20 °C set-point.

For residential use, the TW energy benefits were very reduced (only 5% decrease for 18 °C set-point), and there was even a heating energy consumption increase (+5%) when the set-point was 20 °C, due to nocturnal heat losses through the TW device.

Regarding the TW device parametric study, the main conclusions could be summarized as follows:

- An increase in the original TW air cavity thickness (10 cm) did not show any thermal performance improvement, and a decrease in the average indoor air temperature was found (−0.9 °C and −1.2 °C).
- Increasing the dimensions of the interior upper/lower TW vents (50 × 10 cm / 50 × 5 cm) allowed to slightly increase their thermal performance (+0.4 °C and +0.5 °C).
- An increase in the original thermal storage wall thickness (5 cm) did not show any thermal performance improvement, and a decrease in the average indoor air temperature was obtained (−0.7 °C and −1.0 °C).
- Changing the material of the storage wall (water) reduced the thermal performance of the TW device, originating a decrease in the average indoor air temperature (−0.4 °C and −0.8 °C).

In short, a TW device could, in fact, significantly improve the thermal behavior of an LSF compartment and reduce heating energy consumption during winter in a Csb Köppen–Geiger [47] Mediterranean climate. However, there were many factors that could influence the TW thermal performance, with adequate design and control to mitigate nocturnal heat losses very important. Otherwise, their thermal performance and energy efficiency improvement could be very insignificant and even decreased.

As most of the research studies, this work also had some limitations, including the assessment of only one climate/location, only one TW orientation (south exposed), only one isolated small compartment (not an entire building) without any window, only one construction system (LSF), only the heating mode during the winter season was evaluated (not an entire year), etc. Thus, in real buildings, thermal behavior and energy performance are much more complex, depending on many more factors. Nevertheless, the obtained results and conclusions could be very useful to identify the main benefits and possible drawbacks of a solar passive TW device in an LSF compartment, as well as to enhance the importance of the indoor set-point temperature and the occupation schedule of the compartment.

Author Contributions: All the authors participated equally in this work. All authors have read and agreed to the published version of the manuscript.

Funding: This research work was supported by ISISE (Institute for Sustainability and Innovation in Structural Engineering) and funded by FEDER funds through the Competitivity Factors Operational Programme—COMPETE and by national funds through FCT—Foundation for Science and Technology within the scope of the project POCI-01-0145-FEDER-032061.

Acknowledgments: The Trombe wall prototype was manufactured by CoolHaven company, and the experimental modules were built with the support of the following companies: Urbimagem; Fachaimper; CoolHaven; Forbo flooring systems; Weber (Saint-Gobain); Termolan; Bifase; Sociveda; Falper, and FibroPlac.

Conflicts of Interest: The authors declare no conflict of interest.

Acronyms

2D	Two-Dimensional
CFD	Computational Fluid Dynamics
DEC	Department of Civil Engineering
DEM	Department of Mechanical Engineering
EPBD	Energy Performance of Buildings Directive
EPS	Expanded Polystyrene
ETICS	External Thermal Insulation Composite System
EU	European Union
FEM	Finite Element Method
GPB	Gypsum Plasterboard
GPS	Global Positioning System
GWTW	Glass-Water Trombe Wall
HFM	Heat Flow Meter
ISO	International Standards Organization
LSF	Lightweight Steel Frame
MW	Mineral Wool
N	North
nZEB	nearly Zero-Energy Buildings
OSB	Oriented Strand Board
PhD	Doctor of Philosophy
PVC	Polyvinyl Chloride
RMSE	Root Mean Square Error
S	South
SHGC	Solar Heat Gain Coefficient
TC	Thermocouple
TTW	Traditional Trombe Wall
TW	Trombe Wall
W	West
WTW	Water Trombe Wall
XPS	Extruded Polystyrene

References

1. European Union. Directive (EU) 2018/844 of the European Parliament and of the Council of 30 May 2018 amending Directive 2010/31/EU on the energy performance of buildings and Directive 2012/27/EU on energy efficiency. In *Official Journal of the European Union*; European Parliament: Luxembourg, 2018; pp. 75–91.
2. European Union. Directive 2010/31/EU of the European Parliament and of the Council of 19 May 2010 on the energy performance of buildings. In *Official Journal of the European Union*; European Parliament: Luxembourg, 2010; pp. 13–35.
3. Brandão De Vasconcelos, A.; Pinheiro, M.D.; Manso, A.; Cabaço, A. EPBD cost-optimal methodology: Application to the thermal rehabilitation of the building envelope of a Portuguese residential reference building. *Energy Build.* **2016**, *111*, 12–25. [CrossRef]

4. Attia, S.; Eleftheriou, P.C.; Xeni, F.; Rodolphe, M.; Ménézo, C.; Kostopoulos, A.; Betsi, M.; Kalaitzoglou, I.; Pagliano, L.; Cellura, M.; et al. Overview and future challenges of nearly zero energy buildings (nZEB) design in Southern Europe. *Energy Build.* **2017**, *155*, 439–458. [CrossRef]
5. D'Agostino, D.; Parker, D. A framework for the cost-optimal design of nearly zero energy buildings (NZEBs) in representative climates across Europe. *Energy* **2018**, *149*, 814–829. [CrossRef]
6. Hu, Z.; He, W.; Ji, J.; Zhang, S. A review on the application of Trombe wall system in buildings. *Renew. Sustain. Energy Rev.* **2017**, *70*, 976–987. [CrossRef]
7. Saadatian, O.; Sopian, K.; Lim, C.H.; Asim, N.; Sulaiman, M.Y. Trombe walls: A review of opportunities and challenges in research and development. *Renew. Sustain. Energy Rev.* **2012**, *16*, 6340–6351. [CrossRef]
8. Wang, D. Classification, experimental assessment, modeling methods and evaluation metrics of Trombe walls. *Renew. Sustain. Energy Rev.* **2020**, *124*, 109772. [CrossRef]
9. Briga-Sá, A.; Martins, A.; Boaventura-Cunha, J.; Lanzinha, J.C.; Paiva, A. Energy performance of Trombe walls: Adaptation of ISO 13790:2008(E) to the Portuguese reality. *Energy Build.* **2014**, *74*, 111–119. [CrossRef]
10. Li, W.; Chen, W. Numerical analysis on the thermal performance of a novel PCM-encapsulated porous heat storage Trombe-wall system. *Sol. Energy* **2019**, *188*, 706–719. [CrossRef]
11. Lin, Y.; Ji, J.; Zhou, F.; Ma, Y.; Luo, K.; Lu, X. Experimental and numerical study on the performance of a built-middle PV Trombe wall system. *Energy Build.* **2019**, *200*, 47–57. [CrossRef]
12. Zhou, L.; Huo, J.; Zhou, T.; Jin, S. Investigation on the thermal performance of a composite Trombe wall under steady state condition. *Energy Build.* **2020**, *214*, 109815. [CrossRef]
13. Hu, Z.; Zhang, S.; Hou, J.; He, W.; Liu, X.; Yu, C.; Zhu, J. An experimental and numerical analysis of a novel water blind-Trombe wall system. *Energy Convers. Manag.* **2020**, *205*, 112380. [CrossRef]
14. Hong, X.; Leung, M.K.H.; He, W. Effective use of venetian blind in Trombe wall for solar space conditioning control. *Appl. Energy* **2019**, *250*, 452–460. [CrossRef]
15. Abbassi, F.; Dimassi, N.; Dehmani, L. Energetic study of a Trombe wall system under different Tunisian building configurations. *Energy Build.* **2014**, *80*, 302–308. [CrossRef]
16. Santos, P.; Simões da Silva, L.; Ungureanu, V. *Energy Efficiency of Light-Weight Steel-Framed Buildings*, 1st ed.; European Convention for Constructional Steelwork (ECCS), Technical Committee 14—Sustainability & Eco-Efficiency of Steel Construction: Mem Martins, Portugal, 2012; ISBN 978-92-9147-105-8.
17. Santos, P.; Martins, C.; Simões Da Silva, L. Thermal performance of lightweight steel-framed construction systems. *Metall. Res. Technol.* **2014**, *111*, 329–338. [CrossRef]
18. Murtinho, V.; Ferreira, H.; Antonio, C.; Simoes da Silva, L.; Gervasio, H.; Santos, P. Architectural concept for multi-storey apartment building with light steel framing. *Steel Constr.* **2010**, *3*, 163–168. [CrossRef]
19. Santos, P. Chapter 3—Energy Efficiency of Lightweight Steel-Framed Buildings. In *Energy Efficient Buildings*; Eng Hwa, Y., Ed.; InTech: London, UK, 2017; pp. 35–60.
20. Gervásio, H.; Santos, P.; Martins, R.; Simões da Silva, L. A macro-component approach for the assessment of building sustainability in early stages of design. *Build. Environ.* **2014**, *72*, 256–270. [CrossRef]
21. Gervásio, H.; Santos, P.; Simões da Silva, L.; Lopes, A.M.G. Influence of thermal insulation on the energy balance for cold-formed buildings. *Adv. Steel Constr.* **2010**, *6*, 742–766.
22. Santos, P.; Martins, R.; Gervásio, H.; Silva, L.S. Assessment of building operational energy at early stages of design—A monthly quasi-steady-state approach. *Energy Build.* **2014**, *79*, 58–73. [CrossRef]
23. Martins, C.; Santos, P.; Simoes da Silva, L. Lightweight steel-framed thermal bridges mitigation strategies: A parametric study. *J. Build. Phys.* **2016**, *39*, 342–372. [CrossRef]
24. Roque, E.; Santos, P. The Effectiveness of Thermal Insulation in Lightweight Steel-Framed Walls with Respect to Its Position. *Buildings* **2017**, *7*, 13. [CrossRef]
25. Soares, N.; Gaspar, A.R.; Santos, P.; Costa, J.J. Multi-dimensional optimization of the incorporation of PCM-drywalls in lightweight steel-framed residential buildings in different climates. *Energy Build.* **2014**, *70*, 411–421. [CrossRef]
26. Özdenefe, M.; Atikol, U.; Rezaei, M. Trombe wall size-determination based on economic and thermal comfort viability. *Sol. Energy* **2018**, *174*, 359–372. [CrossRef]
27. Duan, S.; Jing, C.; Zhao, Z. Energy and exergy analysis of different Trombe walls. *Energy Build.* **2016**, *126*, 517–523. [CrossRef]

28. Soares, N.; Santos, P.; Gervásio, H.; Costa, J.J.; Simões da Silva, L. Energy efficiency and thermal performance of lightweight steel-framed (LSF) construction: A. review. *Renew. Sustain. Energy Rev.* **2017**, *78*, 194–209. [CrossRef]
29. LSF System B(A)a, by 'Balthazar Aroso Arquitectos, Lda', Manufactured by Urbimagem—Sistemas de Arquitectura e Construção, Lda, 2019. Available online: www.urbimagem.com/en/empresa/ (accessed on 10 March 2019).
30. Norbord. Technical Brochure—OSB3 Zero. 2019. Available online: www.norbord.co.uk/our-products/sterlingosb-zero/sterlingosb-zero-osb3/ (accessed on 10 March 2020).
31. Termolan. Technical Brochure—PA30 Mineral Wool, 2019. 2019. Available online: http://termolan.pt/wp-content/uploads/2018/06/pa30-pt.pdf (accessed on 10 March 2020).
32. Weber(Saint-Gobain). Technical Brochure—EPS100, Weber.Therm Classic, 2019. Available online: www.pt.weber/files/pt/2019-04/FichaTecnica_sistema_webertherm_classic.pdf (accessed on 10 March 2020).
33. Danopren. Technical Brochure—XPS TL-P 50, 2019. Available online: https://portal.danosa.com/danosa/CMSServlet?node=483103&lng=4&site=3 (accessed on 10 March 2020).
34. TEC. Technical Brochure—XPS RoofTEC SL/FloorTEC 300, 2019. Available online: www.sotecnisol.pt/resources/777f40511b178afb7f9e2c1a7a9e55af/fichas_tecnicas/dop_rooftec.floortec_50_2016_por.pdf (accessed on 10 March 2020).
35. Danopol. Technical Brochure—Water-Proofing PVC Membrane: HSF 1.5 Light Grey, 2019. Available online: https://portal.danosa.com/danosa/CMSServlet?node=210302&lng=4&site=3 (accessed on 10 March 2020).
36. Rosa, N.C.F. Study of structural and thermal performance of lightweight steel framing (LSF) modular construction. Ph.D. Thesis, University of Coimbra, Coimbra, Portugal, 2018.
37. DesignBuilder. DesignBuilder Software Version 5.5.0.012 and Materials Database, 2018. Available online: https://designbuilder.co.uk/ (accessed on 14 February 2019).
38. Santos, C.; Matias, L. *ITE50—Coeficientes de Transmissão Térmica de Elementos da Envolvente dos Edifícios (in Portuguese)*; LNEC—Laboratório Nacional de Engenharia Civil: Lisboa, Portugal, 2006.
39. Sarlon. Technical Brochure—Vinyl Acoustic Flooring: Sarlon Traffic 19dB, 2019. Available online: www.forbo.com/flooring/en-aa/products/acoustic-flooring/sarlon-trafic-19-db/blbfkp (accessed on 10 March 2020).
40. Vicaima. Technical Brochure—Wooden Door: Portaro-SBD-EI30. 2019. Available online: www.vicaima.com/files/files/Vicaima-FT-Portaro-SBD-EI30.pdf (accessed on 10 March 2020).
41. ISO 6946. *Building Components and Building Elements—Thermal Resistance and Thermal Transmittance—Calculation Methods*; International Organization for Standardization: Geneva, Switzerland, 2017.
42. THERM. Software Version 7.6.1. Lawrence Berkeley National Laboratory, United States Department of Energy, 2017. Available online: https://windows.lbl.gov/software/therm (accessed on 14 February 2019).
43. DEM Weather Station. ADAI@DEM.UC Weather Station Located in the Dep. of Mechanical Engineering, University of Coimbra, Portugal, 2020. Available online: www.wunderground.com/dashboard/pws/ICOIMBRA14 (accessed on 15 February 2020).
44. CoolHaven. CoolHaven—Modular Constructions, 2020. Available online: www.cool-haven.com (accessed on 15 February 2020).
45. Davis Instruments. Wireless Vantage Pro2 Plus Automatic Weather Station, Including UV and Solar Radiation Sensors, from Davis Instruments, 2019. Available online: www.davisinstruments.com/product/wireless-vantage-pro2-plus-including-uv-solar-radiation-sensors/ (accessed on 15 February 2019).
46. Delta-T Devices. Sunshine Sensor BF5 (pyranometer), from Delta-T Devices, 2019. Available online: www.delta-t.co.uk/product/bf5/ (accessed on 15 February 2019).
47. Kottek, M.; Grieser, J.; Beck, C.; Rudolf, B.; Rubel, F. World Map of the Köppen-Geiger climate classification updated. *Meteorol. Zeitschrift* **2006**, *15*, 259–263. [CrossRef]
48. Tinytag. Tinytag Ultra 2—TGU-4500, TGU-4500—Indoor Temperature and Relative Humidity Data Logger with Built-in Sensors, 2019. Available online: https://www.geminidataloggers.com/data-loggers/tinytag-ultra-2/tgu-4500 (accessed on 15 February 2019).
49. Santos, P.; Gonçalves, M.; Martins, C.; Soares, N.; Costa, J.J. Thermal Transmittance of Lightweight Steel Framed Walls: Experimental Versus Numerical and Analytical Approaches. *J. Build. Eng.* **2019**, *25*, 100776. [CrossRef]
50. Santos, P.; Lemes, G.; Mateus, D. Analytical Methods to Estimate the Thermal Transmittance of LSF Walls: Calculation Procedures Review and Accuracy Comparison. *Energies* **2020**, *13*, 840. [CrossRef]

51. Santos, P.; Lemes, G.; Mateus, D. Thermal Transmittance of Internal Partition and External Facade LSF Walls: A Parametric Study. *Energies* **2019**, *12*, 2671. [CrossRef]
52. Tyre4BuildIns. Research Project Tyre4BuildIns—'Recycled Tyre Rubber Resin-Bonded for Building Insulation Systems towards Energy Efficiency', Funded by FEDER European and FCT National Funds. Reference: POCI-01-0145-FEDER-032061. University of Coimbra, Portugal, 2020. Available online: www.tyre4buildins.dec.uc.pt (accessed on 15 January 2020).
53. PRT_Coimbra.085490_IWEC. International Weather for Energy Calculation (IWEC), Weather Data for Coimbra (WMO Station 085490), EnergyPlus Weather Database, 2009. Available online: https://energyplus.net/weather-location/europe_wmo_region_6/PRT//PRT_Coimbra.085490_IWEC (accessed on 15 February 2019).
54. REH. *Portuguese Regulation for Energy Performance of Residential Buildings (in Portuguese), Approved by Decree-Law n. 118/2013 of 20 August*; N. 159; Diário da República – I série (in Portuguese): Lisboa, Portugal, 2013; pp. 4988–5005.
55. Jaber, S.; Ajib, S. Optimum design of Trombe wall system in mediterranean region. *Sol. Energy* **2011**, *85*, 1891–1898. [CrossRef]
56. Hong, X.; He, W.; Hu, Z.; Wang, C.; Ji, J. Three-dimensional simulation on the thermal performance of a novel Trombe wall with venetian blind structure. *Energy Build.* **2015**, *89*, 32–38. [CrossRef]

© 2020 by the authors. Licensee MDPI, Basel, Switzerland. This article is an open access article distributed under the terms and conditions of the Creative Commons Attribution (CC BY) license (http://creativecommons.org/licenses/by/4.0/).

Article

Design and Construction of a New Metering Hot Box for the In Situ Hygrothermal Measurement in Dynamic Conditions of Historic Masonries

Mirco Andreotti [1], Marta Calzolari [2], Pietromaria Davoli [3], Luisa Dias Pereira [3,*], Elena Lucchi [4] and Roberto Malaguti [1]

1. Istituto Nazionale di Fisica Nucleare, Sezione di Ferrara, 44122 Ferrara, Italy; mirco.andreotti@fe.infn.it (M.A.); malaguti@fe.infn.it (R.M.)
2. Department of Engineering and Architecture of the University of Parma, 43124 Parma, Italy; marta.calzolari@unipr.it
3. Architettura>Energia Research Centre, Department of Architecture-University of Ferrara, 44121 Ferrara, Italy; pietromaria.davoli@unife.it
4. Eurac Research, 39100 Bolzano, Italy; elena.lucchi@eurac.edu
* Correspondence: dsplmr@unife.it; Tel.: +39-0532-293631

Received: 29 May 2020; Accepted: 6 June 2020; Published: 9 June 2020

Abstract: The main purpose of the HeLLo project is to contribute to data available on the literature on the real hygrothermal behavior of historic walls and the suitability of insulation technologies. Furthermore, it also aims at minimizing the energy simulation errors at the design phase and at improving their conservation features. In this framework, one of the preliminary activities of the study is the creation of a real in situ hot box to measure and analyze different insulation technologies applied to a real historic wall, to quantify the hygrothermal performance of a masonry building. Inside this box, 'traditional' experiments can be carried out: recording heat flux, surface temperature, and air temperatures, as well as relative humidity values through the use of a new sensing system (composed of thermocouples and temperature/relative humidity combined sensors). Within this paper, the process of development, construction, and validation of this new metering box is exhibited. The new hot box, specifically studied for historic case studies, when compared to other boxes, presents other advantages compared to previous examples, widely exemplified.

Keywords: metering hot box; in situ; hygrothermal measurement; dynamic conditions; historic masonries; HeLLo

1. Introduction

Energy refurbishment of existing buildings is one of the priorities of the European policies to reduce fuel consumption, starting from the recognition of the 'exemplary role of public bodies' buildings (art.5 2012/27/UE) [1] to activate effective strategies in the private building stock. Existing buildings in the European Union are, indeed, responsible for 40% of final energy consumption [2] and for 36% of carbon dioxide (CO_2) emissions [3]. Approximately 35% of the buildings are more than 50 years old [3]. Considering the low rate of new buildings construction, 3% in Europe [4], and 2% in the USA [5], energy efficiency in existing, historical, and historic buildings is one of the greatest opportunities towards a sustainable future.

Besides the social and cultural value of all historic buildings, the specific value of heritage assets in Italy strongly justifies the origin of the current research: according to the Italian Ministry for Cultural Heritage and Activities, there are more than 20,000 historic centers of different ages. In light of such numbers, it is evident that many Italian cities are largely made up of historic buildings, which almost often require a greater commitment in the design of conservative and improving interventions than those devoted to the process of new construction. Nonetheless, this is also verified in many other European heritage city centers. Two examples could be pointed out: Edinburgh (Scotland) and Antwerp (Belgium). In 1995, UNESCO added the Old and New Towns of Edinburgh as a World Heritage Site [6]. In this site, 75% of the 4500 individual buildings are listed for their special architectural or historic interest. This city's latest management plan concerning the heritage site (2017–2022) has 6 main objectives and 39 actions, of which stands out" strengthening care and maintenance of buildings and streets' and the 'sustainable re-use of underused and unused buildings" [6]. In Antwerp, instead, there are several listed buildings, such as the Vleeshuis Museum located in the historic center. This significant building has been the object of study [7] in various domains (e.g., evaluation of brick masonry or the assessment of hygrothermal parameters and conservation of important housed collection).

The interest in historic and historical buildings has been gaining cultural and social strategic roles. One important way of preserving built heritage for the future is to keep it in use and to accommodate new uses, avoiding its transformation into a 'museum' and preserving its cultural memory. In order to make this operation successful, it is mandatory for their adaptation to today's comfort requests for indoor human activities. Moreover, promoting the control of hygrothermal parameters and indoor air quality in such buildings also means assuring better conservation of the decorative features that make them distinguishable and enhance their architectural quality.

The building envelope plays an important role in terms of energy transmission. Particularly, the opaque surface in historic buildings constitutes the largest surface of the envelope, and heat losses through this element are, therefore, of most importance [8–10]. In fact, some authors defend that in historic buildings heat loss through windows is only 10%, while walls and roof account for 60% (35% and 25%, respectively) [11,12]. This means that the intervention aiming to enhance the energy performance of the building should involve the envelope's components to reach a high level of efficiency. As well-known, sometimes, it is impossible due to the presence of architectural features to be preserved, and the project has to focus on different strategies. In other cases, the envelope's insulation is possible operating only on the inner façade of the building. Unfortunately, also in these situations, other difficulties may occur, hindering the good result of the operation.

One of the most significant issues in the field of efficiency topics is the buildings energy consumption gap [13–15] between design and post-occupancy phase [16,17]. In many cases, it has been verified that this gap is due to occupants' behavior [18,19], but it can also be justified by erroneous decisions or values accepted at the design phase (i.e., poor practice or uncertainty in building energy simulation—BES [20]). Many authors have been demonstrating the limitations of traditional BES tools and procedures for the estimation of energy performance of historical buildings [21–25]. This topic reaches a significant dimension in historic buildings refurbishment, once the real wall composition of such buildings is frequently unknown [26] and, for practical matters, in many occasions, several projects and estimations are based on general assumptions [27].

The calibration of the hygrothermal models with measured data is very important to avoid irreparable damage to historic buildings. The combination of several hygrothermal variables [e.g., heat flux (φ), surface temperature (Ts), air temperature (Ta), and relative humidity (RH)] should lead to more reliable models.

2. Aims and Methodology

The main purpose of this study was to contribute to data available on literature on the real hygrothermal behaviour of historic walls and the suitability of insulation technologies, also aiming at minimizing the energy simulation errors at the design phase and at improving their conservation features (i.e., avoiding risks of condensation or damaging their structure). These errors can become very significant. For example, the wrong definition of the thermal behaviour of a thick and heavy external wall, a very common situation in historic buildings (e.g., "the results divergence in thermal mass simulation using different tools" [28] (p. 74) or simply the use of different modelling tools [23]), might lead not only to the definition or to the choice of inappropriate insulation solutions (e.g., risking at generating condensation by changing the original hygrothermal wall behaviour), but also it can lead to mistaken thermal spatial requirements or to over dimensioning of HVAC (heating, ventilation and air conditioning) systems. The negative implication of miss-sized systems and the corresponding increased energy consumption has being recognized in the scientific community [29–31]. Furthermore, moisture reduces thermal performance and causes deterioration of insulation materials [7–9,25].

On the basis of these assumptions, the energy refurbishment of historic heritage with testimonial value is an asset. Given the impossibility to remove samples to be tested in the laboratory, and the likely unknown hygrothermal behaviour of historic walls, in situ measurement methods must be more frequently implemented, expeditious, simple to operate. Though each historic building presents unique features, the developed methods should be, preferably, replicable and repeatable.

In this framework, the HeLLo research project [32] created a real in situ laboratory of measurement to analyse different insulation technologies when applied on a real historic building to quantify the hygrothermal performance of a masonry building. As a first step of the research, the authors developed a version of a revised in situ metering hot box, topic of the present paper, perfectly thought for historic buildings, to adapt the standard in situ measurement techniques to the historic case study. The paper presents the main characteristics and uses of this hot box for in situ hygrothermal tests.

3. State of the Art

The literature shows two different kinds of in situ tests: (i) test for determining the thermal performance of building elements, in terms of thermal resistance (R-value or R), thermal conductance (C-value or C), or thermal transmittance (U-value or U) [33]; and (ii) hygrothermal monitoring for determining the hygrothermal behaviour of the various wall layers [7,34].

First, commonly used standard tests to experimentally determine the thermal performances of walls [35] were divided in two groups: (i) In situ tests measurements based on the use of the heat flow meter (HFM) method [36–38] or the quantitative infrared thermography testing (ITT) [39,40]; and (ii) laboratory tests performed on hot box chambers [41,42]. Soares et al. [33] and Bienvenido-Huertas et al. [43] have performed two of the most significant literature reviews on this subject. HFM method is a non-destructive testing (NDT) for determining the thermal transmission properties (R, C, or U values) of an existing building directly in situ. The apparatus was composed of a data-logger equipped with two thermal sensors and one heat flux plate for gathering the internal and external Ta or Ts and the φ through the element. The international standard ISO 9869 [37] defined the calibration and the installation procedures, the data processing techniques, the methodology for correcting systematic errors, and the reporting format. In parallel, the literature presents several methods to solve meteorological and practical issues to reduce the errors and the uncertainties due to the measurement location [44,45], the influence of the boundary conditions [44], [45], or the presence of non-homogeneity, high thermal inertia [44], or moisture content [45] in the structure. In addition, the quantitative ITT permits to measure directly in situ the R-value of a masonry, avoiding the problems related to non-correct locations, non-homogeneity in the walls, or the influence of the boundary conditions [46]. Otherwise, ITT was also used in a qualitative way to measure the thermal pattern of walls [47]. Laboratory tests permit to measure the thermal properties of building components in steady-state or dynamic controlled conditions. The guarded hot plate (GHP) measures the steady-state thermal conductivity (λ-value or λ) of homogeneous flat walls [46,47]. The international standard ISO 8302 [48] and the ASTM C177 [49] defined the minimum requirements for designing the apparatus and the testing procedure. The main problem was related to the errors connected to gaps and edge losses. Several studies proposed analytical calculation models for reducing this error [50–52]. The hot box apparatus measures the steady-state and the dynamic thermal performance (R, C, and U values, Ts, internal T, and RH) of inhomogeneous samples. Basically, it is composed of two climatic chambers maintained at different temperatures that simulate the internal and external conditions. The building element under measurement was inserted between the two chambers, and the thermal performance was obtained, measuring the power required to keep the hot chamber at a constant temperature. The ISO standard 8990 [41], the American ASTM C1363 [53], the European EN 1934 [54], and the Russian GOST 26602.1 [55] defined the minimum requirements for designing the hot box apparatus and the measurement procedure. Two alternative methods are available: the guarded hot box (GHB) and the calibrated hot box (CHB). GHB is composed of a climatic chamber for simulating the exterior temperature, a metering chamber heated to simulate the indoor conditions, and a guard chamber for minimizing the lateral heat flows at the edges of the metering chamber [41,53]. CHB is composed only by a climatic and a metering chamber, surrounded by a "temperature-controlled space" to reduce the errors generated by the apparatus [41,53]. Concerning the hot box method, many researchers have developed their own compact facility, but only a very few correspond to in situ affectations. A significant majority of the examples found in the literature correspond to variations of the hot box method, e.g., facilities for laboratory tests more targeted at wall/materials sample testing [44,56]. In [57], the authors showed the new design of a compact hot box apparatus used for determining properties of wall samples, developed according to ISO 8990 [41]. Though upfront and useful in laboratory, this tool was not developed for in situ measurement and the test rig dimensions are ruled by the sample size requirement. One variation of these models are full scale boxes simulating entire ambiences/buildings [58,59], among which are distinguished outdoor test boxes solutions for building envelope experimental characterisation [60]. Once again, these intend to study new materials/walls, and not existing building construction solutions, for example: window shutters [61], heat insulation solar glass (a type of multifunction PV module) [62], glazed façades with water film [63], multilayer, inhomogeneous, and massive walls [64,65]. More common instead is the use of a combined strategy for data comparison, as for example the in situ testing coupled with computer modelling and steady-state testing in a GHB [66] or, the comparison of steady-state and in situ testing of high R walls incorporating vacuum insulation panels [66].

On the other hand, only a few scientific studies combine both methodologies, and solely for measuring the thermal performances of building elements: The 'chamber'/box and the HFM. In 2008, Peg and Wu [67] approached this strategy by designating an entire room of an apartment situated in a new residential development district in Nanjing as a 'test chamber' where in situ measuring method for the R-value of buildings was tested (defining 'measuring points' arrangement in several walls), but no box was in fact generated. In 2015, in their turn, authors had verified the feasibility of a new developed simple hot box-HFM method (SHB-HFM) to address an in situ measurement of wall thermal transmittance [68]. This SHB-HFM was preceded by another experiment developed by Chinese researchers in 2012, designated Temperature Control Box-HFM method (TCB-HFM) [69] cited in [70]. However, the authors of [70] (p. 748) described this TCB-HFM as not suitable for the in situ measurement, also noticing that measurement thermal transmittance results obtained in [69] were "55% higher than the design thermal transmittance and that the measurement error was attributed to high moisture", denoting the problem of not controlling for humidity in the test.

Besides the final aim of monitoring hygrothermal parameters instead of exclusively the thermal transmittance, the most significant difference between the boxes presented in [68] or [70] and the new one now presented lies in the dimension—none of the SHB-HFM boxes surpasses 0.90 m × 0.90 m × 0.30 m. Further developments on this topic are presented in Section 4.2.

In situ monitoring can be very significant in the case of historic buildings, since: (i) walls samples cannot be examined in the lab (for cultural heritage protection issues, no samples can be removed from original sites); (ii) many historic buildings are abandoned or not in use, and, therefore, are not heated; (iii) many of these building present particular features as high ceilings/volumes and therefore the traditional 1 m × 1 m lab measured surface might not be representative enough of the vertical heat stratification of a historic wall.

The hygrothermal monitoring of heritage buildings can be divided into: environmental monitoring and contact monitoring used, respectively, to assess the environmental condition of a room and the hygrothermal performance inside to a building element [34]. Skills and procedures for the environmental monitoring of Ta and RH are defined by several standards that focus particularly on cultural heritage (CH) [71,72], in order to avoid damage and risks for CH object and surface and users' discomfort [73–78]. Contact monitoring is used to quantify damage already occurred and to predict the presence of potential hygrothermal risks for CH [79]. The methodologies used can be divided into: (i) surface monitoring of Ts and RHs; and (ii) monitoring of T and RH inside the walls [34]. No standard procedures have been developed for the surface monitoring of CH building elements [79]. As a matter of fact, the procedure normally used for new and existing buildings without any heritage value cannot be applied to historic surfaces as risks and losses of historic materials should be avoided.

Moisture content within walls has proved difficult to measure because several variables are unknown, including the influence of the probe on the test results [80]. Moisture content inside the walls can be measured in two ways using: (i) direct methods based on the gravimetric analysis; and (ii) indirect methods based on the drilling of wooden dowels inserted into the building element. The gravimetric analysis consists in the measurement of water content in a building material sample, weighing its mass with analytical scales in a range of controlled wet and dry conditions [81,82]. Standard CEN EN 16,682 [81] and UNI 11,085 [82] define the operative procedure. This process involves the drill of samples at various heights and widths across the area being tested and thus, is not always suitable for CH building elements. Indirect methods have been categorized according to measurement principles in resistance, voltage, capacitance, thermal-based, and innovative (e.g., neutron probes, nuclear magnetic resonance, medical ECG electrodes, and fibre optic sensor) methods [80]. Resistance-based moisture methods are widely used, thanks to the variation of the electrical resistance of the materials under different moisture contents [80]. Particularly, this method has been successfully used mainly in timber construction [79,80,83], and, most recently, in solid brick walls [80]. No standards procedures are defined because several factors affect the electrical resistance, such as the timber species, the speed of growth, the origin, and the storage [80,83]. Otherwise, calibration factors exist for different timber species [80]. However, this method has proven to be stable for slow and long-term moisture measurements, with examples of sensors working for a minimum of 20 years [83]. The results obtained in shorter monitoring periods are not accurate.

Herein, a new approach is suggested: to assess in situ the hygrothermal performance of historic walls (aiming at testing future indoor insulation solutions), a new metering hot box is proposed in combination with T-RH sensors (and eventually added thermocouples if desired), through a low cost and simplified data acquisition system [34]. To the best of the authors' knowledge, the new box suggested within the next sections is the first of its kind, totally addressed to historic buildings in situ measurement. Moreover, the developed experiment allows long-term monitoring, against 'punctual' measurement in laboratories or short-time HFM measurements as proposed in [69], not addressed to historic material. Commonly, most studies of this kind and in this field involve the thermal behaviour of walls solely. Alike [56,84,85], the hygrothermal performance assessment is also intended.

4. Case-Study Presentation and Experimental Methodology

4.1. Contextualization and Configuration of the Tested Wall

The in situ test was being performed in Palazzo Tassoni Estense in Ferrara, Italy. This 15th century listed building is part of a UNESCO site [86], with characteristics representative of many historic buildings. Since 1997, the Palazzo has been the subject of several studies, which resulted in an architecture project and a scientific restoration intervention [87]. The complex of the Palace is located in the NW part of a block, currently housing almost exclusively the Department of Architecture of the University of Ferrara, near the ancient walls of the city.

In order to provide a proper background, it is opportune to recall "that it was built within the Borso Addition (an area of urban expansion wanted by Borso d'Este, who was then the Duke of Ferrara)" during the mid-15th century, then "confiscated by Ercole I d'Este and gifted to the Tassoni Counts in 1476" [87] (p. 129). By 1491, in a letter to the Duke, "the architect Biagio Rossetti affirmed being in charge of the renovation works of the palace." It "housed the Tassoni Estense family until 1858, when it was designated as the seat of the Provincial Psychiatric Hospital. (...) The mental institution remained active until the 1970s" [87] (p. 129).

The Palace was built in masonry bricks and it has considerable architectural interest, e.g., (i) "the main entrance from the street is made of decorated white marble"; (ii) "the perron, in the upper floor, has been restored and it preserves only partially its original features"; (iii) "the access doors to the main hall are still the original and exquisite renaissance artifacts" [34] (p. 10).

The room (700 m^3) and the wall under-study are part of this complex and are located on the ground floor of an area that has not been refurbished yet, currently unoccupied and without any HVAC system (Figures 1 and 2).

Figure 1. Ground floor plan of Palazzo Tassoni and location of the room where the experiment is carried out. External views of the surrounding buildings.

Facing the challenge of assessing the hygrothermal behaviour of the historic wall, the authors' option was twofold: (i) conditioning and buffering the openings of a 700 m^3 space; or (ii) building an in situ chamber that simulated the conditions of a smaller room that still had as an external boundary, the original historic wall. The authors opted for the second hypothesis, both because of the sustainability of the experiment itself (less energy is required/wasted) either before the risks of the operation (limiting the intervention on the historic building, reduces the risks and impact on the heritage features).

Though the HFM method was probably the most internationally recognized and widely used method, it presented several disadvantages for the intended experiment. On one side, this method suggested high-temperature differences between the indoor and outdoor air (ΔT_a). As a matter of fact, "[...] the increase in the measurement temperature difference between the indoor and outdoor environment can weaken the influence of the temperature fluctuation and decrease the test error" [68] (p. 49) but, unfortunately, this difference cannot always be guaranteed in a real field situation (outdoor climate cannot be controlled). On the other side, aiming at the authors' future intention of testing indoor thermal solutions, the RH parameter could not be neglected, and the experience could not be limited to wall U-value measurement.

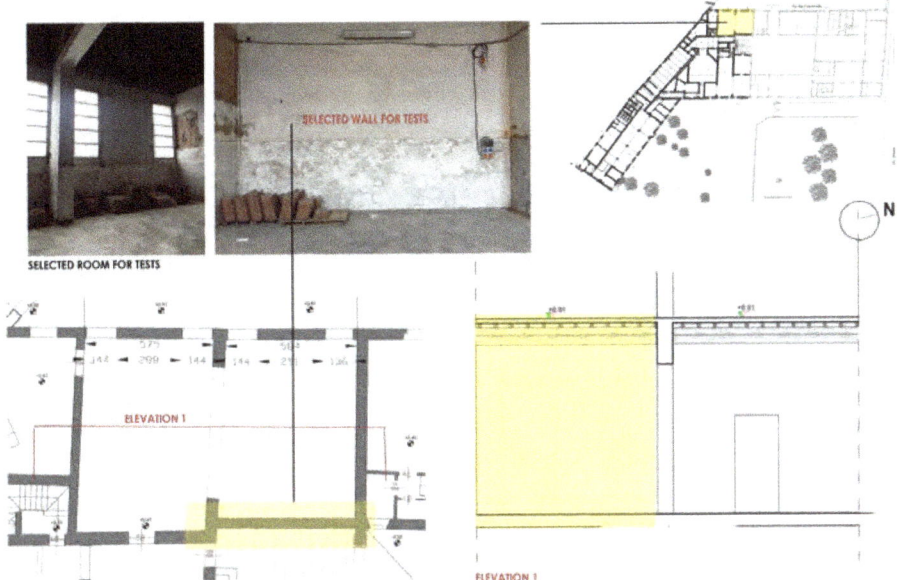

Figure 2. Ground floor plan of Palazzo Tassoni and signaling of the location of the room and wall on which the experiment is carried out (in yellow). Internal views of the room. Inside elevation of the wall (in yellow).

Due to all these premises, field restraints, and the final research goal, an in situ acclimatized box was built, aiming at simulating a 'standard' indoor environment ($T_a \approx 20\ °C$, $RH \approx 55\%$; in accordance with several standards/guidelines, e.g., EN ISO 7730 [88], EN ISO 13,788 [89], ISO 17772-1 [90]) that potentially guaranteed a satisfying ΔT_a between the indoor face of the monitored wall and the external site condition (outdoor climate), 'business as usual' and that always allowed the collection of hygrothermal data of the wall behaviour (and, therefore, using combined T and RH sensors) [34].

4.2. Design of the New Metering Hot Box

As stated earlier, in order to overcome field experiment restraints, the authors proposed a combined strategy between the in situ monitoring and hot box method to enhance robust measurement and reliable data acquisition. As in [70] (p. 747), the idea was that this simplified solution "[...] avoids the heavy equipment of the hot box method and overcomes outdoor and indoor thermal environment limitation of the HFM Method". Moreover, it was also worth mentioning that often in situ measurement was not done because using the standard method ISO 9869, a measurement period of more than 10 days was normally required [91]. Using the proposed method, the monitoring campaign can be performed almost continuously or with very limited interruption periods.

One of the common characteristics of most metering boxes is their mobile base design. Anticipating future studies on different case studies, risks on the selected room where the monitoring campaign was initially foreseen, or given its own weight ($\approx$700 kg), alike in [92], the newly developed metering box was intentionally provided with wheels thus that it could be more easily moved. Moreover, it was built of a modular timber structure to be more simply dismantled or size adjusted in the event it had to be moved to another room or building with other specifications. This feature, i.e., the box possibility of re-assemblage and re-usability, emphasized the experiment's sustainability.

According to [70] (p. 752), for the SHB-HFM, the minimum box dimension should vary according to the different measurement walls. Considering this, for a wall thickness of 0.30 m it was recommended that the minimum box dimensions were about 1 m: i.e., for a wall thickness of 0.24 m or 0.360 m, for a 'preferred' temperature difference of 25 °C, the box dimension may vary between 0.7 m and 1.3 m [70] (Table 3, p. 755).

In the current study, authors have taken into account all these assumptions, but also the specificity of the field of the current and future case-studies: historic buildings, many times characterized by internal volumes with significant high ceilings. For this reason, the newly proposed box had dimensions significantly bigger than the minimum suggested by [70], closer to a climatic chamber than exactly a hot box, aiming at reproducing a fraction of a typical indoor volume of a Palazzo, for example.

The newly developed metering hot box used to perform the hygrothermal tests is depicted in Figure 3. This gross box size is 2.50 × 2.50 × 4.01 m, built with 'platform system' circa 0.13 m thick walls composed by two 0.018 oriented strand board (osb) panels mounted on a timber structure made of elements 0.09 × 0.09 m. To avoid thermal losses and maintain the setup temperature and humidity values, the 5 walls faces (including the pavement) were provided of 0.10 m high-density stone wool insulation material, then protected with a vapor barrier. The net size was 2.04 × 2.42 × 3.55 m (volume of the chamber).

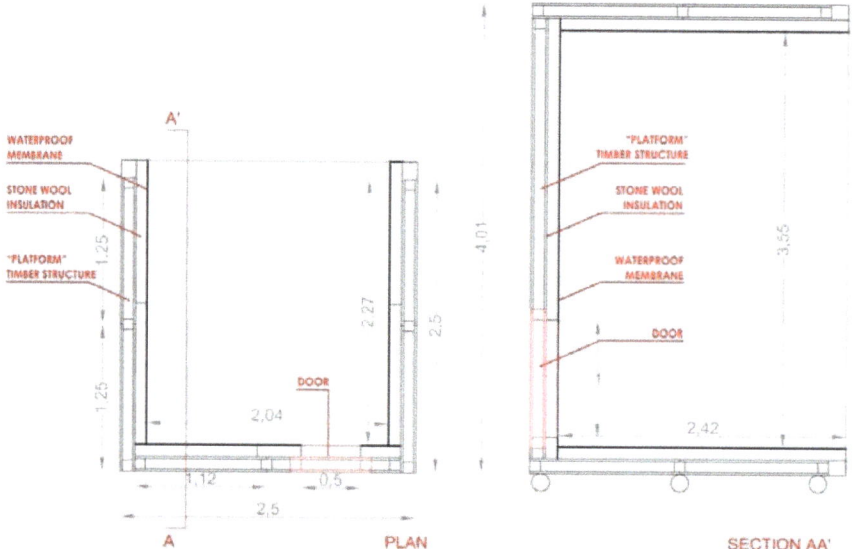

Figure 3. Drawing of the metering box (horizontal plan and vertical section). Measurements expressed in meters.

Box dimensions were determined not only by the anticipated study of probable vertical heat stratification common in historic buildings often with high ceilings (see also Section 5.1), but also due to the favoured and anticipated study of at least 2 insulation systems put in parallel, as depicted in Figure 4, alike the experimental study developed by Kloseiko et al. [93]. In [93], the hygrothermal performance of an internally insulated brick wall was studied, with different insulation systems, measuring 1.00 m width each (this way, by placing sensors in the middle, a 0.50 m distance from each material border was assured).

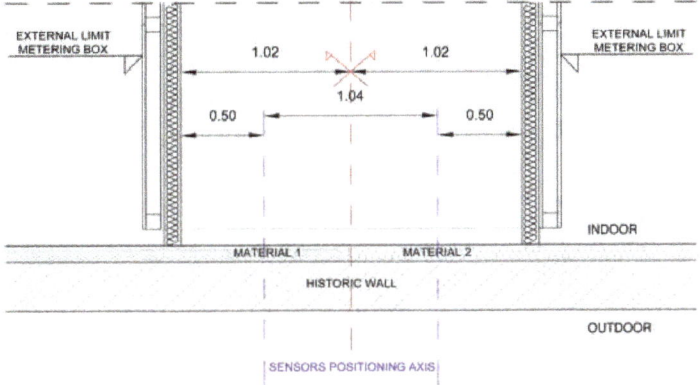

Figure 4. Metering hot box positioning (plan). Relation between the box and sensor location if two insulation materials are tested.

4.3. Construction of the Metering Box

One of the main objectives of the HeLLo project, besides the general scientific final goal to make actors of buildings sector aware of strengths and weaknesses of the most common energy retrofit technical solutions when applied to historic buildings [32], was the development of a very wide and structured program of dissemination The idea was to open the door of the laboratory to other different stakeholders and involve them in the project activities. Among this open Labs program, in the one called SchoolLab activity, in a unique didactical approach, students of the 2nd year of the Degree of Architecture were involved in the activity of the box constructing—Figure A1 in Appendix A unveils some of the steps of the box construction. During this phase, only the 'outer-shell' was executed (the platform frame structure), being later internally coated with 10 cm high-density stone wool, covered by a vapor barrier, Figure 5.

Figure 5. Box finishing: Thermal insulation and vapor barrier application.

4.4. Monitoring System of the Metering Box

Monitoring the hygrothermal behaviour of historic building components was slightly more complex than in existing non-historic ones. As stated in [94] (p. 97), "[...] common mounting systems for long-term surface measurements are risky to original surfaces in historic buildings", e.g., standard installation methods (e.g., adhesive bonds and sensors fixed to walls with holders and/or screws) might damage original surfaces when sensors are later removed. This assumption relates to cultural heritage protection requirements [95] of NDT or methods with the least damages [94].

The metering box was provided with a 2000 W heating convector (with 3 power levels), locally controlled by its own sensor (PID, as described below) and 2 ultrasonic humidifiers, (argo HYDRO digit), 30W/each, self-regulated, which guaranteed indoor air parameters at the desired setup conditions (T $\approx$ 20 °C, RH $\approx$ 55%), Figure 6.

Figure 6. Box indoor hygrothermal control equipment.

In [34], authors presented an innovative measuring method for the hygrothermal assessment of historic walls. In the current study, the same low-cost and conservation compatible technology was also used to control the hygrothermal parameters of the metering box system. The air temperature and RH inside and outside the metering box were controlled by T-RH combined sensors. "These sensors are based on a capacitive polymer RH sensor and a PTA (Proportional to Absolute) integrated temperature sensor (Telaire T9602; Amphenol). They were IP67 certified to guarantee protection in a harsh environment. These sensors used a PDM output signal, and a low pass RC filter was needed to have a voltage signal to acquire hygrothermal data" [34] (p. 7). The sensors of the metering box were connected and managed by a data acquisition system based on a Master Slave configuration.

The initial version of the developed remote sensing technology [34] was upgraded and tuned to fit the current requirements of the HeLLo project. The T-RH measurement system was unchanged, and it was still based on Amphenol probes coupled with an RC lowpass filter, and readout of the analog values was performed by Analog Input Seneca devices with Modbus communication. Old thermostatic heating control was replaced with a more sophisticated Seneca module based on retroactive PID (Proportional, Integrative, Derivative) algorithm and coupled with a triac solid state relay. The temperature probe of the PID module was a PT100 class B. In order to keep the temperature constant inside the metering box, the PID control works on cycles of 120 seconds. Temperature trends were evaluated in terms of temperature integral of previous cycles and heat was activated for a fraction of cycle. The PID module was connected on the same Modbus net and can be configured and monitored by the same software that acquires T-RH values.

The acquisition software was updated, including readout and control of PID modules, and in order to have a configurable number of probes and readout modules, dedicated features were introduced. Substantial updates were done with the main control software. Some of the newly implemented features are listed below:

- Email notification is sent hourly if everything works with a summary of all measurements, otherwise, every critical change in status is notified, with detailed information;
- An hourly backup copy of raw data is performed on a local drive;
- Daily processing of raw data is performed to produce more usable data files;
- Web pages with a summary of system and measurement details are updated every minute and synchronized with an online web server. This allows us to have a simple view of status available on every device connected to the internet, like a PC and mobile device;
- A protected local folder containing raw data backup and all pdf reports are synchronized with an external cloud.

5. Results and Discussion of the Conditions inside the Box

5.1. Preliminary Test for the Evaluation of Heat Stratifications

A simple test was performed to control vertical heat stratification inside the box. Five hand-made thermocouples (TC) with an accuracy of 0.5 °C (calibrated in the laboratory), were placed on the surface of the historic wall, between 0.90 m and 3.40 m from the floor to circa 0.50 m from the box boundaries (Figure 7a), during a four-day monitoring period. As shown in Figure 7b, between the highest TC (h = 3.40 m, in black) and the lowest one (h = 0.90 m, in pink) there was an average difference of 4 °C. This simple test has confirmed the anticipated heat stratification, common in historic buildings, justifying the height of the box.

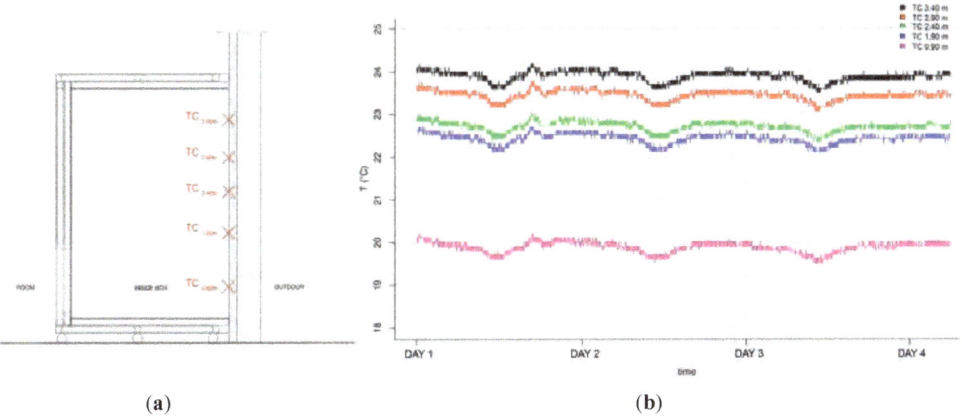

Figure 7. Results of the simplified test to evaluate the heat stratification inside the box: (a) Vertical section of the metering box with the position of the TCs; (b) plot of the monitored T value of the five survey points (TC).

5.2. Validation of the Hygrothermal Set-Up

Figure 8 shows the distribution of the T-RH combined sensors (see Section 4.4) for the monitoring of the following environmental parameters (T_a and RH):

- Of the outdoor climate conditions (T/RH $_{OUT}$);
- Of the non-refurbished and not heated room in which the box is located (T/RH $_{ROOM}$);
- Inside the metering box (T/RH $_{IN\ BOX}$), placed in the center at circa 1.0 m from the top of the box.

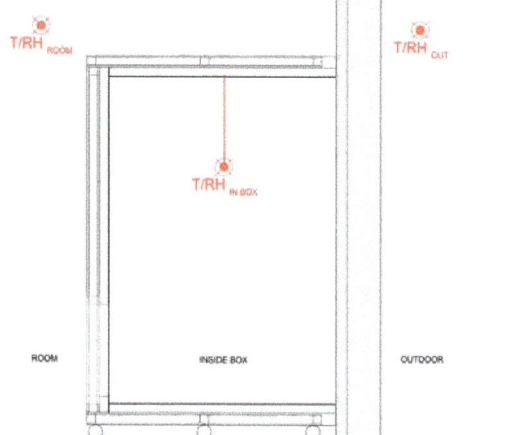

Figure 8. Vertical section of the metering box with the position of the T-RH combined sensors.

The validation of the maintenance of the desired setup conditions was reached after the first period of tests and tuning (27 December 2019–10 January 2020). Figure 9 shows a recently monitored two-week period of a very stable indoor environment. Moreover, in the figure, two small peaks can be observed, brief in time and amplitude, corresponding to the moment of maintenance procedures of the monitoring campaign. In other words, the moments when the door of the box was opened, and the conditions of the air inside the box naturally mixed with those of the room. The insignificance of these events can be further observed in detail in Figure 10.

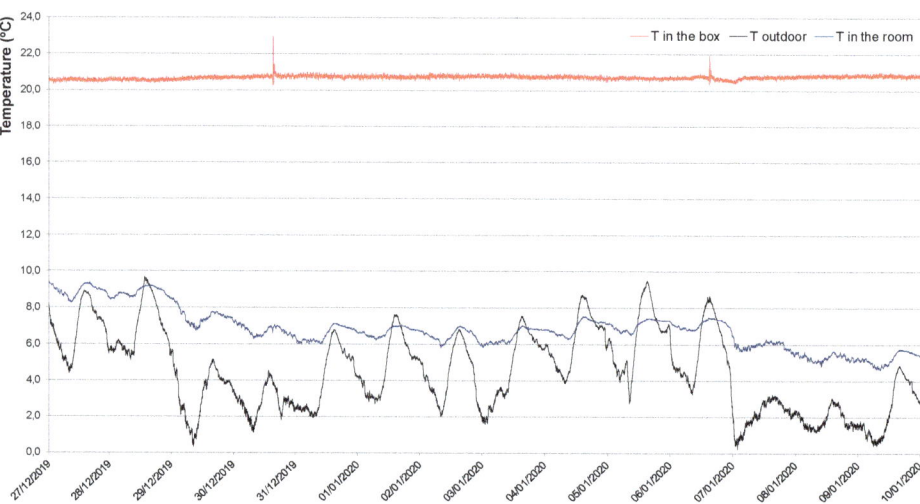

Figure 9. *Cont.*

Relative Humidity

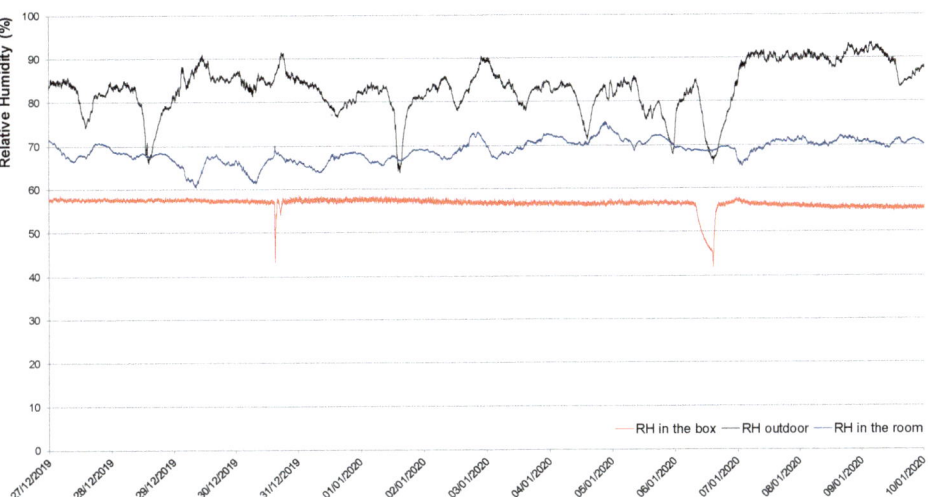

Figure 9. Graphical representation of the monitored parameters [T (°C) and RH (%)] values between 27 December 2019–10 January 2020.

Temperature

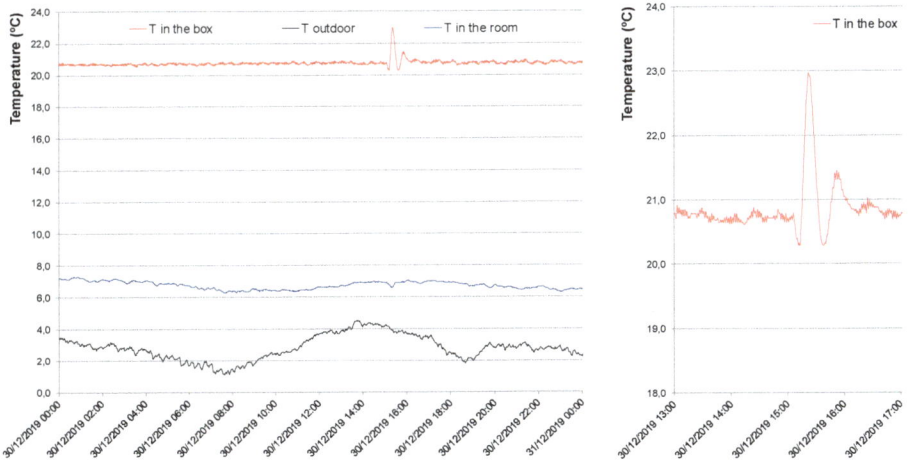

Figure 10. *Cont.*

Relative Humidity

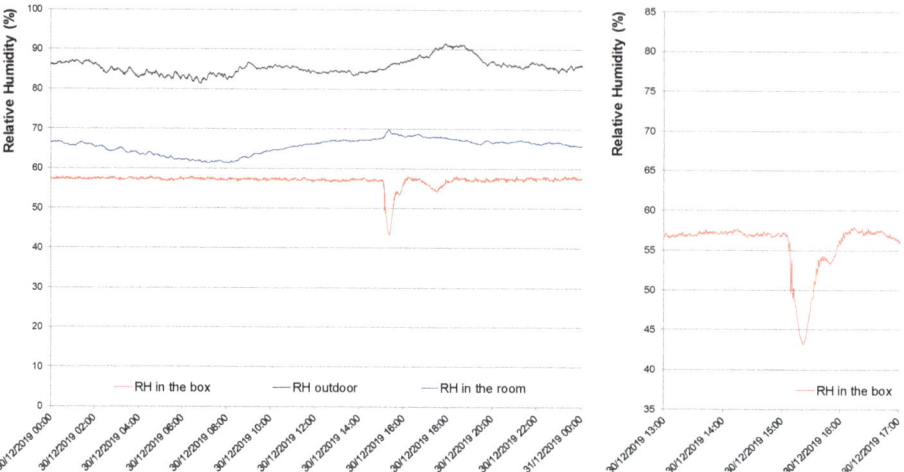

Figure 10. Graphical representation of the monitored parameters [T (°C) an RH (%)] values on 30 December 2019.

As declared, the door of the box was opened twice during this period, on the 30 December 2019 and 6 January 2020, i.e., the experiment might need to be controlled on-site up to once a week. Nonetheless, this action interferes with almost nothing with the continuity and stability of the indoor conditions. Looking at the most noticeable peak, registered on the 30th December, Figure 10, from the moment the door was opened, the indoor temperature suffered a maximum variation of 2.7 °C. Likewise, RH Δmax = 15%.

As shown in Figure 9, the outdoor climate in this winter period is quite varied. The same goes for the conditions of the room where the box is placed, which, as expected, were close to the outdoor conditions, less the thermal influence of the inertia of the historic building envelope. For the entire monitoring period (27 December 2019–10 January 2020), the conditions in the box were definitely stable, average T = 20.7 °C, average RH = 56.5%.

6. Conclusions and Outlook

The new developed device is absolutely disruptive in the field: until this moment, for similar studies, the developed in situ facilities addressed the wall thermal transmittance solely, neglecting the importance of the water vapor permeability factor on the overall wall performance. The feasibility of the new metering hot box has been verified by an in situ measurement for the hygrothermal survey of retrofit wall behaviour in a demonstration MSCA-IF project [32], creating a stable hygrothermal environment by the box. When compared to other boxes the new hot box presents other advantages compared to previous examples:

- It allows long-term monitoring simulating a 'real' indoor environment (e.g., the study in [69] was proposed for short-time HFM measurements);
- It contributes to minimizing biased results since it addresses both the thermal and hygric phenomena of the walls (e.g., in [68] (p. 49) authors recalled that the measured thermal transmittance results obtained in [69] were "55% higher than the design thermal transmittance and that the measurement error was attributed to high moisture");
- It is the first one of its kind addressing historic buildings—probably the type of buildings where data collection in situ is more urgent to be collected, considering that often the characteristics of the materials and their real performance are unknown;

- The singularity of addressing historic buildings justifies the box height. Its width was determined to allow future studies on the hygrothermal performances of internal thermal insulation (two materials can be tested in parallel);
- Similar to a traditional hot box, it was provided with wheels to be more easily moved. In order to minimize the impact on the monitoring, for maintenance purposes, the box was provided with a back door (located on the opposite side where main parameters are being collected);
- The box was built during an educational activity, involving students from a perspective of learning through practice, making the entire scientific process more inclusive.

One other significant advantage could be pointed out: the box re-usability (i.e., enhanced sustainability). As it is provided with wheels, it can be easily moved against another wall in the same room or, more importantly, due to its construction by modules, it can disassemble and used in other case studies. Lastly, it also allows the realization of in situ tests with different settings, for example, 'stress test'.

Author Contributions: Authors are listed in alphabetical order. Conceptualization and methodology, M.C., L.D.P., E.L.; validation and formal analysis, M.A., M.C., L.D.P., R.M.; investigation, M.A., M.C., P.D., L.D.P.; data curation M.A., L.D.P., R.M.; supervision P.D.; writing, review, and editing: M.A., M.C., P.D., L.D.P., E.L. All authors have read and agreed to the published version of the manuscript.

Funding: The results presented in this paper are part of the HeLLo project that has received funding from the European Union's Horizon 2020 research and innovation programme under the Marie Sklodowska-Curie grant agreement No 796712.

Acknowledgments: The authors acknowledge *Giorgi Roberto, Lavorazione Legno* and ROCKWOOL® Italia S.p.A. for the use of their materials and the contribution to the execution of the metering box.

Conflicts of Interest: The authors declare no conflict of interest. The funders had no role in the design of the study; in the collection, analyses, or interpretation of data; in the writing of the manuscript, or in the decision to publish the results.

Nomenclature

BES	Building Energy Simulation	NDT&E	Non-Destructive Testing and Evaluation
CH	Cultural Heritage	OSB	Oriented strand board
CHB	Calibrated Hot Box	PID	Proportional, Integrative, Derivative
EU	European Union	φ	Heat flux (W·m^{-2})
GHB	Guarded Hot Box	SHO	Simple Hot Box
GHP	Guarded Hot Plate	T	Temperature (°C)
HVAC	Heating, ventilation and Air conditioning	Ta	Air Temperature (°C)
HeLLo	Heritage energy Living Lab onsite	TC	Thermocouples
HFM	Heat Flow Meter	TCB	Temperature Control Box
IRT	Infrared Thermography	Ts	Surface temperature
ITT	Infrared Thermography Testing	RH	Relative Humidity (%)
NDT	Non-Destructive Testing	RHa	Air Relative Humidity (%)
NDE	Non-Destructive Evaluation	U-value	Thermal transmittance (W/m^{2}K)

Appendix A

Figure A1. Box construction during SchooLab activity with students (HeLLo [32]).

References

1. European Commission. *Directive 2012/27/Eu of the European Parliament and of the Council*; 2012; 56p, Available online: http://eur-lex.europa.eu/legal-content/EN/TXT/PDF/?uri=CELEX:32012L0027&from=PT (accessed on 28 May 2020).
2. Filippi, M. Remarks on the green retrofitting of historic buildings in Italy. *Energy Build.* **2015**, *95*, 15–22. [CrossRef]
3. European Commission. European Commission > Energy > Topics > Energy Efficiency > Buildings. Available online: https://ec.europa.eu/energy/en/topics/energy-efficiency/energy-performance-of-buildings/energy-performance-buildings-directive (accessed on 8 June 2016).
4. Buildingradar.com. Europe Construction Market Forecast from 2015 to 2020. Available online: https://buildingradar.com/construction-blog/european-construction-market-forecast/ (accessed on 26 June 2016).

5. Phoenix, T. Lessons learned: ASHRAE's approach in the refurbishment of historic and existing buildings. *Energy Build.* **2015**, *95*, 13–14. [CrossRef]
6. Edinburgh World Heritage. *Old and New Towns of Edinburgh World Heritage Site Management Plan 2017–2022*; Edinburgh, UK, 2017; Available online: https://ewh.org.uk/plan/assets/Management-Plan-2018.pdf (accessed on 28 May 2020).
7. Litti, G.; Khoshdel, S.; Audenaert, A.; Braet, J. Hygrothermal performance evaluation of traditional brick masonry in historic buildings. *Energy Build.* **2015**, *105*, 393–411. [CrossRef]
8. Historic England. *Energy Efficiency and Historic Buildings: Insulating Solid Walls*; Historic England: London, UK, 2012; pp. 1–61. [CrossRef]
9. Lucchi, E. Thermal transmittance of historical stone masonries: A comparison among standard, calculated and measured data. *Energy Build.* **2017**, *151*, 393–405. [CrossRef]
10. Lucchi, E. Thermal transmittance of historical brick masonries: A comparison among standard data, analytical calculation procedures, and in situ heat flow meter measurements. *Energy Build.* **2017**, *134*, 171–184. [CrossRef]
11. Peak District National Park. Conserving Your Historic Building. Sustainability and Historic Buildings: A Guide for Owners and Occupiers. Available online: http://www.special-eu.org/assets/uploads/Sustainability_and_Historicbuildings_by_Peak_District.pdf (accessed on 28 May 2020).
12. Raftery, C. Energy Performance in Protected Structures: Planning implications and grants. In *Energy Efficiency in Historic Houses*; Irish Georgian Society: Dublin, Ireland, 2013; pp. 7–18. ISBN 978-0-9545691-4-3. Available online: https://www.igs.ie/uploads/Energy_Efficiency_in_Historic_Houses.pdf (accessed on 28 May 2020).
13. Khoury, J.; Alameddine, Z.; Hollmuller, P. Understanding and bridging the energy performance gap in building retrofit. *Energy Procedia* **2017**, *122*, 217–222. [CrossRef]
14. Zou, P.X.W.; Xu, X.; Sanjayan, J.; Wang, J. Review of 10 years research on building energy performance gap: Life-cycle and stakeholder perspectives. *Energy Build.* **2018**, *178*, 165–181. [CrossRef]
15. Bortoluzzi, D.; Costa, A.; Casciati, S. Reducing Energy Performance Gap in Buildings—Built2Spec Project Solution. *Proceedings* **2017**, *1*, 640. [CrossRef]
16. Bourdeau, M.; Guo, X.; Nefzaoui, E. Buildings energy consumption generation gap: A post-occupancy assessment in a case study of three higher education buildings. *Energy Build.* **2018**, *159*, 600–611. [CrossRef]
17. Built2Spec. D1.2 Performance Gap and Its Assessment Methodology in Built2Spec Project. 2015. Available online: http://built2spec-project.eu/wp-content/uploads/2016/06/Built2Spec_Performance_gap_assessment_methodology_part_1.pdf (accessed on 28 May 2020).
18. Barthelmes, V.M.; Becchio, C.; Fabi, V.; Corgnati, S.P. Occupant behaviour lifestyles and effects on building energy use: Investigation on high and low performing building features. *Energy Procedia* **2017**, *140*, 93–101. [CrossRef]
19. Soares, N.; Bastos, J.; Pereira, L.D.; Soares, A.; Amaral, A.R.; Asadi, E.; Rodrigues, E.; Lamas, F.B.; Monteiro, H.; Lopes, M.A.R.; et al. A review on current advances in the energy and environmental performance of buildings towards a more sustainable built environment. *Renew. Sustain. Energy Rev.* **2017**, *77*, 845–860. [CrossRef]
20. van Dronkelaar, C.; Dowson, M.; Spataru, C.; Mumovic, D. A Review of the Regulatory Energy Performance Gap and Its Underlying Causes in Non-domestic Buildings. *Front. Mech. Eng.* **2016**, *1*, 1–14. [CrossRef]
21. Calzolari, M.; Davoli, P. Instruments for the calculation of energy performance in historical buildings. Limits of applicability and tuning proposal. *SMC Sustain. Mediterr. Constr. L. Cult. Res. Technol. Mag.* **2018**, *1*, 108–114.
22. Adhikari, R.S.; Lucchi, E.; Pracchi, V. Energy modelling of historic buildings: Applicability, problems and compared results. In Proceedings of the 3rd European Workshop on Cultural Heritage Protection (EWCHP): Taking Care of Our Treasures, Bolzano, Italy, 16–17 September 2013; pp. 119–125.
23. Evangelisti, L.; Battista, G.; Guattari, C.; Basilicata, C.; de Lieto Vollaro, R. Analysis of two models for evaluating the energy performance of different buildings. *Sustainability* **2014**, *6*, 5311–5321. [CrossRef]
24. Rye, C. A Short Paper on the Conventions and Standards that Govern the Understanding of Heat Loss in Traditional Buildings. 2012. Available online: http://www.sdfoundation.org.uk/downloads/STBA-Short-Paper-on-heat-loss-FINAL.pdf (accessed on 28 May 2020).
25. Akkurt, G.G.; Aste, N.; Borderon, J.; Buda, A.; Calzolari, M.; Chung, D.; Costanzo, V.; Del Pero, C.; Evola, G.; Huerto-Cardenas, H.E.; et al. Dynamic thermal and hygrometric simulation of historical buildings: Critical factors and possible solutions. *Renew. Sustain. Energy Rev.* **2020**, *118*, 109509. [CrossRef]

26. Calzolari, M. *Prestazione Energetica Delle Architetture Storiche: Sfide e Soluzioni. Analisi dei Metodi di Calcolo per la Definizione del Comportamento Energetico*; Franco Angelli: Milano, Italy, 2016; ISBN 9788891740885.
27. Belpoliti, V.; Bizzarri, G.; Boarin, P.; Calzolari, M.; Davoli, P. A parametric method to assess the energy performance of historical urban settlements. Evaluation of the current energy performance and simulation of retrofit strategies for an Italian case study. *J. Cult. Herit.* **2018**, *30*, 155–167. [CrossRef]
28. Mantesi, E.; Hopfe, C.J.; Cook, M.J.; Glass, J.; Strachan, P. The modelling gap: Quantifying the discrepancy in the representation of thermal mass in building simulation. *Build. Environ.* **2018**, *131*, 74–98. [CrossRef]
29. Pérez-Lombard, L.; Ortiz, J.; Coronel, J.F.; Maestre, I.R. A review of HVAC systems requirements in building energy regulations. *Energy Build.* **2011**, *43*, 255–268. [CrossRef]
30. Dias Pereira, L.; Bispo Lamas, F.; Gameiro da Silva, M. Improving energy use in schools: From IEQ towards energy-efficient planning—method and in-field application to two case studies. *Energy Effic.* **2019**, *12*, 1253–1277. [CrossRef]
31. Rhodes, J.D.; Stephens, B.; Webber, M.E. Using energy audits to investigate the impacts of common air-conditioning design and installation issues on peak power demand and energy consumption in Austin, Texas. *Energy Build.* **2011**, *43*, 3271–3278. [CrossRef]
32. HeLLo EU H2020 MSCA-IF-ES HeLLo Project. 2019. Available online: https://cordis.europa.eu/project/rcn/215475/factsheet/en (accessed on 7 April 2019).
33. Soares, N.; Martins, C.; Gonçalves, M.; Santos, P.; Simões da Silva, L.; Costa, J.J. Laboratory and in-situ non-destructive methods to evaluate the thermal transmittance and behaviour of walls, windows, and construction elements with innovative materials: A review. *Energy Build.* **2019**, *182*, 88–110. [CrossRef]
34. Lucchi, E.; Dias Pereira, L.; Andreotti, M.; Malaguti, R.; Cennamo, D.; Calzolari, M.; Frighi, V. Development of a Compatible, Low Cost and High Accurate Conservation Remote Sensing Technology for the Hygrothermal Assessment of Historic Walls. *Electronics* **2019**, *8*, 1–20. [CrossRef]
35. International Organization for Standardization (ISO). ISO 7345:2018. In *Thermal Performance of Buildings and Building Components—Physical Quantities and Definitions*; ISO: Geneva, Switzerland, 2018.
36. International Organization for Standardization (ISO). ISO 8301:1991. In *Thermal Insulation—Determination of Steady-State Thermal Resistance and Related Properties—Heat Flow Meter Apparatus*; ISO: Geneva, Switzerland, 1991.
37. International Organization for Standardization (ISO). ISO 9869-1:2014. In *Thermal Insulation. Building Elements. In Situ Measurement of Thermal Resistance and Thermal Transmittance—Part 1: Heat Flow Meter Method*; ISO: Geneva, Switzerland, 2014.
38. Santos, P.; Gonçalves, M.; Martins, C.; Soares, N.; Costa, J.J. Thermal transmittance of lightweight steel framed walls: Experimental versus numerical and analytical approaches. *J. Build. Eng.* **2019**, *25*, 100776. [CrossRef]
39. International Organization for Standardization (ISO). ISO 10880:2017. In *Non-Destructive Testing—Infrared Thermography Testing—General Principles*; ISO: Geneva, Switzerland, 2017.
40. International Organization for Standardization (ISO). ISO 6781:1983. In *Thermal Insulation—Qualitative Detection of Thermal Irregularities in Building Envelopes—Infrared Method*; ISO: Geneva, Switzerland, 1983.
41. International Organization for Standardization (ISO). ISO 8990:1994. In *Thermal Insulation—Determination of Steady-State Thermal Transmission Properties—Calibrated and Guarded Hot Box*; ISO: Geneva, Switzerland, 1994.
42. Asdrubali, F.; Baldinelli, G. Thermal transmittance measurements with the hot box method: Calibration, experimental procedures, and uncertainty analyses of three different approaches. *Energy Build.* **2011**, *43*, 1618–1626. [CrossRef]
43. Bienvenido-Huertas, D.; Moyano, J.; Marín, D.; Fresco-Contreras, R. Review of in situ methods for assessing the thermal transmittance of walls. *Renew. Sustain. Energy Rev.* **2019**, *102*, 356–371. [CrossRef]
44. Sala, J.M.; Urresti, A.; Martín, K.; Flores, I.; Apaolaza, A. Static and dynamic thermal characterisation of a hollow brick wall: Tests and numerical analysis. *Energy Build.* **2008**, *40*, 1513–1520. [CrossRef]
45. Giorgio, P.; De Carli, M. A measuring campaign of thermal conductance in situ and possible impacts on net energy demand in buildings. *Energy Build.* **2013**, *59*, 29–36. [CrossRef]
46. Nardi, I.; Lucchi, E.; de Rubeis, T.; Ambrosini, D. Quantification of heat energy losses through the building envelope: A state-of-the-art analysis with critical and comprehensive review on infrared thermography. *Build. Environ.* **2018**, *146*, 190–205. [CrossRef]

47. Lucchi, E. Applications of the infrared thermography in the energy audit of buildings: A review. *Renew. Sustain. Energy Rev.* **2018**, *82*, 3077–3090. [CrossRef]
48. International Organization for Standardization (ISO). ISO 8302:1991. In *Thermal Insulation—Determination of Steady-State Thermal Resistance and Related Properties—Guarded Hot Plate Apparatus*; ISO: Geneva, Switzerland, 1991.
49. ASTM International (American Society for Testing and Materials). ASTM C177—19. In *Standard Test Method for Steady-State Heat Flux Measurements and Thermal Transmission Properties by Means of the Guarded-Hot-Plate Apparatus*; ASTM: West Conshohhocken, PA, USA, 2019.
50. Reddy, K.S.; Jayachandran, S. Investigations on design and construction of a square guarded hot plate (SGHP) apparatus for thermal conductivity measurement of insulation materials. *Int. J. Therm. Sci.* **2017**, *120*, 136–147. [CrossRef]
51. Kobari, T.; Okajima, J.; Komiya, A.; Maruyama, S. Development of guarded hot plate apparatus utilizing Peltier module for precise thermal conductivity measurement of insulation materials. *Int. J. Heat Mass Transf.* **2015**, *91*, 1157–1166. [CrossRef]
52. Sanjaya, C.S.; Wee, T.-H.; Tamilselvan, T. Regression analysis estimation of thermal conductivity using guarded-hot-plate apparatus. *Appl. Therm. Eng.* **2011**, *31*, 1566–1575. [CrossRef]
53. ASTM International (American Society for Testing and Materials). ASTM C1363-19. In *Standard Test Method for Thermal Performance of Building Materials and Envelope Assemblies by Means of a Hot Box Apparatus*; ASTM: West Conshohhocken, PA, USA, 2019.
54. European Committee for Standardization (CEN). *EN 1934:1998 Thermal Performance of Buildings—Determination of Thermal Resistance by Hot Box Method Using Heat Flow Meter—Masonry*; CEN: Brussels, Belgium, 1998.
55. Interstate Standard of Russian Federation (ISRF). Standard GOST 26602.1-99. In *Windows and Doors. Methods of Determination of Resistance of Thermal Transmission*; ISFR: Moscow, Russia, 1999.
56. Vereecken, E.; Roels, S. A comparison of the hygric performance of interior insulation systems: A hot box-cold box experiment. *Energy Build.* **2014**, *80*, 37–44. [CrossRef]
57. Byrne, A.; Byrne, G.; Robinson, A. Compact facility for testing steady and transient thermal performance of building walls. *Energy Build.* **2017**, *152*, 602–614. [CrossRef]
58. Cucumo, M.; De Rosa, A.; Ferraro, V.; Kaliakatsos, D.; Marinelli, V. A method for the experimental evaluation in situ of the wall conductance. *Energy Build.* **2006**, *38*, 238–244. [CrossRef]
59. Han, J.; Lu, L.; Peng, J.; Yang, H. Performance of ventilated double-sided PV façade compared with conventional clear glass façade. *Energy Build.* **2013**, *56*, 204–209. [CrossRef]
60. Cattarin, G.; Causone, F.; Kindinis, A.; Pagliano, L. Outdoor test cells for building envelope experimental characterisation—A literature review. *Renew. Sustain. Energy Rev.* **2016**, *54*, 606–625. [CrossRef]
61. Silva, T.; Vicente, R.; Rodrigues, F.; Samagaio, A.; Cardoso, C. Development of a window shutter with phase change materials: Full scale outdoor experimental approach. *Energy Build.* **2015**, *88*, 110–121. [CrossRef]
62. Young, C.-H.; Chen, Y.-L.; Chen, P.-C. Heat insulation solar glass and application on energy efficiency buildings. *Energy Build.* **2014**, *78*, 66–78. [CrossRef]
63. Qahtan, A.; Keumala, N.; Rao, S.P.; Abdul-Samad, Z. Experimental determination of thermal performance of glazed façades with water film, under direct solar radiation in the tropics. *Build. Environ.* **2011**, *46*, 2238–2246. [CrossRef]
64. Gao, Y.; Roux, J.J.; Teodosiu, C.; Zhao, L.H. Reduced linear state model of hollow blocks walls, validation using hot box measurements. *Energy Build.* **2004**, *36*, 1107–1115. [CrossRef]
65. Kus, H.; Özkan, E.; Göcer, Ö.; Edis, E. Hot box measurements of pumice aggregate concrete hollow block walls. *Constr. Build. Mater.* **2013**, *38*, 837–845. [CrossRef]
66. Baldwin, C.; Cruickshank, C.A.; Schiedel, M.; Conley, B. Comparison of steady-state and in-situ testing of high thermal resistance walls incorporating vacuum insulation panels. *Energy Procedia* **2015**, *78*, 3246–3251. [CrossRef]
67. Peng, C.; Wu, Z. In situ measuring and evaluating the thermal resistance of building construction. *Energy Build.* **2008**, *40*, 2076–2082. [CrossRef]
68. Meng, X.; Gao, Y.; Wang, Y.; Yan, B.; Zhang, W.; Long, E. Feasibility experiment on the simple hot box-heat flow meter method and the optimization based on simulation reproduction. *Appl. Therm. Eng.* **2015**, *83*, 48–56. [CrossRef]

69. Zhu, X.F.; Li, L.P.; Yin, X.B.; Zhang, S.H.; Wang, Y.; Liu, W.; Zheng, L. An in-situ test apparatus of heat transfer coefficient for building envelope. *Build. Energy Effic.* **2012**, *256*, 57–60. Available online: http://en.cnki.com.cn/Article_en/CJFDTotal-FCYY201206018.htm (accessed on 28 May 2020).
70. Meng, X.; Luo, T.; Gao, Y.; Zhang, L.; Shen, Q.; Long, E. A new simple method to measure wall thermal transmittance in situ and its adaptability analysis. *Appl. Therm. Eng.* **2017**, *122*, 747–757. [CrossRef]
71. EN 15758:2010. *Conservation Of Cultural Property—Procedures and Instruments for Measuring Temperatures of the Air and the Surfaces of Objects*; CEN: Brussels, Belgium, 2010.
72. EN 16242. *EN 16242 Conservation of Cultural Heritage—Procedures and Instruments for Measuring Humidity in the Air and Moisture Exchanges between Air and Cultural Property*; CEN: Brussels, Belgium, 2012.
73. Ente Italiano di Normazione (UNI). UNI 10829. In *Beni di Interesse Storico e Artistico—Condizioni Ambientali di Conservazione—Misurazione ed Analisi*; UNI: Milano, Italy, 1999.
74. European Committee for Standardization (CEN). *Standard CEN/TC 346—Conservation of Cultural Heritage*; CEN: Brussels, Belgium, 2009.
75. European Committee for Standardization (CEN). *EN 15757:2010 Conservation of Cultural Property—Specifications for Temperature and Relative Humidity to Limit Climate-Induced Mechanical Damage in Organic Hygroscopic Materials*; 2010; Available online: http://shop.bsigroup.com/ProductDetail?pid=000000000030173518 (accessed on 28 May 2020).
76. European Committee for Standardization (CEN). CEN UNI EN 15759-1 (2012) (English). In *Conservation of Cultural Property—Indoor Climate—Part 1: Guidelines for Heating Churches, Chapels and Other Places of Worship*; CEN: Brussels, Belgium, 2012.
77. Ente Nazionale Italiano di Unificazione (UNI). UNI 10969. In *Beni Culturali. Principi Generali per la Scelta e il Controllo del Microclima per la Conservazione*; UNI: Milano, Italy, 2002.
78. Ente Nazionale Italiano di Unificazione (UNI). UNI 10586. In *Condizioni Climatiche per Ambienti di Conservazione di Documenti Grafici e Caratteristiche Degli Alloggiamenti*; UNI: Milano, Italy, 1999.
79. Camuffo, D. *Microclimate for Cultural Heritage: Conservation, Restoration, and Maintenance of Indoor and Outdoor Monuments*, 2nd ed.; Elsevier: Amsterdam, The Netherlands, 2014; ISBN 9780444632968. [CrossRef]
80. Walker, R.; Pavía, S.; Dalton, M. Measurement of moisture content in solid brick walls using timber dowel. *Mater. Struct.* **2016**, *49*, 2549–2561. [CrossRef]
81. European Committee for Standardization (CEN). Standard prEN 16682. In *Conservation of Cultural Heritage. Guide to the Measurements of Moisture Content in Materials Constituting Movable and Immovable Cultural Heritage*; CEN: Brussels, Belgium, 2013.
82. Ente Nazionale Italiano di Unificazione (UNI). UNI 11085. In *Beni Culturali—Materiali Lapidei Naturali ed Artificiali—Determinazione del Contenuto D'acqua: Metodo Ponderale*; UNI: Milano, Italy, 2003.
83. Odgaard, T.; Bjarløv, S.P.; Rode, C. Interior insulation—Experimental investigation of hygrothermal conditions and damage evaluation of solid masonry façades in a listed building. *Build. Environ.* **2018**, *129*, 1–14. [CrossRef]
84. Walker, R.; Pavía, S. Thermal and moisture monitoring of an internally insulated historic brick wall. *Build. Environ.* **2018**, *133*, 178–186. [CrossRef]
85. Hansen, T.K.; Bjarløv, S.P.; Peuhkuri, R.H.; Harrestrup, M. Long term in situ measurements of hygrothermal conditions at critical points in four cases of internally insulated historic solid masonry walls. *Energy Build.* **2018**, *172*, 235–248. [CrossRef]
86. UNESCO. Ferrara, City of the Renaissance, and its Po Delta. 1995. Available online: http://whc.unesco.org/en/list/733 (accessed on 9 January 2019).
87. Davoli, P. Complexity, information surplus and interdisciplinarity management. The Rehabilitation of Tassoni Estense Palace in Ferrara. In *Conserving Architecture*; Jain, K., Ed.; AADI CENTRE: Ahmedabad, Indian, 2017; pp. 124–145. ISBN 978-81-908528-2-1.
88. International Organization for Standardization (ISO). ISO 7730: 2005. In *Ergonomics of the Thermal Environment. Analytical Determination and Interpretation of Thermal Comfort Using Calculation of the PMV and PPD Indices and Local Thermal Comfort Criteria*; UNI: Geneva, Switzerland, 2005.
89. Ente Italiano di Normazione (UNI). UNI EN ISO 13788. In *Prestazione Igrotermica dei Componenti e Degli Elementi per Edilizia—Temperatura Superficiale Interna per Evitare L'umidita' Superficiale Critica e la Condensazione Interstiziale. Metodi di Calcolo*; UNI: Milano, Italy, 2013.

90. International Organization for Standardization (ISO). ISO 17772-1: 2017. In *Energy Performance of Buildings—Indoor Environmental Quality—Part 1: Indoor Environmental Input Parameters for the Design and Assessment of Energy Performance of Buildings*; ISO: Geneva, Switzerland, 2017.
91. Rasooli, A.; Itard, L.; Ferreira, C.I. A response factor-based method for the rapid in-situ determination of wall's thermal resistance in existing buildings. *Energy Build.* **2016**, *119*, 51–61. [CrossRef]
92. Nardi, I.; Paoletti, D.; Ambrosini, D.; De Rubeis, T.; Sfarra, S. Validation of quantitative IR thermography for estimating the U-value by a hot box apparatus. *J. Phys. Conf. Ser.* **2015**, *655*, 012006. [CrossRef]
93. Kloseiko, P.; Arumagi, E.; Kalamees, T. Hygrothermal performance of internally insulated brick wall in cold climate: A case study in a historical school building. *J. Build. Phys.* **2015**, *38*, 444–464. [CrossRef]
94. Raffler, S.; Bichlmair, S.; Kilian, R. Mounting of sensors on surfaces in historic buildings. *Energy Build.* **2015**, *95*, 92–97. [CrossRef]
95. Council of Europe. Guidelines on Cultural Heritage: Technical Tools for Heritage Conservation and Management. 2012. Available online: https://rm.coe.int/16806ae4a9 (accessed on 28 May 2020).

© 2020 by the authors. Licensee MDPI, Basel, Switzerland. This article is an open access article distributed under the terms and conditions of the Creative Commons Attribution (CC BY) license (http://creativecommons.org/licenses/by/4.0/).

Article

Mechanical and Thermal Performance Characterisation of Compressed Earth Blocks

Elisabete R. Teixeira [1], Gilberto Machado [1], Adilson de P. Junior [1], Christiane Guarnier [2], Jorge Fernandes [1], Sandra M. Silva [1] and Ricardo Mateus [1,*]

[1] Department of Civil Engineering, Institute for Sustainability and Innovation in Structural Engineering (ISISE), University of Minho, 4800-058 Guimarães, Portugal; b8416@civil.uminho.pt (E.R.T.); a73118@alunos.uminho.pt (G.M.); cpjunior.adilson@gmail.com (A.d.P.J.); jepfernandes@me.com (J.F.); sms@civil.uminho.pt (S.M.S.)

[2] Federal Center for Technological Education "Celso Suckow da Fonseca" (CEFET/RJ), Rio de Janeiro-RJ 20271-110, Brazil; christiane.guarnier@cefet-rj.br

* Correspondence: ricardomateus@civil.uminho.pt; Tel.: +351-253-510-200

Received: 30 April 2020; Accepted: 2 June 2020; Published: 10 June 2020

Abstract: The present research is focused on an experimental investigation to evaluate the mechanical, durability, and thermal performance of compressed earth blocks (CEBs) produced in Portugal. CEBs were analysed in terms of electrical resistivity, ultrasonic pulse velocity, compressive strength, total water absorption, water absorption by capillarity, accelerated erosion test, and thermal transmittance evaluated in a guarded hotbox setup apparatus. Overall, the results showed that compressed earth blocks presented good mechanical and durability properties. Still, they had some issues in terms of porosity due to the particle size distribution of soil used for their production. The compressive strength value obtained was 9 MPa, which is considerably higher than the minimum requirements for compressed earth blocks. Moreover, they presented a heat transfer coefficient of 2.66 W/(m^2·K). This heat transfer coefficient means that this type of masonry unit cannot be used in the building envelope without an additional thermal insulation layer but shows that they are suitable to be used in partition walls. Although CEBs have promising characteristics when compared to conventional bricks, results also showed that their proprieties could even be improved if optimisation of the soil mixture is implemented.

Keywords: compressed earth blocks (CEBs); compressive strength; durability; guarded hot box; thermal transmittance

1. Introduction

Earth has been used as a building material since ancient times in several different ways around the world [1–4]. Industrialised building systems and the dissemination of materials, like concrete [5], have replaced earthen construction. Today, earthen construction is associated with poverty [2], and most of this type of construction is located in developing countries. The continuous increase in the energy cost of some building materials (cement and ceramic bricks) and environmental issues are promoting the use of sustainable materials, such as the earthen materials known by their abundance and low-cost production [5–7].

Compressed earth blocks (CEBs) are one of the most widespread earthen building techniques. They represent a modern descendent of the moulded earth block, commonly called as the adobe block [7]. The compaction of earth improves the quality and performance of the blocks [4] but also promotes several environmental, social, and economic benefits [8,9]. Regarding the environmental advantages of using earthen products, a previous study showed that in a cradle-to-gate analysis of different walls, the use of earthen building elements could result in reducing the potential environmental impacts by about 50% when compared to the use of conventional building elements [2].

Earthen construction is known to undergo rapid deterioration under severe weather conditions [10]. If not built adequately, earthen buildings have lower durability and are more vulnerable to extreme weather conditions and rainfall than conventional buildings. This situation means higher maintenance and repair costs during the life cycle of a building [11].

In the last few years, there has been increasing interest in overcoming the mechanical and durability issues related to earthen blocks. Different stabilisation techniques were used to improve durability and compressive strength [6,11–13]. Dynamic compaction alone or together with chemical stabilisation using several additives has been shown to considerable improve the mechanical performance of CEBs [10,13]. In contrast, compaction increases thermal conductivity [14]. The stabilisation process of raw earth refers to any mechanical, physical, physicochemical, or combined methods that enhance its properties [10]. Bahar et al. [15] studied the effect of several stabilisation methods on mechanical properties. The results showed that the combination of compaction and cement stabilisation is an effective solution for increasing the strength of earth blocks. Amoudi et al. [16,17] developed an experimental program to study the mechanical properties of cement-stabilised earth blocks. They verified that cement in the presence of water forms hydrated products that occupy the voids and wrap the soil particles. That process leads to an improvement in compressive strength, water absorption, dimensional stability, and durability. In many countries, CEBs are stabilised with cement or lime, and there are successful examples of their use in the construction of buildings. Several studies highlighted their lower construction cost, simpler construction processes, and the contribution of this material to maintaining a better indoor environment quality when comparing to the use of conventional building materials [8,11,18–20]. Besides that, some studies show that the addition of lime to compressed earth blocks can improve their mechanical and hydrous properties [21,22].

Regarding the thermal proprieties of earthen building products, since they are massive, they contribute to increasing the thermal inertia of the buildings. This feature can have a positive influence on the thermal performance of buildings in certain climates. A previous study showed that in locations with hot summers and temperate winters, such as the Mediterranean areas, earthen construction could provide comfortable indoor temperature by passive means alone [23]. This property can reduce heating and cooling energy needs and therefore contribute to lower life cycle environmental and financial costs. Nevertheless, compared to the number of studies focusing on mechanical properties, there are fewer studies related to the thermal properties of compressed earth blocks [11]. Adam and Jones [24] measured the thermal conductivity of lime and cement stabilised hollow and massive earth blocks using a guarded hot box method. The authors verified that the thermal conductivity was higher on stabilised blocks (0.20 W/m^2·K and 0.50 W/m^2·K). The compressive strength used in the compaction of the blocks, the type of soil, and the additives used can significatively influence the thermal conductivity of a CEB building element [24]. For that reason, in the literature, it is possible to find very different thermal conductivity values for earthen products. For example, according to the Portuguese thermal regulation [25], the thermal conductivity to consider for adobe, rammed earth, and compressed earth blocks is 1.10 W/(m^2·K). At the same time, other studies show quite different values, also depending on the considered earthen building technique—earth materials with fibres (0.42–0.90 W/(m^2·K)), adobe (0.46–0.81 W/(m^2·K)), or rammed earth (0.35–0.70 W/(m^2·K)) [1,26].

When designing a sustainable building, the design team must have comprehensive information regarding the different building products they can use [19]. Information should include that related to the life cycle environmental (e.g., embodied energy and global warming potential), functional (e.g., mechanical and thermal) and economic (e.g., construction and maintenance cost) performances.

Based on this context, this research is within a series of studies that are being developed by the same authors to develop comprehensive information about earthen construction. Past studies include those related to analysing the contribution of this type of construction in improving the indoor environmental quality [23] and reducing the embodied environmental impacts [2].

The present research is focused on an experimental investigation to evaluate the mechanical, durability and thermal performances of compressed earth blocks produced by a Portuguese company. This study aims to analyse the functional quality of the abovementioned product and assess its potential to be used in the construction of buildings.

2. Materials and Methods

The compressed earth blocks tested are a commercial product made by a manufacturer located in the city of Serpa, district of Beja (southern Portugal), which is also a contractor that builds earthen and conventional buildings. This contractor is one of the leading earth building systems builders in Portugal. The share of the earthen building systems corresponds to around 12% of the total company's activity, and during the year 2014, the company produced 338 m^3 of rammed earth and 36 m^3 of compressed earth blocks. Usually, rammed earth is used to build 60-cm-thick walls, and the dimensions of the CEBs produced by this company varies. In this work, 30 cm × 15 cm × 7 cm compressed earth blocks were studied, since it is the most common block produced by the company. Additionally, this size is the most used in the Portuguese construction. The soil mixture is stabilised by using 6% by weight (wt) of hydraulic lime and 1% wt of hydrated lime. The mix also uses water, generally extracted on-site (groundwater) (10% by weight), which evaporates during the drying process. In the majority of cases, earthen building elements are built from soil extracted from the construction site. Additionally, according to the company's data, the compressed earth blocks are made and compacted using a mechanical tapping machine. The company provided compressed earth blocks and the soil used for their production. They were experimentally analysed in different labs of the Department of Civil Engineering of the University of Minho, located in the city of Guimarães, district of Braga (northern Portugal).

2.1. Soil Characterisation

The soil was characterised in terms of particle size distribution, sand equivalent, clay content, cohesion limits, and compaction properties. These properties evaluate the quality of soil to be used in earthen construction. The particle size distribution was determined according to the EN 196:1966 standard [27]. The main goal of the sand equivalent test is to estimate the percentage of sand that exist in a soil fraction with particles with less than 2 mm. This test was done according to EN 933-8:2002 [28]. The methylene blue test allows the quantification of clay content present in a soil sample through the ionic change between the cations that exist in the soil particles and was done according to EN 933-9:2002 [29]. The cohesion limits of soil are fundamental for the final quality of CEBs. The main goal of this test is assessing the liquid limit (L_L), the plastic limit (L_P), and the index of plasticity (I_P) of the soil. The cohesion limits were determined and calculated according to the EN 143:1969 standard [30].

The compaction properties of soil are fundamental in earthen products since there is a direct relation between dry density and compressive strength of a product. A more compact product presents higher strength. The main goal of the Proctor compaction test consists in analysing the optimum water content. This water content corresponds to the water content of a soil that allows it to achieve its dry density for specific energy of compaction. This test was done according to LNEC E 197:1966 standard [31], considering two types of compaction (light and hard) in a small mould.

Table 1 and Figure 1 summarise the characteristics of the soil used for CEBs production. The soil presented a good particle size distribution and showed the four types of particles in significant percentages (15.9% pebble, 47.2% sand, 17.6% silt, and 19.4% clay). As shown in Table 1, the soil has a liquid limit of 29% and a plasticity index of 11%. Therefore, it can be classified as a fair to poor clayed soil (type A6) according to the American Association of State Highway Transportation Officials (AASHTO) system [32]. However, according to the CRATerre group [33], these figures are within the limits of the recommended classes for soil to be used as a construction material. The methylene blue test shows 2.28 g of methylene blue per 100 g of soil, indicating a low degree of expansion, as also confirmed by the plasticity index, suggesting a low clay content in the studied soil [34]. This result is good since expansive soils are affected by humidity variations that change its consistency [35]. The analysed soil has a maximum dry density between 1.95 g/cm³ and 1.99 g/cm³, which means that this soil is classified as "very good" to be used as construction material [33].

Table 1. Particle size distribution and Atterberg limits of the soil used.

Property	Parameter	Value
Particle size distribution	Gravel (>4.75 mm)	15.90
	Sand (2.00–0.06 mm)	47.20
	Clay and silt (<0.06 mm)	36.95
Atterberg limits	Liquid limit L_l (%)	29.00
	Plasticity limit L_p (%)	18.00
	Plasticity index I_p (%)	11.00
Sand equivalent	- (%)	23.49
Fine content by methylene blue test	Methylene blue value (g/100 g soil)	2.28
Modified Proctor test—Light	Optimum water content (%)	12.00
	Maximum dry density (g/cm³)	1.95
Modified Proctor test—Heavy	Optimum water content (%)	11.80
	Maximum dry density (g/cm³)	1.99

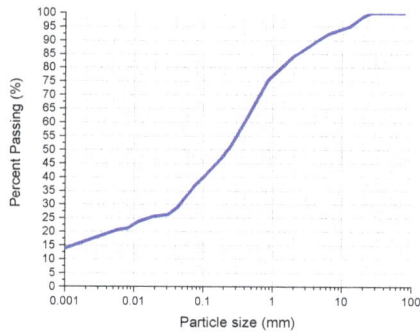

Figure 1. Particle size distribution of the soil, according to the results of the sedimentation test.

2.2. Compressed Earth Blocks Characterisation

2.2.1. Electrical Resistivity

The electrical resistivity was measured using the ResipodProceq equipment, made by Proceq SA (Schwerzenbach, Switzerland), which comprises four equidistant (38 mm) electrodes (Figure 2). During this test, an alternate current was provided between the external electrodes and the electrical potential difference between the internal electrodes was measured. The electrical resistivity was measured through the Ohm's law and computed by the equipment used. The tested samples were the ones used for the water absorption by capillarity test after they achieved the saturation point. Four measurements were done for each saturated compressed earth block sample.

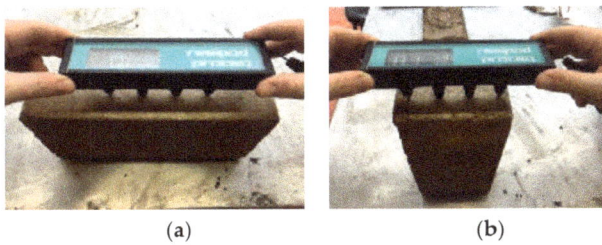

Figure 2. Electric resistivity test: measurements were done in two samples faces—(**a**) width and (**b**) length.

2.2.2. Ultrasonic Pulse Velocity

The Ultrasonic Pulse Velocity (UPV) test evaluates some materials properties, such as elasticity modulus, homogeneity, mechanical resistances, and cracking. It is also possible to calculate the propagation velocity [36]. The UPV tests consists of measuring the time that a given sound pulse takes to pass through a known section of a specimen. This is based on the wave propagation theory, where a sound pulse propagates faster in a dense material and slowly in a porous material. It is therefore possible to calculate the propagation velocity [36], and this test allows the indirect determination of the intrinsic characteristics of a given sample [36]. There is almost nothing in the literature about the use of this test in compressed earth blocks. Nevertheless, there are some studies that have already been performed on rammed earth [37,38] that disclose that there is a relation between the UPV and the compressive strength of earthen products. The UPV measurement was developed according to EN 12504-4:2007 [39] in two directions (direct and indirect—see Figure 3). The measure of UPV in the direct position was obtained with the transmitter and receiver transducers positioned on two opposite sides. The indirect measurements were done by placing the transmitter on one face and the receiver on a perpendicular side. An appropriate coupling gel was applied between the transducers and the sample to prevent the existence of voids in the contact area. Three independent readings were registered for each sample. Equation (1) is used to calculate the UPV, which is the ratio between the distance (*L*) between the transductors (emission and receptor) and the propagation time (*t*).

$$UPV\ (m/s) = \frac{L}{t} \tag{1}$$

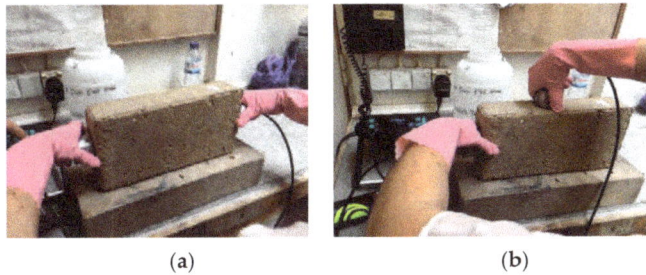

Figure 3. Ultrasonic Pulse Velocity (UPV) test—(**a**) direct method and (**b**) indirect method.

2.2.3. Compressive Strength

The compressive strength test used a hydraulic press machine with a capacity of 3000 kN (Figure 4), coupled with a hydraulic control system, according to NP EN 772-1 [40]. For the test, two transducers were used, one that belongs to the press and another external to measure the vertical displacement

(LVSTs). The test used displacement control, with a regular load velocity of 0.5 kN/s. The experiment consisted of applying an increasing compressive load until the load achieved 40% to 50% of the failure value after registering the maximum load peak. Six samples were tested to assess the compressive strength of CEBs.

Figure 4. Compressive strength test.

2.2.4. Total Water Absorption

The assessment of the total water absorption of a block is essential since it can be used for routine quality checks, classification according to required durability and structural use, and to estimate the volume of voids [4,41]. Usually, the less water a block absorbs and retains, the better its structural performance and durability. Reducing the total water absorption capacity of a block has often been considered as one way of improving its quality [41].

The total water absorption test consists of immersing a block in the water until no further increase in apparent mass is observed. This experiment followed the LNEC E 394:1993 standard [42]. It is considered that there was no increase in the apparent mass when two consecutive measurements did not differ by more than 0.1% by mass. The test was carried out at atmospheric pressure in which three samples were immersed in water for 1, 2, 3, 4, and 5 h. After each period, the surface of the specimen was wiped with a cloth to remove any adsorbed water. Then the samples were weighed. Initially, the test was done for 24 h, as recommended by the standard and observed in other studies [4,41]. However, after 24 h immersion in water, the CEBs disintegrated, and it was not possible to measure its wet weight. The percentage of water absorbed (A) was calculated using Equation (2). W_h is the weight of the specimen after each period of immersion, and W_s is the dry block weight.

$$A\ (\%) = \frac{(W_h - W_s)}{W_s} \qquad (2)$$

2.2.5. Water Absorption by Capillarity

This test consists of quantifying the amount of water absorbed by capillarity in the compressed earth block. The experiment was performed in three entire blocks following the LNEC E 393:1993 standard [43]. This is a Portuguese standard for analysing the water absorption in concrete, and it was used since the procedure is similar to the international standards specific for earthen products [10,40,44]. Before being immersed, each specimen dried for 14 days in an oven at a controlled temperature of 60 ± 5 °C. In the following step, each sample was weighed with a precision of 0.1 g, and then its lower face was immersed in a 5 mm water bath. Samples were left in the bath for 10, 20, 30, 45, 60, and 90 min, and 2, 3, 4, 6, 24, and 72 h, to identify the water saturation point (Figure 5). The water absorption by capillarity coefficient, C_b, was calculated for 10 min, according to UNE 41410:2008 [44] and using Equation (3).

$$C_b = \frac{100 \times (M_1 - M_0)}{S\sqrt{t}} \qquad (3)$$

where

C_b is the water absorption by capillarity coefficient (g/cm^2·min$^{0.5}$);
M_1 is the weight of the block after immersion in water (g);
M_0 is the weight of the block before immersion in water (g);
S is the immersed area (cm^2);
t is the immersion time (min).

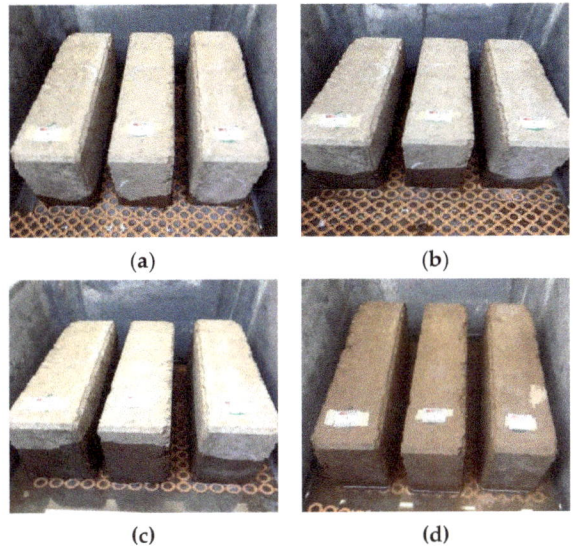

Figure 5. Water absorption by capillarity test: (**a**) after 10 min of water immersion; (**b**) after 30 min of water immersion; (**c**) after 45 min of water immersion; and (**d**) after 60 min of water immersion.

2.2.6. Accelerated Erosion

This test analyses the degradation process of a specimen caused by water falling on it. This experiment verifies the surface resistance to erosion, thus evaluating the durability of the analysed blocks. The test was performed according to NZS 4298:1998 [39], and a rain simulator was used (Figure 6). The climate parameters used were the ones for Penhas Douradas region, Guarda district since it is the Portuguese region with the highest precipitation values (1715 mm). In the experiment, a direct rainfall exposure index (worst scenario) was considered, which means that a flow rate of 14.26 L/min was used in the rainfall simulation. The outlet pressure in the water nozzle was 45 kPa, respecting the conditions recommended for erosion tests in the international standards.

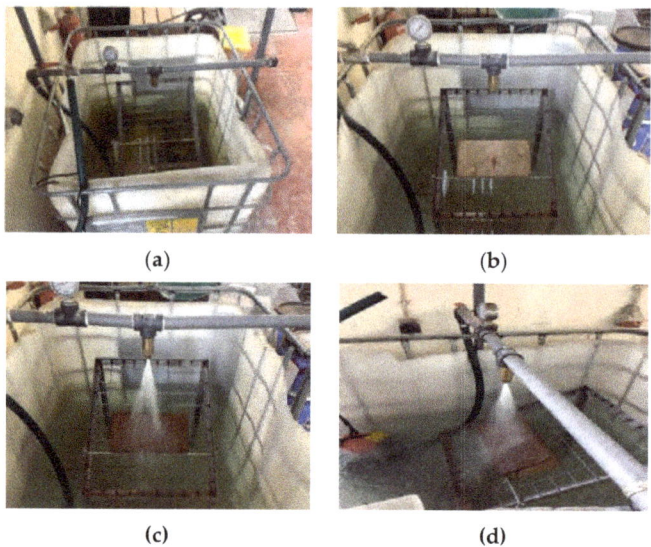

Figure 6. Accelerated erosion test: (**a**) setup; (**b**) sample preparation; (**c**) and (**d**) sample test.

2.2.7. Thermal Transmittance

The characterisation of the thermal properties of the CEBs is based on the analysis of the thermal transmittance (U-value) of the product. The thermal transmittance was measured using a guarded hot box set up apparatus, built for this study in the Department of Civil Engineering of the University of Minho, according to ASTM C1363-11:2011 [45] (Figure 7a). The hot box consists of two five-sided chambers (dimensions: 2.0 m × 1.4 m × 1.6 m), the cold and the hot one. The envelope is well insulated, made of extruded polystyrene (20 cm; U = 0.21 W/(m²·K)), to reduce the heat flux through the envelope and minimise heat losses by conduction. The specimen is placed in the mounting ring placed between the two chambers (Figure 7b). The setup is placed in an indoor environment with a controlled temperature below to the ones in the measurement chambers.

The thermal transmittance of the sample is obtained by measuring the heat flux rate needed to maintain the hot chamber at a steady temperature (in this study 35 ± 5 °C). Two ventilation devices were placed in the back wall of the cold chamber (Figure 7a), which allow the cold air to enter into the chamber and the hot air to exit. The ventilation is necessary to maintain uniform heat flux conditions through the specimen. In the hot chamber, there is a heating system, controlled by a temperature controller, that controls the defined temperature. The temperature in the chambers was measured by four thermocouples (two in each chamber—one in the middle of the chamber, and the other near the sample). A heat flux sensor was installed in the centre of the sample (Figure 7c). Preliminary calibration measurements were carried out successfully to evaluate the heat losses and the heat transfer through a wall with known thermal transmittance.

In this study, the heat flux method was used to determine the U-value. There is a heat flux through a material when there is a temperature difference between two sides. Heat flows from the warmer side to the colder side. It is possible to calculate the U-value of a specimen using the standardised methodology of ISO 9869-1:2014 [46] by assessing the heat flux together with the temperatures in both chambers. In this experiment, the greenTEG gSKIN® U-Value Kit (KIT-2615C) was used to automatically quantify the temperatures, the heat flux through the material and the U-value. The U-value is obtained from the average values of the heat flux through a small CEBs wall sample (composed of three blocks) and the temperature difference, ΔT, between the chambers, using Equation (4). In this experiment, the heat flux was assessed in two points of the CEBs wall.

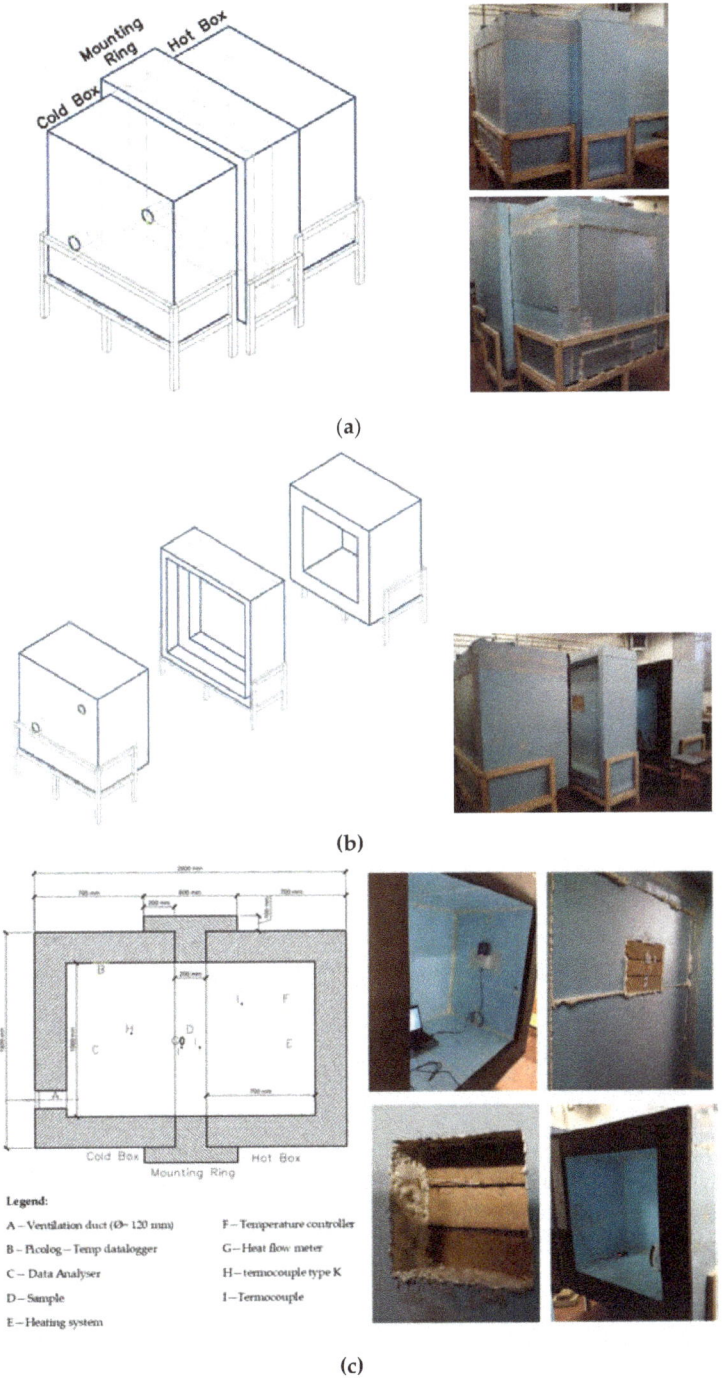

Figure 7. Hot box apparatus. Legend: (**a**) hotbox closed; (**b**) cold chamber, mounting ring, and hot chamber; (**c**) longitudinal plan view of the hotbox apparatus and position of the measurement equipment.

$$U - value = \frac{\sum_{j=1}^{n} \varphi_j}{\sum_{j=1}^{n} \Delta T_j} \left(W/(m^2 \cdot K) \right), \qquad (4)$$

where

n is the total number of data points;
φ is the heat flux in (W/m^2);
ΔT is the temperature (°C) difference between the two sides of the specimen.

3. Results

In this section, the mechanical, durability, and thermal characterisation of the CEBs will be described and discussed.

3.1. Electrical Resistivity

Electrical resistivity was measured to analyse the porosity of CEBs, and the results are presented in Figure 8. CEBs have electrical conductivity mainly because ions can propagate in their body. Electrical resistivity is directly dependent on CEBs permeability. In water-saturated CEBs with higher porosity, the propagation of ions is easier, and therefore, there is a lower electrical resistivity [47]. From Figure 8, it is possible to conclude that the measurements done in the direction of the bigger dimension (length) of the sample showed similar results for all samples. However, in the measurements carried out in the other direction (width), sample 2 presented higher values than samples 1 and 3. This result can be an indication that sample 2 was denser, with fewer pores or with pores with smaller dimensions (meaning reduced permeability and conductibility). These outputs disclose some disparity between the porosity of tested CEB samples.

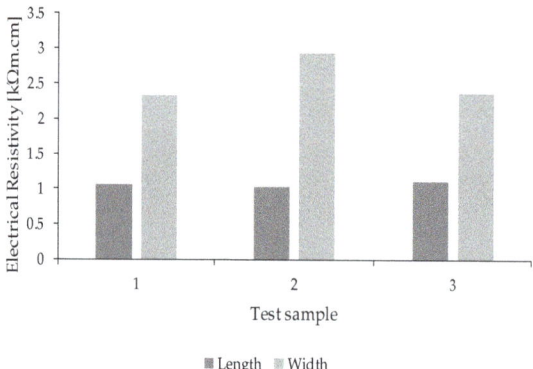

Figure 8. Results of the electrical resistivity tests of the three specimens carried out in the two directions.

3.2. Ultrasonic Pulse Velocity

Figure 9 presents the results for the UPV for each sample, using the same samples used in the electrical resistivity test. Five measurements were carried out for every sample (for each direction, direct and indirect) and results presented in Figure 9 are the average of the results obtained for each sample.

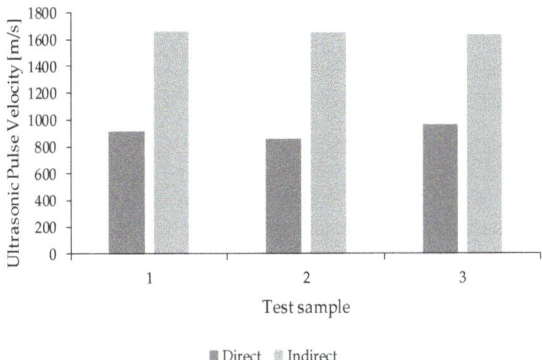

Figure 9. Ultrasonic Pulse Velocity measurements.

A sonic impulse propagates with lower velocity in a porous body and with higher velocity in a denser one. Therefore, according to the analysis of the results, it is possible to conclude that sample 2 presented slightly lower UPV, which can be an indication of a higher number of voids. These results are similar to the electrical resistivity test results. The sample performance differences could be due to the incorrect homogenisation of the soil mixture and/or cracking.

3.3. Compressive Strength

The compressive strength test is considered a reference test for CEBs since it is regarded as an essential indicator of masonry strength. Figure 10 presents the results obtained for the compressive strength of six samples, as well as the average value obtained for this parameter. The results showed a variation on compressive strength between 7.8 MPa and 11.0 MPa, being the average 9.0 ± 1.3 MPa. These values are very good ones since it is known that the minimum compressive strength requirements for CEBs, varying between 1.0 MPa and 2.8 MPa [7,21]. These higher values can be related to the compaction process used since compacting the soil using a press improves the quality of the material. The higher density obtained by compaction significantly increases the compressive strength of the blocks [48]. Another reason for the higher compressive strength is the presence of lime in the mixture. Lime allows the development of calcium silicate hydrate (CSH) together with the formation of minor amounts of calcite, which causes increased strength [21].

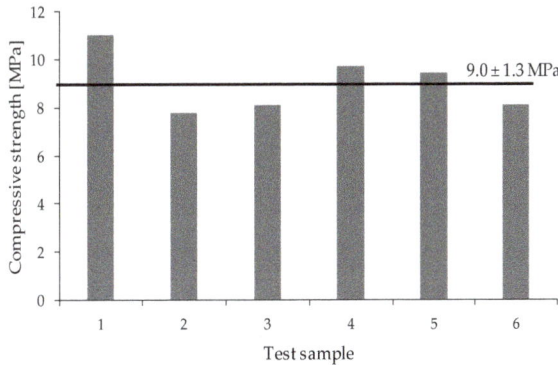

Figure 10. Compressive strength results.

3.4. Total Water Absorption

Analysing Figure 11, the maximum values obtained for the total water absorption for each sample were 8.7%, 11.3%, and 10.0% for samples 1, 2, and 3, respectively. It is possible to conclude that the total water absorption varied between 8.7 and 11.3%, being these values favourable when compared with clay bricks (0–30%), concrete blocks (4–25%), or calcium silicate bricks (6–16%) [4]. Although this result seems good, the fast absorption and desegregation of blocks can influence the durability negatively. Total water absorption is influenced by the granulometry of the soil and compaction pressure. These two aspects have a meaningful impact in the density, mechanical strength, compressibility, permeability, and porosity of CEBs [4,48].

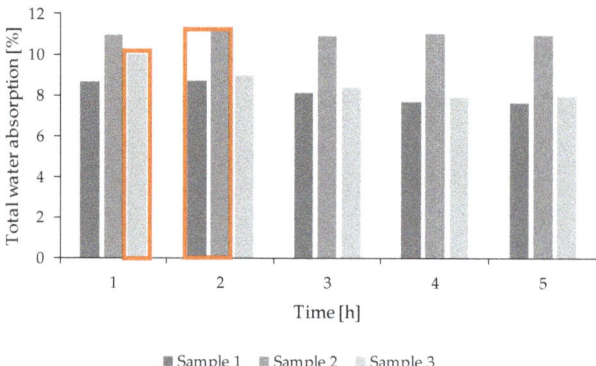

Figure 11. Total water absorption results. The results highlighted by the orange box represents the maximum values obtained in the total water absorption test.

3.5. Water Absorption by Capillarity

CEBs used in a structure may undergo alternating phenomena of absorption and release of water, mainly because of the capillarity effect [49]. The curves shown in Figure 12 illustrate the variation of water absorption by capillarity until the total saturation of each sample is reached, as a function of the square root of test time. Figure 13 shows the variation of the water absorption coefficient (average of the three samples tested) of compressed earth blocks as a function of the square root of test time. This coefficient was determined for all test periods. However, the literature mentions that the value at the end of 10 min (represented with an orange rectangle in Figure 13) is representative of the behaviour of masonry exposed to a violent storm [21]. At the end of this period, the water absorption coefficient value was 34.6, and therefore, the blocks are classified as having high capillarity [21]. This result can be related to the fact that the soils used to manufacture the CEBs have a high percentage of sand (Table 1). In this study, the high presence of bigger particles (in terms of size) in CEBs seems to be an issue related with the quality of the particle size distribution of the soil, used for the CEBs production, than the quality of the soil itself, which did not cause good reaction with lime and resulted in CEBs with high porosity [10].

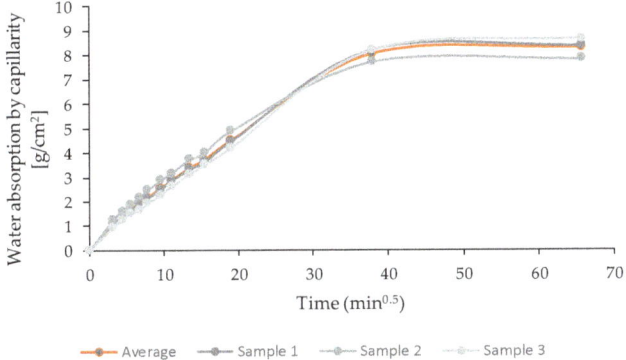

Figure 12. Total water absorption results.

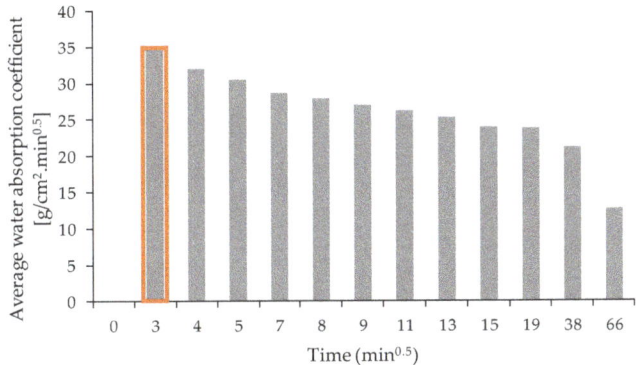

Figure 13. Variation of the average water absorption during the test period.

3.6. Accelerated Erosion

The degradation analysis of CEBs was done, and the results are presented in Figures 14–16. A total of seven blocks were analysed in this test. Initially, three blocks were tested for one hour. Since blocks did not present any type of erosion, this test was repeated in four additional blocks, and the results were the same. Based on these results and according to NP EN 12504-4 [39], CEBs were classified with an erosion index of 1 (erosion depth between 0–20 mm/h), which means that they had very good results in terms of durability.

To understand how these blocks behave if exposed for more time to the erosion test and if there is any relation with the other physical properties mentioned before, the analysis was extended for one additional hour.

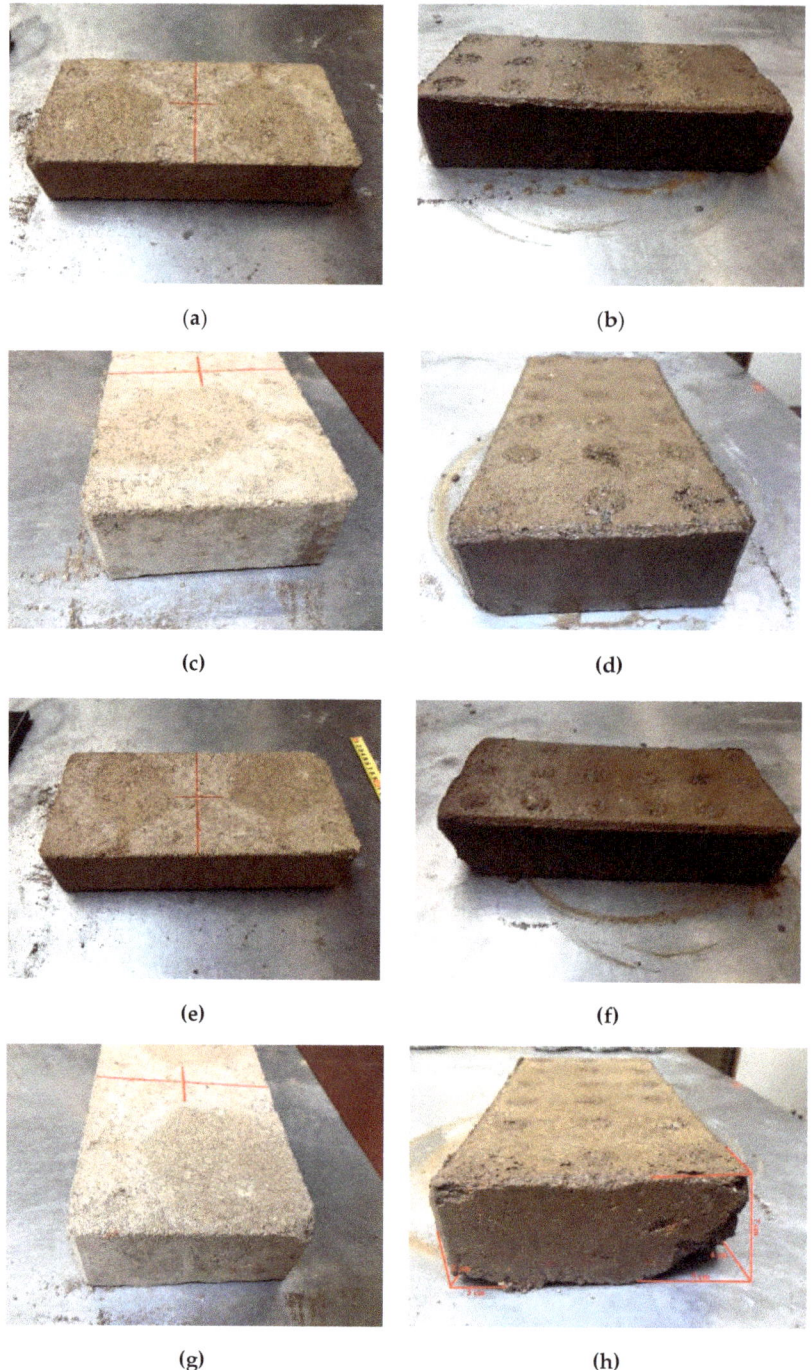

Figure 14. Accelerated erosion test. (**a,c,e,g**)—four sides of the exposed face of sample 1 before the test; (**b,d,f,h**)—four sides of the exposed face of sample 1 after two hours of testing.

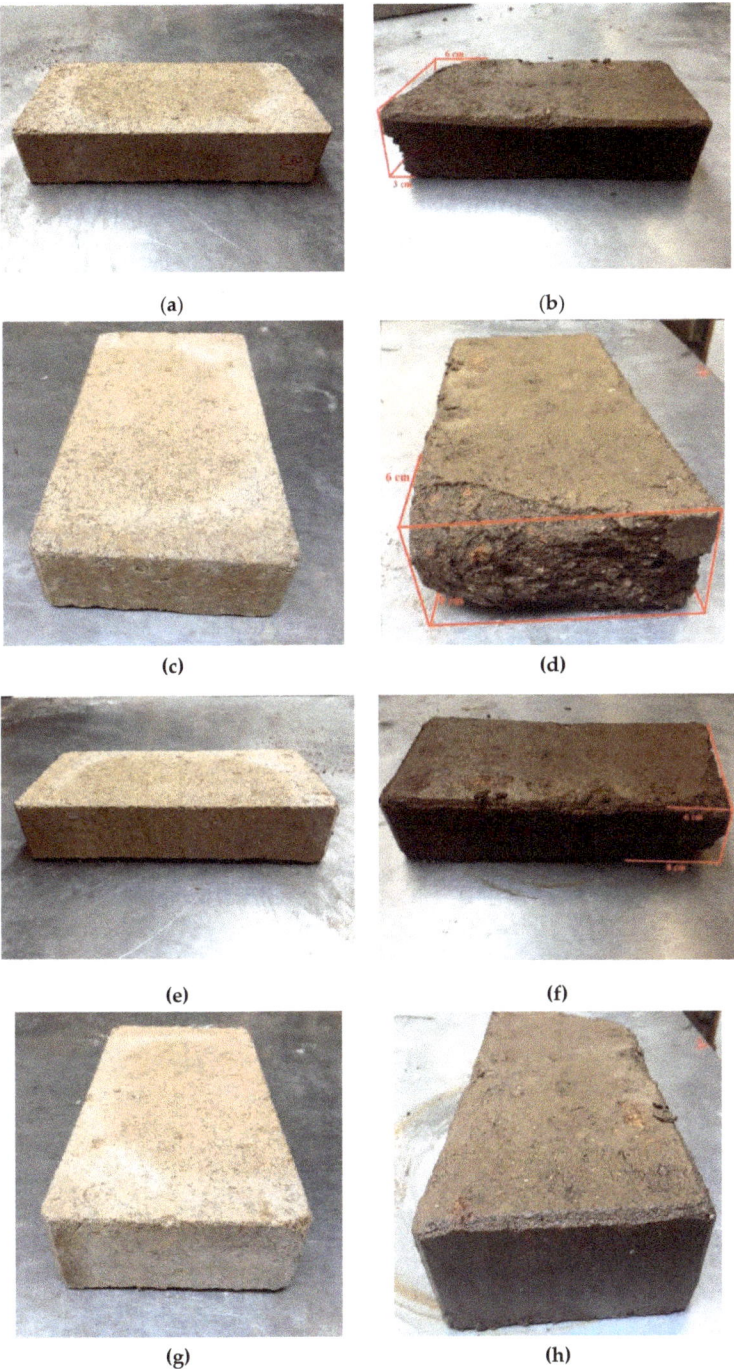

Figure 15. Accelerated erosion test. (**a,c,e,g**)—four sides of the exposed face of sample 3 before the test; (**b,d,f,h**)—four sides of exposure face of sample 3 after two hours testing.

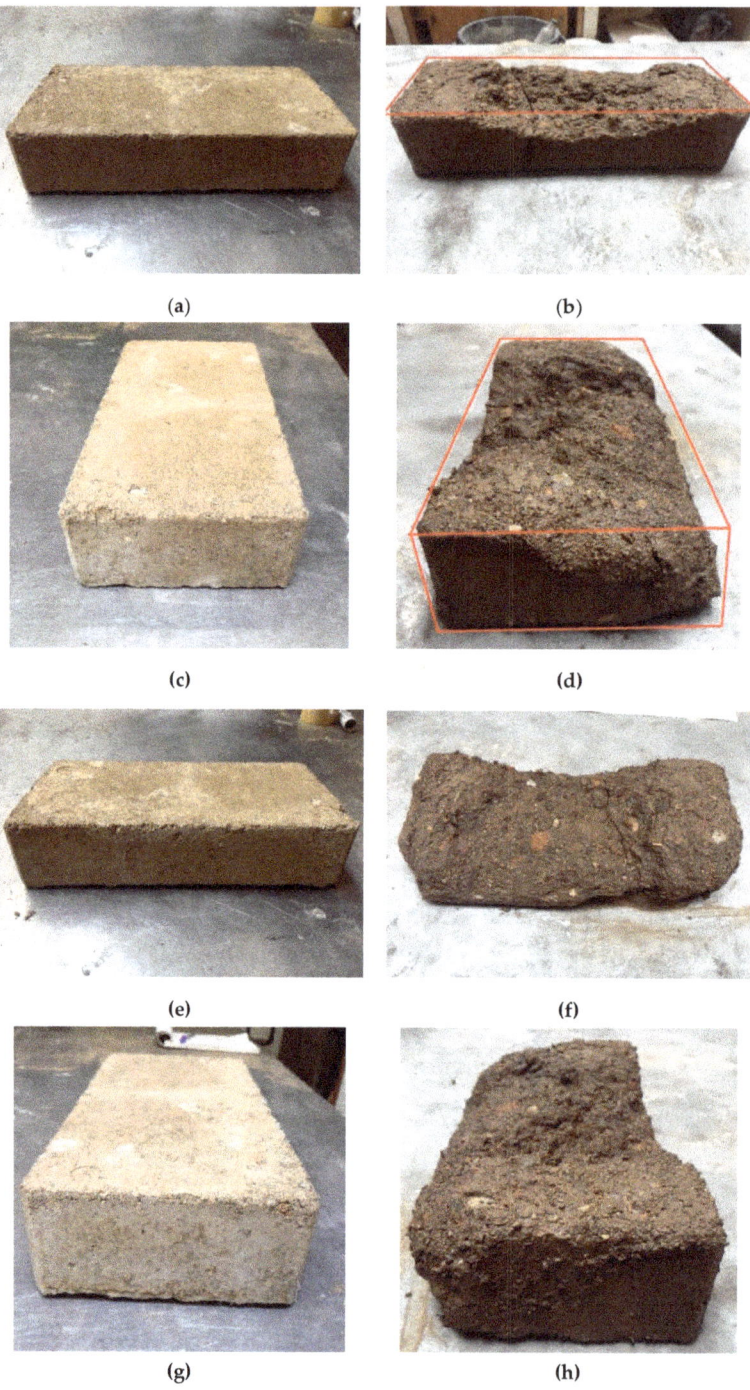

Figure 16. Accelerated erosion test. (**a**,**c**,**e**,**g**)—four sides of the exposed face of sample 6 before the test; (**b**,**d**,**f**,**h**)—four sides of the exposed face of sample 6 after testing.

After two hours of water exposition, CEBs presented different behaviour. Sample 1 did not show significant damage in the majority of faces exposed to water but presented a loss in one part of the block where there was already a small defect before the test (Figure 14). Sample 2 showed similar behaviour to sample 1. Samples 3 (Figure 15), 4 and 5 presented some damage in the two smaller faces, even though the face in contact with the water was the one with a larger area (highlighted with a red cross in Figure 14). Samples 6 (Figure 16) and 7 suffered significant damage, and at the end of two hours, they were almost destroyed. It is then essential to analyse the reasoning behind the different deterioration levels of CEBs since they were manufactured using the same soil, mixture, and compaction process. The most probable explanation for the differences is the inadequate particle size distribution in the soil used. According to the literature in the field of earthen blocks, the soil should only present particles below 5 mm of diameter [48]. Nevertheless, it was possible to see particles with higher dimensions in the most damaged samples (Figure 16d,f,h). These big size particles affected the homogeneity of the mixture, which negatively influenced the porosity and the porous structure of the CEBs. The lack of uniformity in the structure of the blocks worsens their behaviour to water. This problem also explains the results achieved in the total water absorption and water absorption by capillarity tests. The presented porosity and durability issues can be minimised if proper soil preparation and/or selection is considered [18].

3.7. Thermal Transmittance

Regarding the thermal transmittance, Figure 17 shows the measurement results for the XPS wall (Figure 17a) and the compressed earth blocks small wall (Figure 17b,c). From the analysis of Figure 17, it is possible to verify that the values measured for temperature and the heat flux were very stable during the test period, in both cases. The temperature in the hot chamber practically did not change since the heat input to the box was controlled so that the temperature established was maintained (35 °C). The heating system turns on when the temperature drops below 35 °C and turns off when the temperature rises to 40 °C. The heat flux is presented as negative values due to heat flux sensor placement in the sample.

The results of the preliminary calibration measurements carried out, using a reference sample with known thermal transmittance (XPS 20 cm), showed an agreement with the technical data provided by the manufacturer (U-value of 0.21 W/(m^2·K) and thermal conductivity of 0.60 W/(m·K). The measured thermal transmittance of the 15 cm CEBs wall was of 2.65 ± 0.16 W/(m^2·K) (thermal resistance of 0.21 (m^2·K)/W on average).

The results indicate that the thermal conductivity of the CEBs studied is significantly lower than the reference values (1.1 W/(m·K)) listed by the Portuguese National Laboratory of Civil Engineering (LNEC) [25]. Considering the thickness of the CEBs wall analysed (15 cm), the thermal transmittance of the CEB wall would be approximately 3.26 W/(m^2·K) (thermal resistance of 0.31 (m^2·K)/W)) (for external walls). The thermal transmittance measured for the CEB wall sample was lower than the value obtained from technical data of LNEC. This result can be explained due to a higher porosity of these blocks, which increased their thermal resistance. This result is in accordance with the other analysis made in those blocks; as was seen in the accelerated erosion test, the CEBs presented a different soil composition, showing in some samples soil particles with a size larger than 5 mm, which leads to a higher thermal conductivity (Figure 16). Moreover, the results could be better if the granulometry of the soil was optimised, since in the experiments found soil particles larger than 5 mm and some small rocks that increase thermal transmittance.

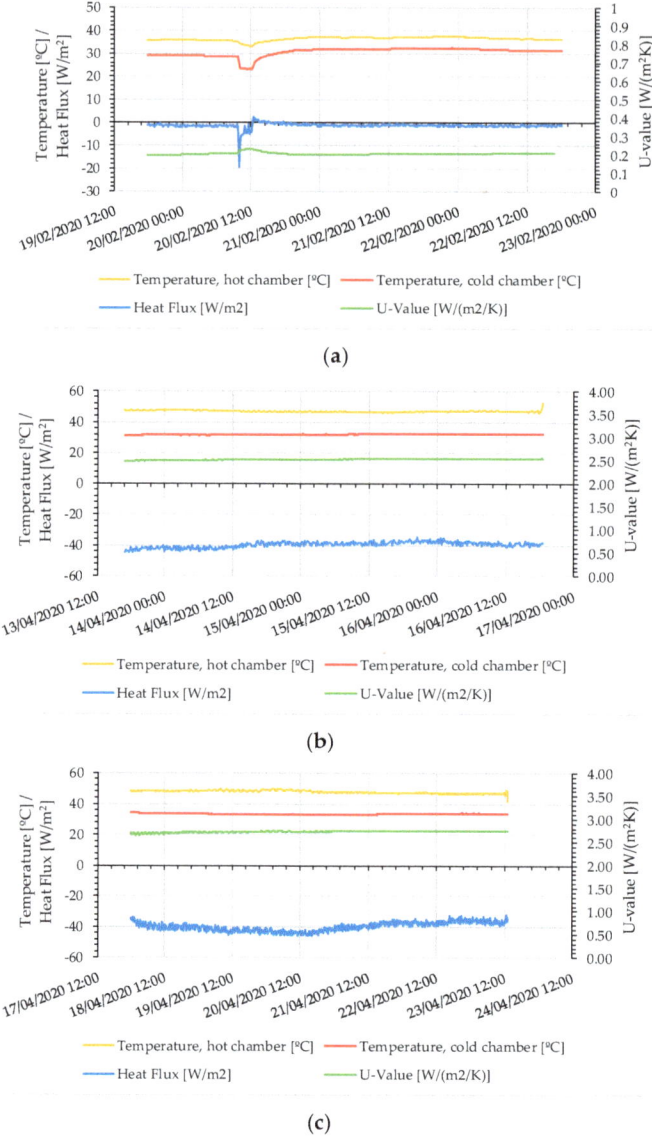

Figure 17. Measurements of the heat flux, hot and cold superficial temperatures and U-value in the XPS wall (**a**) and in the small CEBs wall (**b**,**c**).

4. Discussion

The compressed earth blocks characterisation is summarised in Table 2. Overall, the results are consistent and show that these blocks presented good mechanical and durability properties and better thermal performance than the reference values listed in the technical data for Portugal, but they also present porosity issues. The non-destructive (electrical resistivity and ultrasonic pulse velocity) and the destructive (water absorption) tests for porosity analysis showed similar results. CEBs samples presented heterogeneity on their mixture composition, which led to the production of blocks with different porosities. The high values obtained in the water absorption test highlight that this is the most problematic characteristic of the CEBs. Contrary, to the other results, the compressive strength analysis and the accelerated erosion test presented significant results. The compressive strength obtained was approximately three times higher than the minimum requirement for CEBs. Moreover, these CEBs were classified as erosion index of 1, which means that they have an erosion depth between 0–20 mm/h, are very resistant, and have good durability properties.

Table 2. Thermophysical proprieties of the analysed compressed earth blocks.

Property		Average	Standard Deviation	Coefficient of Variance (%)
Electrical resistivity (kΩm·cm)	Length	1.07	0.07	6.63
	Width	2.54	0.30	11.81
Ultrasonic pulse velocity (m/s)	Direct	908.34	51.97	5.72
	Indirect	1647.21	12.32	0.75
Compressive strength (MPa)		9.01	1.25	13.90
Total water absorption (%)		9.98	1.29	12.96
Water absorption by capillarity (g/cm^2·min$^{0.5}$)		34.62	5.13	14.82
Heat Transfer Coefficient (W/m^2·K)		2.65	0.15	5.59
Thermal resistance (m^2·K/W)		0.21	-	-
Thermal conductivity (W/m·K)		0.60	-	-

One of the essential features of these building elements is thermal resistance. In this study, thermal transmittance was analysed, corresponding to a U-value of 2.66 W/(m^2·K). The results seem to be in accordance with the results obtained for the quality and durability parameters, showing that the results could be better if the granulometry of the soil was optimised, since in the experiments found soil particles larger than 5 mm and some small rocks that increase thermal transmittance. Therefore, the optimisation of the soil particle size distribution in the mixture before CEB production is necessary to increase thermal performance while maintaining high mechanical resistance.

5. Conclusions

The results presented in this study show a strong relationship between soil and mixture preparation and compaction with CEB properties. During the investigation, it was possible to observe that the use of soil with particles with higher dimensions than the ones recommended by the international standards for CEB production had a significant effect in some of CEBs properties, such as: porosity, water absorption, durability and thermal performance. However, even though these blocks did not have the proper production, they did not present mechanical resistance issues. In general, the analysed CEBs are adequate to be used for construction of partition walls. Moreover, it was seen in the literature that these blocks presented several benefits in terms of environmental performance. Taking this into account, optimising the soil particle size distribution in the mixture before CEB production could be a solution to minimise these issues (mainly in thermal performance) while maintaining high mechanical resistance. The optimisation of the distribution will lead to a production of elements with lower porosity, and it is known that porosity has a direct relation to mechanical resistance and durability. However, by reducing porosity, it is expected that thermal transmittance increases. The study of CEB optimisation and the characterisation of the optimised CEB properties will be studied in the future in order to assess the real contribution of this optimisation in CEB mechanical, durability, and thermal performance. In short, optimisation of compressed earth blocks is necessary to improve their functional quality and increase their potential for use in the construction of buildings.

Author Contributions: Conceptualization, E.R.T., J.F., and R.M.; methodology, E.R.T., S.M.S., and R.M.; validation, S.M.S. and R.M.; formal analysis, E.R.T. and J.F.; investigation, E.R.T., G.M., A.d.P.J., and C.G.; resources, R.M.; writing—original draft preparation, E.R.T.; writing—review and editing, E.R.T., A.d.P.J., C.G., J.F., S.M.S., and R.M.; supervision, S.M.S. and R.M.; project administration, R.M.; funding acquisition, R.M. All authors have read and agreed to the published version of the manuscript.

Funding: The authors would like to acknowledge the support granted by the FEDER funds through the Competitively and Internationalization Operational Programme (POCI) and by national funds through FCT (the Foundation for Science and Technology) within the scope of the project with the reference POCI-01-0145-FEDER-029328, and of the Ph.D. grant with the reference PD/BD/113641/2015, which were fundamental for the development of this study.

Acknowledgments: The authors would like to acknowledge the support granted by DANOSA "*Derivados asfálticos normalizados, S.A.*" industry for the hotbox construction by providing all the necessary insulation material.

Conflicts of Interest: The authors declare no conflict of interest.

References

1. Minke, G. *Building with Earth: Design and Technology of a Sustainable Architecture*; Birkhäuser: Basel, Switzerland, 2012; ISBN 9783034608725.
2. Fernandes, J.; Peixoto, M.; Mateus, R.; Gervásio, H. Life cycle analysis of environmental impacts of earthen materials in the Portuguese context: Rammed earth and compressed earth blocks. *J. Clean. Prod.* **2019**, *241*, 118286. [CrossRef]
3. Pacheco-Torgal, F.; Jalali, S. Earth construction: Lessons from the past for future eco-efficient construction. *Constr. Build. Mater.* **2012**, *29*, 512–519. [CrossRef]
4. Taallah, B.; Guettala, A.; Guettala, S.; Kriker, A. Mechanical properties and hygroscopicity behavior of compressed earth block filled by date palm fibers. *Constr. Build. Mater.* **2014**, *59*, 161–168. [CrossRef]
5. Leitão, D.; Barbosa, J.; Soares, E.; Miranda, T.; Cristelo, N.; Briga-Sá, A. Thermal performance assessment of masonry made of ICEB's stabilised with alkali-activated fly ash. *Energy Build.* **2017**, *139*, 44–52. [CrossRef]
6. Ruiz, G.; Zhang, X.; Edris, W.F.; Cañas, I.; Garijo, L. A comprehensive study of mechanical properties of compressed earth blocks. *Constr. Build. Mater.* **2018**, *176*, 566–572. [CrossRef]
7. Mansour, M.B.; Jelidi, A.; Cherif, A.S.; Jabrallah, S.B. Optimizing thermal and mechanical performance of compressed earth blocks (CEB). *Constr. Build. Mater.* **2016**, *104*, 44–51. [CrossRef]
8. Saidi, M.; Cherif, A.S.; Zeghmati, B.; Sediki, E. Stabilization effects on the thermal conductivity and sorption behavior of earth bricks. *Constr. Build. Mater.* **2018**, *167*, 566–577. [CrossRef]
9. Chaibeddra, S.; Kharchi, F. Performance of Compressed Stabilized Earth Blocks in sulphated medium. *J. Build. Eng.* **2019**, *25*, 100814. [CrossRef]
10. Mahdad, M.; Benidir, A. Hydro-mechanical properties and durability of earth blocks: Influence of different stabilisers and compaction levels. *Int. J. Sustain. Build. Technol. Urban. Dev.* **2018**, *9*, 44–60.
11. Zhang, L.; Gustavsen, A.; Jelle, B.P.; Yang, L.; Gao, T.; Wang, Y. Thermal conductivity of cement stabilized earth blocks. *Constr. Build. Mater.* **2017**, *151*, 504–511. [CrossRef]
12. Lavie Arsène, M.I.; Frédéric, C.; Nathalie, F. Improvement of lifetime of compressed earth blocks by adding limestone, sandstone and porphyry aggregates. *J. Build. Eng.* **2020**, *29*, 101155. [CrossRef]
13. Islam, M.S.; Elahi, T.E.; Shahriar, A.R.; Mumtaz, N. Effectiveness of fly ash and cement for compressed stabilized earth block construction. *Constr. Build. Mater.* **2020**, *255*, 119392. [CrossRef]
14. Narayanaswamy, A.H.; Walker, P.; Venkatarama Reddy, B.V.; Heath, A.; Maskell, D. Mechanical and thermal properties, and comparative life-cycle impacts, of stabilised earth building products. *Constr. Build. Mater.* **2020**, *243*, 118096. [CrossRef]
15. Bahar, R.; Benazzoug, M.; Kenai, S. Performance of compacted cement-stabilised soil. *Cem. Concr. Compos.* **2004**, *26*, 811–820. [CrossRef]
16. Al-Amoudi, O.S.B.; Khan, K.; Al-Kahtani, N.S. Stabilization of a Saudi calcareous marl soil. *Constr. Build. Mater.* **2010**, *24*, 1848–1854. [CrossRef]
17. Guettala, A.; Abibsi, A.; Houari, H. Durability study of stabilized earth concrete under both laboratory and climatic conditions exposure. *Constr. Build. Mater.* **2006**, *20*, 119–127. [CrossRef]
18. Kariyawasam, K.K.G.K.D.; Jayasinghe, C. Cement stabilized rammed earth as a sustainable construction material. *Constr. Build. Mater.* **2016**, *105*, 519–527. [CrossRef]

19. Toure, P.M.; Sambou, V.; Faye, M.; Thiam, A. Mechanical and thermal characterization of stabilized earth bricks. *Energy Procedia* **2017**, *139*, 676–681. [CrossRef]
20. Gouny, F.; Fouchal, F.; Pop, O.; Maillard, P.; Rossignol, S. Mechanical behavior of an assembly of wood-geopolymer-earth bricks. *Constr. Build. Mater.* **2013**, *38*, 110–118. [CrossRef]
21. Taallah, B.; Guettala, A. The mechanical and physical properties of compressed earth block stabilized with lime and filled with untreated and alkali-treated date palm fibers. *Constr. Build. Mater.* **2016**, *104*, 52–62. [CrossRef]
22. Guettala, A.; Houari, H.; Mezghiche, B.; Chebili, R. Durability of lime stabilized earth blocks. *Courr. Savoir* **2002**, *2*, 61–66.
23. Fernandes, J.; Mateus, R.; Gervásio, H.; Silva, S.M.; Bragança, L. Passive strategies used in Southern Portugal vernacular rammed earth buildings and their influence in thermal performance. *Renew. Energy* **2019**, *142*, 345–363. [CrossRef]
24. Adam, E.A.; Jones, P.J. Thermophysical properties of stabilised soil building blocks. *Build. Environ.* **1995**, *30*, 245–253. [CrossRef]
25. Pina dos Santos, C.A. *ITE50—Coeficientes de Transmissão Térmica de Elementos da Envolvente dos Edifícios*; National Laboratory of Civil Engineering: Lisboa, Portugal, 2006.
26. Berge, B. *The Ecology of Building Materials*, 2nd ed.; Elsevier: Amsterdam, The Netherlands, 2009; ISBN 9781856175371.
27. LNEC. *196: Solos - Análise Granulométrica*; Especificação do Laboratório Nacional de Engenharia Civil: Lisboa, Portugal, 1966; Volume 10.
28. IPQ. EN, NP. 933-8, Norma Portuguesa (Setembro de 2002). In *Ensaios das Propriedades Geométricas dos Agregados: Determinação do Teor de Finos - Ensaio do Equivalente de Areia*; IPQ: Lisboa, Portugal, 2002.
29. IPQ. EN, NP. 933-9. In *Ensaios das Propriedades Geométricas dos Agregados Parte 9: Determinação de Finos - Ensaio do Azul de Metileno*; IPQ: Lisboa, Portugal, 2002.
30. IPQ. NP, NP. 143: 1969. In *Determinação dos Limites de Consistência*; IPQ: Lisboa, Portugal, 1969.
31. LNEC. *197: Solo-Cimento. Ensaio de Compactação*; Especificação do Laboratório Nacional de Engenharia Civil: Lisboa, Portugal, 1966; Volume xiii.
32. Standard A.S.T.M. *D3282 (2009) Standard Practice for Classification of Soils and Soil-Aggregate Mixtures for Highway Construction Purposes*; ASTM International: West Conshohocken, PA, USA, 2004. [CrossRef]
33. Doat, P.; Hays, A.; Houben, H.; Matuk, S.; Vitoux, F.; CraTerre, F. *Building with Earth*, 1st ed.; The Mud Village Society: New Delhi, India, 1991.
34. Arab, P.B.; Araújo, T.P.; Pejon, O.J. Identification of clay minerals in mixtures subjected to differential thermal and thermogravimetry analyses and methylene blue adsorption tests. *Appl. Clay Sci.* **2015**, *114*, 133–140. [CrossRef]
35. Yukselen, Y.; Kaya, A. Suitability of the methylene blue test for surface area, cation exchange capacity and swell potential determination of clayey soils. *Eng. Geol.* **2008**, *102*, 38–45. [CrossRef]
36. Duarte Gomes, N. Caracterização de Blocos de Terra para Construção de Alvenarias Ecoeficientes. Master's Thesis, University Nova, Lisbon, Portugal, 2015.
37. Canivell, J.; Martin-del-Rio, J.J.; Alejandre, F.J.; García-Heras, J.; Jimenez-Aguilar, A. Considerations on the physical and mechanical properties of lime-stabilized rammed earth walls and their evaluation by ultrasonic pulse velocity testing. *Constr. Build. Mater.* **2018**, *191*, 826–836. [CrossRef]
38. Martín-del-Rio, J.J.; Canivell, J.; Falcón, R.M. The use of non-destructive testing to evaluate the compressive strength of a lime-stabilised rammed-earth wall: Rebound index and ultrasonic pulse velocity. *Constr. Build. Mater.* **2020**, *242*, 118060. [CrossRef]
39. IPQ. *EN, NP. 12504-4. 2007, Ensaios de Betão nas estruturas—Parte 4: Determinação da velocidade de propagação dos ultra-sons*; IPQ: Lisboa, Portugal, 2009.
40. IPQ. *EN, NP. 772-1. 2002, Methods of Test for Masonry Units—Part 1: Determination of Compressive Strength*; IPQ: Lisboa, Portugal, 2002.
41. Kerali, A.G. Durability of Compressed and Cement-Stabilised Buildings Blocks. Ph.D. Thesis, University of Warwick, Coventry, UK, 2001.
42. LNEC. *394, Betões, Determinação da Absorção de Água por Imersão, Ensaio à Pressão Atmosférica*; Laboratório Engenharia Civil: Lisboa, Portugal, 1993.

43. LNEC. *393, Concrete—Determination of the Absorption of Water through Capillarity*; Test at environment pressure (in Portuguese); Laboratório Nacional de Engenharia Civil: Lisboa, Portugal, 1993.
44. UNE 41410. Bloques de tierra comprimida para muros y tabiques Definiciones, especificaciones y métodos de ensayo. *Asoc. Española Norm. Y Certif. —Aenor* **2008**, 26.
45. Standard A.S.T.M. *C1363-11, Standard Test Method for Thermal Performance of Building Materials and Envelope Assemblies by Means of a Hot Box Apparatus*; ASTM International: West Conshohocken, PA, USA, 2011.
46. BSI. *BS ISO 9869-1: 2014: Thermal Insulation–Building Elements–in situ Measurement of Thermal Resistance and Thermal Transmittance—Part 1: Heat Flow Meter Method*; BSI: London, UK, 2014.
47. Motahari Karein, S.M.; Ramezanianpour, A.A.; Ebadi, T.; Isapour, S.; Karakouzian, M. A new approach for application of silica fume in concrete: Wet granulation. *Constr. Build. Mater.* **2017**, *157*, 573–581. [CrossRef]
48. Rigassi, V.; CRATerre-EAG. *Compressed Earth Blocks: Manual of Production*; Deutsches Zentrum für Entwicklungstechnologien—GATE: Eschborn, Germany, 1985; Volume I, ISBN 3528020792.
49. Adam, E.A.; Agib, A.R.A. *Compressed Stabilised Earth Block Manufacture in Sudan*; Graphoprint for UNESCO: Paris, France, 2001.

© 2020 by the authors. Licensee MDPI, Basel, Switzerland. This article is an open access article distributed under the terms and conditions of the Creative Commons Attribution (CC BY) license (http://creativecommons.org/licenses/by/4.0/).

Article

Life-Cycle Assessment of Alternative Envelope Construction for a New House in South-Western Europe: Embodied and Operational Magnitude

Helena Monteiro [1,2,*], **Fausto Freire** [2] and **John E. Fernández** [3,4]

1. Low Carbon & Resource Efficiency, R&Di, Instituto de Soldadura e Qualidade, 4415-491 Grijó, Portugal
2. ADAI, Department of Mechanical Engineering, University of Coimbra, Rua Luís Reis Santos, 3030-788 Coimbra, Portugal; fausto.freire@dem.uc.pt
3. Building Technology Program, Department of Architecture, Massachusetts Institute of Technology, 77 Massachusetts Avenue, Cambridge, MA 02139, USA; fernande@mit.edu
4. Environmental Solutions Initiative, Massachusetts Institute of Technology, Cambridge, MA 02139, USA
* Correspondence: himonteiro@isq.pt; Tel.: +351-963-378-570

Received: 30 June 2020; Accepted: 6 August 2020; Published: 11 August 2020

Abstract: The building envelope is critical to reducing operational energy in residential buildings. Under moderate climates, as in South-Western Europe (Portugal), thermal operational energy may be substantially reduced with an adequate building envelope selection at the design stage; therefore, it is crucial to assess the trade-offs between operational and embodied impacts. In this work, the environmental influence of building envelope construction with varying thermal performance were assessed for a South-Western European house under two operational patterns using life-cycle assessment (LCA) methodology. Five insulation thickness levels (0–12 cm), four total ventilation levels (0.3–1.2 ac/h), three exterior wall alternatives (double brick, concrete, and wood walls), and six insulation materials were studied. Insulation thickness tipping-points were identified for alternative operational patterns and wall envelopes, considering six environmental impact categories. Life-cycle results show that, under a South-Western European climate, the embodied impacts represent twice the operational impact of a new Portuguese house. Insulation played an important role. However, increasing it beyond the tipping-point is counterproductive. Lowering ventilation levels and adopting wood walls reduced the house life-cycle impacts. Cork was the insulation material with the lowest impact. Thus, under a moderate climate, priority should be given to using LCA to select envelope solutions.

Keywords: LCA; environmental impact; house; building envelope; thermal performance

---

1. Introduction

Households represent around 27% of the European Union's (EU) final energy consumption. To address this, EU regulatory efforts have been enacted to promote energy efficiency, and the new EU Green Deal roadmap aims to encourage that EU building stock (new and existing buildings) become energy and resource efficient. To support new building developments, a life-cycle perspective is recommended since reducing operational energy through improved building envelopes is likely to affect the impact of other life-cycle phases of new buildings.

Life-cycle assessment (LCA) has been extensively used to study residential buildings [1,2], building options [3,4], and building construction [5]; however, most studies have focused on primary energy and/or CO_2 emissions, disregarding other environmental impacts. Review articles on LCA of buildings [6,7] agree that comparing different studies is not linear because building characteristics (size, shape, construction, and occupation) vary with location and climate, and the studied methodological assumptions (functional unit, lifespan, life-cycle impacts, exclusions) widely vary [8,9]; although some trends can be climate specific, each study helps to explain a climatic and regional context.

Studies covering cold climate houses in developed countries have concluded that the operational phase has a preponderant weight in the total life-cycle of the building [1,10]. Moreover, studies of conventional buildings in different countries (Sweden [11], Kazakhstan [12], Alaska, USA [13], Spain [14], Portugal [15]) have showed that operational energy is dominant, representing 60–90% of the total environmental impacts. Thus, reducing heating and cooling is essential. Interestingly, a study that provided an LCA benchmark for dwellings in North Italy and Denmark [16] showed that, in North Italian case studies, operational impacts accounted for 69–76%, and embodied impacts accounted for 24–31% of the overall impact, whereas in Danish cases, the impacts per life-cycle phase are reversed due to the low impact of the future Danish energy grid. This shows that life-cycle results are also highly sensitive to specific regional conditions other than climate, such as the energy mix.

Dylewski [17] studied the environmental impact of diverse thermal insulation materials for exterior walls in Poland, considering alternative heat sources, in order to find the optimal insulation thickness considering both economic and ecological net present value of insulation (as an investment). Results showed that significantly higher thicknesses were recommended when considering environmental data as compared to economic data—for instance, 0.46–0.52 m of expanded polystyrene (EPS) insulation considering brick walls and a heat pump system. Some LCA studies assessed low energy and passive houses [18–20], in which operational energy is substantially reduced. Generally, when operational energy is reduced, the relative contribution of embodied impact rises and therefore a life-cycle perspective is essential [21].

In low energy houses, embodied energy can amount to 50–70% of the life-cycle energy [22], and the building envelope is accountable for a significant share of embodied impacts. Consequently, alternative building options must be carefully assessed in new dwellings and, again, a life-cycle assessment study to support decision-making at an early building design stage is desirable.

Some studies covered South European dwellings [14,21,23–27]. However, a trend regarding which life-cycle phase has the most impact in new houses located in South Europe under a mild climate was not determined. Embodied and operational impacts are both significant, but their life-cycle contributions appear to be highly sensitive to construction options, energy systems, operation/occupation behavior, and regional aspects. The electric production mix (share of renewable) is essential to characterize the environmental impacts of the use phase [14,28,29]. Additionally, operational heating and cooling behavior can significantly affect a study outcome [23,30].

Thus, the prevailing strategies for cold climate houses should not be directly transposed onto other building or climatic contexts [31] because, depending on the local context, the embodied impact may surpass the operational impact. Studying alternative passive architecture measures and their influence on operational energy of buildings in Spain, a recent study [32] concluded that, for some climate regions, a few passive strategies could reduce operational energy to the passive house level: north–south orientation, small window-to-wall ratio (<20%), insulated envelope (U = 0.35 W/m^{2}K). Nevertheless, the authors recognize that user behavior remains unaddressed. Furthermore, as a life-cycle perspective was not considered, the embodied energy of the building measures was not assessed.

In South-Western Europe, many houses are exposed to a moderate Mediterranean warm climate, and interior comfort (operational patterns) may be dependent on user behavior (influenced by cultural heating habits and economic constraints). Thus, typical operational energy levels of these houses are lower than in most North and Central European countries [28]. According to Lavagna et al. [10], a considerable part of the environmental life-cycle impacts of EU building stock is associated with single family houses located in moderate climates, and new houses in this climate have not been widely assessed considering user behavior.

Regarding building components, exterior walls comprise a significant part of the construction embodied impact [29,33], and roofs were also identified as significant [34], especially for top-floor dwellings. A recent life-cycle study has assessed 114 flat roof alternatives for a Portuguese apartment located in Lisbon considering environmental, energy, and economic criteria [34]. The functional unit assessed was 1 m² of roof used during a 50-year lifetime. The study concluded that, with an identical insulation layer, the roof impacts can vary widely among alternatives (e.g., the best non-accessible roof can lower CO_2 emissions by 30%).

The goal of this study is to assess the life-cycle environmental influence of key building envelope options (with varying thermal performance) for a South-Western European compact house located in Portugal in a moderate Mediterranean climate. This research investigates how operational and embodied impacts of a house vary with building envelope alternatives in order to identify the alternatives with the lowest impacts. LCA and building dynamic simulation were integrated to assess the following envelope construction options throughout the walls and roof: five insulation thickness levels (0–12 cm), four total ventilation levels (0.3–1.2 ac/h), and six insulation materials. In addition, since exterior walls represent most of the building envelope area, three exterior wall construction alternatives (double brick, concrete, and wood walls) were also considered.

2. Materials and Methods

LCA methodology [35] was used to assess the environmental impact of building envelope alternatives for a new South-Western European house located in a mild Mediterranean warm climate in Coimbra, Central Portugal (1460 heating degree days). An attributional LCA approach [36] was selected since it was not expected that the flows within the supply chains would change as a consequence of the adoption of the alternatives assessed. A process-based life-cycle inventory was built based on previous research [23,28,37] and using average background data. The functional unit selected was to build and use a house (for a 4-person family) during its lifespan. A lifespan of 50 years was assumed since it is a common lifespan considered for buildings in the literature [8,37]. The life-cycle study included three life-cycle phases: construction, operation (heating and cooling), and maintenance of the building and envelope alternatives. Furthermore, these phases are considered the most significant and amount to the majority of a building's life-cycle impacts (82–98%, based on [16]).

The construction phase included material production, transport to the construction site, and on-site construction processes (considering a 5% material waste factor). Materials and techniques commonly used in Portugal during the last few decades were assumed. The environmental impacts of building material production and transport were aggregated by average construction product or process and assessed based on European background data from ecoinvent v3.2 [38], using SimaPro 8.3 software [39].

Maintenance activities that preserve the physical characteristics of the building during its lifespan (painting, vanishing, and roof water-proof layer replacement) were taken into account based on data from local construction material producers [40,41]. Detailed information regarding the maintenance activities schedule can be found in [28]. Their environmental impact was assessed based on background data from ecoinvent v3.2.

The annual heating and cooling loads for the house and the various building alternatives were obtained by thermal simulation in DesignBuider © v3.0 [42], which is a dynamic thermal simulation tool based on the Energyplus calculation engine (tested and validated under the comparative standard method of test BESTEST and ANSI/ASHRAE Standard 140–2011). Operational patterns were considered to better represent mild climate modest energy (heating and cooling) use, typical of Portuguese dwellers. In the LCA, the heating and cooling electric energy requirements obtained by thermal simulation were converted to life-cycle environmental impacts using inventory data for the Portuguese electricity generation mix in 2012 [43]. In the last few years, Portugal has consistently had a large share of electricity generated from renewable energy sources when compared to other European countries, which influences the operational life-cycle impact.

Information regarding the case study definition, namely construction and alternative construction scenarios considered, can be found in Section 2.1, while operational phase details and the operational patterns considered are presented in Section 2.2.

In the life-cycle impact assessment (LCIA) stage, two well-known LCIA methods were used. These were the cumulative energy demand (CED) method, to account for the non-renewable primary energy (NRPE), and the CML 2001 method, to account for the following environmental impacts [44]: abiotic depletion, global warming potential (GWP), acidification, eutrophication, photochemical oxidation, and ozone layer deletion (OLD).

Given the comparative nature of this LCA study, the life-cycle model implemented assumed a few simplifications, which are identified and explained in Table 1.

Table 1. Life-cycle model simplifications.

Simplifications and Processes Out of the Scope	Reason
Energy used on construction site	It is considered of minor importance in other studies [1,45].
Furniture, plumbing, sanitary equipment, heat distribution pipes, change in land use	These are not affected by the alternative building envelope options and do not affect the comparative nature of the findings. Hence, embodied impacts are underestimated in the life-cycle model.
Appliances and domestic hot water use, lighting	These needs are not dependent on envelope options. Improvements are independent of the building and mainly related to available technology (appliances efficiency) and user behavior.
Insulation materials' thermal properties were assumed to remain the same throughout the lifespan	Though the EU standards recommend considering the aging process of construction products to estimate the decay of thermal properties, overtime was out of the scope of our study.
End-of-life phase	Expected to have a small life-cycle magnitude, representing less than 4% in Mediterranean dwellings (Nemry et al., 2010). Additionally, to predict waste treatment scenarios for such distant future (50 years) encompasses high uncertainty and waste treatment processes can change.

2.1. Construction: Base Case House and Envelope Alternatives

The house under study is a Portuguese household occupied by a 4-person family. A single-family house was selected because it is the most common residential building type in Portugal. The compact building shape, typology, and area are representative of an average Portuguese house based on statistical data [46,47]: it has two floors, 133 m^2, and a 3-bedroom typology. Table 2 describes the main building components of the base case house; axonometric drawings of the building can be found in [37].

Table 2. Base case house building components description.

Building Component	Area (m^2)	Units	Description
Roof	74.4		Gravel (0.05 m); polypropylene felt; extruded polystyrene (XPS) insulation (0.06 m); bitumen layer (0.005 m); anhydrite screed (0.05 m); reinforced concrete slab (0.15 m); lime mortar (0.02 m); U = 0.39 W/m^{2}K.
Slab	76.4		Wooden flooring (0.04 m square joists, air-layer, 0.02 m planks); anhydrite screed (0.03 m); reinforced concrete slab (0.15 m); lime mortar (0.02 m).
Ground floor	80		Wooden flooring (0.04 m square joists, air-layer XPS) 0.02 m planks); lightweight anhydrite screed (0.05 m); reinforced concrete (0.12 m); gravel (0.20 m) on ground; U = 0.56 W/m^{2}K.

Table 2. Cont.

Building Component	Area (m²)	Units	Description
Structure			Beams, columns, foundations: reinforced concrete
Exterior walls	220		Base plaster painted; hollow-brick masonry (0.11 m); air-cavity with XPS (0.06 m); hollow-brick masonry (0.15 m); base plaster; painting; U = 0.33 W/m²K.
Interior walls	110		Hollow-brick masonry (0.11 m); base plaster (0.02 + 0.02 m); painting.
Windows	1	11	Aluminum-frame with thermal break; double-glazing U = 1.1 W/(m² K); exterior plastic shutters
Doors (interior)	1.6	8	Wooden doors, varnished.
Exterior door	2	1	Wooden doors, varnished (U = 1.8 W/(m²K).

A parametric analysis of the alternative construction options studied (presented in Table 3) was performed for the following: five envelope insulation levels (0–12 cm), five insulation materials, four total ventilation levels (including infiltration), and three exterior wall systems.

Table 3. Envelope construction alternatives and base case.

Passive Construction	Alternatives Studied			Base Case
Envelope extruded polystyrene (XPS) insulation level (cm) [1,2]	0; 3; 6; 9; 12			6
Total ventilation level, including infiltration (ac/h) [1]	0.3; 0.6; 0.9; 1.2			0.6
Exterior wall construction type	Double hollow-brick masonry (XPS insulation)	Concrete block masonry[2] (EPS insulation)	Wood walls (XPS insulation)	Double hollow-brick masonry (XPS insulation)
Insulation material [1](equivalent U-value)	XPS; XPS CO_2; EPS; Cork; Polyurethane rigid foam (PUR); Rock wool			XPS

[1] Measures applied both to façades and roof. [2] Instead of the exterior thermal insulation composite system (ETICS), the hypothetical non-insulated concrete wall (0 cm) has a base plaster finish.

A hypothetical non-insulated scenario (0 cm), which does not meet the legal thermal requirements, was considered with the sole purpose of better showing how operational and embodied impacts vary with the insulation level (i.e., allowing us to draw in results figures which are representative of the polynomial trend-line from 0 cm through the following insulation thicknesses). Nevertheless, in the analysis, a focus is given to insulated alternatives (3–12 cm).

2.2. Building Operational Conditions

Operational energy consumption is directly affected by the building characteristics and by the construction options studied. The operational phase included the impact of heating and cooling the house with a 10 kW air-water heat pump system (2.8 $COP_{heating}$ and 2.0 $EER_{cooling}$). Table 4 summarizes the energy building simulation settings used to assess the house with alternative building construction alternatives in DesignBuilder © v3.0. A window-shutter schedule, presented in Table 5, was assumed to account for typical use of the window-shutters to benefit from solar gains during the heating season and avoid them during cooling season.

Table 4. Building simulation settings, OP100.

Building Simulation Settings	Description
3D build-up model	
Living area (m^2)	133.2
Conditioned volume	360
Heating set-point air temperature (with no set-back)	20 °C
Cooling set-point air temperature (with no set-back)	25 °C
Heating Ventilation and Air Conditioning (HVAC)schedule; gains schedule	0:00–24:00 (24 h/7 a year)
Location	Coimbra, Portugal
Latitude/longitude (°)	40.2°/−8.4°
Elevation above sea (m)	140
Hourly weather data	PRT_Coimbra_IWEC
Internal gains (lumped into a single value)	4 W per m^2 of living area; as recommended by [48]
Air-tightness (infiltration)	Dependent on total ventilation scenario
Gains schedule	0:00–24:00 (24 h/7 a year)

Table 5. Window shading schedule.

Annual Period	Days	Shutters Open	Shutters Closed (Shading)
30 September to 30 June	weekdays	7 h–19 h	19 h–7 h
	weekends	9 h–19 h	19 h–9 h
30 June to 30 September	weekdays	7 h–8.5 h	8.5 h–7 h
	weekends	9 h–12 h	12 h–9 h

A continuous operational pattern (OP100) that reflects continuous interior comfort conditions and occupation (identified in Table 3) was initially used to thermally assess the residential building performance. However, in mild climates, users do not heat and cool continuously, nor do they heat all the rooms simultaneously. Due to this fact, the final energy results were significantly higher when compared to statistical data on energy consumption in Portuguese houses. For instance, comparing the thermal energy requirements for an equivalent existing house with identical shape/construction (based on Portuguese building stock characteristics [47]) and the average real heating energy consumption per square meter in houses in Portugal (inferred from statistical data; [49]), a continuous operational pattern reveals a significant gap [9,28]. Portuguese real household consumption can be 75% lower than simulated energy needs for maintaining continuous comfort conditions. This gap, called the prebound effect [30], represents the way in which user behavior can reduce expected energy consumption levels. It seems that Portuguese dwellers heat their homes partially, or at cooler temperatures, or have their heating on for less time than assumed in the simulated continuous operational pattern. This is possible because winter climate conditions are not as harsh as in North and Central European locations and the summer climate is not hot but warm. Furthermore, occupants tend to use heating more economically in houses that are thermal underperformers [30,50]. Consequently, the prebound effect percentage might change with the thermal performance of the building, decreasing the benefit of energy efficiency measures. As real operational energy consumption data are limited, in this LCA study, two alternative operational pattern scenarios were used to inform heating and cooling habits:

- OP25, which represents a low occupancy and modest and partial heating and cooling level, reinforced by Portuguese statistical data; it holds 25% of the energy requirements of simulated continuous operational pattern.
- OP50, which assumes the average occupancy of a working-out family and medium heating and cooling level, holding 50% of the simulated heating and cooling energy requirements for OP100.

This study did not intend to assess the specific effect of dynamic (zoned and intermittent) operational patterns, which widely vary with the household. Stazi et al. [51] covered these aspects and the effect of thermal mass (inertia) in three super-insulated multifamily buildings both for hot and cold climates. They concluded that, in such highly insulated envelopes, thermal mass had a low influence on operational energy savings (marginal benefit). Additionally, thermal mass (masonry alternative) had a stronger effect on comfort levels (less discomfort hours for intermittent cooling) but it had 20% higher environmental life-cycle impacts (for ecoindicator'99).

3. Results

The main LCA results are presented for two operational patterns (OP25 and OP50). Firstly, the influence of alternative ventilation and insulation levels was assessed for the base case house (house with double hollow-brick walls and double-glazing windows, using heat pump system). Later, the influence of alternative exterior wall systems and insulation level were assessed. Lastly, alternative insulation materials were considered. When assessing alternative insulation levels, trend-lines (polynomial, order 4) are shown in the figures to clearly indicate the influence of varying insulation levels from a hypothetical 0 cm insulation.

3.1. Influence of Ventilation Level vs. Insulation Level

Four total ventilation levels (0.3–1.2 ac/h) and five insulation levels (0–12 cm) were considered for the base case house. Life-cycle results are presented for non-renewable primary energy (NRPE) in Section 3.1.1. and for six environmental impact categories in Section 3.1.2.

3.1.1. Non-Renewable Primary Energy

The construction phase of insulated house alternatives (3–12 cm) was the most important phase (Figure 1) in terms of life-cycle NRPE, representing 63–82% in OP25 and 49–76% in OP50, whereas the operational phase represented 8–28% and 14–43% of NRPE in OP25 and OP50, respectively. Insulation thickness tipping-points, for which NRPE was reduced, were identified: these were 3–6 cm for OP25 and 6–9 cm for OP50. However, in OP50, the total life-cycle benefit of having more insulation than 6 cm was less than 1% for all ventilation scenarios. The insulation tipping-point did not change significantly with the ventilation level.

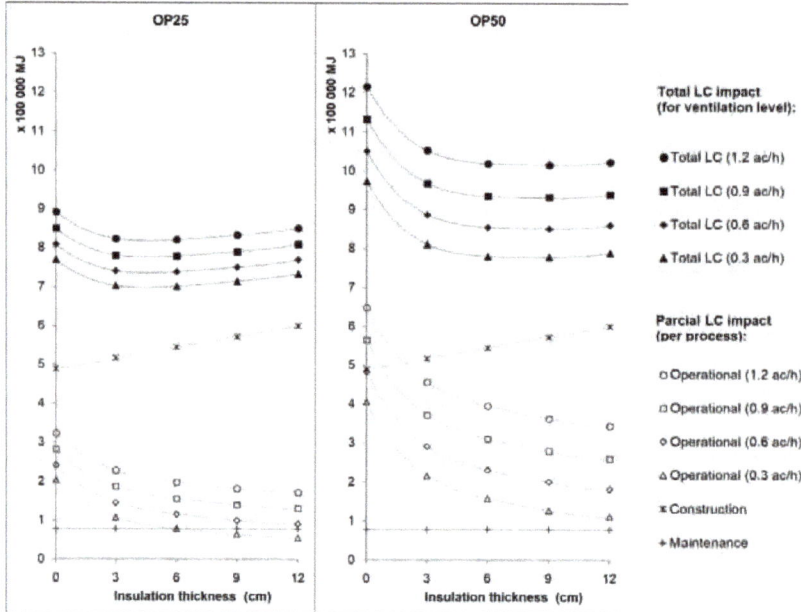

Figure 1. NRPE results for base case house with a heat pump for OP25 and OP50: ventilation level vs. insulation level.

In a well-insulated (6 cm) and air-tight (0.3 ac/h) house with modest energy use (OP25), maintenance had a similar impact to operational energy. When operational energy is reduced, other life-cycle phases' relative contributions are increased.

Compared with a hypothetical non-insulated house (0 cm, 1.2 ac/h), a 6 cm XPS layer reduced operational NRPE by 39–61% (from OP25 to OP50) but it only achieved a life-cycle reduction of 8–9% (OP25) or 16–20% (OP50). Lowering the overall ventilation level from 1.2 to 0.3 ac/h reduced operational NRPE by 38–68% (from OP25 to OP50) and life-cycle NRPE by 14–15% (OP25) or 20–23% (OP50). Assessing the joint effect of the measures (6 cm insulation; 0.3 ac/h ventilation), maximum NRPE reductions of 21% (OP25) and 36% (OP50) were achieved compared to the hypothetical worst scenario. The base case house (6 cm; 0.6 ac/h) yielded a 17% (OP25) and a 30% (OP50) NRPE reduction.

3.1.2. Environmental Impact Assessment

LCIA results are presented for OP25 (Figure 2) and OP50 (Figure 3) to determine whether a broader environmental impact assessment results in the same conclusions as the NRPE analysis. Results show that abiotic depletion, acidification, and GWP correlate with NRPE (Figure 1). In OP25, the insulation tipping-point was between 3 and 6 cm for most categories (exceptions: eutrophication and OLD), whereas in OP50, the tipping-point varied widely: 3–6 cm for GWP and photochemical oxidation; 9–12 cm for abiotic depletion and acidification. For eutrophication, the tipping-point was above 12 cm even in OP25, since the insulation material used (XPS) had relatively low impact in this category. Regarding OLD, the impact of construction (87–99%) surpassed, by far, operational impacts in insulated alternatives. Construction materials, especially XPS insulation, had a significant contribution to OLD. The high impact of XPS is justified by the extrusion process that uses a hydrofluorocarbon (HFC-134a).

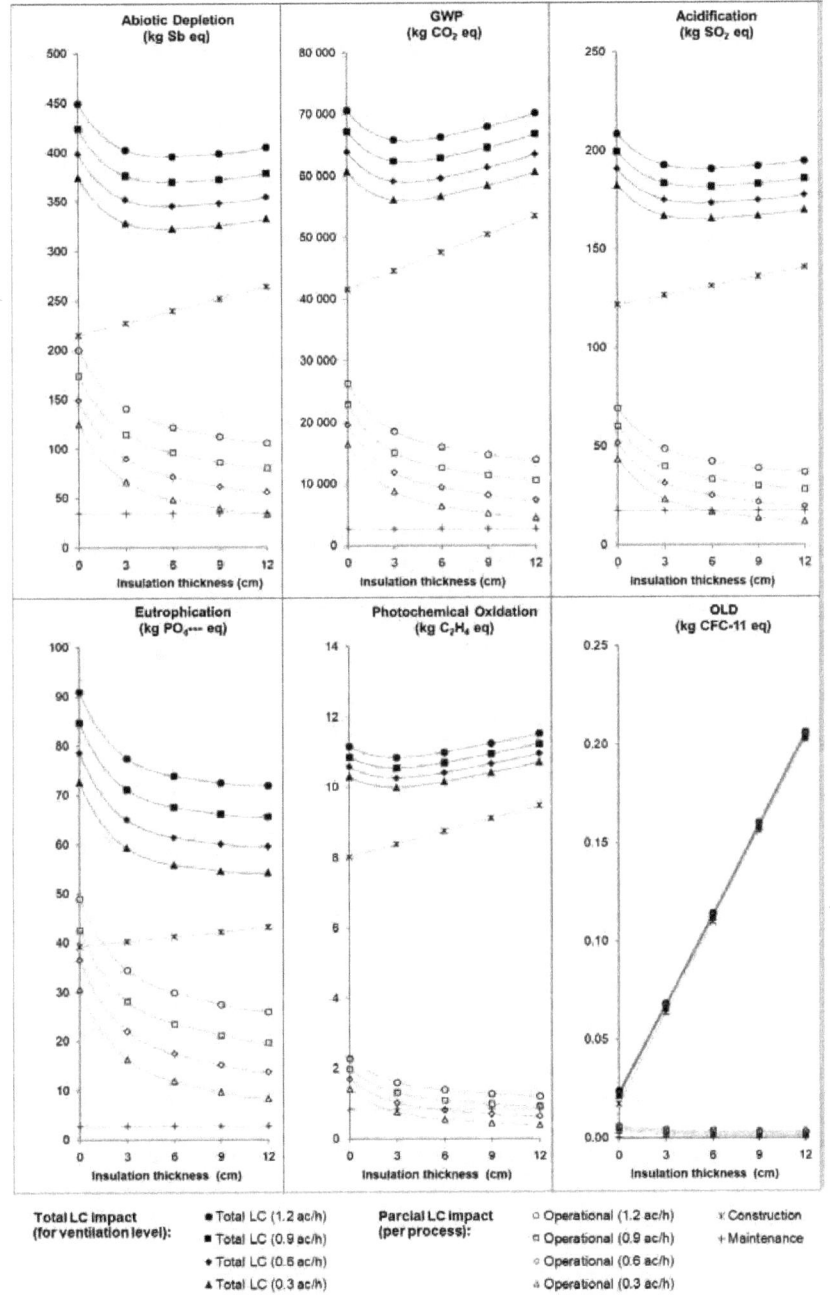

Figure 2. LCIA results for base case house with a heat pump for OP25: ventilation level vs. insulation level.

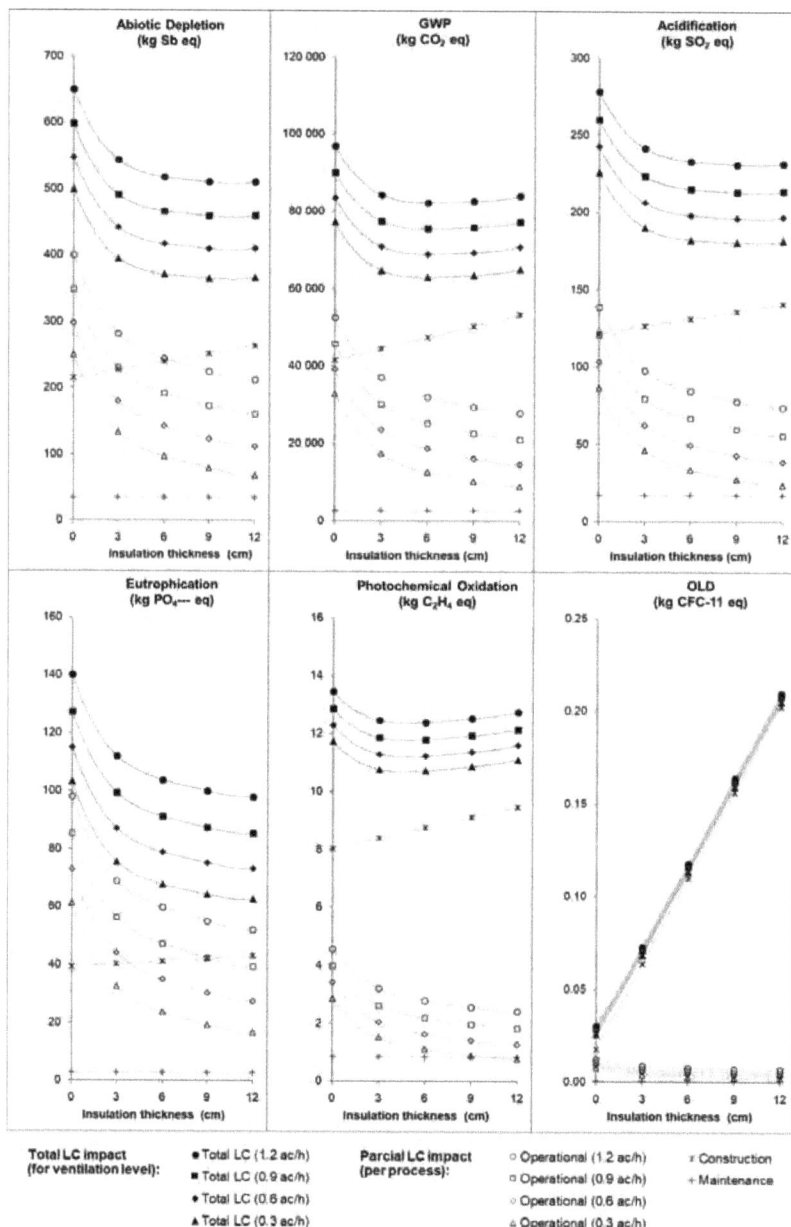

Figure 3. LCIA results for base case house with a heat pump for OP50: ventilation level vs. insulation level.

In OP25, construction was the most significant life-cycle phase for all categories in insulated house alternatives. Furthermore, in photochemical oxidation, construction had a significant impact (77–88%). In OP50, the most significant phase (construction or operation) varies with the insulation and ventilation levels. For the house with two simple passive construction measures (6 cm XPS and 0.6 ac/h), embodied impacts had a life-cycle contribution above 67%.

3.2. Influence of Exterior Wall Construction Alternatives vs. Insulation Level

In this subsection, three exterior wall alternatives—double brick, lightweight concrete, and wooden wall—were assessed jointly with different envelope insulation levels. Results are presented for the base case house with 0.6 ac/h ventilation level.

3.2.1. Primary Energy

Figure 4 presents NRPE for OP25 and OP50. Results show that the operational energy of the three exterior wall house alternatives is similar and mostly dependent on the envelope insulation level. Embodied energy (NRPE) surpassed operational energy for all insulated alternatives, amounting to 62–78% in OP25 and 52–70% in OP50, whereas operation varied from 12% to 25% in OP25 and 21% to 40% in OP50.

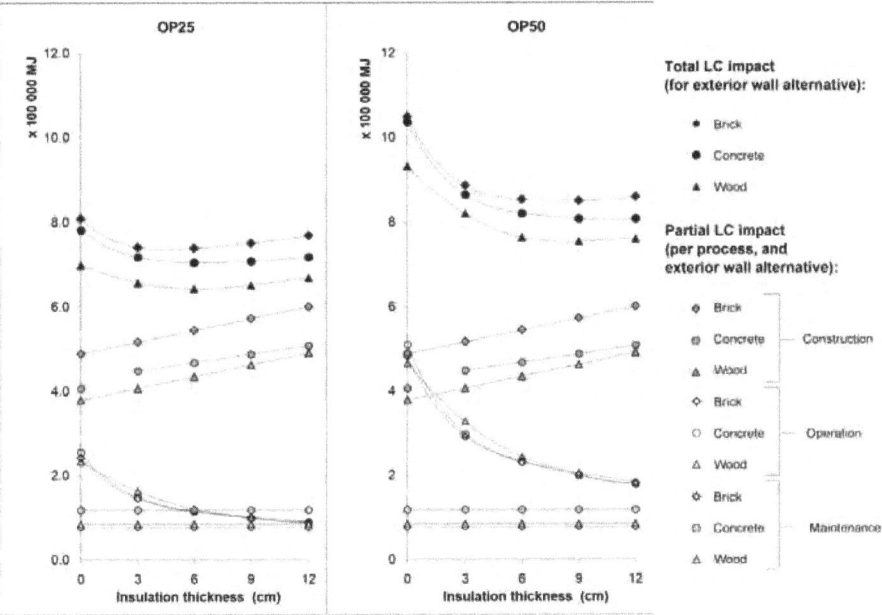

Figure 4. NRPE for OP25 and OP50 for exterior wall house alternatives (brick, concrete, and wood) vs. insulation level.

The double brick wall construction had the highest embodied energy. Comparatively, the concrete wall construction alternative had 13–15% lower embodied NRPE (depending on insulation level), and the wood wall alternative had 22–18% lower embodied NRPE. In the CED method, wood is considered a renewable source of energy and has low embodied NRPE. Thus, the wood wall house had the lowest NRPE, with a reduction of 11–14% (OP25) or 7–11% (OP50) NRPE when compared with the base case brick house. The concrete wall house had a NRPE reduction of 3–7% (OP25) or 1–6% (OP50) since the embodied energy reduction was partially offset by the higher maintenance requirements. Maintenance of a concrete wall house results in a higher NRPE than the other exterior wall alternatives, mainly due to the acrylic plaster finishing of ETICS (exterior thermal insulation composite system).

The insulation tipping-point varied both with the exterior wall alternative and with the operational patterns. For OP25, tipping-points were 6 cm for concrete and wood wall houses and 3 cm for the base case brick house. For OP50, the tipping-points were around 12 cm for concrete wall, 9 cm for wood wall, and 6 cm for brick wall house.

3.2.2. Environmental Impact Assessment

Figures 5 and 6 present the LCIA results for OP25 and OP50, respectively. Acidification closely correlates with NRPE. Abiotic depletion had a slightly higher operational relative contribution. Other environmental categories present some differences in the life-cycle phase contributions, insulation tipping-points, and specific insulation material impacts.

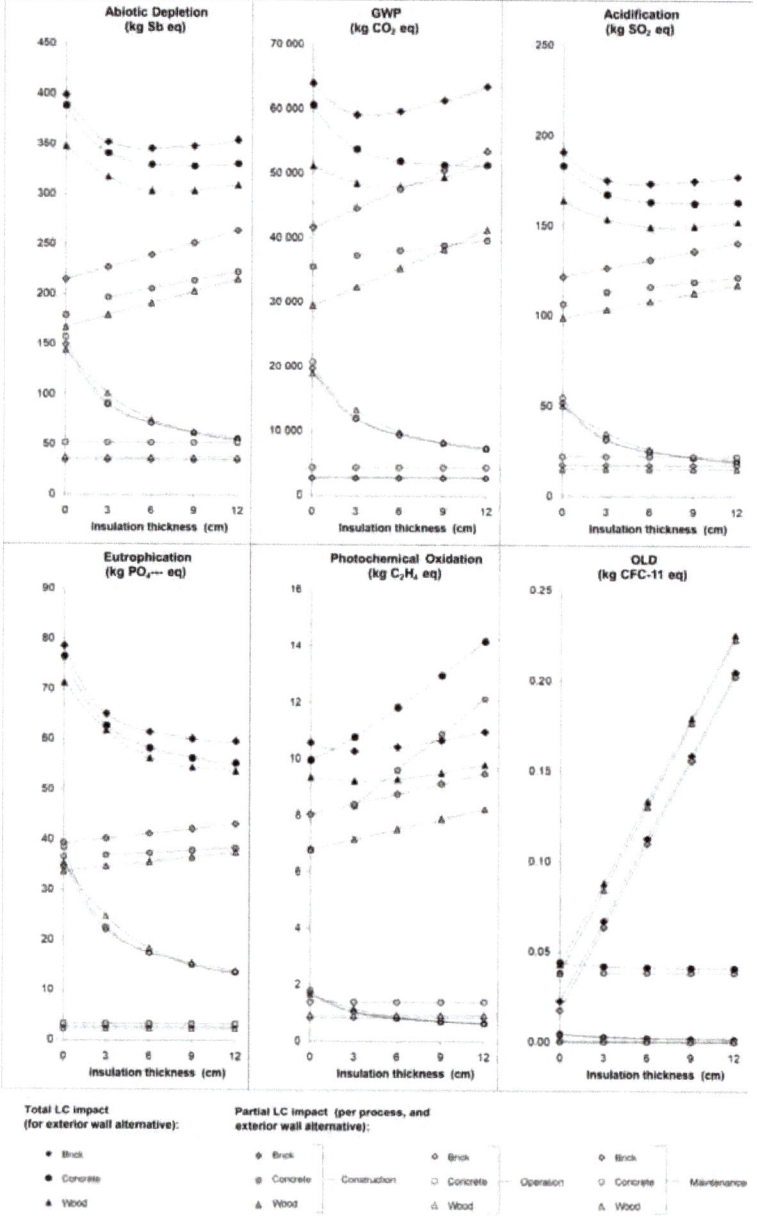

Figure 5. LCIA house results for OP25: exterior wall construction alternatives vs. insulation level.

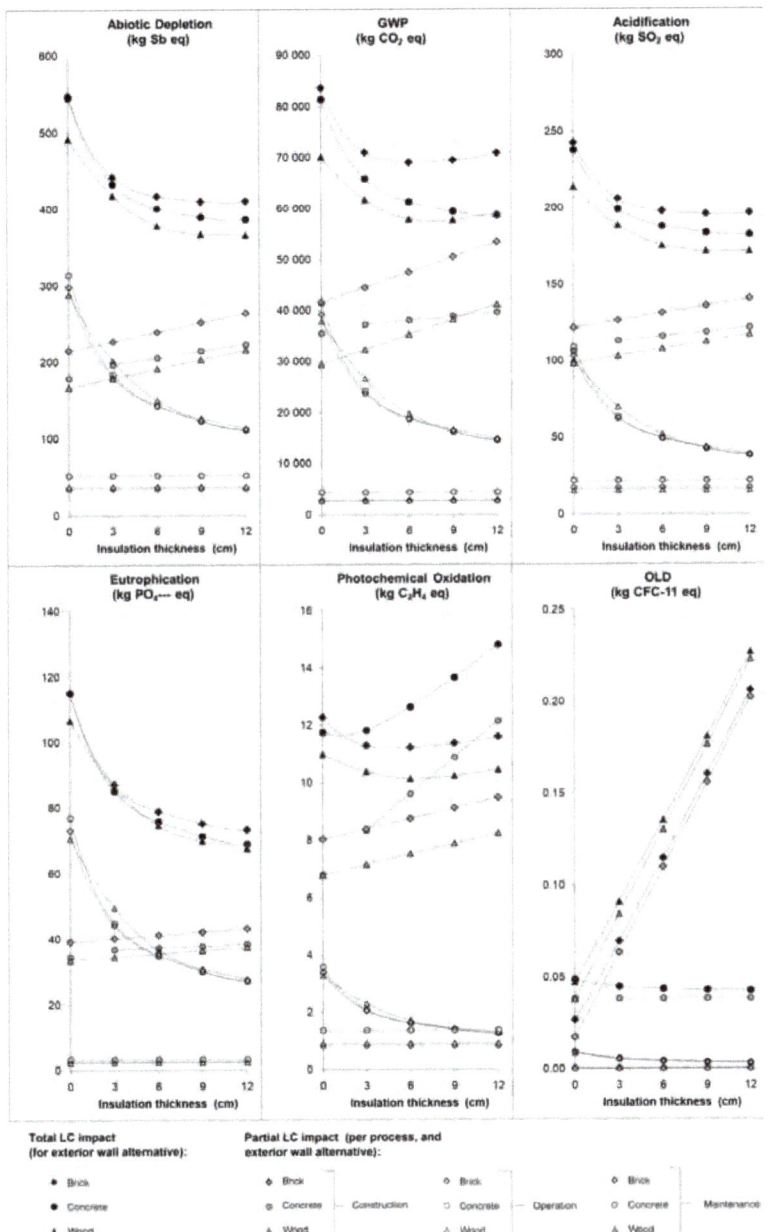

Figure 6. LCIA house results for OP50: exterior wall construction alternatives vs. insulation level.

In GWP, photochemical oxidation, and OLD, dissimilar embodied impacts are associated with XPS (brick and wood walls) and EPS (concrete wall ETICS). XPS had a 3.8 times higher GWP impact and 2000 times higher OLD impact than EPS for the same insulation thickness, whereas EPS had a 3.5 times higher photochemical oxidation impact than XPS. The OLD impact magnitude of XPS insulation is due to HFC-134a being used during the extrusion process, as explained in Section 3.2.1.

The insulation thickness tipping-point varied with exterior wall alternative, operational patterns, and impact categories. In OP25, the tipping-points for most environmental categories were as follows: between 3 and 6 cm for the brick wall alternative and 6 cm for wood wall alternative (exceptions: eutrophication, OLD); for the concrete wall, it was above 12 cm for three categories (GWP, eutrophication, OLD), 9 cm for abiotic depletion and acidification, and 0 cm for photochemical oxidation. Results show that the acrylic plaster finishing of ETICS had a high photochemical oxidation impact that surpasses operational energy savings due to insulation. In OP50, the tipping-points for both the brick and the wood wall were as follows: nearly 6 cm for GWP and photochemical oxidation; 9–12 cm for abiotic depletion and acidification; above 12 cm for eutrophication. The tipping-points for the concrete wall alternative were above 12 cm for five categories, except for photochemical oxidation.

Comparing the three exterior wall alternatives, the double brick wall had the highest embodied and total life-cycle impacts in four categories (abiotic depletion, GWP, acidification, eutrophication). Wood wall construction was the alternative with the lowest impacts in five categories, presenting a reduction of 7–20% in contrast to the brick wall alternative. In OP25, the embodied impacts of construction held most of the life-cycle impact in all environmental categories for insulated house alternatives. In OP50, the same was valid in four categories (except in abiotic depletion and eutrophication).

Assuming a 6 cm insulation level, which is a likely insulation level for a new house, the construction phase accounts for most of the house life-cycle impacts both in OP25 and OP50. In OP25, construction amounts to 62–84% of life-cycle impacts, operation 7–33%, and maintenance accounted for 5–16% in five categories (except OLD, which is explained below). Meanwhile, in OP50, construction accounted for 48–78% of life-cycle impacts, operation 13–49%, and maintenance accounted for 3–13%. OLD is a particular category in which embodied impacts were responsible for almost all impacts (88–98%), especially in the wall alternatives that incorporated XPS insulation (96–98% of life-cycle impacts), as explained in Section 3.2.1.

3.3. Influence of Insulation Material

To assess the specific influence of the selected insulation material, Table 6 presents how the embodied impact of the construction stage of the house varies with alternative insulation materials. The insulation materials' thicknesses were defined to have an equivalent insulation level to the base case house, which means that the building envelope delivers the same U-values of the base case (with 6 cm XPS).

Table 6. Influence of alternative insulation materials on the life-cycle impacts of the house compared to the base case house (XPS insulation).

Insulation (Thermal Conductivity [1] W/m^{2}K)		NRPE	AD	GWP	AP	EP	PO	OLD
XPS CO$_2$	0.035	1.4%	1.4%	−7.4%	−0.5%	0.2%	3.2%	−96.9%
EPS	0.038	3.5%	3.6%	−6.4%	0.9%	−0.2%	45.4%	−96.9%
Cork	0.038	−6.8%	−6.3%	−10.4%	−1.1%	−0.5%	−3.3%	−96.9%
PUR	0.04	1.3%	0.7%	−6.4%	3.2%	14.0%	4.4%	−96.7%
Rock Wool	0.025	−1.7%	−2.1%	−7.9%	0.1%	2.0%	−3.1%	−97.0%
No insulation [2]	0.035	−10.4%	−10.3%	−12.7%	−7.3%	−5.6%	−8.0%	−97.0%

[1] The base case XPS thermal conductivity was 0.030 W/m^{2}K; [2] The non-insulation scenario allows us to account for the embodied impact of base case thermal insulation.

Results clearly show that changing the insulation material from XPS to cork panels can reduce the house embodied impact in the construction stage in all categories while ensuring the same operational impact. In fact, if cork insulation is considered, comparing the embodied with the operational LC impact (presented for alternative insulation levels in previous figures), the cork thickness tipping-point for the brick house would be between 12 cm (for OP25) and 16 cm (for OP50), being able to reduce the overall life-cycle NRPE by around 5.6–7.8% (in OP25-OP50). Results show that cork insulation is preferred compared to the other materials; the only downside would be the higher space that it takes to ensure the same performance (e.g., to ensure the same envelope U-value, cork insulation must be 1.33 times thicker than base case XPS).

4. Discussion

This study's results reinforce the idea that LCA is crucial not only to avoid problem-shifting but also to identify the most significant life-cycle processes, materials, and hotspots for improvement in new houses. Additionally, they highlight that under mild South European climates (e.g., Coimbra) and in the Portuguese context, even a lightly insulated house can have higher embodied impacts than the operational (for heating and cooling), whereas a new house (base case with 6 cm and 0.6 ac/h) is likely to have more embodied impacts in all environmental impact categories.

This finding may be surprising when compared to other studies, even for South European houses (Table 7), because both user behavior and climate widely vary. For instance, Italy and Spain are South European countries but they may have higher heating requirements or higher cooling requirements than houses in temperate, warm, summer, Mediterranean climates (Csb Köppen–Geiger climate classification) depending on the specific location of the buildings studied. Furthermore, users may heat and cool their houses differently (continually or partially) and this was shown to affect the operational energy magnitude in this study. Thus, operational patterns should reflect patterns of inhabiting and acclimatizing a house (typical user behavior). Assessing operational patterns more intensively than actual consumption might result in higher embodied energy (than needed) and be counterproductive.

Table 7. Comparison of case study and literature life-cycle results for GWP.

		Life-Cycle GWP (kg CO_2eq/m^2.year)			
Location	OP (C/P)	Operation HVAC	Construction	Maintenance	Total
Spain, Barcelona [14]	C	1.7 heating 10.7 cooling	4.5	2.9	49.4
Spain, Zaragoza [26]	C	10.2 HVAC	10,3	-	25 [1]
Spain, Lleida [2] [27]	C	53.2 heating 21.1 cooling	60.5	-	134.8
Italy, Piedmont [21]	C	0.78 HVAC	10.8	-	17.4 [1]
Portugal, Coimbra: base case house [1]	P	2.5 heating 0.3 cooling	7.5	0.4	10.4

Legend: OP (operational pattern): C—continuous; P—partial. [1] Other operational energy needs were accounted for beyond heating and cooling; [2] the case study is a house-like cubicle with similar construction to the base case.

Other reasons that may justify such differences are the following:

(a) Design-related: the fairly compact building, north–south oriented, with a low window-to-wall ratio. Some of these passive design measures were identified as being important to reduce operational energy in a Mediterranean climate [32]. Nevertheless, it would be interesting to assess the influence of different building designs for this climatic and operational context from a life-cycle perspective.
(b) The heavyweight building components (exterior and interior brick walls, concrete structure, roof and slabs) are known to incorporate high embodied impact (e.g., both brick and concrete production involve high energy consumption processes) [12,51];
(c) The high performance of the heating and cooling system adopted (heat pump). As shown in other studies, heating systems can play a key role in reducing environmental impacts [28,52];

(d) The Portuguese electricity mix, which has a substantial share of renewables [43]. In the last few years (and likely in the next few years), the electric mix should continue to have an increased contribution of renewable energy, which is expected to have lower environmental impact. Thus, it is even more likely that the operation phase has lower overall significance in new houses. Therefore, it is important to assess the embodied impacts in construction materials in order to arrive at construction alternatives with lower overall environmental impacts and consider those impacts at the project level jointly with operational environmental impacts at the local scale to avoid problem shifting.

Regarding the base case house, this study showed that reducing the ventilation level to 0.3 ac/h without compromising indoor air quality reduced life-cycle impacts by 4–14%, while adopting an alternative wood wall construction instead of the brick one reduced LC impact by 7–20%. These two measures are more beneficial passive solutions than increasing XPS insulation beyond 6 cm thickness, which only marginally reduced the overall impact (NRPE). Increasing insulation levels results in gradually lower NRPE savings and can even generate higher NRPE (when insulation is above the tipping-point), since embodied energy requirements offset operational energy savings.

This should hold true for new houses with a fairly compact shape and small window-to wall ratios, such as the base case, using a heat pump system, under similar climate conditions.

Operational impact was more affected by the insulation thermal resistance and thickness than by the varying construction of the exterior wall. This can be justified because all house alternatives had high thermal inertia, due to the heavyweight core of the house (concrete structure and brick interior walls), which remained unchanged. In this case, for the same insulation level, the life-cycle differences among exterior wall alternatives were mainly due to embodied impacts and maintenance procedures typical of different construction types.

The study also identified other material issues for improvement, namely the following:

(a) Cork insulation had the lowest life-cycle impacts when compared with other insulation materials;
(b) The base case XPS insulation had a high impact on OLD. This impact is justified by the extrusion process that used hydrofluorocarbons (HFC-134a). Recently, XPS producers started to use CO_2 and acetone or HCF-152a as alternative blowing agents to replace HFC-134a. An LCA study of insulation materials [53] that assumed this replacement showed that new production methods can drastically reduce XPS OLD impact (from 1.64×10^{-4} to 7.27×10^{-8} kg CFC-11eq, per kg of XPS) and, in that case, the insulation tipping-point would be above the 12 cm thickness for both OP25 and OP50.
(c) The acrylic plaster used in ETICS concrete walls was associated with a high impact for photochemical oxidation, so alternative production methods for this finishing layer should be studied

5. Conclusions

An LCA of a house located in Coimbra (in mild, warm, Mediterranean climate) was performed, considering two operational patterns (OP25 and OP50). The influence of the following alternative building envelope options were assessed: insulation thickness levels (0–12 cm); ventilation levels (0.3–1.2 ac/h); insulation materials; exterior envelope solutions (double brick, concrete, and wood walls). The results showed that combining two simple passive construction measures, a good envelope insulation level (6 cm), and an air-tight envelope (0.3 ac/h ventilation level) may lead to important LC primary energy savings of 21% (for OP25) to 36% (for OP50) when compared to a hypothetical uninsulated house (0 cm; 1.2 ac/h). Increasing the base case XPX insulation thickness has only marginal life-cycle benefits and can even increase the overall life-cycle impacts (depending on operational patterns). Thus, to avoid problem-shifting, LCA is critical to assess the balance between embodied and operational impact. Insulation tipping-points (with reduced life-cycle impact) were identified for the various environmental categories ranging between 3 and 6 cm for OP25 and 6 and 9 cm for OP50 for the brick wall house with XPS insulation.

Regarding the base case house (brick wall; 6 cm; 0.6 ac/h), two measures were identified to have more benefit than increasing XPS thickness: (a) the replacement of brick walls by wood walls (achieved a LC reduction of 7–20%); (b) increases in envelope air-tightness and reductions in total ventilation level to 0.3 ac/h (achieved a LC reduction of 4–14%). Regarding alternative insulations, cork panels resulted in the lowest embodied impact for an equivalent U-value envelope. Furthermore, for this material, the tipping-point thickness was around 12–16 cm, and it enabled a reduction in the life-cycle NRPE impact of the base case house by around 5–8%.

This study showed that construction represents a significant share (62–81%) of the LC impacts of new houses with fairly simple construction measures, using a heat pump system to satisfy current modest Portuguese operational user demands. This is a surprising result alongside other comparable studies, especially of buildings in Mediterranean countries because LCA impacts are strongly influenced by the climate and cultural local conditions (how to build and inhabit a house) and energy mix. Embodied impacts are currently not routinely considered in building energy performance certification [54]. However, as new buildings are expected to be very low energy in operation, neglecting embodied impacts may lead to problem-shifting, having higher embodied impacts in upfront construction than the avoided impacts in operation.

Thus, the adoption of construction options with lower embodied impact is highly important. To further reduce the environmental impact of buildings under mild climates, data on the environmental impact embodied in materials should be freely available in the marketplace—for instance, through widespread environmental product declaration (EPD) or product environmental footprint (PEF) schemes. This would greatly benefit architects, engineers, and households as they take into account the environmental impacts of their decision-making. Finally, to assess the overall sustainability of a wide range of building alternatives, future research work should further examine building life-cycle costs at higher resolutions and a greater range of the associated social impacts.

Author Contributions: Conceptualization, H.M. and F.F.; methodology, formal analysis and investigation, visualization, writing—original draft preparation, funding acquisition: H.M.; writing—review and editing, H.M., F.F and J.E.F.; supervision, F.F., J.E.F. and H.M. All authors have read and agreed to the published version of the manuscript.

Funding: This work partially results from the funding that H.M. received from the Portuguese Fundação para a Ciência e a Tecnologia (SFRH/BD/33736/2009) and from the Energy for Sustainability (EfS) initiative from University of Coimbra.

Acknowledgments: H.M. acknowledges MIT Portugal Program and MIT for providing assistance and allowing H.M. to perform part of this research at the MIT Building Technology Lab.

Conflicts of Interest: The authors declare no conflict of interest. The funders had no role in the design of the study; in the collection, analyses, or interpretation of data; in the writing of the manuscript, or in the decision to publish the results.

References

1. Nemry, F.; Uihlein, A.; Colodel, C.M.; Wetzel, C.; Braune, A.; Wittstock, B.; Hasan, I.; Kreißig, J.; Gallon, N.; Niemeier, S.; et al. Options to reduce the environmental impacts of residential buildings in the European Union—Potential and costs. *Energy Build.* **2010**, *42*, 976–984. [CrossRef]
2. Invidiata, A.; Lavagna, M.; Ghisi, E. Selecting design strategies using multi-criteria decision making to improve the sustainability of buildings. *Build. Environ.* **2018**, *139*, 58–68. [CrossRef]
3. Göswein, V.; Rodrigues, C.; Silvestre, J.D.; Freire, F.; Habert, G.; König, J. Using anticipatory life cycle assessment to enable future sustainable construction. *J. Ind. Ecol.* **2019**, *24*, 178–192. [CrossRef]
4. Iribarren, D.; Marvuglia, A.; Hild, P.; Guiton, M.; Popovici, E.; Benetto, E. Life cycle assessment and data envelopment analysis approach for the selection of building components according to their environmental impact efficiency: A case study for external walls. *J. Clean. Prod.* **2015**, *87*, 707–716. [CrossRef]

5. Chau, C.K.; Leung, T.; Ng, W. A review on life cycle assessment, life cycle energy assessment and life cycle carbon emissions assessment on buildings. *Appl. Energy* **2015**, *143*, 395–413. [CrossRef]
6. Cabeza, L.F.; Rincón, L.; Vilarino, V.; Pérez, G.; Castell, A. Life Cycle Assessment (LCA) and life cycle energy analysis (LCEA) of buildings and the building sector: A review. *Renew. Sustain. Energy Rev.* **2014**, *29*, 394–416. [CrossRef]
7. Karimpour, M.; Belusko, M.; Xing, K.; Bruno, F. Minimising the life cycle energy of buildings: Review and analysis. *Build. Environ.* **2014**, *73*, 106–114. [CrossRef]
8. Buyle, M.; Braet, J.; Audenaert, A. Life cycle assessment in the construction sector: A review. *Renew. Sustain. Energy Rev.* **2013**, *26*, 379–388. [CrossRef]
9. Soares, N.; Bastos, J.; Pereira, L.D.; Soares, A.R.; Amaral, A.; Asadi, E.; Rodrigues, E.; Lamas, F.; Monteiro, H.; Lopes, M.; et al. A review on current advances in the energy and environmental performance of buildings towards a more sustainable built environment. *Renew. Sustain. Energy Rev.* **2017**, *77*, 845–860. [CrossRef]
10. Lavagna, M.; Baldassarri, C.; Campioli, A.; Giorgi, S.; Valle, A.D.; Castellani, V.; Sala, S. Benchmarks for environmental impact of housing in Europe: Definition of archetypes and LCA of the residential building stock. *Build. Environ.* **2018**, *145*, 260–275. [CrossRef]
11. Gustavsson, L.; Joelsson, A. Life cycle primary energy analysis of residential buildings. *Energy Build.* **2010**, *42*, 210–220. [CrossRef]
12. Tokbolat, S.; Nazipov, F.; Kim, J.; Karaca, F. Evaluation of the environmental performance of residential building envelope components. *Energies* **2019**, *13*, 174. [CrossRef]
13. Hossain, Y.; Marsik, T. Conducting Life Cycle Assessments (LCAs) to determine carbon payback: A case study of a highly energy-efficient house in rural Alaska. *Energies* **2019**, *12*, 1732. [CrossRef]
14. Ortíz-Rodriguez, O.O.; Castells, F.; Sonnemann, G. Life cycle assessment of two dwellings: One in Spain, a developed country, and one in Colombia, a country under development. *Sci. Total Environ.* **2010**, *408*, 2435–2443. [CrossRef]
15. Rodrigues, C.; Freire, F. Environmental impact trade-offs in building envelope retrofit strategies. *Int. J. Life Cycle Assess.* **2016**, *22*, 557–570. [CrossRef]
16. Rasmussen, F.N.; Ganassali, S.; Zimmermann, R.K.; Lavagna, M.; Campioli, A.; Birgisdóttir, H. LCA benchmarks for residential buildings in Northern Italy and Denmark—Learnings from comparing two different contexts. *Build. Res. Inf.* **2019**, *47*, 833–849. [CrossRef]
17. Dylewski, R. Optimal thermal insulation thicknesses of external walls based on economic and ecological heating cost. *Energies* **2019**, *12*, 3415. [CrossRef]
18. Stephan, A.; Crawford, R.H.; De Myttenaere, K. A comprehensive assessment of the life cycle energy demand of passive houses. *Appl. Energy* **2013**, *112*, 23–34. [CrossRef]
19. Dodoo, A.; Gustavsson, L.; Sathre, R. Lifecycle carbon implications of conventional and low-energy multi-storey timber building systems. *Energy Build.* **2014**, *82*, 194–210. [CrossRef]
20. Kylili, A.; Ilic, M.; Fokaides, P.A. Whole-building Life Cycle Assessment (LCA) of a passive house of the sub-tropical climatic zone. *Resour. Conserv. Recycl.* **2017**, *116*, 169–177. [CrossRef]
21. Blengini, G.; Di Carlo, T. The changing role of life cycle phases, subsystems and materials in the LCA of low energy buildings. *Energy Build.* **2010**, *42*, 869–880. [CrossRef]
22. Dodoo, A.; Gustavsson, L.; Sathre, R. Life cycle primary energy implication of retrofitting a wood-framed apartment building to passive house standard. *Resour. Conserv. Recycl.* **2010**, *54*, 1152–1160. [CrossRef]
23. Monteiro, H.; Freire, F. Life-cycle assessment of a house with alternative exterior walls: Comparison of three impact assessment methods. *Energy Build.* **2012**, *47*, 572–583. [CrossRef]
24. Bastos, J.; Batterman, S.A.; Freire, F. Life-cycle energy and greenhouse gas analysis of three building types in a residential area in Lisbon. *Energy Build.* **2014**, *69*, 344–353. [CrossRef]
25. Sierra-Pérez, J.; Boschmonart-Rives, J.; Gabarrell, X. Environmental assessment of façade-building systems and thermal insulation materials for different climatic conditions. *J. Clean. Prod.* **2016**, *113*, 102–113. [CrossRef]
26. Bribian, I.Z.; Uson, J.A.A.; Scarpellini, S. Life cycle assessment in buildings: State-of-the-art and simplified LCA methodology as a complement for building certification. *Build. Environ.* **2009**, *44*, 2510–2520. [CrossRef]

27. Llantoy, N.; Chàfer, M.; Cabeza, L.F. A comparative Life Cycle Assessment (LCA) of different insulation materials for buildings in the continental Mediterranean climate. *Energy Build.* **2020**, *225*, 110323. [CrossRef]
28. Monteiro, H.; Fernández, J.E.; Freire, F. Comparative life-cycle energy analysis of a new and an existing house: The significance of occupant's habits, building systems and embodied energy. *Sustain. Cities Soc.* **2016**, *26*, 507–518. [CrossRef]
29. Rossi, B.; Marique, A.-F.; Reiter, S. Life-cycle assessment of residential buildings in three different European locations, case study. *Build. Environ.* **2012**, *51*, 402–407. [CrossRef]
30. Sunikka-Blank, M.; Galvin, R. Introducing the prebound effect: The gap between performance and actual energy consumption. *Build. Res. Inf.* **2012**, *40*, 260–273. [CrossRef]
31. Gaspar, P.; Santos, A.L. Embodied energy on refurbishment vs. demolition: A southern Europe case study. *Energy Build.* **2015**, *87*, 386–394. [CrossRef]
32. Fernandez-Antolin, M.-M.; Del Río, J.M.; Costanzo, V.; Nocera, F.; Gonzalez-Lezcano, R.A. Passive design strategies for residential buildings in different Spanish climate zones. *Sustainability* **2019**, *11*, 4816. [CrossRef]
33. Monteiro, H.; Freire, F. Environmental life-cycle impacts of a single-family house in Portugal: Assessing alternative exterior walls with two methods. *Gazi Univ. J. Sci.* **2011**, *24*, 528–534.
34. Jean, S.-S.; Lee, P.-I.; Hsueh, P.-R. Treatment options for COVID-19: The reality and challenges. *J. Microbiol. Immunol. Infect.* **2020**, *53*, 436–443. [CrossRef] [PubMed]
35. ISO. *ISO 14044: Environmental Management—Life Cycle Assessment—Requirements and Guidelines*; International Organization of Standardization: Geneva, Switzerland, 2006.
36. Weidema, B.P.; Pizzol, M.; Schmidt, J.; Thoma, G. Attributional or consequential Life Cycle Assessment: A matter of social responsibility. *J. Clean. Prod.* **2018**, *174*, 305–314. [CrossRef]
37. Monteiro, H. Comprehensive Life Cycle Assessment of New Houses in Portugal: Building Design, Envelope, and Operational Conditions. PhD Thesis, Faculty of Sciences and Technology of University of Coimbra (FCTUC), Coimbra, Portugal, 2017.
38. Wernet, G.; Bauer, C.; Steubing, B.; Reinhard, J.; Moreno-Ruiz, E.; Weidema, B.P. The ecoinvent database version 3 (part I): Overview and methodology. *Int. J. Life Cycle Assess.* **2016**, *21*, 1218–1230. [CrossRef]
39. PRé Sustainability SimaPro. Available online: https://simapro.com/about/ (accessed on 7 February 2020).
40. Kellenberger, D.; Althaus, H.-J.; Jungbluth, N.; Künniger, T.; Lehmann, M.; Thalmann, P. *Life Cycle Inventories of Building Products*; Final Report Ecoinvent Data v2.0 No.7; Ecoinvent: Dübendorf, Switzerland, 2007.
41. Künzel, H.; Künzel, H.M.; Sedlbauer, K. Long-term performance of external thermal insulation systems (ETICS). *Archit. Z. Für Gesch. Der Baukunst* **2006**, *5*, 11–24.
42. *DesignBuilder* ©, version 3.0; DesignBuilder Software Ltd.: London, UK, 2017.
43. Garcia, R.; Marques, P.; Freire, F. Life-cycle assessment of electricity in Portugal. *Appl. Energy* **2014**, *134*, 563–572. [CrossRef]
44. Hischier, R.; Weidema, B.; Althaus, H.-J.; Bauer, C.; Doka, G.; Dones, R.; Frischknecht, R.; Hellweg, S.; Humbert, S.; Jungbluth, N.; et al. *Implementation of Life Cycle Impact Assessment Methods Data v2.2.*; Ecoinvent Rep. No. 3; Ecoinvent: Dübendorf, Switzerland, 2010; Volume 176.
45. Gervásio, H.; Santos, P.; Martins, R.; Da Silva, L.S. A macro-component approach for the assessment of building sustainability in early stages of design. *Build. Environ.* **2014**, *73*, 256–270. [CrossRef]
46. INE. PORDATA Housing Confort and Living Conditions. Available online: https://www.pordata.pt/en/Subtheme/Portugal/Accommodation-53 (accessed on 20 September 2002).
47. INE. *O Parque Habitacional e a sua Reabilitação—Análise e Evolução 2001-2011*; LNEC, Ed.; INE: Lisboa, Portugal, 2013; ISBN 978-989-25-0246-5.
48. RCCTE. *Regulamento das Características de Comportamento Térmico dos Edifícios (Code of the Buildings Thermal Behaviour Characteristics)*; DL: Lisboa, Portugal, 2006; pp. 2468–2513.
49. INE-I.P./DGEG. *Inquérito ao Consumo de Energia no Sector Doméstico 2010*; INE: Lisboa, Portugal, 2011.
50. Galvin, R.; Sunikka-Blank, M. Economic viability in thermal retrofit policies: Learning from ten years of experience in Germany. *Energy Policy* **2013**, *54*, 343–351. [CrossRef]
51. Stazi, F.; Tomassoni, E.; Bonfigli, C.; Di Perna, C. Energy, comfort and environmental assessment of different building envelope techniques in a Mediterranean climate with a hot dry summer. *Appl. Energy* **2014**, *134*, 176–196. [CrossRef]
52. Pineau, D.; Rivière, P.; Stabat, P.; Hoang, P.; Archambault, V. Performance analysis of heating systems for low energy houses. *Energy Build.* **2013**, *65*, 45–54. [CrossRef]

53. Pargana, N.; Pinheiro, M.D.; Silvestre, J.D.; De Brito, J. Comparative environmental life cycle assessment of thermal insulation materials of buildings. *Energy Build.* **2014**, *82*, 466–481. [CrossRef]
54. Toftum, J.; Baxter, V. Nearly-zero energy buildings. *Sci. Technol. Built Environ.* **2016**, *22*, 883–884. [CrossRef]

© 2020 by the authors. Licensee MDPI, Basel, Switzerland. This article is an open access article distributed under the terms and conditions of the Creative Commons Attribution (CC BY) license (http://creativecommons.org/licenses/by/4.0/).

Article

Energy Analyses of Serbian Buildings with Horizontal Overhangs: A Case Study

Danijela Nikolic [1,*], Slobodan Djordjevic [1], Jasmina Skerlic [2] and Jasna Radulovic [1]

1 Faculty of Engineering, University of Kragujevac, 34000 Kragujevac, Serbia; sdjordjevic@energetika-kragujevac.com (S.D.); jasna@kg.ac.rs (J.R.)
2 Faculty of Technical Sciences, University of Pristina temporarily settled in Kosovska Mitrovica, 38220 Kosovska Mitrovica, Serbia; jskerlic@gmail.com
* Correspondence: danijela1.nikolic@gmail.com

Received: 24 July 2020; Accepted: 1 September 2020; Published: 3 September 2020

Abstract: It is well known that nowadays a significant part of the total energy consumption is related to buildings, so research for improving building energy efficiency is very important. This paper presents our investigations about the dimensioning of horizontal overhangs in order to determine the minimum annual consumption of building primary energy for heating, cooling and lighting. In this investigation, embodied energy for horizontal roof overhangs was taken into account. The annual simulation was carried out for a residential building located in the city of Belgrade (Serbia). Horizontal overhangs (roof and balcony) are positioned to provide shading of all exterior of the building. The building is simulated in the EnergyPlus software environment. The optimization of the overhang size was performed by using the Hooke Jeeves algorithm and plug-in GenOpt program. The objective function minimizes the annual consumption of primary energy for heating, cooling and lighting of the building and energy spent to build overhangs. The simulation results show that the building with optimally sized roof and balcony overhangs consumed 7.12% lessprimary energy for heating, cooling and lighting, compared to the house without overhangs. A 44.15% reduction in cooling energy consumption is also achieved.

Keywords: building; overhangs; energy consumption; optimization; GenOpt; EnergyPlus

1. Introduction

Reduction in energy consumption is globally of great importance as the combustion of fossil fuels emits significant amounts of greenhouse gases, primarily carbon dioxide. Fossil fuels are also a limited resource which isdecreasing in Nature and should be very cautiously used. In order to reduce the primary energy consumption, it is essential to focus on reducing energy consumption in buildings. Building energy consumption is related to the exploitation conditions, where the largest consumers are the heating, cooling and domestic hot water systems, appliances, etc. Reducing energy consumption can be achieved by the construction of energy efficient buildings which have lower total energy consumption and lower greenhouse gas emissions. It is very important to apply as many measures to design energy efficient buildings as possible, primarily in the passive design of buildings. An application of passive energy elements on buildings, which include elements of shading by horizontal roof overhangs improves thermal indoor comfort, reduces primary energy consumption and hence reduces greenhouse gas emissions. Sometimes, a building design strives to insulate the building from outside influences, and thus to reduce energy exchange. At the other hand, it is necessary to utilize energy from the environment in the best way in order to achieve even better results. The implementation of these principles at the building design stage is the most effective way to achieve good results in the reduction of the energy required for heating, cooling, and lighting.

Many studies were carried out to analyze the impact of shading elements to energy consumption and most authors found a reduction in energy consumption for cooling due to shadowing. Cooling load due to solar gain represents about half of the total cooling load of residential buildings [1]. Solar radiation through the building windows can be decreased with different shading devices installed on the exterior side of building windows [2]. Skias and Kolokotsa analyzed the office building energy consumption for cooling in Athens (Greece) and ways of reducing it during the summer period by placing shadowing elements [3]. Their investigation was carried out in TRNSYS 16, and the application of the horizontal roof overhangs on the south side of the building yielded building energy savingsthat ranged from 7.2% to 17.5%. Kim et al. studied energy saving for cooling with the IES_VE program for a shaded building located in South Korea [4]. They found that by building with horizontal overhangs on the south façade, it is possible to achieve energy savings for cooling of 11%. Raeissi and Taheri investigated the energy consumption for heating and cooling in a family home with horizontal roof overhangs located in Shiraz (Iran), at an altitude of 1491 m [5]. Analyses were performed for cooling and heating periods. The optimization of the primary energy consumption for building cooling and heating achieved a reduction in energy consumption for cooling by 12.7% and increased the energy consumption for heating by 0.63%. Bojic at al. in their paper [1] analyzed the primary energy consumption in residential building with overhangs during the summer season. The obtained results showed that in the case of a house without optimized overhangs, there is an increase in primary energy consumptionby 3.36% and in that case, the operative energy consumption is lower. Imessad at al. investigated a building with horizontal overhangs, located in Algiers, where there is a temperate Mediterranean climate [6]. This analysis was carried out in TRNSYS software, and results showed that horizontal overhangs in combination with natural ventilation can and improve thermal comfort and reduce cooling energy demand in the summer periodby 35%. Datta in his study [7] analyzed building with external fixed shading device for windows, in different cities in Italy (north to south). With a simulation model in the TRNSYS program, he optimized shading device size with the aim to minimize annual primary energy consumption in buildings. The results showed that with optimum shading a 70% solar gain can be avoided in Milan during the summer season. An air-conditioned office in England with fixed external overhang was investigated with simulations in the DOE-2 modeling program [8]. The obtained results showed that energy savings depend on latitude, so in Scotland it was between 1% and 9%. With moveable external shading the highest energy savings can be achieved. Yao [9] simulated a high residential building in Ningbo, China, which has movable solar shading devices in south-facing rooms, in the EnergyPlus software. The simulation results showed that movable solar shading devices can reduce building energy consumption by 30.87% and improve visual comfort for about 20%. Atzeri et al. [10] in their paper investigated an open-space office located in Rome (Italy). They used the EnergyPlus software, and compared the influence of indoor and outdoor shading devices on primary energy consumption, thermal and visual comfort. The main conclusion was that external shading devices can reduce cooling needs and increased heating load. Florides at al. [11] modeled and simulated a modern building with the aim to reduce its thermal load. They recommended a window overhang length of 1.5 m, with which it is possible to save 7% of annual cooling energy consumption for a building with single walls and without roof insulation, and 19% of annual cooling energy consumption for the buildings with walls and roof with 50 mm insulation [11]. Liu et al. [12] investigated shading devices on opaque facades of public buildings in Hong Kong and the possibilities for energy savings with them. They varied the length, the number and the tilt angle of the different configurations of shading devices and found optimal values for west-oriented overhangs, with an energy saving potential of up to 8%.

Aldawoud simulated the energy behavior of an office building with external shading devices and electrochromic glazed windows, located in Phoenix, a city in Arizona, USA, which has a very hot and dry climate. Simulations were carried out in the Design Builder software. Among the other energy performance factors, great attention was paid to the energy consumption for heating, cooling and interior lighting. The simulation results showed that electrochromic glazing provided the greatest reduction of solar heat gains during hot summer days. Also, well-designed overhangs allow a significant reduction in cooling load [13].

Mandalaki et al. [14] analyzed the energy needed for heating, cooling and lighting for office rooms with shading devices, located in the cities of Athens and Chania (Greece). The aim of the analyses was to determine the optimal size of shading devices with integrated south-facing PV panels, which generate electricity for lighting. The results showed that shading devices decrease the building energy consumption. Stamatakis et al. applied multi-criteria analyses of monocrystalline PV panels mounted on south-facing shading devices on office buildings in the Mediterranean region [15]. A novel design of energy-efficient shading devices with amorphous panels was investigated by parametric modeling [16]. Objective functions were the minimal value of total energy consumption and useful daylight illuminance. The achieved savings in total building energy consumption was 14%, with a daylight level above 50%.

Bellia et al. have provided an overview of lighting analysis, energy analysis, HVAC system energy requirements and comprehensive analyses of thermal, visual and energetic aspects for buildings with fixed, movable and others shading systems [17]. Also, a review of simulation modeling for different type of shading devices which are implemented in modern buildings today was given by Kirmtat et al. [18]. The effects of horizontal and vertical louver shading devices, applied to different building façades at different locations, on building energy consumption are analyzed in [19] using the TRNSYS software. Obtained results showed significant energy savings in comparison to a building without shading devices. Valladares-Rendón et al. investigated solar protection and building energy saving in buildings with balanced daylighting and visibility and optimal orientation for façade shading systems [20]. The investigated buildings were in the subtropical zone, at 59 different locations. The results showed that passive strategies can reduce energy consumption by 4.64% to 76.57%. Numerical simulations showed that 58.62% of the locations should apply east oriented, 24.13% northeast oriented, 12.06% west oriented and 5.17%southeast oriented optimal designs. Al-Masarni and Al-Obaidi theoretically and experimentally analyzed current applications and trends of dynamic shading systems [21]. Their outcomes give a classification of shading models and analysis of their performance, with some recommendations for improving dynamic shading systems' performance, which can be very useful for architects. Tabadkani et al. reviewed studies with automatic shading control methods for balancing comfort and energy savings in buildings [22]. They concluded that existing studies investigated only automatic shading controls such as roller shades or venetian blinds, which can contribute to the reduction of energy consumption.

Serbia is among the countries that has the lowest level of energy efficiency in Europe and is therefore located at the bottom of the list of energy-efficient countries. This information is fully illustrated by the fact that in Serbia there are an estimated 300,000–400,000 energy-inefficient residential buildings (single family houses) which have no thermal insulation and with an annual final energy consumption of 220 kWh/m^2 [23], while the European annualenergy consumption ranges from 55 kWh/m^2 in Malta and 70 kWh/m^2 in Portugal, to 300 kWh/m^2 in Romania [24].

Energy consumption in buildings at the global level is 20–40% of total energy consumption, while in Serbia it is at the 35%level [25]. This energy consumption is related to the exploitation conditions of buildings. In the structures of total energy consumption of Serbian building, about 60% of the energy consumption is related to the space heating [23], or approximately 65 million MWh per annum [26]. About 76% of this consumption pertains to single family houses and 24% to multifamily houses [26].

Residential buildings represent the biggest part of national building stock of Serbia, and more than 90% of them are single family houses. Most of these residential buildings (58.78%) are older buildings that were built in the 1960s, 1970s and 1980s, and are characterized by excessive energy consumption, due to the absence or poor thermal insulation, whether due to inefficient doors and windows, etc. In accordance with the national residential buildings typology in Serbia, these buildings belong to the groups D1, E1 and F1 [26], and they are usually two-storey, free standing, single family buildings.

In the last 20 years, some basic energy-saving measures have been implemented in these buildings in order to improve their energy efficiency—application of thermal insulation on the external walls, roof and floors, replacement the old inefficient windows, doors, etc. In that way, a certain energy savings is achieved, but, it is also necessary to implement some other measures and find other ways for minimizing building energy consumption, especially for heating and cooling.

In the literature, there is almost no investigation of how the installation of overhangs influences the common consumption of energy for heating, cooling and lighting in Serbian buildings. This paper reports numerical investigations about how shading by horizontal roof and balcony overhangs influences the primary and final energy consumption for heating, cooling and lighting of residential buildings in Serbia through the year. Analyzed buildings are modeled in accordance with the national residential buildings typology in Serbia, and they represent typical buildings which were built in the period from 1960 until 1990, with thermal insulation on the external walls and energy-efficient windows. In this paper, optimal size of the horizontal roof overhangs, which are placed over east, west, north, and south wall, are obtained by simultaneous operation of the two programs EnergyPlus and GenOpt. The optimization is performed to minimize the primary energy consumption for heating, cooling and lighting. In these processes, the embodied energy of concrete horizontal roof overhangs was taken into account [27].

The primary energy saving and cooling energy consumption results obtained with numerical simulations and optimizations are within the frame of research results of the other authors who have conducted similar studies, but in some other regions of Europe. Serbia lies in the central part of the Balkan Peninsula, and has a moderate continental climate, characterized by cold winters, warm summers, and well-distributed rainfall, like in the other northern and central parts of the Balkans. The results of this study are not merely useful for the study of the methods for improving building energy efficiency aimed at optimizing overhangs and minimizing of Serbian building energy consumption, but above all, they could represent useful information for similar studies conducted in other parts of Europe that share the same or similar characteristics in terms of climate and topography.

2. Materials and Methods

2.1. Description of Modeled Buildings

In this research, the energy consumption is investigated for three buildings shown in Figure 1 as models in the EnergyPlus software [28]. These buildings are detached with two-floors. They have almost the same characteristics.

They only differ in their overhang characteristics such as type and dimensions. They were: (1) a basic building, (2) a building with optimized roof overhangs (ORO building), and (3) a building with optimized balcony overhangs (OBO building). Two types of overhangs were studied—roof and balcony. The roof overhangs were parts of roof that acted as overhangs for the second floor apartments. The balcony overhangs acted as overhangs for the first floor apartments and as the balconies for the second floor apartments. The basic building had roof overhangs with depths of 0.2 m. The ORO building had optimized roof overhangs. The OBO building had optimized balcony overhangs. The balcony overhangs were balconies of the second floor apartments that acted as overhangs for the first floor apartments. This building had also roof overhangs with the same depths as in the ORO building. All overhangs are thermally insulated with polystyrene (0.05 m) to avoid or minimize the appearance of thermal bridges. The cross-section of the building in Figure 1d shows the distribution of rooms on the first and second floor. Each floor has four rooms of identical size of 23 m^2: kitchen, living room, bedroom 1 and bedroom 2. Each of them was air-conditioned and illuminated by an average brightness of 500 lux. Additionally, there were a toilet and corridor.

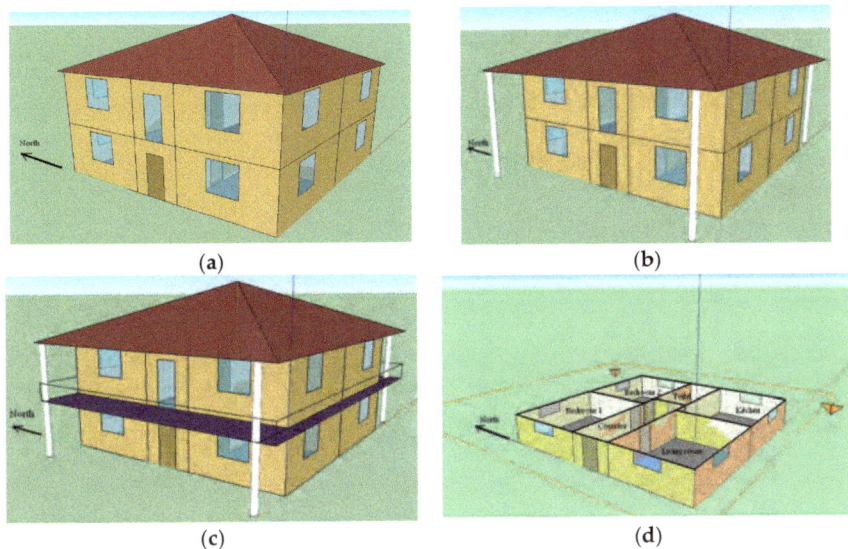

Figure 1. The house geometry: (**a**) basic building, (**b**) ORO building, (**c**) OBO building, and (**d**) the cross section of the first story of these buildings.

To study the impact of shadowing more in details, the influence of tenant activities in buildings is excluded, although in practice this is not situation. The overhang geometry is shown in Figure 2 for the ORO and OBO building.

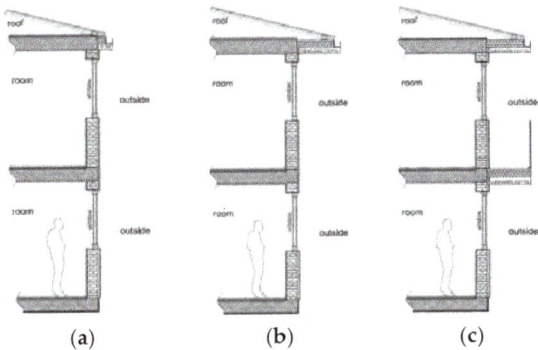

Figure 2. Overhang geometry: (**a**) basic building, (**b**) ORO building and (**c**) OBO building.

Each investigated building had a total floor area of 234 m², of which F = 186 m² were cooled and heated. The constructions used in the envelope of each house are shown in Table 1. These building materials and constructions are usual in Serbian buildings and correspond to typical Serbian construction materials. The windows were double glazed with the air gap of 15 mm, and the U-value of 2.72 W/(m²K). Inward opening side-hung windows are implemented in modeled buildings. The ratio of the areas of glass surface to that of the external wall surface was 13.96%. Then, the total area of the exterior walls was 224 m² (with the roof of 358 m²) and that of the windows was 32 m².

Table 1. Materials used in the envelope of the buildings.

Construction	Layers	Material	Thickness [m]	Conductivity [W/m·K]	Density [kg/m³]	Specific Heat [J/kg·K]
External wall	Outside Layer	Cementmortar	0.015	0.81	1600	1050
	Layer 2	Polystyrene	0.15	0.041	20	1260
	Layer 3	Clay block	0.19	0.52	1200	920
	Layer 4	Lime mortar	0.015	0.81	1600	1050
Inner wall	Outside Layer	Lime mortar	0.015	0.81	1600	1050
	Layer 2	Clay block	0.19	0.52	1200	920
	Layer 3	Lime mortar	0.015	0.81	1600	1050
Ceiling panel	Outside Layer	Cement screed	0.04	1.4	2100	1050
	Layer 2	Glass wool	0.08	0.04	50	840
	Layer 3	Monta block	0.16	0.6	1200	920
	Layer 4	Lime mortar	0.015	0.81	1600	1050
Floor (parquet)	Outside Layer	Sand	0.2	0.81	1700	840
	Layer 2	Concrete	0.15	0.93	1800	960
	Layer 3	PVC foil	0.00015	0.19	1460	1100
	Layer 4	Stirodure	0.05	0.03	33	1260
	Layer 5	Cement screed	0.04	1.4	2100	1050
	Layer 6	Parquet	0.02	0.21	700	1670
Floor (tiles)	Outside Layer	Sand	0.2	0.81	1700	840
	Layer 2	Concrete	0.15	0.93	1800	960
	Layer 3	PVC foil	0.00015	0.19	1460	1100
	Layer 4	Stirodure	0.05	0.03	33	1260
	Layer 5	Cement screed	0.04	1.4	2100	1050
	Layer 6	Ceramic tiles	0.015	0.87	1700	920
Roof	Outside Layer	Roof tiles	0.03	0.99	1900	880
	Layer 2	Air gap/wood	0.035	0.14	550	2090
	Layer 3	Glass wool/wood	0.08	0.04	50	840
	Layer 4	Gypsum board	0.012	0.19	800	1090

The installed windows and doors on the building envelope provide the infiltration of 0.5 ach. The infiltration parameter has been adopted for load calculations to ensure minimum outdoor fresh air for building zones without any forced ventilation. It was assumed that these rooms would have almost the same occupancy, lighting, and small power schedule (see Figure 3).

The heating and cooling are assumed to operate according to the schedules, during the entire year, to meet the temperature heating and cooling setpoints given in Table 2.

Table 2. Setpoint Schedule.

Heating			Cooling		
Period	15 October to 15 April		Period	15 May to 15 September	
06.00–22.00	20 °C	22 °C	06.00–22.00	24 °C	
22.00–06.00	18 °C	18 °C	22.00–06.00	30 °C	
Zone	Kitchen, living room, bedroom 1, bedroom 2	Toilet	Zone	Kitchen, living room, bedroom 1, bedroom 2	

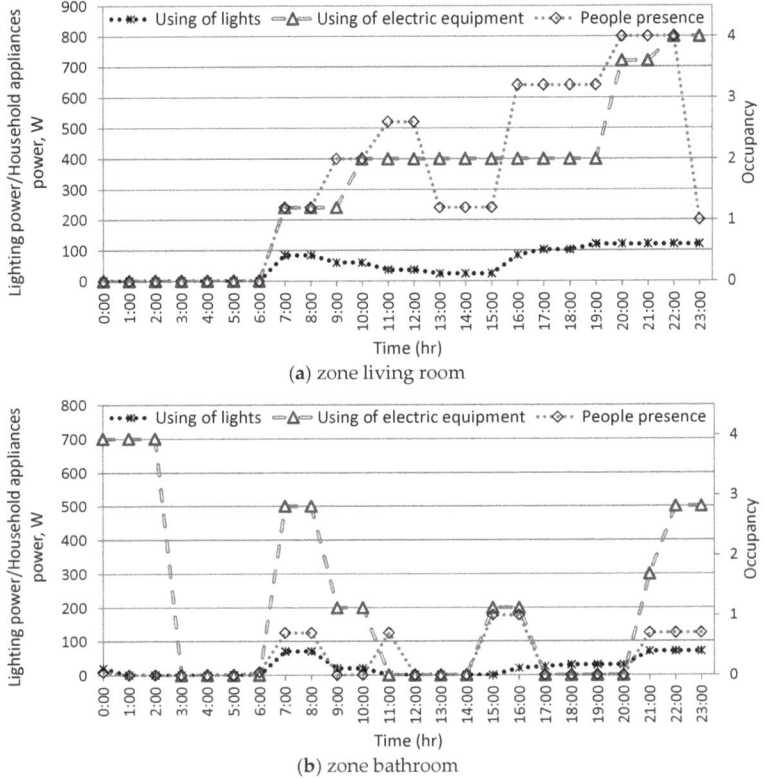

Figure 3. Schedules of the people presence, the use of lighting devices and the use of electric equipment (**a**) in a living room and (**b**) in a bathroom.

2.2. Location and Climate

The investigated residential buildings were located in the city of Belgrade (Republic of Serbia). Its average height above sea level is about 117 m, its latitude is 44°82′ N and longitude 20°28′ E. The time zone for Belgrade is GMT+1 h. Belgrade has a moderate continental climate with four defined seasons (winter, spring, summer, autumn). In the city of Belgrade summers are very warm and humid, while the winters are cool and snowy. The EnergyPlus uses weather data from its own database file, which contains a large variety of parameters: dry bulb temperatures (minimum and maximum), relative humidity, air pressure etc. Figure 4 represents monthly averages weather data for Belgrade—maximum and minimum air temperature and relative air humidity [28].

The EnergyPlus software also calculates solar radiation for every day in the year. Daily average solar radiation for Serbia is different in different parts of country: it is about 1.1 kWh/m^2 at the north and 1.7 kWh/m^2 at the south in January; in July it is about 5.9 kWh/m^2 at the north and 6.6 kWh/m^2 at the south of Serbia. Annually average solar radiation in Serbia is from 1200 kWh/m^2 for north-west to 1800 kWh/m^2 at the south of Serbia [29]. Solar radiation is dependent on the time of day and the sun's angle toward Earth. This angle varies by latitude and longitude, and season. Also, atmospheric conditions can affect radiation levels—clouds, air pollution and the hole in the ozone layer. These factors cause typical radiation levels to differ. Figure 5 presents average monthly values of solar radiation (direct, diffuse and global) for Belgrade, obtained from EnergyPlus' own weather file [29]. In accordance with the EnergyPlus software, direct solar radiation is measured as beam normal solar irradiance, while global and diffuse solar radiation are measured at a horizontal plane. That's the reason why global solar radiation is not equal to the sum of direct and diffuse solar radiation.

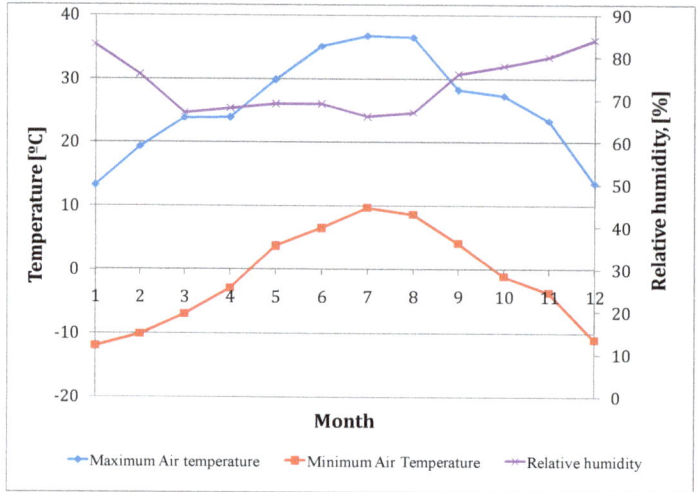

Figure 4. Monthly weather data for Belgrade, from EnergyPlus weather file.

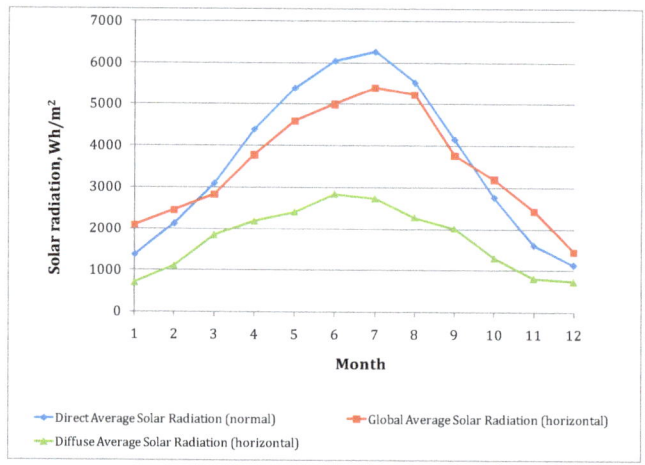

Figure 5. Solar radiation for Belgrade, from EnergyPlus weather file.

2.3. Software–Simulation and Optimization

In this research, two software packages were used: EnergyPlus [28] and GenOpt [30]. With these packages, energy research was performed for three buildings: the basic building, the ORO building and the OBO building, shown in Figure 1. For the basic building, energy simulations were done by using EnergyPlus. By using the EnergyPlus and Genopt software, energy optimizations were performed for the ORO building and the OBO building.

The basic house shown in Figure 1a and the ORO house shown in Figure 1b were modeled by using idf files of EnergyPlus. Then, an idf model template which was made included four variables: the depth of south roof overhang (s_r), the depth of north roof overhang (n_r), the depth of west roof overhang (w_r), and the depth of roof east overhang (e_r). These variables are simultaneously varied between their minimum (0.2 m) and maximum values (3 m). For variation, the GPS HookeJeeves algorithm was used. The objective function (the minimum of E_{tot}) was programmed by Genopt code.

The OBO house shown in Figure 1c was modeled also by using an idf file of EnergyPlus, which included also four variables: the depth of south balcony overhang (s_b), the depth of north balcony overhang (n_b), the depth of west balcony overhang (w_b), and the depth of east balcony overhang (e_b). The objective function (the minimum of E_{tot}) was also programmed by using Genopt with HookeJeeves algorithm. At this idf model template, the values of (s_r), (n_r), (w_r), and (e_r) were put constant with the optimal values obtained for the ORO house. The simulation resultsare obtainedin the output file: theoptimumhorizontal roof overhangssize, power consumptionof heating, air conditioning, lighting and total primaryenergy consumption.

To simulate the energy performance of a building, the EnergyPlus software was used, in which the architecture and all system parameters that correspond to its physical condition are set. To ensure adequate thermal comfort in winter, electric heaters are used. This is not a typical heating system used in Serbian buildings. The most significant advantageous of using electric heaters are that they have the ability to fine-tune the temperature in the room and maintain a minimum temperature (as protection against freezing). In the space that is used periodically, it can be used for heating one room or the whole house.

Heating thermostats are set to the appropriate temperature during winter. In summer to maintain proper thermal comfort in rooms air conditioners are used with the appropriate thermostats. The room air conditioners are operated by electricity. To maintain an appropriate light level, the combined impact of daylight and electric lighting is investigated by entering the appropriate parameters in a given time interval (using the DayLightingControls function implemented in EnergyPlus) [28].

Finding the optimal size of the horizontal roof overhangs was done with the Hooke Jeeves optimization method [31] with the help of GenOpt [30]. The objective function minimizes the consumption of primary energy for heating, cooling and lighting of the building and energy spent to build a horizontal roof overhangs. The program GenOpt operates with fixed parameters, and with variable parameters in which the optimization is performed. Its Ini file defines the objective function and all necessary parameters and variables that are required for optimization. The command file is given as the pattern of the traits that are necessary for the execution of the optimization algorithm.

2.4. Energy Analyses of Modeled Buildings

In these investigations, energy analyses were performed with the aim to minimize primary energy consumption of modeled buildings with optimized size of overhangs. Also there were calculated some environmental performances of the buildings, like energy payback time and greenhouse substitution time.

2.4.1. Primary Operating Energy Consumption

The annual primary operating energy consumption of a house was calculated by equation:

$$E_p = (E_{ac} + E_{eh} + E_{eq} + E_{el})K_{ec}/F, \tag{1}$$

Here, E_{ac} is annual electricity consumption by the air conditioners, E_{eh} is annual electricity consumption by the electric heaters, E_{eq} is annual electricity consumption for the electric equipment, E_{el} is annual electricity consumption for lighting, K_{ec} is primary energy factor and F is total conditioned floor area. The K_{ec} is defined as the ratio of the total primary energy consumption by energy sources and the total supplied electricity, and for Serbia K_{ec} = 3.04 [32].

2.4.2. Annualized Embodied Energy

The annualized embodied energy (AEE) for horizontal roof overhangs depends on the overhang size (width, depth, and thickness) and material.

For the geometry of roof and its overhangs (see Figure 6), AEE_r (annualized embodied energy for ORO building) is calculated as

$$AEE_r (f_n F) = \rho_c \delta_c s_{ec} A_c + \rho_{ct} \delta_{ct} s_{et} (A_{ct} - A_0), \qquad (2)$$

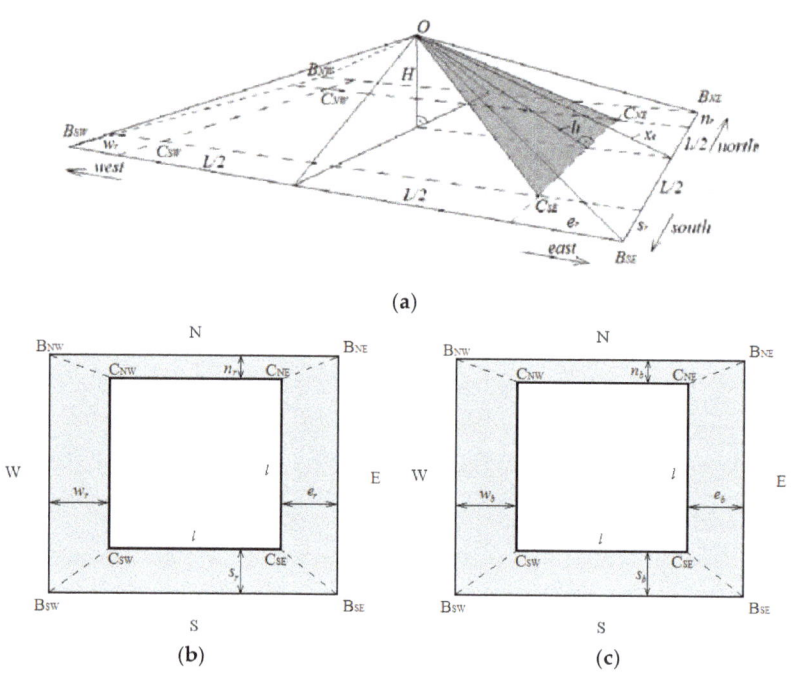

Figure 6. Sketch of roof construction: (**a**) 3d view, (**b**) view from the top of the ORO building and (**c**) view from the top of the OBO building.

Here, ρ stands for the material density for roof overhangs (concrete, $\rho_c = 2150$ kg/m^3 [1,2]; clay tile, $\rho_{ct} = 1900$ kg/m^3 [1,2]), $\delta_c = 0.18$ m stands for the thickness of the roof overhangs, $\delta_{ct} = 0.014$ m stands for the thickness of the roof clay tile, s_e stands for the roof overhangs specific embodied energy (concrete, $s_{ec} = 1.924$ MJ/kg; clay tile, $s_{et} = 6.5$ MJ/kg) [33], and f_n stands for the roof overhangs lifecycle (20 years, [1,34]). Variable A_c stands for the area of the roof overhangs made by using concrete. From Figure 6, this surface is obtained when the rectangle area (CNW CNE CSE CSW CNW) is subtracted from square area (BNW BNE BSW BSE BNW). Variable A_0 stands for the area of the tile roof surface without overhangs. This surface represents a sum of four roofs triangular surfaces of the same size A ("O CNW CNE O", "O CNE, CSE O", "O CSC, CSW O", "O CSW, CNW O"). Variable A_{ct} stands for the areas of the tile roof surface with overhangs (optimization). This surface represents a sum of four roofs triangular surfaces A_n "O BNW BNE O", A_e "O BNE, BSE O", A_s "O BSC, BSW O", A_w "O BSW, BNW O".

When $(A_{ct} - A_0)$ is multiplied by δ_{ct}, the volume of tiles is obtained because of increase in the roof area with established overhangs. The area of the roof overhangs made by using concrete (see Figure 6b) is given as:

$$A_c = L (e_r + w_r + n_r + s_r) + e_r n_r + e_r s_r + w_r s_r + w_r n_r, \qquad (3)$$

In Figure 6b, e_r stands for the depth of roof overhang at the east building side, w_r stands for the depth of roof overhang at the west building side, n_r stands for the depth of roof overhang at the north building side and s_r stands for the depth of roof overhang at the south building side, L = 10.8 m stands for length of the buildings wall.

The difference between the areas of the tile roof surface with and without overhangs is given as:

$$(A_{ct} - A_0) = [(A_{e,r} - A) + (A_{w,r} - A) + (A_{n,r} - A) + (A_{s,r} - A)], \quad (4)$$

When the equations from (3) and (4) are substituted into (2), the following equation which describes the annualized embodied energy for ORO building is obtained:

$$AEE_r(f_nF) = \{\rho_c \delta_c s_{sc}[L(e_r + w_r + n_r + s_r) + e_r n_r + e_r s_r + w_r s_r + w_r n_r] + \rho_{ct} \delta_{ct} s_{sct}[(A_{e,r} - A) + (A_{w,r} - A) + (A_{n,r} - A) + (A_{s,r} - A)]\}, \quad (5)$$

The variable A (area) is calculated as:

$$A = Lh/2, \quad h = [(L/2)^2 + H^2]^{1/2}, \quad A = L[(L/2)^2 + H^2]^{1/2}/2, \quad (6)$$

Here, $h_r = ((L/2)^2 + H^2)^{1/2} = 5.99$ m = const., H = 2.6 m stands for the height of the roof and the values of $x_{e,r}, x_{w,r} x_{n,r} x_{s,r}$ in $A_{i,r}, A_{w,r}, A_{n,r}$, and $A_{s,r}$ are the following:

$$x_{e,r} = [(L/2 + e_r)^2 + H^2]^{1/2}, \quad x_{w,r} = [(L/2 + w_r)^2 + H^2]^{1/2}$$
$$x_{n,r} = [(L/2 + n_r)^2 + H^2]^{1/2}, \quad x_{s,r} = [(L/2 + s_r)^2 + H^2]^{1/2}, \quad (7)$$

Finally, the areas $A_{e,r}, A_{w,r}, A_{,r}$, and $A_{s,r}$ are the following:

$$A_{e,r} = x_{e,r}(L + s_r + n_r)/2 = (L + s_r + n_r)[(L/2 + e_r)^2 + H^2]^{1/2}/2$$
$$A_{w,r} = x_{w,r}(L + s_r + n_r)/2 = (L + s_r + n_r)[(L/2 + w_r)^2 + H^2]^{1/2}/2$$
$$A_{n,r} = x_{n,r}(L + e_r + w_r)/2 = (L + e_r + w_r)[(L/2 + n_r)^2 + H^2]^{1/2}/2$$
$$A_{s,r} = x_{s,r}(L + e_r + w_r)/2 = (L + e_r + w_r)[(L/2 + s_r)^2 + H^2]^{1/2}/2, \quad (8)$$

When these values are substituted in (5) then:

$$AEE_r(f_nF) = \{\rho_c \delta_c s_{sc}[L(e_r + w_r + n_r + s_r) + e_r n_r + e_r s_r + w_r s_r + w_r n_r] + \rho_{ct} \delta_{ct} s_{sct} 1/2((L + s_r + n_r)[(L/2 + e_r)^2 + H^2]^{1/2} + (L + s_r + n_r)[(L/2 + w_r)^2 + H^2]^{1/2} + (L + e_r + w_r)[(L/2 + n_r)^2 + H^2]^{1/2} + (L + e_r + w_r)[(L/2 + s_r)^2 + H^2]^{1/2} - 4L[(L/2)^2 + H^2]^{1/2}\}, \quad (9)$$

The annualized embodied energy (AEE_b) for horizontal balcony overhangs (Figure 6c) depends on overhang size (width, depth, and thickness) and material. AEE_b is calculated as:

$$AEE_b(f_b F) = \rho_c \delta_b A_b s_{ec}, \quad (10)$$

where $\delta_b = 0.18$ m stands for the thickness of the balcony overhangs, $s_{ec} = 1.924$ MJ/kg stands for the balcony overhangs embodied energy [8], f_b stands for the roof overhangs lifecycle (20 years). L = 10.8 m stands for the length of the buildings wall, h_b stands for the depth of the balcony overhangs.

The area of the roof overhangs made by using concrete (see Figure 6b) is given as:

$$A_b = L(e_b + w_b + n_b + s_b) + e_b n_b + e_b s_b + w_b s_b + w_b n_b, \quad (11)$$

In Figure 6c, e_b stands for the depth of balcony overhang at the east building side, w_b stands for the depth of balcony overhang at the west building side, n_b stands for the depth of balcony overhang at the north building side and s_b stands for the depth of balcony overhang at the south building side, L = 10.8 m stands for the length of the buildings wall.

2.4.3. Partial Annualized Primary Energy Consumption

The partial annualized primary energy consumption is equal to the sum of the primary operating energy consumption E_p and annualized embodied energy (AEE_r, AEE_b):

$$E_{tot} = E_p + AEE_r + AEE_b, \qquad (12)$$

This equation is the objective function for the optimization routine. For the ORO building, the optimization is performed in respect to the four depths of the roof overhangs e, w, n, and s (when $AEE_b = 0$). For the OBO building, the optimization is performed in respect to the four depths of the balcony overhangs e_b, w_b, n_b, and s_b (when $AEE_r = 0$).

2.4.4. Primary Operating Energy Savings

When the optimized overhangs are installed, the achieved primary operating energy savings for heating, cooling and lighting in buildings (in percents) is given as:

$$e_{psav} = 100\ (E_{p,0} - E_{p,opt})/E_{p,0}, \qquad (13)$$

Here, $E_{p,opt}$ stands for primary operating energy consumption after installation of optimized overhangs, $E_{p,0}$ stands for primary operating energy consumption without roof overhangs.

2.4.5. Energy Payback Time

Energy payback time (EPBT) is time, in years, required to primary energy savings disannul the primary energy spent to overhangs building, and it is given in next equation [35]:

$$EPBT = (AEE\ (f_n))/(E_{p,0} - E_{p,opt}), \qquad (14)$$

The energy recovery (ER) is defined as number of time cycles due to primary energy saving (generated during whole lifecycle) is more than the primary energy which is needed for overhangs building. The energy recovery (ER) is given by:

$$ER = (E_{p,0} - E_{p,opt})/AEE, \qquad (15)$$

2.4.6. Greenhouse Substitution Time for Horizontal Overhangs

Greenhouse substitution time for horizontal overhangs (GHGST) is defined as the time period (in years) required for substituting the entire amount of CO_2 emitted during the construction of horizontal overhangs due to the effect of emission reductions from the operation of the same horizontal overhangs. The amount of CO_2 emitted in a process of production, transportation, building and installation of a horizontal roof overhangs (in t CO_2) [1,2] is:

$$G_{CO2} = AEE_r(f_n)\ GHG_c, \qquad (16)$$

where GHG_c stands for CO_2 emissions intensity of the production of concrete in tCO_2/t concrete. The annual decrease of the emission of CO_2 due to the application of horizontal roof overhangs is:

$$S_{CO2} = (E_{p,0} - E_{p,opt})\ k_{CO2,ec}, \qquad (17)$$

where $k_{CO2,ec}$ stands for the equivalent to CO_2 emissions for an energy mix for electricity production. Then, CO_2 substitution time is given as:

$$GHGST = \rho\ \delta\ l\ (h_E + h_S + h_W + h_N)\ GHG_c/[(E_{p,0} - E_{p,opt})\ k_{CO2,ec}], \qquad (18)$$

3. Results and Discussion

Obtained optimal depths of the roof overhangs are listed in Table 3. A0 represents the results for the basic building, A1 represent the results for the ORO building (with the optimized roof overhangs) and A2 represents the results for the OBO building (with optimized depths of balconies used as overhangs, the depths of roof overhangs are the same as that for the ORO building).

Table 3. Results with implemented cooling, heating and lighting control.

	Depth of Overhangs (m)			
	East	South	West	North
A0	0.2	0.2	0.2	0.2
A1	2.1	0.95	1.9	0.2
A2	2.6	0.7	2.4	0.4

Basic building A0 has no overhangs, so the values of 0.2 m represent only roof protrusions on the second floor. For ORO building, with optimization routine it was obtained the maximum value of east roof overhang (2.1 m) and west roof overhang (1.9 m), while the south roof overhang was 0.95 m. For the OBO building (with optimal values for roof overhangs), a maximum value of the east balcony overhang of 2.6 m and a west balcony overhang of 2.4 m were obtained, while the south balcony overhang was 0.7 m and north balcony overhang was 0.4 m. These values can be explained by the small angle of incidence of the Sun in the morning and in the afternoon during the summer period, so solar gains during that period can be significant. With implementation of overhangs overheating can be avoided through the summer months, with a reduction of solar gains.

Values of solar radiation through the windows and its reduction, in the basic and OBO buildings (monthly) are presented in Figure 7. It is not difficult to conclude that solar gains through the windows at the OBO building are significantly lower than the solar gains through the windows of the basic building. Optimized horizontal overhangs provide great protection from the solar radiation (especially during the March–October period), preventing overheating and thus reducing cooling energy consumption, i.e., the total building energy consumption.

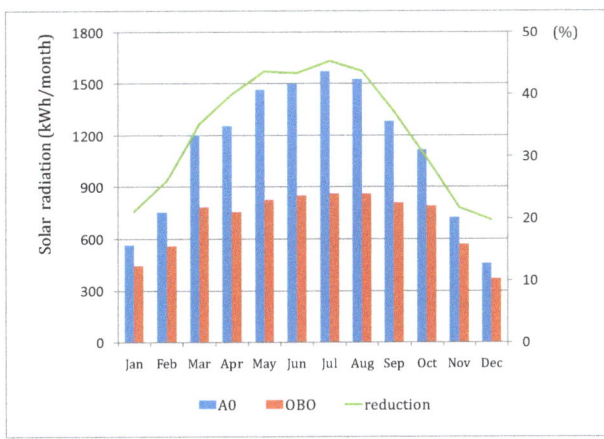

Figure 7. Solar radiation through windows and its reduction in OBO house.

The total value of the annual solar radiation through the windows in the basic building is 13,404.7 kWh (57.29 kWh/m^2), and in the OBO building with optimized overhangs it is 8451.11 kWh (36.12 kWh/m^2). The annual difference in solar radiation of these two analyzed buildings is 4953.59 kWh (21.17 kWh/m^2) and solar radiation is reduced by 36.95% (annually). The monthly reduction of solar radiation through the windows is lower in winter months, while it has a greater value during spring, summer and autumn months.

The specific final energy consumption for lighting, electric equipment, heating and air-conditioners for cooling (in kWh/m^2) for all analyzed buildings is presented in Figure 8. With the implementation of overhangs, a small increase in annual lighting energy was observed for the ORO and OBO buildings compared to the basic building (0.23 kWh/m^2 and 0.53 kWh/m^2, respectively). Due to overhangs, a smaller amount of daylight enters these buildings, so electric lighting is used more during some time intervals (according to the software's DayLightingControls function). In the winter period there are small solar gains through the windows of the ORO and OBO buildings, so the specific final heating energy consumption increases; for ORO building the increase of heating energy consumption is 2.63 kWh/m^2; for OBO building the increase of heating energy consumption is 4.98 kWh/m^2. During the summer period, overhangs prevent building overheating, so the amount of cooling energy is significantly reduced. For the ORO building the decrease of annual cooling energy consumption is 6.18 kWh/m^2; for the OBO building the decrease of annual cooling energy consumption is 11.9 kWh/m^2. The specific final energy consumption for electric equipment is the same for all the buildings, 21.98 kWh/m^2.

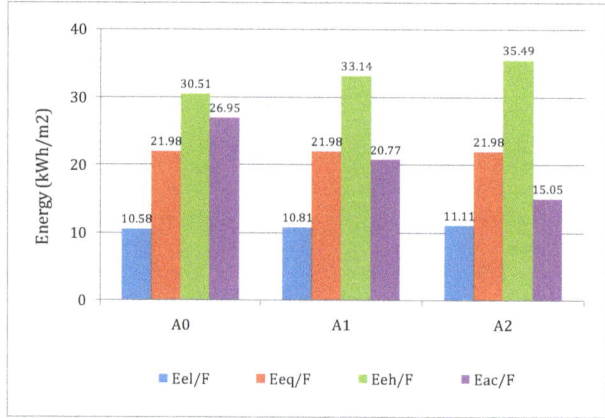

Figure 8. Specific final energy consumption for lighting, electric equipment, heating and cooling in the analyzed buildings.

The total final and primary energy consumption for the analyzed buildings are shown in Table 4. The highest annual energy consumption corresponds to the basic building, case A0, (90.02 kWh/m^2 of final energy and 273.7 kWh/m^2 of primary energy), then the ORO building, case A1, (86.07 kWh/m^2 of final energy and 263.5 kWh/m^2 of primary energy), while the lowest energy consumption corresponds to the OBO building, case A2, (83.63 kWh/m^2 of final energy and 254.2 kWh/m^2 of primary energy). Annual energy savings are 3.69% for the ORO building, and 7.12% for the OBO building, compared to the basic building without horizontal overhangs.

Table 4. Final and primary energy consumption in analyzed buildings.

	E$_{final}$	E$_{prim}$
A0	90.02	273.7
A1	86.70	263.6
A2	83.63	254.2

The percentages of primary energy reduction and specific final energy reduction for the ORO and OBO buildings, compared to a basic house, are shown in Figure 9. Total primary energy consumption in the ORO building with the optimized roof overhangs (A1), was 3.69% lower, compared to the basic building without overhangs. Cooling energy consumption was 22.95% lower, while heating energy increased by 8.62%, in the form of energy consumption for lighting (2.1%). Significantly greater energy saving is achieved in OBO buildings with the optimized roof overhangs and optimized depths of balconies used as overhangs (A2)—total primary energy consumption in the OBO building is 7.12% lower, compared to the basic building. A greater cooling energy saving is also obtained (44.15%), while the primary heating energy was increased by 16.33%, mainly as energy consumption for lighting (4.98%). These values represent very significant energy savings.

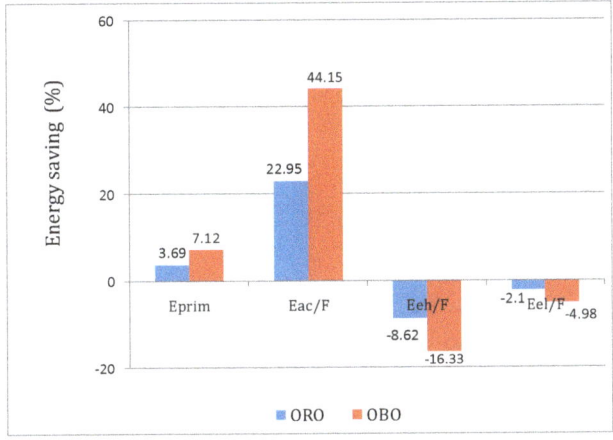

Figure 9. Primary and specific final energy savings in analyzed buildings (in percents, compared to basic building without overhangs).

The duration required for any primary energy savings to compensate for the primary energy needed to build overhangs, in accordance with Equation (14) is EPBT = 6.44 years for the ORO house, and EPBT = 6.60 years for the OBO house. The number of time primary energy savings (Equation (14)) are generated during the lifecycle by using the optimal overhangs more than the primary energy needed to build overhangs is ER= 3.11 for the ORO, and 3.03 for the OBO house (Table 5).

Table 5. Embodied energy, AEEr, EPBT and ER.

	Embodied Energy			EPBT Years	ERTimes
	AEEr (kWh/m^2)	AEErConcrete (kWh/m^2)	AEErClay Tile (kWh/m^2)		
A0	/	/	/	/	/
A1	3.28	2.66	0.62	6.44	3.11
A2	6.44	5.82	0.62	6.60	3.03

The CO_2 substitution time (GHGST) is the time required to substitute the entire amount of CO_2 emitted during the construction of a technical system due to the effect of emission reductions resulting from operation of the system. The amount of CO_2 emitted during the construction of concrete horizontal roof overhangs is GCO_2 = 3.03 t CO_2. Then, the CO_2 emissions intensity of the concrete production is taken as GHGc = 0.13 t CO_2/t concrete from [35,36]. The CO_2 emission reductions resulting from the application of horizontal roof overhangs is annually S_{CO2} = 2.07 t CO_2 where equivalent CO_2 emissions for EPS $kCO_{2,ec}$ = 3.1, taken from [32]. Finally, the CO_2 substitution time is GHGST = 1.47 years.

Validation of the Results

The average annual specific primary energy consumption for the buildings which belong to the groups D1, E1 and F1, according to the national residential buildings typology in Serbia [26] (with applied thermal insulation, replaced old inefficient windows and no overhangs), is 281 kWh/m² [37]. This value is near the annual specific primary energy consumption for the basic building without overhangs, analyzed in these investigations (273.7 kWh/m²). The investigated basic building, like the ORO and OBO buildings, represents typical buildings from the D1, E1 and F1 groups, with thermal insulation on external walls and energy efficient windows. Having in mind these facts, it can be said that the data obtained by simulations and optimizations are valid.

4. Sensitivity to the Accuracy of the Input Data

In these investigations, optimization is performed for the horizontal roof overhangs. The overhangs are made by using concrete of the specific embedded energy (due its production process of materials, construction process, manufacture and installation) s_{ec} = 1.924 MJ/kg. As there are different conditions of concrete production and construction, specific embedded energy of concrete as input data may be different. An analysis, which presents how these changes influence to the output simulation results, for ORO and OBO building, is given below (Tables 6 and 7):

$$e_{p\%} = 100 \, (E_{\text{prim} +/-20\%} - E_{\text{primref}})/E_{\text{primref}}, \tag{19}$$

Table 6. Sensitivity to the accuracy of the specific embodied energy for ORO house.

	ORO Building									
	Depth of Horizontal Roof Overhangs				Energy					
	EAST (h_E)	SOUTH (h_S)	WEST (h_W)	NORTH (h_N)	E_{prim}	E_{ac}	E_{eh}	E_{eq}	E_{el}	$e_{p\%}$
	m	m	m	m	kWh/m²	kWh/m²	kWh/m²	kWh/m²	kWh/m²	%
S_{ec}	0.2	0.2	0.2	0.2	273.7	26.95	30.51	21.98	10.58	/
$S_{ec+20\%}$	1.9	0.95	1.8	0.2	263.7	21.02	33.03	21.98	10.80	0.08
S_{ecref}	2.1	0.95	1.9	0.2	263.5	20.77	33.14	21.98	10.81	ref.
$S_{ec-20\%}$	2.1	0.95	2.2	0.51	262.8	20.50	33.24	21.98	10.81	−0.27

Table 7. Sensitivity to the accuracy of the specific embodied energy for OBO house.

	OBO Building									
	Depth of Horizontal Balcony Overhangs				Energy					
	EAST (h_E)	SOUTH (h_S)	WEST (h_W)	NORTH (h_N)	E_{prim}	E_{ac}	E_{eh}	E_{eq}	E_{el}	$e_{p\%}$
	m	m	m	m	kWh/m²	kWh/m²	kWh/m²	kWh/m²	kWh/m²	%
S_{ec}	0.2	0.2	0.2	0.2	273.66	26.95	30.51	21.98	10.58	/
$S_{ec+20\%}$	2.2	0.7	2.1	0.2	254.64	15.43	35.32	21.98	11.04	0.16
S_{ecref}	2.6	0.7	2.4	0.4	254.24	15.05	35.49	21.98	11.11	ref.
$S_{ec-20\%}$	2.6	0.7	2.4	0.5	254.16	15.00	35.51	21.98	11.11	−0.03

During optimization for ORO building, if the specific embodied energy s_{ec} increases by 20%, the depth of east and west roof overhangs decreases, increasing the primary energy consumption by 0.08%. In this case cooling energy increases, while heating energy decreases. If the specific embodied energy s_{ec} decreases by 20%, depth of west and north roof overhang increases, with a decreasing of the primary energy consumption by 0.27%. The amount of cooling energy decreases in this case, while heating energy increases.

During optimization for the OBO building, if the specific embodied energy s_{ec} increases by 20%, the depth of east, west and north roof overhangs decreases, increasing the primary energy consumption by 0.16%. In these simulations, cooling energy has a small increase, while heating energy has a small decrease. If the specific embodied energy s_{ec} decreases by 20%, depth of north roof overhang increases, decreasing the primary energy consumption by 0.03%. In this case, the amount of cooling energy has a small decrease, while amount of heating energy has a very small increase. The obtained results show a very small deviation (in the range of −0.03–0.16%) from the main results obtained by simulations and optimization process.

5. Conclusions

This paper represents a numerical investigation about how shading by horizontal roofs and balcony overhangs influences the annual primary and final energy consumption for heating, cooling and lighting in residential building in Serbia. Energy consumption is investigated for three buildings modeled in the EnergyPlus software in accordance with national residential buildings typology in Serbia. These buildings were detached with two-floors and they had almost the same characteristics. They only differed in their overhang characteristics such as type and dimensions. They were: (1) a basic building, (2) a building with optimized roof overhangs (ORO building), and (3) a building with optimized balcony overhangs (OBO building). The basic building had roof overhangs with depths of 0.2 m, the ORO building had optimized roof overhangs and the OBO building had optimized balcony overhangs (besides the roof overhangs with the depths as in the ORO building).

The optimal sizes of the horizontal overhangs, which are placed over the east, west, north, and south walls, are obtained by simultaneous operation of the two programs EnergyPlus and GenOpt. The aim of the optimization was to minimize the primary energy consumption for heating, cooling and lighting (embodied energy of concrete horizontal roof overhangs was taken into account).

Simulation results performed in this paper showed that horizontal overhangs on the analyzed buildings can reduce annual solar radiation through the windows by 36.95%. The reduction of the solar radiation through windows is less through the winter months than in summer months, so there is the possibility of optimizing the size of overhangs separately, for each side. This optimization process can achieve even greater energy savings and reduced primary energy consumption.

For the ORO building, the optimal dimensions of roof overhang depths are 2.1 m facing east, 0.95 m facing south, 1.9 m facing west and 0.2 m facing north. The reduction of heat gains due to solar radiation decreases the energy consumption for cooling by 22.95%, while the energy consumption for heating and lighting increases by 8.62 and 2.1%, respectively. Total primary energy consumption is reduced by 3.69%.

For the OBO building, the optimal dimensions of the depths of balconies used as overhangs, (the depths of roof overhangs are the same as that for the ORO house) are 2.6 m facing east, 0.7 m facing south, 2.4 m facing west and 0.4 m facing north. The reduction of heat gains due to solar radiation decreases the energy consumption for cooling by 44.15%, while the energy consumption for heating, and lighting are increased by 16.33% and 4.98%, respectively. The total primary energy consumption is reduced by 7.12%.

The time needed for the primary energy savings to compensate for the primary energy needed to build overhangs is 6.44 years for the ORO house and 6.60 years for the OBO house. The number of time cycles of primary energy savings generated during the lifecycle by using the optimal overhangs more than the primary energy needed to build overhangs is ER = 3.11 for the ORO and 3.03 for the OBO building. The CO_2 substitution time GHGST is 1.47 years.

In the sensitivity analyses, we investigated how changes of the specific embedded energy of concrete, as input data, influence to the output simulation results for the ORO and OBO buildings. The obtained results show a very small deviation from the main results obtained by simulations and optimization process (in the range of −0.03–0.16%). Energy Plus is a software package which is intensively validated and has been tested using the IEA HVAC BESTEST E100–E200 series of tests [25,32]. Regardless of the high accuracy of the obtained results, all software tools, no matter how good and powerful they may be, can give a certain deviation in terms of the accuracy of the results.

Implementation of the roof overhangs on the existing building involves nailing rafter extensions onto existing rafters. When the rafter extensions are installed, then sheathing and roofing is started. If the roof is near the end of its life cycle, it is a good time to reroof the entire house. In that case, the overhangs will be built together with the roof. The case of implementation of balcony overhangs on an existing building can be very difficult job. First, concrete pillars must be added at the building construction, after that connection elements between the building and pillars have to be installed, and finally the balconies can be built. This process is more complex than implementation of roof overhangs.

Our future research may deal with analyses of different types of roof (flat and sloped), different roof construction types used in Serbia and different kinds of shading elements. Then, the optimization results may be compared if the roof is built using concrete, steal, laminated wood, or classic wood. In addition, the economics should be analyzed to show how these solutions are acceptable in practice and what any eventual energy penalty of this acceptance is.

Author Contributions: Conceptualization, D.N. and S.D.; methodology, D.N. and J.S.; software, S.D. and J.S.; validation, J.S., S.D. and J.R.; formal analysis, D.N., S.D. and J.R.; investigation, D.N., S.D., J.S.; resources, D.N., J.S. and J.R.; data curation S.D. and J.R.; writing—original draft preparation, D.N., S.D., J.S. and J.R.; writing—review and editing, D.N. and J.S.; visualization, D.N. and J.R.; supervision, D.N. and J.R.; project administration, J.S. and S.D.; funding acquisition, D.N. All authors have read and agreed to the published version of the manuscript.

Funding: This paper presents results obtained within realization of two projects TR 33015 and III 42006, funded by Ministry of Education, Science and Technological Development of the Republic of Serbia.

Conflicts of Interest: The authors declare no conflict of interest.

References

1. Bojic, M.; Cvetkovic, D.; Bojic, L. Optimization of geometry of horizontal roof overhangs during a summer season. *Energy Effic.* **2017**, *10*, 41–54. [CrossRef]
2. Djordjevic, S.; Bojic, M.; Cvetkovic, D.; Malesevic, J.; Miletic, M. Influence of house shadowing to the consumption of primary energy for heating, cooling and lighting. In Proceedings of the 7 IQC 2013, 7th International Quality Conference, Kragujevac, Serbia, 24 May 2013.
3. Skias, I.; Kolokotsa, D. Contribution of shading in improving the energy performance of buildings. In Proceedings of the 2nd PALENC Conference and 28th AIVC Conference on Building Low Energy Cooling and Advanced Ventilation Technologies in the 21st Century, Crete Island, Greece, 27–29 September2007.
4. Kim, G.; Lim, H.S.; Lim, T.S.; Schaefer, L.; Kim, J.T. Comparative advantage of an exterior shading device in thermal performance for residential buildings. *Energy Build.* **2012**, *46*, 105–111. [CrossRef]
5. Raeissi, S.; Taheri, M. Optimum Overhang Dimensions for Energy Saving. *Build. Environ.* **1998**, *33*, 293–302. [CrossRef]
6. Imessad, K.; Derradji, L.; AitMessaoudene, N.; Mokhtari, F.; Kharchi, R. Impact of passive cooling techniques on energy demand for residential buildings in a Mediterranean climate. *Renew. Energy* **2014**, *71*, 589–597. [CrossRef]
7. Datta, G. Effect of fixed horizontal louver shading devices on thermal performance of building by TRNSYS simulation. *Renew. Energy* **2001**, *23*, 497–507. [CrossRef]
8. Littlefair, P.; Ortiz, J.; Bhaumik, C.D. A simulation of solar shading control on UK office energy use. *Build. Res. Inform.* **2010**, *38*, 638–646. [CrossRef]
9. Yao, J. An investigation into the impact of movable solar shades on energy, indoor thermal and visual comfort improvements. *Build. Environ.* **2014**, *71*, 24–32. [CrossRef]
10. Atzeri, A.; Cappelletti, F.; Gasparella, A. Internal versus external shading devices performance in office buildings. *Energy Procedia* **2014**, *45*, 463–472. [CrossRef]
11. Florides, G.A.; Tassou, S.A.; Kalogirou, S.A.; Wrobel, L.C. Measures used to lower building energy consumption and their cost effectiveness. *Appl. Energy* **2002**, *73*, 299–328. [CrossRef]
12. Liu, S.; Kwok, Y.T.; Lau, K.; Chan, P.W.; Ng, E. Investigating the energy saving potential of applying shading panels on opaque façades: A case study for residential buildings in Hong Kong. *Energy Build.* **2019**, *193*, 78–91. [CrossRef]

13. Aldawoud, A. Conventional fixed shading devices in comparison to an electrochromic glazing system in hot, dry climate. *Energy Build.* **2013**, *59*, 104–110. [CrossRef]
14. Mandalaki, M.; Zervas, K.; Tsoutsos, T.; Vazakas, A. Assessment of fixed shading devices with integrated PV for efficient energy use. *Sol. Energy* **2012**, *86*, 2561–2575. [CrossRef]
15. Stamatakis, A.; Mandalaki, M.; Tsoutsos, T. Multi-criteria analysis for PV integrated in shading devices for Mediterranean region. *Energy Build.* **2016**, *117*, 128–137. [CrossRef]
16. Kirimtat, A.; Krejcar, O.; Ekici, B.; Tasgetiren, M.F. Multi-objective energy and daylight optimization of amorphous shading devices in buildings. *Sol. Energy* **2019**, *185*, 100–111. [CrossRef]
17. Bellia, L.; Marino, C.; Minichiello, F.; Pedace, A. An overview on solar shading systems for buildings. *Energy Procedia* **2014**, *62*, 309–317. [CrossRef]
18. Kirimtat, A.; Koyunbaba, B.K.; Chatzikonstantinou, I.; Sariyildiz, S. Review of simulation modeling for shading devices in buildings. *Renew. Sustain. Energy Rev.* **2016**, *53*, 23–49. [CrossRef]
19. Palmero-Marrero, A.I.; Oliveira, A.C. Effect of louver shading devices on building energy requirements. *Appl. Energy* **2010**, *87*, 2040–2049. [CrossRef]
20. Valladares-Rendón, L.G.; Schmid, G.; Lo, S.L. Review on energy savings by solar control techniques and optimal building orientation for the strategic placement of façade shading systems. *Energy Build.* **2017**, *140*, 458–479. [CrossRef]
21. Al-Masrani, S.M.; Al-Obaidi, K.M. Dynamic shading systems: A review of design parameters, platforms and evaluation strategies. *Autom. Constr.* **2019**, *102*, 195–216. [CrossRef]
22. Tabadkani, A.; Roetzel, A.; Li, H.X.; Tsangrassoulis, A. A review of automatic control strategies based on simulations for adaptive facades. *Build. Environ.* **2020**, *175*. [CrossRef]
23. Nikolic, D.; Skerlic, J.; Radulovic, J. I Energy efficient buildings—Legislation and design. In Proceedings of the 2nd International Conference on Quality of Life, Kragujevac, Serbia, 8–10 June 2017.
24. Gaglia, A.G.; Tsikaloudaki, A.G.; Laskos, C.M.; Dialynas, E.N.; Argiriou, A.A. The Impact of the Energy Performance Regulations' updated on the construction technology, economics and energy aspects of new residential buildings: The case of Greece. *Energy Build.* **2017**, *155*, 225–237. [CrossRef]
25. Bojic, M.; Nikolic, N.; Nikolic, D.; Skerlic, J.; Miletic, I. A simulation appraisal of performance of different HVAC systems in an office building. *Energy Build.* **2011**, *43*, 2407–2415. [CrossRef]
26. Jovanovic-Popovic, M.; Kavran, J. Energy Efficiency and Renewal of Residential Buildings Stock. *Int. J. Contemp. Archit. New ARCH* **2014**, *1*, 93–100.
27. Dixit, M.K.; Fernandez-Solis, J.L.; Lavy, S.; Culp, C.H. Identification of parameters for embodied energy measurement. *Energy Build.* **2010**, *42*, 1238–1247. [CrossRef]
28. Department of Energy (DOE). EnergyPlus Software, Version 8.0. 2015. Available online: https://energyplus.net/ (accessed on 20 May 2020).
29. Pavlovic, T.; Milosavljevic, D.; Radonjic, I.; Pantic, L.; Radivojevic, A.; Pavlovic, M. Possibility of electricity generation using PV solar plants in Serbia. *Renew. Sustain. Energy Rev.* **2013**, *20*, 201–218. [CrossRef]
30. Wetter, M. *GenOpt—Generic Optimization Program*; Technical Report LBNL-54199 User Manual; Lawrence Berkeley National Laboratory: Berkeley, CA, USA, 2004.
31. Hooke, R.; Jeeves, T.A. Direct search solution of numerical and statistical problems. *J. Assoc. Comput. Mach.* **1961**, *8*, 212–229. [CrossRef]
32. Bojic, M.; Djordjevic, S.; Malesevic, J.; Miletic, M.; Cvetkovic, D. A simulation appraisal of a switch of district to electric heating due to increased heat efficiency in an office building. *Energy Build.* **2012**, *50*, 324–330. [CrossRef]
33. Peng, J.; Lun, L.; Yang, H. Review on life cycle assessment of energy payback and greenhouse gas emission of solar photovoltaic systems. *Renew. Sustain. Energy Rev.* **2013**, *19*, 255–274. [CrossRef]
34. Cabeza, L.F.; Barreneche, C.; Miro, L.; Martınez, M.; Fernandez, A.I.; Urge-Vorsatz, D. Affordable construction towards sustainable buildings: Review on embodied energy in building materials. *Curr. Opin. Environ. Sustain.* **2013**, *5*, 229–236. [CrossRef]
35. Kim, H.C.; Fthenakis, V.M. Life cycle energy demand and greenhouse gas emissions from an Amonix high concentrator photovoltaic system. In Proceedings of the IEEE 4th World Conference on PV Energy Conversion, Waikoloa, HI, USA, 7–12 May 2006.
36. Goggins, J.; Keane, T.; Kelly, A. The assessment of embodied energy in typical reinforced concrete building structures in Ireland. *Energy Build.* **2010**, *42*, 735–744. [CrossRef]

37. Novikova, A.; Csoknyai, T.; Jovanovic-Popovic, M.; Stankovic, B.; Zivkovic, B.; Ignjatovic, D.; Sretenovic, A.; Szalay, Z. *The Typology of the Residential Building Stock in Serbia and Modelling Its Low-Carbon Transformation*, 1st ed.; The Regional Environmental Center for Central and Eastern Europe (REC): Szentendre, Hungart, 2015; pp. 16–45.

 © 2020 by the authors. Licensee MDPI, Basel, Switzerland. This article is an open access article distributed under the terms and conditions of the Creative Commons Attribution (CC BY) license (http://creativecommons.org/licenses/by/4.0/).

Article

Effect on the Thermal Properties of Mortar Blocks by Using Recycled Glass and Its Application for Social Dwellings

Vicente Flores-Alés [1], Alexis Pérez-Fargallo [2], Jesús A. Pulido Arcas [3] and Carlos Rubio-Bellido [1,*]

1. Department of Building Construction II, Universidad de Sevilla, 41012 Seville, Spain; vflores@us.es
2. Department of Building Science, University of Bío-Bío, Concepción 1202, Chile; aperezf@ubiobio.cl
3. Center for Research and Development of Higher Education, Graduate School of Arts and Sciences, College of Arts and Sciences, The University of Tokyo, Meguro City, Tokyo 153-8907, Japan; jpulido@g.ecc.u-tokyo.ac.jp
* Correspondence: carlosrubio@us.es; Tel.: +34-686-135-595

Received: 25 September 2020; Accepted: 29 October 2020; Published: 31 October 2020

Abstract: Including recycled waste material in cement mixes, as substitutes for natural aggregates, has resulted in diverse research projects, normally focused on mechanical capacities. In the case of recycled glass as an aggregate, this provides a noticeable improvement in thermal properties, depending on its dosage. This idea raises possible construction solutions that reduce the environmental impact and improves thermal behavior. For this research, an extended building typology that is susceptible to experiencing the risk of energy poverty has been chosen. The typology is typical for social housing, built using mortar blocks with crushed glass. First, the basic thermophysical properties of the mortars were determined by laboratory tests; after that, the dynamic thermal properties of representative constructive solutions using these mortars were simulated in seven representative climate zones in Chile. An analysis methodology based on periodic thermal transmittance, adaptive comfort levels and energy demand was run for the 21 proposed models. In addition, the results show that thermal comfort hours increases significantly in thermal zones 1, 2, 3 and 6; from 23 h up to 199 h during a year. It is in these zones where the distance with respect to the neutral temperature of the m50 solution reduces that of the m25 solution by half; i.e., in zone 1, from −429 °C with the m25 solution to −864 °C with the m50. This research intends to be a starting point to generate an analysis methodology for construction solutions in the built environment, from the point of view of thermal comfort.

Keywords: crushed glass; periodic thermal transmittance; energy demand; adaptive comfort; social housing

1. Introduction

The average glass recycling rate in the European Union (28 member countries) has reached a 76% threshold for the first time. This means that more than 12.4 million tons of glass were collected throughout the European Union in 2017, 2% more than in 2016 [1,2]. The proportion of recycled glass in the US is currently estimated to be around 35%. There is very little reliable data available for other countries. In the case of China, the recycling rate for container glass is currently still below 20% and in South Africa it is over 41% [3]. It is important to highlight the impact of domestic recycling in a recovery chain of simple and safe containers. Glass is inert and maintains its inherent properties regardless of how many times it has been recycled. If suitably collected, it can be recycled ad infinitum in a closed circuit, hence repeatedly using this waste will help preserve the natural resources of the Earth, minimizing landfill spaces and saving energy and money [4].

1.1. Effect of Crushed Glass on Thermal and Mechanical Properties of Mortars

Cement mixes have traditionally been researched as construction products, in particular incorporating inert aggregates with sufficient resistance capacity to be substitutes of the virgin aggregate, with the ultimate goal of reducing the environmental impact associated with gravel and sand extraction. The chemical composition of glass mainly has a formless siliceous nature, which makes it compatible with natural aggregates, although when a very small size particle is used (<20 μm) [5], it has a reactive nature induced by the high alkalinity of cement [6]. The alkalinity of glass causes the breakdown of the matrix favoring, by its formless nature, the formation of calcium silicate hydrate, improving the cementitious performance [7]. The fine particles of glass have a high specific surface [8] and therefore favor high pozzolanic reaction kinetics due to the strong reaction between the alkali in the cement and the reactive silica in the glass [9]. It must also be considered that glass can intervene in alkali aggregate reactions, although the participation of recycled material in this reaction depends on the particle size, with an expansive phenomena of sizes above 1 mm being favored [10]. In fact, the research made shows that fine waste glass has a mitigating capacity of the ASR [11]. Finally, it is worth mentioning that there are incipient studies which address the capacity that incorporating glass into cement mortar provides to improve the resistance to the penetration of chlorides [12,13], and to develop bactericidal features [14,15].

Regarding resistance, several researchers have shown that crushed glass increases compression resistance [16,17], although the data has a significant spread in the results, obtaining optimal improvement percentages of between 10% and 30% compared to the reference mortar, depending on mortar type and the maximum size of the glass [18]. Regardless of the aggregate substitution, the dose conditions, and especially the w/c ratio, are determining factors in the mechanical behavior of the product [19]. In addition, Castro and de Brito (2013), regarding the mechanical behavior, showed a general improvement in terms of resistance to the carbonatation of concretes that contain glass waste (size <4 mm) as a natural fine aggregate [20].

Likewise, previous research has highlighted the capacity of glass aggregate as a substitute for sand to substantially reduce the thermal conductivity of mortar [21], demonstrating that energy savings can be achieved when using a glass aggregate instead of the sand alternative [22]. Sikora et al. have shown that substituting fine sand by WG can improve the thermal properties of cement mortars, while maintaining an acceptable mechanical strength [23].

1.2. Low-Cost Materials and Its Application to Energy-Efficient Dwellings

In the last few years, the building industry has been striving to reduce its energy consumption, and one of its strategies for this focuses on using recycled products as a construction material. In this sense, since crushed glass has beneficial effects on the thermal conductivity of mortars, is easy to obtain and has a low price, it has emerged as a viable option. Since crushed glass has a controversial effect on the mechanical properties, its use in structural elements is out of the question, being rather applied to coatings and finishing without structural function, but where thermal insulation becomes crucial to reduce the energy demands of the building.

Research on mortar blocks brings the opportunity of introducing an affordable material that also allows for a better insulation, as suggested by other authors [24], and finds a specific application in projects with limited financial resources, such as those comprising social dwellings. The use of recycled glass has wider implications, it is an environmentally friendly material, an affordable constructive solution and is technically feasible, even in countries where manpower has limited technical skills [25]. Previous research by the authors has focused on clarifying the feasibility of such materials on the basis of its chemical, mineralogical, physical, thermal and mechanical properties, and former studies claim that a percentage of recycled glass between 25% and 50% results in mortars with a lower thermal conductivity and higher density that, at the same time, have mechanical capacities comparable to mortars with natural aggregates [26,27].

Up to date, the great majority of studies in the field, as well as building codes, use the static thermal transmittance as a proxy for assessing the insulation capacity of a given material. Nevertheless, this approach ignores the complex interplay among other variables, such as the thermal inertia [28], which, in combination with the former, exerts a remarkable influence on the energy performance of buildings [29]. Static thermal transmittance relies on a simplification, where the temperature gap between the interior and the exterior of the building is constant, disregarding the effect that warm and cold climates with wide thermal oscillations might have on the energy demand of the building [30]. Recently, the UNE-EN ISO 13786 Standard [31] has taken the lead in incorporating the so-called dynamic thermal properties of building materials. In brief, this document assumes that heat flow between both sides of the envelope depends on the dynamic variation of the temperature gap through time. Starting with the calculation procedure as per the UNE-EN ISO 6946 [32], this standard considers both the static and the dynamic thermal transmittance.

Being this a novel approach, research on this area is still limited, but some researchers have already shed light on this, suggesting that periodic thermal transmittance can lead to a reduction in the energy demand of the building [33]; the authors have also made a contribution in this field, claiming that not static thermal transmittance, but thermal inertia, is the driver to improve indoor thermal comfort in social dwellings located in the Central-South area of Chile [34]. However, research in this field is still scarce and fragmented. Plenty of studies deal with the development of a new construction material with improved static thermal transmittance that, after determination of its properties by laboratory test, is applied to common constructive solutions. In the case of dynamic thermal properties, the process is more complex and there is still a research gap in considering studies that comprise the whole process: Determination of the properties by laboratory tests, implementation of the material in constructive solutions, and analysis of the effect on different aspects of energy demand and building comfort.

This study aims at filling-in this research gap by presenting a study comprising all these steps. In turn, the results of this study are expected to be applied to the design and construction of social dwellings in Chile. Subsidized housing always represents a challenge for designers and builders, as it needs to balance the constraints of a limited construction budget with the need of providing the lower strata of society with decent living standards. Chile is a representative case study for two reasons: First, this country encompasses a great variety of climates, including hot dry deserts in the North and cold steppes in the South. Second, this country has had a continuous and solid program of social dwellings, with 3.6 million subsidies granted between 1.964 and 2.015, and an estimated investment of 19 billion Euros since 1.990 [35]. At the present time, Supreme Decrees 01 and 49 establish the basic technical standards for social dwellings [36–39], which consist of predefined typologies with standardized constructive solutions: Built surfaces are between 36 and 55 m^2 and usually have a living-dining area, a kitchen, a bathroom and two or three bedrooms. Considerations of energy efficiency were introduced only after the enactment of the General Urbanism and Constructions Ordinance (OGUC, in Spanish) in 2007 [40], which was the first legislation that established the limits for the U-values of the external envelope in Chilean buildings. After this, the government has put much effort into improving the benchmark for energy efficiency by releasing technical guidelines that, although not mandatory, have started to impregnate professional practice in the country: The Standards for the Sustainable Construction of Housing, published in 2014, raised the benchmark for thermal envelopes but is still not mandatory [41].

This background describes a country that, in spite of an increasing awareness about energy efficiency in buildings, still has a long way to go. Chilean researchers have clarified how low insulated houses can have unacceptable low temperatures in winter, as low as 14 °C [42], which, in turn, may lead to a higher prevalence of respiratory illnesses [43]. The thermal adaptation of users in central-southern Chile has its own particularities, but currently, from the adaptive thermal comfort models included in the standards, the model that is part of ASHRAE 55-2017 is the one with the most similarities to those of the users [42,44].

In sum, there is a need for comprehensive research on how recycled glass incorporated into construction blocks can improve both static and thermal properties of the thermal envelope, and on how these constructive solutions may find an application in social dwellings in Chile.

The article is organized in three sections. First, the methodology used in the research is described, analyzing the following aspects: (i) types of construction solutions considered and stationary transmittance; (ii) considerations about the periodic thermal transmittance calculation of UNE-EN ISO 13786; (iii) definition of the case study and the thermal modeling; (iv) analysis of the studied climate zones; and (v) the approach of the comfort and energy demand analysis. Second, the results are presented and discussed. Finally, the main conclusions of the results obtained in the study are summarized.

2. Materials and Methods

2.1. Thermopyshical Properties of the Material

In previous works, the potential viability of the materials being studied has been clarified [26,27,45], starting from the evaluation of the chemical, mineralogical, physical, thermal and mechanical characteristics, as well as the correlation with thermal conductivity coefficients [46]. The thermal conductivity results have confirmed that the doses with 25% and 50% of glass aggregate have thermal conductivity coefficients that are noticeably lower than those of the reference material, and also higher densities, maintaining a sufficient mechanical capacity compared with natural aggregate mortars (Table 1).

Table 1. Sand and glass composition of mortars used in the study, density of the end product (ρ), and thermal conductivity (δ) at 30 °C.

	Cement	% Sand	% Glass	ρ (g/cm^3)	δ (W/Km)
mR	1	3	0	1.60	1.2884
m25	1	2.25	0.75	1.95	1.0589
m50	1	1.5	1.5	2.04	0.8662

The particles were classified according to their size as per UNE EN 933-1:2012 standard (0.063, 0.125, 0.250, 0.50, 1.00, 2.00 mm) [30] and continuous particle size with maximum compactness were prepared in accordance with Fuller's curve (Figures 1 and 2) [28]. Thermal conductivity was determined by differential scanning calorimetry (MDSC) at 30 °C, thermal diffusivity (cm^2/s) and thermal conductivity (W/mK) were measured on Linseis measuring equipment (LFA 1600) and a DSC Q20-TA [35].

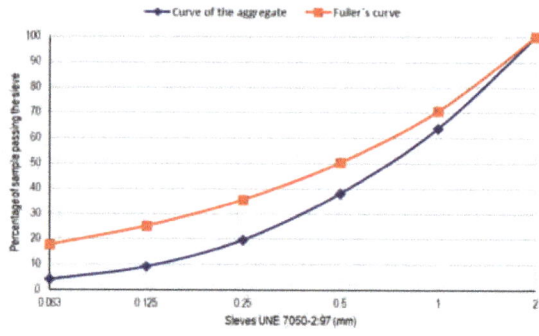

Figure 1. Aggregate particle size of crushed glass. Reprinted from Flores-Alés, V., Alducin-Ochoa, J. M., Martín-del-Río, J. J., Torres-González, M., and Jiménez-Bayarri, V. (2020). Physical-mechanical behaviour and transformations at high temperature in a cement mortar with waste glass as aggregate. Journal of Building Engineering, 29, 101158.

Figure 2. Crushed glass and mortar based on recycled glass as aggregate.

2.2. Constructive Solutions and Static Thermal Properties

This study considered three constructive solutions, which are commonplace in the construction of dwellings. Mortar blocks with a thickness of 11.5 cm constitute the core of the external walls; they are coated by a layer of cement mortar on the outside and plaster on the inside. The three solutions differ in the percentage of crushed glass used to elaborate the mortar blocks (mR, m25 and m50) which, in turn, modified their static properties: Thermal conductivity, gross density and specific heat capacity (Table 2). It is remarkable how higher percentages of crushed glass brings lower thermal conductivities and higher densities, while maintaining a nearly constant specific heat capacity.

Table 2. 15 cm brick thermal transmittance: mR, m25, m50.

	Layer Name	Thermal Conductivity l (W/mK)	Gross Density r (kg/m^3)	Spec. Heat Capacity C (J/kgK)	Layer Thickness d (m)	R (m^{2}K/W)
mR	Rsi (int. heat transfer resistance)					0.13
	Plaster	0.570	1150.0	1000	0.0150	0.026
	Glass mortar mR brick	1.288	1603.0	659	0.1150	0.089
	Cement mortar	1.400	2000.0	800	0.0200	0.014
	Rse (ext. heat transfer resistance)					0.04
			U-value:	3.3349	W/m^{2}K	
m25	Rsi (int. heat transfer resistance)					0.13
	Plaster	0.570	1150.0	1000	0.0150	0.026
	Glass mortar mR25 brick	1.059	1951.0	671	0.1150	0.109
	Cement mortar	1.400	2000.0	800	0.0200	0.014
	Rse (ext. heat transfer resistance)					0.04
			U-value:	3.1329	W/m^{2}K	
m50	Rsi (int. heat transfer resistance)					0.13
	Plaster	0.570	1150.0	1000	0.0150	0.026
	Glass mortar mR50 brick	0.866	2039.0	687	0.1150	0.133
	Cement mortar	1.400	2000.0	800	0.0200	0.014
	Rse (ext. heat transfer resistance)					0.04
			U-value:	2.9124	W/m^{2}K	

2.3. Assumption of Periodic Thermal Transmittance and Dynamic Thermal Properties

The UNE-EN ISO 13786 standard defines the analytical calculation procedure and the parameters related to the dynamic thermal behavior of building envelopes [31]. The theoretical basis of this was primarily established by Carslaw and Jaeger [47], who analyzed the sinusoidal ratio between the heat flow and the indoor and outdoor temperatures. The oscillation period (T) of the temperatures can be an hour, a day or a year, and this research assumes that the sinusoidal variation has a period of 1 day, which corresponds to the daily oscillations of the external temperatures [31]. After this, three static properties of all the considered materials must be known: thermal conductivity (λ), density (ρ), and specific thermal capacity (c). In this case, they were already determined by laboratory tests. Thermal bridges do not have to be considered due to their low impact on dynamic thermal properties [31].

The calculation procedure uses the thermal transference matrices for each one of the homogeneous layers of the wall (Z_{mn}). The elements of a transference matrix are defined as (Equation (1)):

$$Z_{mn} = \begin{pmatrix} Z_{11} & Z_{12} \\ Z_{21} & Z_{22} \end{pmatrix}$$
$$Z_{11} = Z_{22} = cosh(\xi)cos(\xi) + j \cdot senh(\xi)sen(\xi)$$
$$Z_{12} = -\frac{\delta}{2\lambda}\{senh(\xi)cos(\xi) + cosh(\xi)sen(\xi) + j \cdot [cosh(\xi)sen(\xi) - senh(\xi)cos(\xi)]\}$$
$$Z_{21} = -\frac{\lambda}{\delta}\{senh(\xi)cos(\xi) - cosh(\xi)sen(\xi) + j \cdot [cosh(\xi)sen(\xi) + senh(\xi)cos(\xi)]\}$$
(1)

where δ (m) is the periodic penetration depth of a thermal wave on the layer's material (Equation (2)), and ξ (dimensionless) is the ratio between d and δ (Equation (3)).

$$\delta = \sqrt{\frac{\lambda T}{\pi \rho c}}$$
(2)

$$\xi = \frac{d}{\delta}$$
(3)

The transference matrix of a wall (Z) is defined as the multiplication of the matrices of the different layers (Z_i) from the outside ($i = N$) to the inside ($i = 1$):

$$Z = \begin{pmatrix} Z_{11} & Z_{12} \\ Z_{21} & Z_{22} \end{pmatrix} = \prod_{i=N}^{1} Z_i$$
(4)

The different periodic variables can be determined by operating the elements of the matrix. The periodic variables defined in UNE-EN ISO 13786, and considered in the study, are: (i) periodic thermal transmittance ($|Y_{12}|$) (W/(m²K)), which is the module of the complex number defined as the complex amplitude of the heat flow density through the indoor component's surface, divided by the complex amplitude of the temperature of the outdoor area, when the indoor temperature is constant (Equation (5)); (ii) time shift periodic thermal transmittance (φ) (h), which is the period of time between the maximum amplitude of a cause and the maximum amplitude of its effect, related to the periodic thermal transmittance (Equation (6)); (iii) decrement factor (f) (dimensionless), which is the quotient between the periodic thermal transmittance module and the U-value (Equation (7)); (iv) internal thermal admittance ($|Y_{11}|$) (W/(m²K)), which is the complex number module defined as the complex amplitude of the heat flow density through the surface of the component adjoining the indoor area, divided by the complex amplitude of the temperature in the same area when the indoor temperature is kept constant (Equation (8)); (v) time shift internal side (φ_{11}) (h), which is the period of time between the maximum amplitude of a cause and the maximum amplitude of its effect, related to the internal thermal admittance (Equation (9)); (vi) external thermal admittance ($|Y_{22}|$) (W/(m²K)), which is the complex number module defined as the complex amplitude of the heat flow density through the surface of the component adjoining the outdoor area, divided by the complex amplitude of the temperature in the same area when the outdoor temperature remains constant (Equation (10)); and (vii) time shift external side (φ_{22}) (h), which is the period of time between the maximum amplitude of a cause and the maximum amplitude of its effect related to the external thermal admittance (Equation (11)).

$$Y_{12} = -\frac{1}{Z_{12}} \tag{5}$$

$$\varphi = \frac{T}{2\pi} arg(Z_{12}) \tag{6}$$

$$f = \frac{|Y_{12}|}{U} \tag{7}$$

$$Y_{11} = -\frac{Z_{11}}{Z_{12}} \tag{8}$$

$$\varphi_{11} = \frac{T}{2\pi} arg(Y_{11}) \tag{9}$$

$$Y_{22} = -\frac{Z_{22}}{Z_{12}} \tag{10}$$

$$\varphi_{22} = \frac{T}{2\pi} arg(Y_{22}) \tag{11}$$

2.4. Case Study and Thermal Model

Figure 3 shows housing model for the study.

A representative case-study was selected from the Project Bank of the Bio-Bio Region's Housing and Urbanization Service (SERVIU, in Spanish) [37]; a detached house with a higher surface of thermal envelope (external walls, roof and slabs) would have, a priori, a worse energy performance in this climate [38]. The prototype was modelled in the EnergyPlus® simulation software [36] and parametric simulations were done considering the thermal constructive properties of Table 3. The base-case scenario considered U-values as per the Chilean Building code, and the thermal properties of three different walls were introduced, using the three different mortars considered (Table 2). Ventilation rates were adjusted as per the minimum values recommended by the Chilean standard. In addition, the values for internal loads, such as occupation, lighting, equipment and ventilation, were considered (Table 4), as well as their schedules (Figure 4).

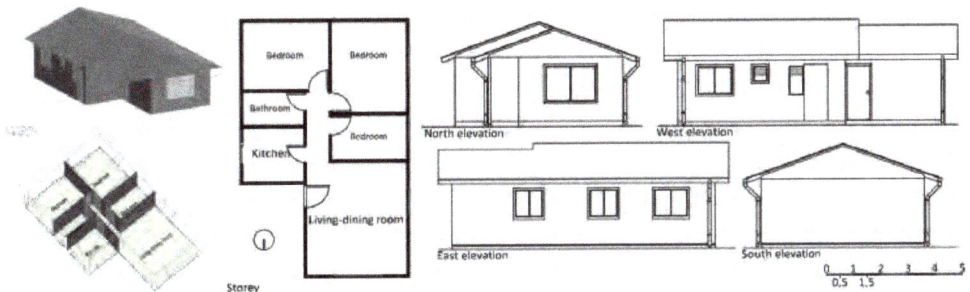

Figure 3. Housing model for the study.

Table 3. Thermal constructive properties of the case studies with the three bricks.

| Case | U Openings (W/m² K) | U envelope (W/m² K) | | | Ventilation l/(s*Person) | | | | Infiltrations (ACh) |
		Roof	Walls	Floor (m²K/W) × 100	Time	Months 5, 6, 7, 8	Months 4, 9, 10	Months 1, 2, 3, 11, 12	
1	mR		3.916–0.600 *						
2	m25	1.94–3.16 *	0.25–0.84 *	3.651–0.593 *	45	24 h	5.2	5.2	5.2	1
3	m50		3.367–0.585 *						

* Depends on the climate zone considered.

Table 4. Internal heat loads for the models.

	Living-Dining Room	Kitchen	Bedroom	Bathroom	Corridor
Illumination (W/m^2)	23	13	12	13	5
Occupation (W/m^2)	8.9	8.9	8.9	8.9	8.9
Equipment (W/m^2)	12.40	12.40	12.40	-	12.40

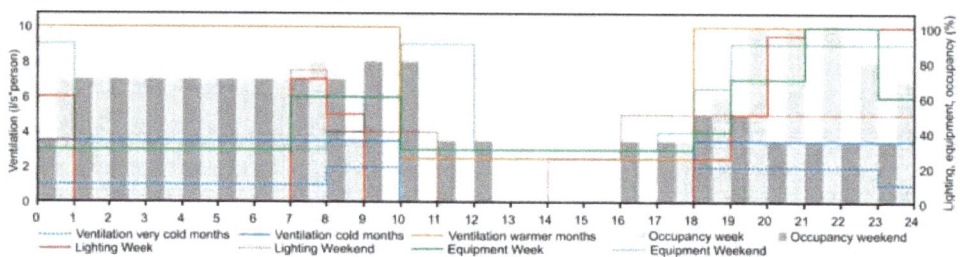

Figure 4. Occupation, lighting and equipment and ventilation schedules. Source: [34].

2.5. Climate Zones

Chile's climate varies greatly, covering the climatic variants B (arid and semi-arid), C (template) and E (cold) of the Köppen–Geiger classification. According to the current standard, Chile is divided into 7 thermal zones considering the annual heating degree days (Table 5) [48]. The Thermal Regulation (RT, in Spanish) for housing came into force in 2000. In a first stage, maximum thermal transmittance requirements were defined for roofs and, in a second stage in 2007, requirements were established for walls, ventilated floors and windows [49], which are mainly based on heating degree days. A representative city has been chosen for each of the 7 thermal zones, which also allows for an easy classification as per Köppen–Geiger (Table 5). An EPW weather file was considered for each city to model the external conditions.

Table 5. Selected locations from Chilean Climatic Zoning.

Zone	Location	Koppen–Geiger Classification	Latitude	Longitude	Elevation	Average January	Oscil. January	Average July	Oscil. July	Heating Degree Day Based on 15 °C
1	Antofagasta	BWk	23.43 °S	70.43 °W	120 m	20.5	7.3	14.9	5	≤500
2	Valparaíso	CSbn	30.03 °S	71.48 °W	41 m	17.8	9.2	11.4	7.3	>500–≤750
3	Santiago	CSb	33.38 °S	70.78 °W	474 m	20.7	17	7.9	11.3	>750–≤1000
4	Concepción	CSbn's	36.77 °S	73.05 °W	16 m	16.6	14.2	8.7	8.5	>1000–≤1250
5	Temuco	CFb	38.75 °S	72.63 °W	120 m	18.0	17.2	6.3	7.6	>1250–≤1500
6	Lonquimay	CFb	38.43 °S	71.23 °W	925 m	15.5	20.9	1.5	9.4	>1500–≤2000
7	Punta Arenas	BSk's	53.00 °S	70.85 °W	37 m	11.2	8.2	2.2	4.7	>2000

2.6. Thermal Comfort and Energy Analysis

The influence of using mortar bricks with glass in housing will be determined starting from the simulation results from EnergyPlus®, using two thermal comfort indicators and four energy demand indicators. The comfort indicators will be obtained by simulating the dwelling in the seven climates with the three brick types (mR, m25 and m50), in free oscillation during the entire year. With the results of the hourly operational temperatures in free oscillation, the number of annual hours where the operational temperatures are within the adaptive thermal comfort (ATC) limits of the model defined in the ASHRAE 55-2017 [44] standard, will be quantified, as will the distance in hourly operational temperature degrees to the thermal neutrality to quantify the reduction or increase of extremely hot or cold temperatures. In this case, a thermal acceptability limit of 80%, as per ASHRAE 55-2017, was considered; Chile, as many other countries, still does not have its own standard for adaptive thermal comfort, thus international documents are adopted. ATC models were originally developed for office buildings [50], whereas they find also application in residential buildings, considering that occupants may change their clothes and operate windows to achieve thermal comfort [51]. Besides, previous studies also support the fact that adaptive comfort finds an application in naturally cooled houses, and finds applicability in social dwellings [42].

The ASHRAE adaptive model is governed by Equation 12, which defined the neutral temperature inside the building (T_n) as a function of the $T_{pma(out)}$; a range of ±3.5 °C gives an acceptability of 80% and ±2.5 °C gives a 90%.

$$T_n = 0.31 \times T_{pma(out)} + 17.8 \quad (12)$$

$T_{pma(out)}$ is a weighted average of the mean external temperatures of the previous 7 days (Equation (13)). $T_{e(d-1)}$ is the average outdoor temperature of the previous day, $T_{e(d-2)}$, the average outdoor temperature of two days prior and so on and so forth; and α is a constant that depends on the thermal oscillation of the local climates, assuming $\alpha = 0.8$ [44].

$$T_{pma(out)} = (1-\alpha) \times \left(T_{e(d-1)} + \alpha \times T_{e(d-2)} + \alpha^2 \times T_{e(d-3)} + \alpha^3 \times T_{e(d-4)} + \cdots\right) \quad (13)$$

($T_{pma(out)}$) must be within 10.0 °C and 33.5 °C so that this standard find application. If ($T_{pma(out)}$) falls outside those limits, the neutral temperature will be a constant, as the standard assumes that when it is too cold or too hot, active cooling or heating becomes necessary, thus internal temperatures are decoupled from the external oscillations. In that case, this study assumed a heating setpoint temperature of 20 °C and a cooling setpoint of 26 °C, as per EN 16798 standard, Category II [52]. If necessary, the dwelling would have to resort to heating or cooling devices and therefore the energy demand was also recorded. Those variables were simulated for the 3 constructive solutions (Table 2) and the 7 climate zones of Chile (Table 5).

3. Results and Discussion

3.1. Periodical Thermal Properties

Taking the data as a base for the static thermal transmittance (Table 2), and following the calculation procedure described in Section 2.2, the periodic thermal properties of the 3 constructive solutions considered were calculated. Thermal conductivity (λ), density (ρ) and specific heat capacity (c) were already known for each solution; the calculation period for the thermal oscillation was 24 h for all cases, which is the recommended value for daily meteorological variations and temperature setback. The dynamic thermal properties were calculated (Table 6) using the tool to calculate thermal mass [53].

Table 6. Calculation results according to EN ISO 13786.

Parameter	Unit	mR	m25	m50
external thermal admittance	W/(m²K)	7.351	7.732	7.615
time shift external side	h	2.64	2.60	2.61
internal thermal admittance	W/(m²K)	4.068	4.132	4.085
time shift internal side	h	1.05	1.18	1.28
periodic thermal transmittance	W/(m²K)	2.890	2.548	2.280
time shift periodic thermal transmittance	h	−2.68	−3.26	−3.63
external areal heat capacity	kJ/(m²K)	101.822	110.720	111.165
Internal areal heat capacity	kJ/(m²K)	47.083	53.592	56.158
decrement factor f		0.866	0.813	0.783

Static thermal transmittance could be reduced by around 13% on using recycled glass aggregates, and similar effects can be seen in the dynamic thermal properties, though the discussion there is more complex. Dynamic thermal transmittance was reduced by around 22%; the decrement factor was reduced from 0.86 to 0.78 (−10%), and the time shift was increased by one hour, which means that the thermal oscillation amplitude is reduced conjointly with a delay in the transmission of heat from the outside to the inside. This could be expected because of the greater heat capacity of the proposed material.

However, to fully grasp the real implications of these properties, additional data were needed. Since dynamic thermal properties are highly dependent on temperature oscillation, which, in turn, is a function of the local climate, it was deemed necessary to clarify how these solutions would work in all the climate zones of Chile. For this purpose, external temperature variations were simulated during a 24-h cycle, by approximating the oscillation to a cosine function in the form of:

$$t(h) = t_{avg} + t_{amp} \times (\cos(t - t_{max})) \quad (14)$$

where t_{avg} is the daily average temperature, t_{amp} is the daily temperature amplitude, t is time in hours and t_{max} is the time of the day when the outdoor temperature reaches its maximum. As a result, this function delivers an output, $t(h)$, which is the hourly external temperature for 24 h. Additional data were needed to calculate the temperature oscillation inside the building: The thermal resistance of the external air layer and the static thermal transmittance are obtained from Table 2; the external thermal admittance and the time shift for the external side are obtained from Table 6. A calculation routine was written in Matlab®, where the properties of the m50 solution (Table 6) and the climate data for each location (Table 5) were input, giving as a result the indoor and outdoor temperature oscillation for the coldest and hottest months of the year. No HVAC systems were considered, so the mere effect of the walls on the indoor environment could be clarified. Data were depicted graphically; the x axis was extended to 36 h and the scales of the y axis were unified for an easier comparison (Figure 5).

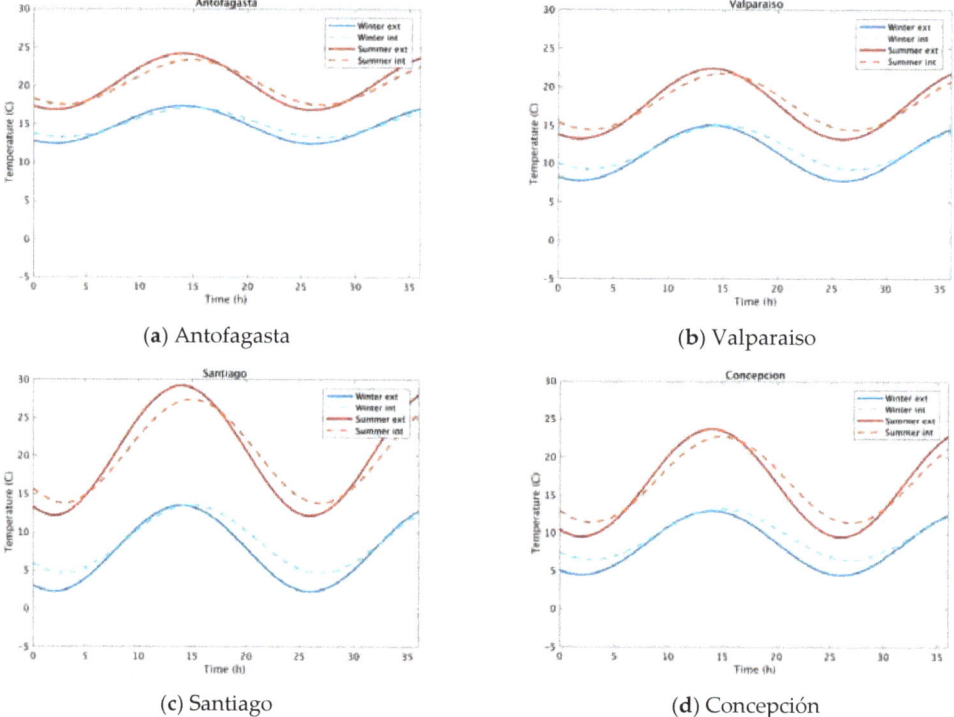

(a) Antofagasta

(b) Valparaiso

(c) Santiago

(d) Concepción

Figure 5. *Cont.*

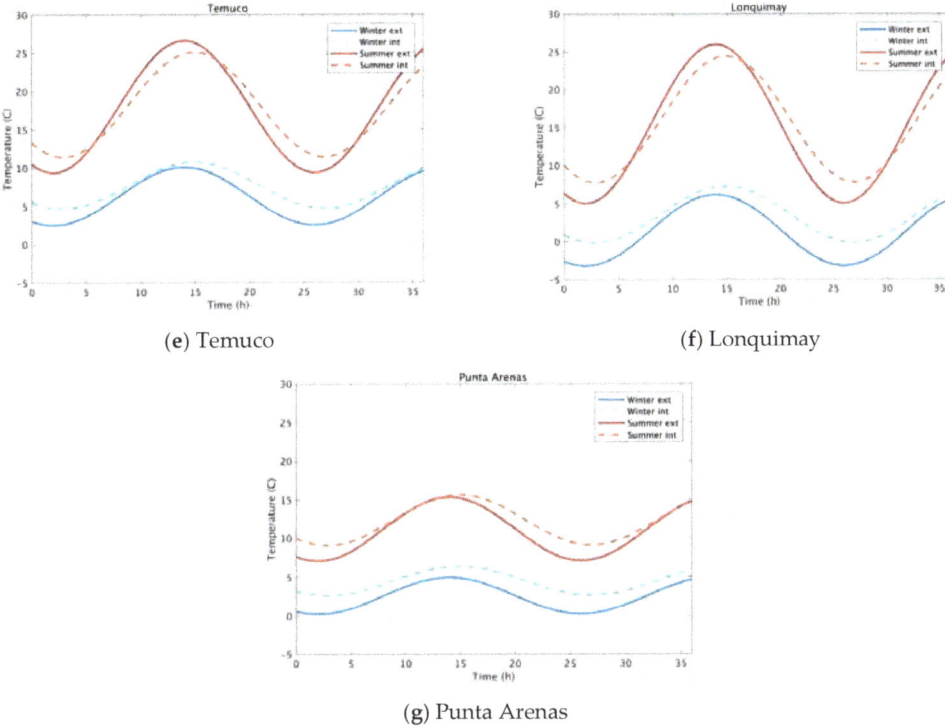

Figure 5. Thermal oscillation of m50 solution, in winter and summer, for representative cities of the seven climate zones of Chile.

Two aspects of the results from these simulations can be commented on: Time shift and decrement factor. In general, this solution works better in climates with large thermal oscillations between day and night, such as Lonquimay or Santiago. During the cold season, indoor temperatures are never lower than outdoor temperatures; while on the contrary, during the hot season, thermal inversion is observed, and the walls can mitigate the low temperatures during the first few hours in the morning and also help in weathering the peak in temperature during the middle of the day. In regions with low thermal oscillation, whether we are talking about cold warm (Antofagasta) or cold climates (Punta Arenas), the effect is not so evident, although indoor temperatures are always a couple of degrees above outdoor temperatures during the cold season.

3.2. Thermal Comfort and Energy Saving Analysis

Dynamic simulations have been developed to establish the impact of glass aggregate mortars on the thermal behavior of social housing in the different thermal zones. The substitution of conventional mortar bricks for glass aggregate mortars can signify anywhere between a 1% and 14% reduction in the enclosure's thermal transmittance, maintaining the same total thickness and insulation (Table 7). This reduction in thermal transmittance, along with a higher thermal inertia, means that the use of said material in zones Z1, Z2, Z3 and Z6, results in an increase in thermal comfort hours when the dwelling operates under free oscillation. These are higher for the m50 mortar, ranging from 65 to 199 h, while they range from 23 and 103 for the m25 mortar. Therefore, the increase in recycled glass percentage has a direct relationship on the increase in hours in comfort for climates where the increase in thermal inertia this material provides, can be taken advantage of.

Table 7. Hours in thermal comfort and distance in respect to the neutral temperature by zones considering the mortar used.

ZONE	Mortar	Transmittance (W/m²K)	Reduction (%)	Hours in Comfort (h)	Difference (h)	Distance to Tn (°C)	Difference (°C)
Z1	mR	3.916		6351		21,157	
	m25	3.651	7%	6454	103	20,728	−429
	m50	3.367	14%	6550	199	20,293	−864
Z2	mR	3.004		4128		35,404	
	m25	2.846	5%	4151	23	34,995	−409
	m50	2.670	11%	4198	70	34,583	−822
Z3	mR	1.903		4390		32,583	
	m25	1.839	3%	4437	47	32,327	−255
	m50	1.764	7%	4464	74	32,083	−500
Z4	mR	1.701		5031		28,727	
	m25	1.649	3%	4998	−33	28,897	170
	m50	1.589	7%	5056	25	28,544	−183
Z5	mR	1.599		4238		35,646	
	m25	1.553	3%	4238	0	35,888	242
	m50	1.499	6%	4275	37	35,523	−122
Z6	mR	1.101		4296		36,363	
	m25	1.079	2%	4330	34	36,096	−267
	m50	1.053	4%	4361	65	35,875	−488
Z7	mR	0.600		2676		55,647	
	m25	0.593	1%	2605	−71	56,178	531
	m50	0.585	3%	2618	−58	56,048	401

There are thermal zones, such as Z4 and Z5, where it is necessary to exceed 25% in the substitution of aggregate for glass so that there is a positive impact; however, this impact is not as important as in zones Z1, Z2, Z3 and Z6. For zone 7 (Punta Arenas), none of the glass aggregate mortars have an increase of comfort hours due to their low daily oscillation during the coldest months, just as was commented in Section 3.1. However, in zone 1 (Antofagasta), there is a significant increase in hours, despite this not being a climate with major thermal oscillations. This is due to the proximity of outdoor hours to the comfort limits (See Table 5).

The distance regarding neutral temperature marked by the ASHRAE adaptive comfort model has also seen important reductions in zones 1, 2, 3 and 6, with the m50 mortar representing almost double the reduction compared to the m25 in these zones. Zone 7, just as in the difference in the results of comfort hours, does not see a reduction in this indicator, as the distance increases by 401 °C for m50 and 531 °C for m25. Zones Z4 and Z5 only have a reduction of the distance when the m50 mortar is used, and end up being −183 °C and −122 °C, respectively (See Table 7).

These results are associated with a higher thermal comfort and lower overheating issues in dwellings for zones 1, 2, 3 and 6 when these operate under free oscillation. Said results can also assume a reduction in respect to heating and cooling consumptions, as well as in demand peaks due to a lower thermal oscillation inside the dwelling, linked with a higher thermal inertia of the enclosures. In Table 8, it can be seen that heating peak demands can be reduced by between 0.1% and 7.6%, with zones Z1 and Z2 providing a higher reduction (2.6–7.6%). The same occurs with the heating demand, with reductions of between 3.4% and 15.3%. In these zones, the use of an m50 mortar assumes a 50% increase in consumption reduction compared to the m25, with Z1 passing from 2667 to 2559 KWh/year and Z2 from 7047 to 6557 KWh/year.

The cooling demand and peaks also see important reductions in zones Z1, Z2 and Z3 (in the rest of the zones there is no cooling demand) (See Table 8). The reductions of the cooling peak oscillate between 4.8% and 12.2%, falling in the most favorable zone, Z1, from 4.14 to 3.64 kWh. The reductions for cooling demand are between 7.3% and 9.8%, with reductions seeing a noticeably similar percentage among the three zones. However, the kWh reduction oscillates between 150 and 366 KWh/year, with the zones providing a better performance, with glass aggregate mortar being in zones 2 and 3, respectively.

Table 8. Heating and cooling demand and peak by zones, depending on mortar used.

ZONE	Mortar	Heating				Cooling			
		Peak Demand (KWh)	Peak Reduction (%)	Total (KWh/Year)	Annual Reduction (%)	Peak Demand (KWh)	Peak Reduction (%)	Total (KWh/Year)	Annual Reduction (%)
Z1	mR	2.97		2667		4.14		1871	
	m25	2.85	4.3%	2460	7.8%	3.78	8.7%	1721	8.0%
	m50	2.75	7.6%	2259	15.3%	3.64	12.2%	1689	9.7%
Z2	mR	3.39		7047		6.45		3746	
	m25	3.31	2.6%	6806	3.4%	6.14	4.8%	3462	7.6%
	m50	3.22	5.1%	6557	7.0%	6.01	6.8%	3380	9.8%
Z3	mR	2.74		5611		4.47		2953	
	m25	2.70	1.5%	5477	2.4%	4.23	5.4%	2739	7.3%
	m50	2.66	2.8%	5339	4.8%	4.13	7.6%	2693	8.8%
Z4	mR	0.92		1403		0.00		0.00	
	m25	0.92	0.6%	1391	0.8%	0.00	-	0.00	-
	m50	0.91	1.3%	1364	2.7%	0.00	-	0.00	-
Z5	mR	1.07		1902		0.00		0.00	
	m25	1.07	0.4%	1891	0.6%	0.00	-	0.00	-
	m50	1.06	1.0%	1863	2.1%	0.00	-	0.00	-
Z6	mR	1.03		1755		0.00		0.00	
	m25	1.03	0.3%	1756	0.0%	0.00	-	0.00	-
	m50	1.03	0.8%	1739	0.9%	0.00	-	0.00	-
Z7	mR	1.10		3176		0.00		0.00	
	m25	1.10	0.1%	3187	−0.4%	0.00	-	0.00	-
	m50	1.09	0.3%	3178	−0.1%	0.00	-	0.00	-

It can be seen by these results that the thermal and energy benefits of using glass aggregate mortars are closely tied to the percentage of glass incorporated, the thermal zone and the dwelling's type of operation. The m50 mortar can imply an increase in the hours of comfort and lower indoor thermal oscillations to avoid overheating in zones 1, 2, 3 and 6. When the dwelling operates with air-conditioning systems, the substantial differences of using these mortars would be in zones 1, 2 and 3, with zone 2 producing the highest saving, with a total reduction of 857 kWh/year, bearing in mind both heating and cooling. However, it should also be remarked that this material might have a controversial effect in the coldest zones, giving as a result a slight increment in the heating demand.

4. Conclusions

In this research, mortar blocks with doses of 25% (m25) and 50% (m50) of recycled glass aggregate were analyzed from a thermal point of view. This assumes a tangential vision to the traditional analysis based on mechanical behavior. The approach to the analysis focuses on social housing, using 21 models located in the 7 thermal representative zones of Chile. Using a methodology based on analyzing periodic thermal transmittance, adaptive comfort levels and energy demand, the thermal analysis is addressed holistically, making it possible to extrapolate this methodology to different construction solutions.

On considering thermal transmittance results, it can be pointed out that with the m50 solutions, this is reduced by 22%, the decrement factor is reduced by 10%, and the time shift increases by approximately one hour. Meanwhile, locations with a higher thermal amplitude between day and night like 6 (Lonquimay) and 3 (Santiago), both in winter and in summer, see significant reductions when compared with traditional solutions without a recycled glass aggregate.

Regarding thermal comfort, we can conclude that an increase is detected in the hours of comfort in all zones except in zone 7 (Punta Arenas), increasing 199 h in zone 1 (Antofagasta) with the m50 solution. The extension effect of hours of comfort and the reduction of distance, compared to the neutral temperature, is accentuated in zones 1 (Antofagasta), 2 (Valparaiso), 3 (Santiago) and 6 (Lonquimay). The m50 solution increases hours of comfort by almost 50% when compared with m25 and reduces the distance in respect to the neutral temperature by half.

Regarding energy demands, the most relevant finding is the reduction of the peak demand in all thermal zones, from 0.1% (heating) with m25 in zone 7, to 12.2% (cooling) with m50 in zone 1. The annual energy demand reduction is seen in zones 1 (Antofagasta), 2 (Valparaiso) and 3 (Santiago), with up to 15%. On this occasion, the m50 reduces the demand by around 50%, compared to the m25.

This research is a starting point to be valued in decision-making when it comes to implementing a construction solution with recycled materials in a country. In future work, the simulation results should be validated with actual prototypes. Further research is needed to generate an implementation methodology of new construction solutions which consider energy behavior, bearing in mind the energy poverty levels of social housing and that this cannot be done in any other way than with low environmental impact solutions.

Author Contributions: All the authors contributed equally to the present research. All authors have read and agreed to the published version of the manuscript.

Funding: This research as well as The APC were funded by National Agency for Research and Development (ANID) of Chile, grant number 1200551.

Acknowledgments: The authors would also like to acknowledge that this paper is part of the project "Conicyt Fondecyt Regular 1200551 - Energy poverty prediction based on social housing architectural design in the central and central-southern zones of Chile: an innovative index to analyze and reduce the risk of energy poverty" funded by the National Agency for Research and Development (ANID). In addition, we would like to acknowledge to the research group "Confort ambiental y pobreza energética (+CO-PE)" of the University of the Bío-Bío for supporting this research. Authors also wish to express their gratitude to the General Research Center at the University of Seville (CITIUS).

Conflicts of Interest: The authors declare no conflict of interest.

References

1. European Federation of Glass Packaging Makers Record Collection of Glass Containers for Recycling hits 76% in the EU—FEVE. Available online: https://feve.org/record-collection-of-glass-containers-for-recycling-hits-76-in-the-eu/ (accessed on 17 May 2020).
2. European Federation of Glass Packaging Makers Glass Recycling Statistics Year 2017. Available online: http://feve.org/about-glass/statistics/ (accessed on 17 May 2020).
3. Harder, J. Glass recycling—Current market trends. *Recovery* **2018**, *5*.
4. Hogland, W. Remediation of an old landfill site. *Environ. Sci. Pollut. Res.* **2002**, *9*, 49–54. [CrossRef] [PubMed]
5. Saillio, M.; Frohard, F.; Chaussadent, T.; Divet, L.; Tagnit-Hamou, A. Durability of concrete with alternative supplementary cementitious materials. In Proceedings of the 10th ACI/RILEM International Conference on Cementitious Materials and Alternative Binders for Sustainable Concrete, Montreal, QU, Canada, 2–4 October 2017.
6. Aliabdo, A.A.; Abd Elmoaty, A.E.M.; Aboshama, A.Y. Utilization of waste glass powder in the production of cement and concrete. *Constr. Build. Mater.* **2016**, *124*, 866–877. [CrossRef]
7. Salim, M.N.; Jenal, R.B.; Ismail, M.P. Reflected guided wave in plate with flaw using civa simulation software. In Proceedings of the 6th International Engineering Conference (ENCON 2013), Kuching Sarawak, Malaysia, 4 July 2013; pp. 978–981.
8. Chen, Z.; Wang, Y.; Liao, S.; Huang, Y. Grinding kinetics of waste glass powder and its composite effect as pozzolanic admixture in cement concrete. *Constr. Build. Mater.* **2020**, *239*, 117876. [CrossRef]
9. Chen, G.; Lee, H.; Young, K.L.; Yue, P.L.; Wong, A.; Tao, T.; Choi, K.K. Glass recycling in cement production-an innovative approach. *Waste Manag.* **2002**, *22*, 747–753. [CrossRef]
10. Idir, R.; Cyr, M.; Tagnit-Hamou, A. Use of fine glass as ASR inhibitor in glass aggregate mortars. *Constr. Build. Mater.* **2010**, *24*, 1309–1312. [CrossRef]
11. Carles-Gibergues, A.; Cyr, M.; Moisson, M.; Ringot, E. A simple way to mitigate alkali-silica reaction. *Mater. Struct. Constr.* **2008**, *41*, 73–83. [CrossRef]
12. Cassar, J.; Camilleri, J. Utilisation of imploded glass in structural concrete. *Constr. Build. Mater.* **2012**, *29*, 299–307. [CrossRef]

13. Shayan, A.; Xu, A. Value-added utilisation of waste glass in concrete. *Cem. Concr. Res.* **2004**, *34*, 81–89. [CrossRef]
14. Sikora, P.; Augustyniak, A.; Cendrowski, K.; Horszczaruk, E.; Rucinska, T.; Nawrotek, P.; Mijowska, E. Characterization of mechanical and bactericidal properties of cement mortars containingwaste glass aggregate and nanomaterials. *Materials (Basel)* **2016**, *9*, 701. [CrossRef] [PubMed]
15. Poon, C.S.; Cheung, E. NO removal efficiency of photocatalytic paving blocks prepared with recycled materials. *Constr. Build. Mater.* **2007**, *21*, 1746–1753. [CrossRef]
16. Ajdukiewicz, A.; Kliszczewicz, A. Influence of recycled aggregates on mechanical properties of HS/HPC. *Cem. Concr. Compos.* **2002**, *24*, 269–279. [CrossRef]
17. Mohajerani, A.; Vajna, J.; Cheung, T.H.H.; Kurmus, H.; Arulrajah, A.; Horpibulsuk, S. Practical recycling applications of crushed waste glass in construction materials: A review. *Constr. Build. Mater.* **2017**, *156*, 443–467. [CrossRef]
18. Kumarappan, N. Partial replacement cement in concrete using waste glass. *Int. J. Eng. Res. Technol.* **2013**, *2*, 1880–1883.
19. McNeil, K.; Kang, T.H.K. Recycled concrete aggregates: A review. *Int. J. Concr. Struct. Mater.* **2013**, *7*, 61–69. [CrossRef]
20. De Castro, S.; de Brito, J. Evaluation of the durability of concrete made with crushed glass aggregates. *J. Clean. Prod.* **2013**, *41*, 7–14. [CrossRef]
21. Lu, J.X.; Zhou, Y.; He, P.; Wang, S.; Shen, P.; Poon, C.S. Sustainable reuse of waste glass and incinerated sewage sludge ash in insulating building products: Functional and durability assessment. *J. Clean. Prod.* **2019**, *236*, 117635. [CrossRef]
22. Alani, A.; MacMullen, J.; Telik, O.; Zhang, Z.Y. Investigation into the thermal performance of recycled glass screed for construction purposes. *Constr. Build. Mater.* **2012**, *29*, 527–532. [CrossRef]
23. Sikora, P.; Horszczaruk, E.; Skoczylas, K.; Rucinska, T. Thermal properties of cement mortars containing waste glass aggregate and nanosilica. *Procedia Eng.* **2017**, *196*, 159–166. [CrossRef]
24. Xiao, Z.; Ling, T.C.; Poon, C.S.; Kou, S.C.; Wang, Q.; Huang, R. Properties of partition wall blocks prepared with high percentages of recycled clay brick after exposure to elevated temperatures. *Constr. Build. Mater.* **2013**, *49*, 56–61. [CrossRef]
25. Khan, M.N.N.; Saha, A.K.; Sarker, P.K. Reuse of waste glass as a supplementary binder and aggregate for sustainable cement-based construction materials: A review. *J. Build. Eng.* **2020**, *28*, 101052. [CrossRef]
26. Flores-Alés, V.; Jiménez-Bayarri, V.; Pérez-Fargallo, A. Influencia de la incorporación de vidrio triturado en las propiedades y el comportamiento a alta temperatura de morteros de cemento. *Boletín Soc. Española Cerámica Vidr.* **2018**, *57*, 257–265. [CrossRef]
27. Flores-Alés, V.; Alducin-Ochoa, J.M.; Martín-del-Río, J.J.; Torres-González, M.; Jiménez-Bayarri, V. Physical-mechanical behaviour and transformations at high temperature in a cement mortar with waste glass as aggregate. *J. Build. Eng.* **2020**, *29*, 101158. [CrossRef]
28. Yilmaz, Z. Evaluation of energy efficient design strategies for different climatic zones: Comparison of thermal performance of buildings in temperate-humid and hot-dry climate. *Energy Build.* **2007**, *39*, 306–316. [CrossRef]
29. Rodrigues, E.; Fernandes, M.S.; Gaspar, A.R.; Gomes, Á.; Costa, J.J. Thermal transmittance effect on energy consumption of Mediterranean buildings with different thermal mass. *Appl. Energy* **2019**, *252*, 113437. [CrossRef]
30. Dodoo, A.; Gustavsson, L.; Sathre, R. Effect of thermal mass on life cycle primary energy balances of a concrete-and a wood-frame building. *Appl. Energy* **2012**, *92*, 462–472. [CrossRef]
31. AENOR. *UNE-EN ISO 13786:2011. Prestaciones Térmicas de los Productos y Componentes para Edificación. Características Térmicas Dinámicas. Métodos de Cálculo*; Asociacion Espanola de Normalizacion: Madrid, Spain, 2011.
32. AENOR. *UNE-EN ISO 6946:2012. Componentes y Elementos para la Edificación. Resistencia Térmica y Transmitancia Térmica. Método de Cálculo*; Asociacion Espanola de Normalizacion: Madrid, Spain, 2012.
33. Stazi, F.; Ulpiani, G.; Pergolini, M.; Di Perna, C. The role of areal heat capacity and decrement factor in case of hyper insulated buildings: An experimental study. *Energy Build.* **2018**, *176*, 310–324. [CrossRef]

34. Rubio-Bellido, C.; Pérez-Fargallo, A.; Pulido-Arcas, J.A.; Trebilcock, M. Application of adaptive comfort behaviors in Chilean social housing standards under the influence of climate change. *Build. Simul.* **2017**, *10*, 933–947. [CrossRef]
35. MINVU Estadisticas Históricas. Available online: http://www.observatoriohabitacional.cl/ (accessed on 28 April 2016).
36. MINVU. *MINVU DS 01. Reglamento del Sistema Integrado de Subsidio Habitacional*; MINVU: Santiago de Chile, Chile, 2011.
37. MINVU. *MINVU DS 49. Reglamento del Programa Fondo Solidario de Elección de Vivienda*; MINVU: Santiago de Chile, Chile, 2011.
38. MOP Decreto con Fuerza de Ley N° 2 Sobre Plan Habitacional. Available online: http://www.sii.cl/pagina/jurisprudencia/legislacion/basica/dfl2_1.htm (accessed on 16 November 2016).
39. MINVU. *MINVU DS 47. Ordenanza General de la Ley General de Urbanismo y Construcciones*; MINVU: Santiago de Chile, Chile, 1992.
40. MINVU (Ed.) Artículo 4.1.10. Exigencias de acondicionamiento térmico de la Ordenanza general de urbanismo y construcciones. In *DS 47 Ordenanza General de la Ley General de Urbanismo y Construcciones*; MINVU: Santiago, Chile, 2007.
41. MINVU and Building Research Establishment. *Estándares de Construcción Sustentable para Viviendas en Chile*; MINVU and Building Research Establishment: Santiago de Chile, Chile, 2018.
42. Pérez-Fargallo, A.; Pulido-Arcas, J.A.; Rubio-Bellido, C.; Trebilcock, M.; Piderit, M.B.; Attia, S. Development of a new adaptive comfort model for low income housing in the central-south of chile. *Energy Build.* **2018**, *178*, 94–106. [CrossRef]
43. Porras-Salazar, J.A.; Contreras-Espinoza, S.; Cartes, I.; Piggot-Navarrete, J.; Pérez-Fargallo, A. Energy poverty analyzed considering the adaptive comfort of people living in social housing in the central-south of Chile. *Energy Build.* **2020**, *223*, 110081. [CrossRef]
44. American Society of Heating, R.A.C.E. (ASHRAE) (Ed.) *ASHRAE Standard 55-2017. Thermal Environmental Conditions for Human Occupancy*; ASHRAE Inc. American Society of Heating, Refrigerating and Air Conditioning Engineers: Atlanta, GA, USA, 2017; ISBN 1041-2336.
45. Flores-Alés, V.; Martín-del-Río, J.J.; Alducin-Ochoa, J.M.; Torres-González, M. Rehydration on high temperature-mortars based on recycled glass as aggregate. *J. Clean. Prod.* **2020**, *275*, 124139. [CrossRef]
46. Blumm, J.; Lindeman, A.; Niedrig, B. Measurement of the thermophysical properties of an NPL thermal conductivity standard Inconel 600. *High Temp. Press.* **2003**, *35/36*, 621–626. [CrossRef]
47. Carslaw, H.S.; Jaeger, J.C. *Conduction of Heat in Solids*, 2nd ed.; Oxford Clarendon Press: Oxford, UK, 1959.
48. Ministerío de Vivienda y Urbanísmo. *Artículo 4.1.10. Exigencias de Acondicionamiento Térmico de la Ordenanza General de Urbanismo y Construcciones*; Ministerío de Vivienda y Urbanísmo: Santiago, Chile, 2011.
49. Ministerío de Vivienda y Urbanísmo. *Artículo 4.1.10. Manual de Aplicación. Reglamentación Térmica. Ordenanza General de Urbanísmo y Construcciones*; Ministerío de Vivienda y Urbanísmo: Santiago, Chile, 2016.
50. Nicol, J.F.; Humphreys, M.A. Adaptive thermal comfort and sustainable thermal standards for buildings. *Energy Build.* **2002**, *34*, 563–572. [CrossRef]
51. Escandón, R.; Suárez, R.; Sendra, J.J.; Ascione, F.; Bianco, N.; Mauro, G.M. Predicting the impact of climate change on thermal comfort in a building category: The Case of Linear-type Social Housing Stock in Southern Spain. *Energies* **2019**, *12*, 2238. [CrossRef]
52. CEN. *Indoor Environmental Input Parameters for Design and Assessment of Energy Performance of Buildings Addressing Indoor Air quality, Thermal Environment, Lighting and Acoustics*; CEN: Brussels, Belgium, 2017; Volume 44, EN 16798.
53. HTflux—Hygrig and Thermal Simulation Software. Available online: https://www.htflux.com/en/ (accessed on 17 May 2020).

Publisher's Note: MDPI stays neutral with regard to jurisdictional claims in published maps and institutional affiliations.

© 2020 by the authors. Licensee MDPI, Basel, Switzerland. This article is an open access article distributed under the terms and conditions of the Creative Commons Attribution (CC BY) license (http://creativecommons.org/licenses/by/4.0/).

Article

Prediction of Cooling Energy Consumption in Hotel Building Using Machine Learning Techniques

Marek Borowski * and **Klaudia Zwolińska**

Faculty of Mining and Geoengineering, AGH University of Science and Technology, 30-059 Kraków, Poland; kzwolinska@agh.edu.pl
* Correspondence: borowski@agh.edu.pl; Tel.: +48-12-6172068

Received: 22 October 2020; Accepted: 24 November 2020; Published: 26 November 2020

Abstract: The diversification of energy sources in buildings and the interdependence as well as communication between HVAC installations in the building have resulted in the growing interest in energy load prediction systems that enable proper management of energy resources. In addition, energy storage and the creation of energy buffers are also important in terms of proper resource management, for which it is necessary to correctly determine energy consumption over time. It is obvious that the consumption of cooling energy depends on meteorological conditions. Knowing the parameters of the outside air and the number of users, it is, therefore, possible to determine the hourly energy consumption of a cooling system in a building with some accuracy. The article presents models of cooling energy prediction in summer for a hotel building in southern Poland. The paper presents two methods that are often used for energy prediction: neural networks and support vector machines. Meteorological data, time data, and occupancy level were used as input parameters. Based on the collected input and output data, various configurations were tested to identify the model with the best accuracy. As the analysis showed, higher prediction accuracy was obtained thanks to the use of neural networks. The best of the proposed models was characterized by the *WAPE* and *CV* coefficients of 19.93% and 27.03%, respectively.

Keywords: energy consumption; heating and cooling system; optimization and management; energy use prediction; neural network; support vector machine

1. Introduction

Nowadays, people spend the majority of time indoors, which leads to increased costs associated with maintaining comfort conditions in buildings. Internal installations such as heating, cooling, and ventilation systems therefore play a key role and thus constitute the main source of costs. Hence, the combination of economic and environmental factors is an important task for manufacturers and designers, contributing to the development of new solutions to provide comfort conditions for small operating and investment outlays. In 2018, the industry sector accounted for about 32% of total global energy consumption. The transport sector accounted for 28% of the energy use. The building construction and operations accounts for the largest share of global final energy use—36%, of which 8% was connected with operation of the non-residential building, 22% with residential building operation and 6% with the construction industry [1]. Global final energy consumption in buildings [2] in 2018 increased 1% from 2017, and about 7% since 2010. According to the Energy efficiency indicators Report published by IEA, nearly 50% of building consumption is related to space heating and 4% with cooling. Increasing energy consumption in the European Union led to appearance of the general regulation in this field—Directive 201013/EU of the European Parliament and of the Council on the energy performance of buildings [3]. According to Article No. 8, Member States should set system requirements for the purpose of optimizing the energy use of technical systems in the buildings. Regulations cover at least heating, hot water, air-conditioning, and large ventilation systems. Furthermore, Member States may encourage the use of active control systems such as automation, control, and monitoring systems that aim to save energy.

In general, methods for estimating energy use have two purposes: design or optimization of the building and HVAC systems (forward modeling), and calculating retrofit savings or implementing model predictive control in existing buildings (data-driven modeling). Behavior of the system is described by a mathematical model, which includes input variables, system structure (physical description), and output variables (reaction to the input variables). Data-driven modeling may be divided into three groups: "Black-Box" (Empirical), Calibrated Simulation, and Gray-Box Approach, which differ in data requirements, time, and effort required to develop the appropriate models. In the first method, a simple or multivariate regression model that describes a relationship between measured energy use and the various input parameters is constructed. A Calibrated Simulation Approach uses a simulation computer program to evaluate existing buildings' energy consumption and then calibrates the physical input parameters to the program. The Gray-Box Approach formulates a physical model and identifies important parameters by statistical analysis. This method is a mixture of physics-based and data-driven methods, and it could be implemented for fault detection and diagnosis (FDD) and online control [4,5].

Building energy consumption prediction is crucial to appropriate energy management, thus improving energy efficiency of systems and performance of the buildings. Generally, for building energy consumption prediction, two techniques are used: statistical methods and artificial intelligence methods. In recent years, artificial intelligence methods have become very popular. This technique is often applied to the prediction of energy consumption due to good, accurate prediction results [6]. Among the most popular data-driven prediction models using empirical approach modeling are artificial neural networks (ANNs) and support vector machines (SVM). One of the popular techniques is also decision tree (DT) and random forest (RF), which generates multiple decision trees that operate as an ensemble [7]. To improve their solutions, many authors use various methods mentioned above and choose the results of the best one [8–10]. The literature contains numerous interesting solutions using different methods. Table 1 illustrates applications of certain algorithms in literature.

Table 1. A review of predictive models for building energy consumption.

Author	Type of Building	Inputs	Outputs	Methods
Sendra-Arranz et al., 2020 [11]	Solar house MagicBox	outdoor temperature, relative humidity, irradiance, indoor CO_2 level, indoor temperature, reference temperature (set by the user)	Power consumption by an HVAC system	Long Short-Term Memory Neural Networks (LSTM)
Wang et al., 2020 [12]	Educational Building (2 buildings)	meteorological data (including outdoor dry bulb temperature, wet bulb temperature, relative humidity, wind direction, wind speed, air pressure, horizontal total radiation), time variable (hour of the day, day type), historical data (energy consumption at the same time as the previous day).	Building energy consumption	RF, GBDT, XGBoost, SVR, and kNN models (Random Forest, Gradient Boosted Decision Tree, Extreme Gradient Boosting, Support Vector Machine, and K-Nearest Neighbor)
Casteleiro-Roca J.L. et al., 2019 [13]	Hotel Building	the energy demand in the previous 24 h, the mean temperature of the previous day and the occupancy rate of the hotel.	Power demand	Hybrid model: clustering, LS-SVR, ANN (MLP, Tan-Sigmoid function), ARIMAX
Jaber et al., 2019 [14]	University Building	time, outdoor dry-bulb temperature, orientation of the building, overall heat transfer coefficient, space volume and window to wall ratio	Hourly cooling energy	feed forward artificial neural network (ANN)
Nasruddina, 2019 [15]	University Building	cooling set point, RH set point, starting delay, stopping delay, supply air flow rate, window area, wall thickness, supply air temperature, supply radiant temperature, supply radiant flow rate.	Annual energy consumption, PPD	Combination of artificial neural network (ANN-MLP) and MOGA
Sha et al., 2019 [16]	Large Retail Building	degree-day, day type, and month type	Cooling/heating Energy HVAC electricity consumption	SVM, ANN, and MLR
Wei et al., 2019 [17]	Office Building	electricity consumption of appliances, number of occupants, electricity consumption of lighting, solar radiation, electricity consumption of the fresh-air system, outdoor temperature barometric pressure, dry-bulb temperature, relative humidity, average wind speed, mean wind direction, mean wind direction indoor temperature and indoor relative humidity	Electricity consumption by an air-conditioning system	FFNN, ELM with BSI
Zhong et al., 2019 [18]	Office Building	dew point temperature, mean wind direction, mean wind velocity, outdoor temperature, precipitation intensity, precipitation quantity, relative humidity, school holiday time, working time schedule	Cooling load	Novel Vector Field-based SVR model
Koschwitz et. al., 2018 [19]	200 non-residential buildings (district scale)		Hourly heating and cooling load	NARX RNN SVM-R (RBF Kernel and polynomial Kernel)
Manjarres et al., 2017 [20]	Office Building	door, outdoor temperatures and relative humidities and occupancy levels, and operation of HVAC	Thermal consumption for heating, electrical consumption	RF regression
Muhammad Waseem Ahmad 2017 [21]	Hotel Building	outdoor air temperature, dew point temperature, relative humidity, windspeed, hour of the day, day of the week, month of the year, number of guests for the day, number of rooms booked	HVAC electricity consumption	feed forward backpropagation neural network (FFNN), RF
Jovanović et al., 2015 [22]	University Campuses	mean daily outside temperature, mean daily wind speed, total daily solar radiation, minimum daily temperature, maximum daily temperature, relative humidity, day of the week, month of the year, heating consumption of the previous day.	Heating consumption	FFNN, radial basis function network (RBFN) and adaptive neuro-fuzzy interference system (ANFIS)

Jovanović et al. [22] presented a prediction of heating energy using various neural networks: feed forward backpropagation neural network (FFNN), radial basis function network (RBFN), and adaptive neuro-fuzzy interference system (ANFIS). The subject of analysis was the university buildings in Norway. The authors have predicted building energy use based on the input feature: mean daily outside temperature, mean daily wind speed, total daily solar radiation, minimum daily temperature, maximum daily temperature, relative humidity, day of the week, month of the year, and heating consumption of the previous day. Data for the working days in the cold period for three years was used. For model FFNN, all input variables mentioned above are used; for model RBFN, seven most influencing parameters, and for model ANFIS, only three of them. The results showed that all three models have very good agreement with measured values. Sha et al. [16] presented a simplified energy prediction method based on the three input features: degree-day, day type, and month type. Their study adopted three machine learning algorithms: MLR, SVR, and ANN. The results showed that ANN and SVR methods have better performance than the MLR model. The authors mentioned that all of the methods do not have sufficient quality in heating energy prediction, due to the size of the training dataset. Manjarres et al. [20] proposed to implement the HVAC energy management system in a separate part of the office building in Spain. Solution includes a two-way communication system, enhanced database management system, and a set of machine learning algorithms based on random forest (RF) regression techniques. The proposed optimizer included information such as: indoor and outdoor temperatures, relative humidity, and occupancy level. The simulations took into account different modes of operation of HVAC systems. Data were collected from 63 days in summer and 46 days in winter, which was the basis during the simulation phase. The solution assumes the ON/OFF operation of the HVAC system and the operation of the mechanical ventilation system in accordance with the proposed solutions, which is to ensure minimization of energy consumption while maintaining the assumed temperature inside the rooms. The authors estimate that the implementation of the described system will contribute to the reduction of heat demand by 48% and 39% for cooling consumption. Ahmad et al. [21] presented a comparison between feed-forward back-propagation artificial neural network and random forest for HVAC electricity consumption of a hotel building in Madrid. Results showed that ANN performed marginally better than RF. Generally, both models have comparable quality of prediction and could be implemented in the building system.

In order to increase the accuracy of forecasting, as well as to expand the possibility of implementing solutions, many authors decide to modify classic prediction models or combine several approaches. The biggest problem of these patterns is the nonlinearity of relationships. Zhong et al. [18] proposed using a novel vector field-based support vector regression method. The purpose of the method is to define the optimal feature space by modifying the input data space. The resulting algorithm is then used to build a predictive model. The implementation of such a solution in an office building in China gave very good results compared to commonly used methods. The algorithm proposed in this article is used to determine the refrigeration load forecasting model based on the integrated data set. The input parameters are both external and internal. Casteleiro-Roca et al. [15] described an intelligent hybrid model to predict the short-term energy demand in a hotel, including three techniques: clustering, MLP, and LS-SVR. It was used for predicting the power load of the building for each hour in a 24-h horizon. The authors identified three input variables: the energy demand in the previous 24 h, the mean temperature of the previous day, and the occupancy rate of the hotel. The obtained results were compared with conventional forecast techniques based on ARIMAX modeling and a method based on tree models. The hybrid appeared to have better accuracy (lower mean absolute error) than the above-mentioned models.

Artificial Intelligence Algorithms are also widely used for electricity consumption forecasting, including district public consumption [23]. Güngör et al. [24] presented electricity consumption prediction for a variety of households using different prediction algorithms (Holt-Winters, ARIMA, LSTM i TESLA). Results showed that TESLA performance is better than other prediction methods. The authors have used five different classifiers (Logistic Regression, Stochastic Gradient Descent,

K-Nearest Neighbors, Random Forest, Support Vector Machine). Another example of such algorithms is the forecast of meteorological conditions for the needs of the HVAC system proposed by Işik and Inalli. A back-propagation neural network perceptron model with seven inputs was used to predict temperature, solar radiation, and relative humidity. A comparison with the ANFIS model showed that the ANN model has better accuracy in terms of forecasting meteorological data for HVAC. [25]

In this study, cooling energy consumption predictive models based on an artificial neural network and support vector machine algorithms are presented. The paper includes a statistical analysis of historical data of hotel building. The cooling load and proposed variables during the summer season are considered. The main contribution of this paper is the development of high-accuracy predictive models and comparison neural network-based model with the support vectors approach. The structure of this study was organized as follows. Firstly, an introduction to energy prediction is presented. A short review of the prediction method was included. Section 2 provided the case of the study and used methods description, including the details of data collection, input variables, and proposed models. In the Results section, the main results obtained from models are presented. In these sections, real energy consumption and predicted load are compared and analyzed. Finally, conclusions and remarks are given in Section 5.

2. Materials and Methods

2.1. Case of Study

The Turówka hotel building located in Wieliczka near Kraków (the south-central part of Poland) is used as a case of study for this paper. The five-story building is a reconstruction of a historic salt store. The hotel has a floor area of 5525.00 m^2 and a volume of 19,300 m^3. Facilities include 50 double rooms, a hotel bar, a restaurant, a drink bar, a conference room, and a pool. A detailed description of the building with an analysis of the cooling and heating load was described in the previous article [26]. Based on the analysis of the summer period in the hotel building, it was found that the cooling system is responsible for 50–60% of the total energy load. Additionally, the analysis showed a clear relationship between the outside temperature and the cooling load; therefore, the presented predictive model applies to the cooling energy prediction during the summer. The data used in the research mainly include three parts: meteorological data, load data, and data related to operational conditions. The input variables used for the predictive models for the calculation of the cooling demands include weather conditions, occupancy level in the hotel, hour, and day of the week. The cooling energy load data is provided by meters systems installed in the building cooling system i.e., feed and return of the high and, depending on demand, the low parameter of the refrigerant. Data transmitted via a serial communications protocol—MODBUS RTU—is stored in a recording system. The measurement system consists of MULTICAL heat meters by Kamstrup and flow sensors submitted to a type of approval according to EN 1434 [27], which includes the 2400-h measurement stability test of the flow sensors. The meteorological data used in this paper are obtained from the National Research Institute—Polish Institute of Meteorology and Water Management; only the outside temperature was measured directly at the hotel area. The data used in predictive models are an hourly time series collected in a summer season in Poland, where the cooling load was observed. The hourly meteorological data were calculated as an average value based on the measurement with sampling time set to 10 min. Data directly from the analyzed object, i.e., outdoor temperature and cooling energy consumption, were collected with a sampling time of 1 min. The data used for the models were calculated as an arithmetic average of measurement from an hourly period of time. For this study, the measurement season was selected from 15 May to 15 September 2019. The framework of the case study is shown in Figure 1.

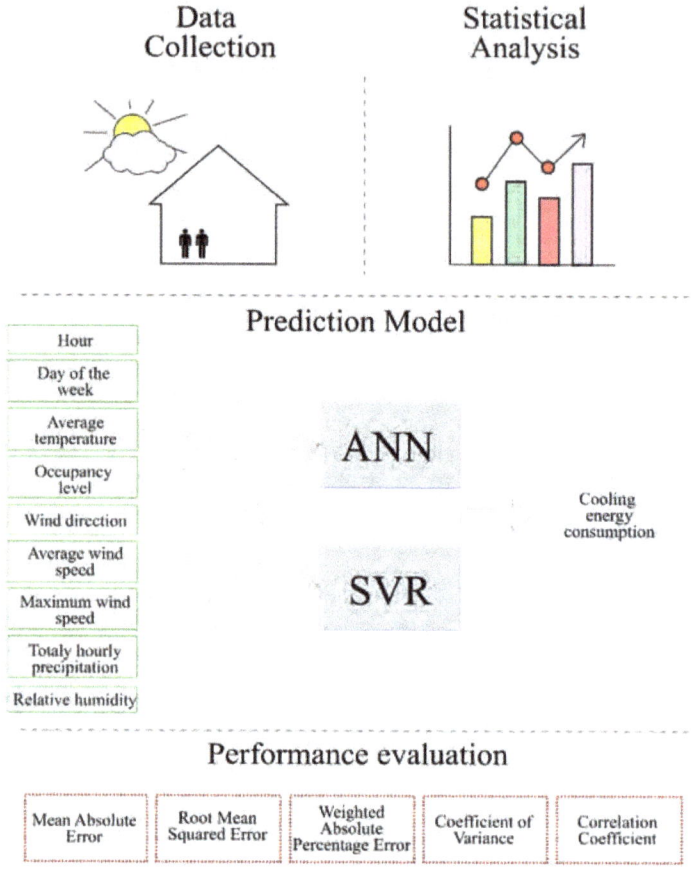

Figure 1. The framework of the case study.

2.2. Methodology

2.2.1. Methodology of Artificial Neural Networks

There are many types of artificial neural networks (ANN) including simple feed forward networks, recurrent neural networks, and spiking neural networks, RBFNs. One of the most commonly used models is the multi-layer back-propagation neural network (BPNN). The BPNN architecture includes three types of the layers: an input layer (variable), an output layer (predicted value), and a hidden layer. A basic processing unit in this model is a neuron. A schematic diagram of an artificial neural network structure which consists of all three layers is shown in Figure 2.

For Multi-layer Perceptron (MLP), neurons in the input layer distribute the input signals x_i to neurons in the hidden layer. Each neuron j in the hidden layer sums up its input signals x_i after weighting them with the strengths of the respective connections w_{ji} from the input layer and computes its output y_j as a function f of the sum [28]:

$$y_j = f\left(\sum w_{ji} x_i\right) \quad (1)$$

In this study, the Statistica Artificial Neural Network Package was used. A partition of the data into training (70%), validation (15%), and test (15%) is carried out.

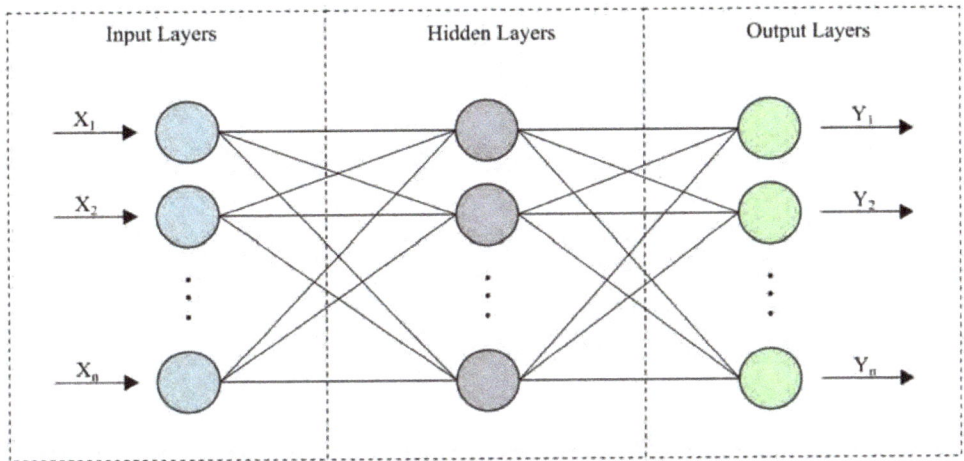

Figure 2. An illustration of a typical ANN topology.

2.2.2. Methodology of Support Vector Machines

Support Vector Machine is a supervised learning method increasingly used in solving nonlinear problems. This method is commonly used for classification, regression, and clustering. One of the main features of this model is the lack of local minima. It consists of two main parts: universal linear learning algorithm and a specific kernel that calculates the inner product of input points in feature space [29].

The universal linear function is as Equation (2):

$$f(x) = w^T \varphi(x) + b \tag{2}$$

where f(x) means the forecasting values and the coefficients w and b are adjustable.

The main aim of SVM is to find the optimal hyperplane between classes, with the maximal margin. The margin is defined as the distance between the closest point in each class and hyperplane. For this purpose, the ε-insensitive loss function is used. Minimizing the overall errors is expressed by Equation (3) [30]:

$$\min_{w,b,\xi^*,\xi} R(w, \xi^*, \xi) = \frac{1}{2} w^T w + C \sum_{i=1}^{N} (\xi_i^* + \xi_i) \tag{3}$$

with the constraints:

$$y_i - w^T \varphi(x_i) - b \leq \varepsilon + \xi_i^* \tag{4}$$

$$-y_i + w^T \varphi(x_i) + b \leq \varepsilon + \xi_i \tag{5}$$

$$\xi_i^*, \xi_i \geq 0 \tag{6}$$

The method of operation of the support vector machine is shown in Figure 3.

There are four types of kernel function linear, polynomial, radial basis function (RBF), and sigmoidal function [29]. The most used kernel functions are the Gaussian RBF with a function of kernel defined by Equation (7):

$$K(x_i, x_j) = \exp(-\gamma |x_i - x_j|^2) \tag{7}$$

where x_i and x_j are vectors in the input space and γ is kernel parameter. An equivalent definition involves a σ parameter, where $\gamma = 1/2\sigma^2$.

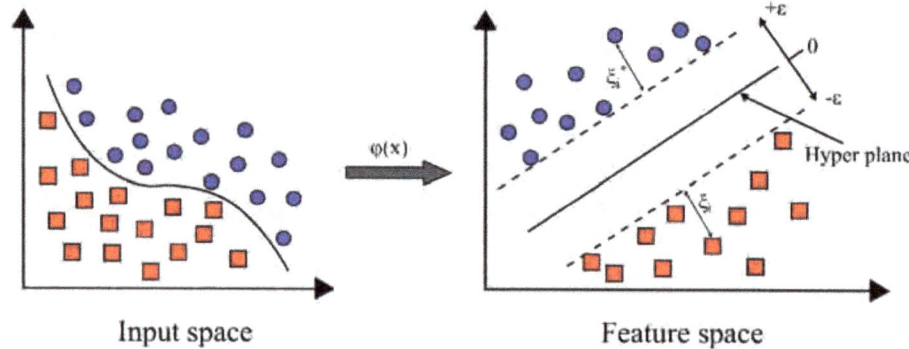

Figure 3. An illustration of a typical ANN topology.

The forecasting accuracy of the SVM model is affected by hyperparameters. In the ε-SVR, three proper parameters need to be determined: C, ε, and the kernel parameter (γ). Parameter ε represents the width of the ε-insensitive loss function and can affect the number of support vectors in the model. The higher ε value results in fewer support vectors and more flat estimates. Parameter C is related to model complexity and the degree of the deviations larger than ε which are tolerated. In models with high C parameter values, the main objective is to minimize the empirical risk only, without regard to model complexity part in the optimization formulation. Parameters σ and γ describe the Gaussian function width [31]. In this paper, the radial basis function RBF function is used as a kernel function to estimate the cooling load of the hotel building. A grid-search technique was applied to find the optimal parameter values. The kernel parameter γ was selected in priori based on the literature. According to the Limsvm-2.6 [32], the kernel parameter is defined as γ = 1/n, where n means the number of input variables. For d-dimensional problems, Cherkassy and Ma [31] noticed that the width parameter σ depends on the number of input variables d in accordance with the formula $\sigma^d \sim (0.2-0.5)$. On this basis, four values of γ parameters were found: 0.1, 0.3, 0.5, 0.7. The ten-fold cross-validation was applied to reduce the error of the model. The dataset was randomly divided into two parts: learning samples (75%) and testing samples (25%). The feasible ranges of the parameters are set as follows: C ∈ [0.5, 150] and ε ∈ [0.01, 0.5].

2.3. Model Evaluation Index

The performance of the proposed models is evaluated by the mean absolute error (*MAE*), root mean square error (*RMSE*), weighted absolute percentage error (*WAPE*), coefficient of variance (*CV*), and coefficient of determination (*r*) [11,12,18,33]. The indicators are calculated as follows:

$$MAE = \frac{1}{n}\sum_{i=1}^{n}|E_A - E_P| \qquad (8)$$

$$RMSE = \sqrt{\frac{1}{n}\sum_{i=1}^{n}(E_A - E_P)^2} \qquad (9)$$

$$WAPE = \frac{\sum_{i=1}^{n}|E_A - E_P|}{\sum_{i=1}^{n}E_A} \qquad (10)$$

$$CV = \frac{\sqrt{\frac{1}{n}\sum_{i=1}^{n}(E_A - E_P)^2}}{\overline{E_A}} \qquad (11)$$

$$r = \frac{cov(E_A, E_P)}{\sigma_{E_A} \sigma_{E_P}} \tag{12}$$

where n denotes the entire number of observations, E_A is the actual value, $\overline{E_A}$ denotes the mean of actual values, and E_P represents the predicted value.

3. Results

3.1. Preliminary Statistical Analysis

The first step in the analysis was a preliminary statistical analysis of the dataset. From the available data, nine parameters were selected that could potentially have an impact on cooling energy consumption. Apart from the most obvious values, such as temperature, air humidity, wind speed, and relative humidity, it was decided to use time variables such as the hour and day of the week as well as the occupancy level. Table 2 presents the information summary of the data used in this work.

Table 2. Descriptive statistics for input and output variables.

Variable	Unit	Minimum	Maximum	Mean	Median
Hour	-		From 0:00 to 23:00		
Day of the week	-		From Monday to Sunday		
Average temperature	°C	6.24	36.16	20.22	19.75
Occupancy level	%	18.90	100.00	77.35	83.00
Wind direction	°		From 12.50° to 341.00°		
Average wind speed	m/s	0.00	5.07	0.63	0.40
Maximum wind speed	m/s	0.00	13.12	2.22	2.07
Total hourly precipitation	mm	0.00	15.00	0.12	0.00
Relative humidity	%	16.65	98.06	70.93	74.60
Cooling energy consumption	kWh/h	0.00	141.30	51.35	42.34

In the beginning, the relationship between cooling energy consumption and the values of the variables was determined. Graphs for each of the variables are presented in Figure 4.

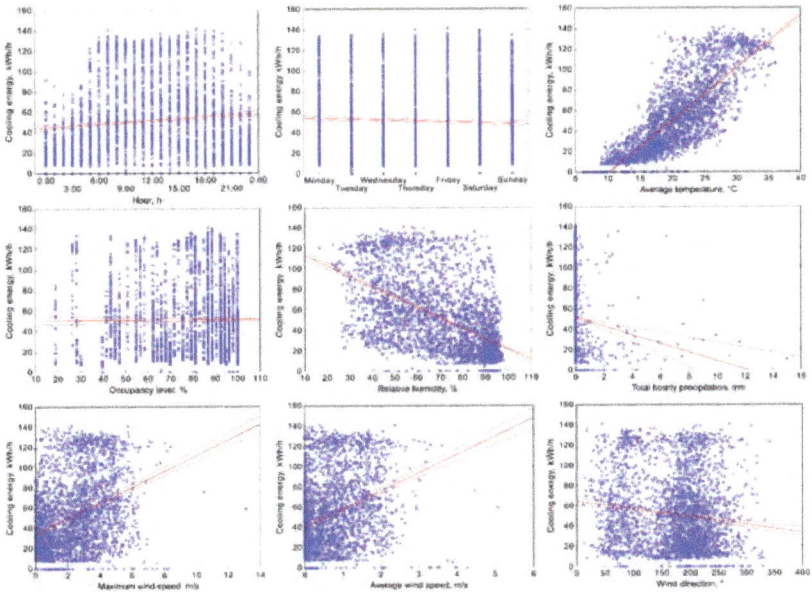

Figure 4. Relationship between cooling energy consumption and selected variables.

The Pearson correlation coefficient was the key parameter determining the acceptance of a given variable for analysis. Coefficients for each between the studied variables and the predicted output were provided. The results are summarized in Figure 5.

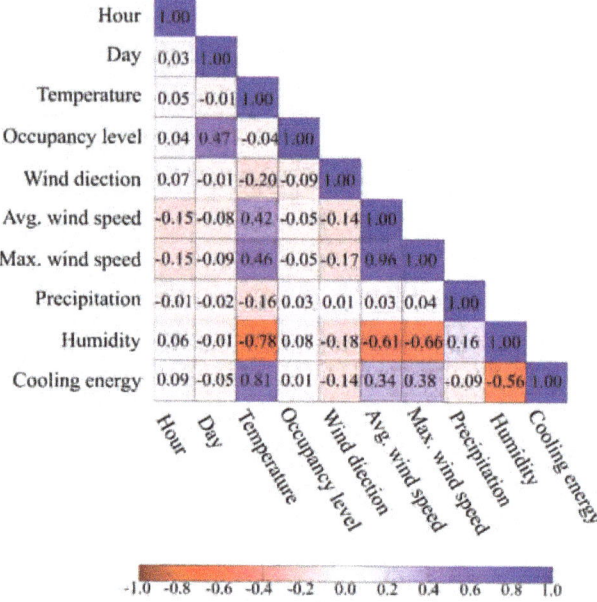

Figure 5. Matrix of the Pearson correlation coefficients.

Variables with low correlation coefficients can be significant in more complex predictive models. Based on the analysis of the correlation coefficient, the rejection of two parameters was initially planned: day and occupancy level. Due to the low correlation coefficients of the variables which, according to the authors, may have a significant impact on the predicted values, a more detailed analysis of the cooling load variability overtime was performed. Figure 6 shows plots of variation over time for the raw data collected with a sampling time of 1 min, and also after taking the hourly mean which was then accepted for prediction models.

As shown in Figure 6, there is a clear relationship between the demand for cooling energy and the time it was recorded. This relationship is not linear, hence the low correlation coefficient in the previous analysis step. The adoption of averaged values for further analysis may of course affect the accuracy of prediction, due to the unstable variability of the demand value; however, due to practical aspects, it was decided that the hourly variables were more efficient and could give sufficient effects in the prediction models. Despite the low correlation coefficients for time parameters and the occupancy level, it was decided that these parameters may be of great importance for a hotel facility due to the variability of the energy load depending on the use by guests.

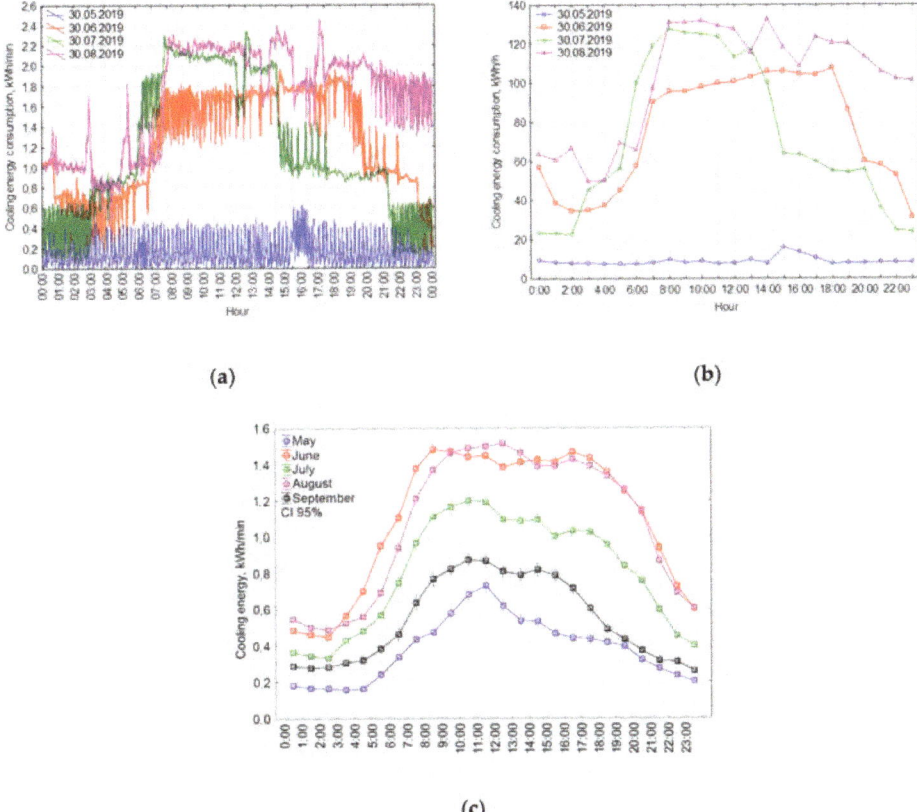

Figure 6. Hourly variation of cooling energy consumption: (**a**) for the sample days (30 May, 30 June, 30 July, 30 August) based on data sampled every minute; (**b**) for example days (30 May, 30 June, 30 July, 30 August) on the basis of hourly average data; (**c**) mean values grouped by hour and categorized by month based on data sampled every minute.

3.2. Prediction Models

3.2.1. Artificial Neural Networks

As mentioned in Section 2.2.1, the Statistica Artificial Neural Network Package was used to prepare predictive models. The data was divided into three groups. In addition, 70% of the data was used for the design of the network, 15% was used for validation, and 15% for testing. The hidden layer activation-functions, output layer activation-function, and the number of hidden neurons are selected using the methodology based on statistical tests and least-squares estimation. Models were created by combining different types of activation functions and a different number of hidden neurons. The selection of the function was made from identity, logistic, hyperbolic tangent, and exponential function. To find the optimum number of hidden neurons, various numbers of neurons were examined. The various combination of activation functions and numbers of hidden neurons as described above were tested. Sum-of-squares was selected as the error function during the network training process. Training Algorithm BFGS (Broyden–Fletcher–Goldfarb–Shanno) was chosen for this work. Five models with the best accuracy were selected and described in Table 3.

Table 3. Network configurations tested.

Model Name	Hidden Layer Activation-Function	Number of Hidden Units	Output Layer Activation-Function	Correlation Coefficients
MLP 1	Logistic	90	Logistic	0.912
MLP 2	Hyperbolic tangent	68	Logistic	0.925
MLP 3	Logistic	83	Logistic	0.909
MLP 4	Hyperbolic tangent	71	Logistic	0.903
MLP 5	Hyperbolic tangent	54	Logistic	0.902

Figure 7 shows the schemes of regressions for all data sets according to each of the MLP models. The plots explain the correlation between the real values and the MLP model output. The solid line in each plot represents the best linear fit between the output and target values.

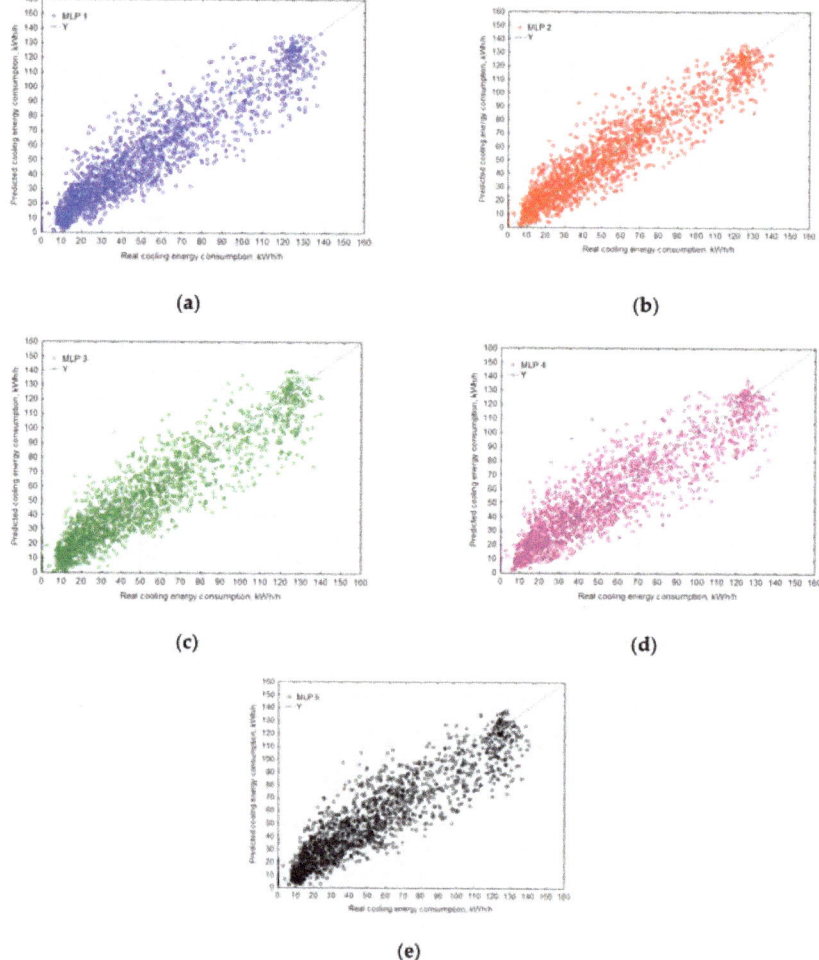

Figure 7. Comparison between real cooling energy consumption and prediction of SVM model: (**a**) for the model MLP-1; (**b**) for the model MLP-2; (**c**) for the model MLP-3; (**d**) for the model MLP-4; (**e**) for the model MLP-5.

In Table 4, sensitivity coefficients are presented. The analysis describes the change in the system's outputs due to variations in the parameters that affect the system. Performing the sensitivity analysis consists of controlling how the network error behaves in the case of fluctuations of the independent variables. For each input variable, its values are converted to the mean (from the training set) so that it does not contribute any information to the model. After supplying such modified data to the network input, the final prediction error is checked. A larger error value means that the model depends on the proposed variable. The higher the value of the sensitivity analysis coefficients, the greater the importance of a given variable for a good fit of the model.

Table 4. Sensitivity analysis of inputs.

Variable	MLP 1	MLP 2	MLP 3	MLP 4	MLP 5
Hour	2.04	2.78	2.14	1.96	1.73
Day of the week	2.01	2.53	1.89	1.58	1.71
Average temperature	9.97	8.74	7.04	5.93	6.84
Occupancy level	1.64	2.82	1.48	1.30	1.42
Wind direction	1.39	1.42	1.26	1.28	1.20
Average wind speed	1.78	1.93	1.72	1.44	1.31
Maximum wind speed	1.93	1.97	1.83	1.55	1.51
Total hourly precipitation	1.04	1.04	1.04	1.04	1.03
Relative humidity	2.30	2.84	2.33	2.01	1.97

According to the statistical analysis and the analysis of the sensitivity of neural networks, a clear impact on energy consumption is noticeable in the case of parameters such as relative humidity, maximum wind speed, occupancy level, hour, and day of the week

3.2.2. Support Vectors Machines

As mentioned above, ranges of gamma, C, and epsilon parameters were defined and tested in the search for optimal values. Below, in Table 5, models with given characteristic parameters are defined which were characterized by the best fit during the tests.

Table 5. Network configurations tested.

Model Name	Gamma	C	Epsilon	No. of Vectors	Correlation Coefficients
SVM 1	0.1	10	0.20	884	0.853
SVM 2	0.3	10	0.20	809	0.867
SVM 3	0.5	10	0.17	877	0.873
SVM 4	0.7	10	0.13	979	0.879

In Figure 8, the relationship between the real energy consumptions and the predicted values was plotted. The line of a theoretical perfect 1:1 match line was marked in red.

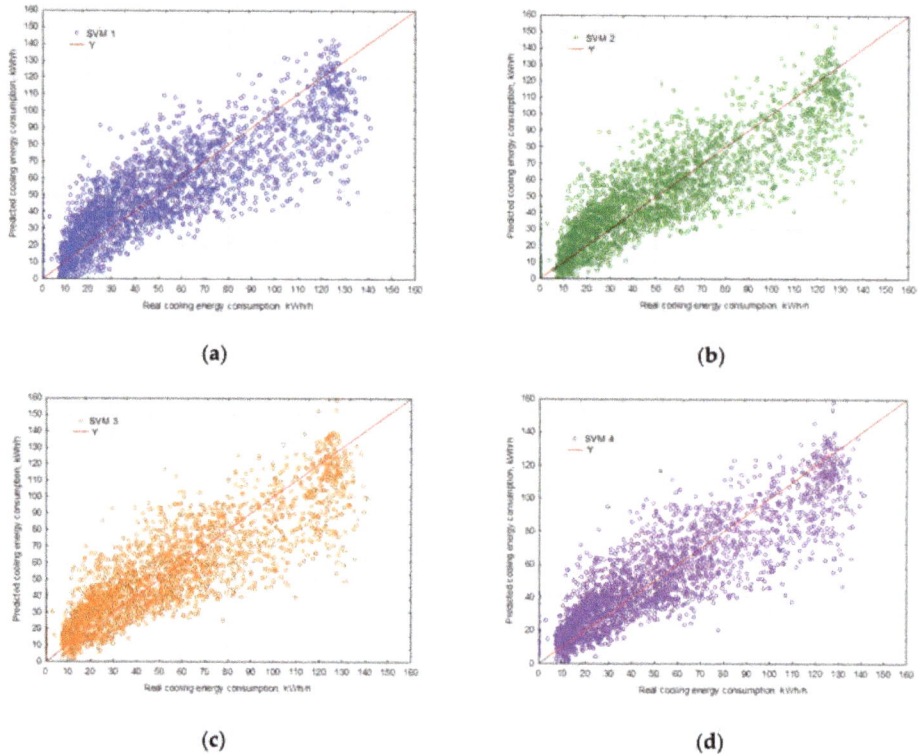

Figure 8. Comparison between real cooling energy consumption and prediction of SVM model: (**a**) for SVM 1 model; (**b**) for SVM 2 model; (**c**) for SVM 3 model; (**d**) for SVM 4 model; performance of models.

Based on the coefficients defined in Section 2.3, selected models of neural networks were compared. As previously mentioned, five matching indexes were selected for this purpose: *MAE*, *RMSE*, *WAPE*, *CV*, and correlation coefficient (r). The results for cooling consumption are presented in Table 6.

Table 6. Performance evaluation of different ANN models for cooling energy consumption.

Indicators	MLP 1	MLP 2	MLP 3	MLP 4	MLP 5
MAE	11.09	10.27	11.46	11.69	11.80
RMSE	15.05	13.88	15.35	15.71	15.78
WAPE	21.54%	19.93%	22.34%	22.77%	22.77%
CV	29.29%	27.03%	29.88%	30.59%	30.72%
r	0.91	0.93	0.91	0.90	0.90

Similarly, for selected SVM models, the coefficient values are summarized in Table 7.

Table 7. Performance evaluation of different SVM models for cooling energy consumption.

Indicators	SVM 1	SVM 2	SVM 3	SVM 4
MAE	14.46	13.79	13.48	13.13
RMSE	18.84	18.10	17.80	17.44
WAPE	28.04%	26.98%	26.20%	25.45%
CV	36.69%	35.24%	34.65%	33.95%
r	0.85	0.87	0.87	0.88

Figure 9 shows comparison of the real values and predicted results of the MLP and SVM models with the high accuracy; it is MLP 2 and an SVM 4 model.

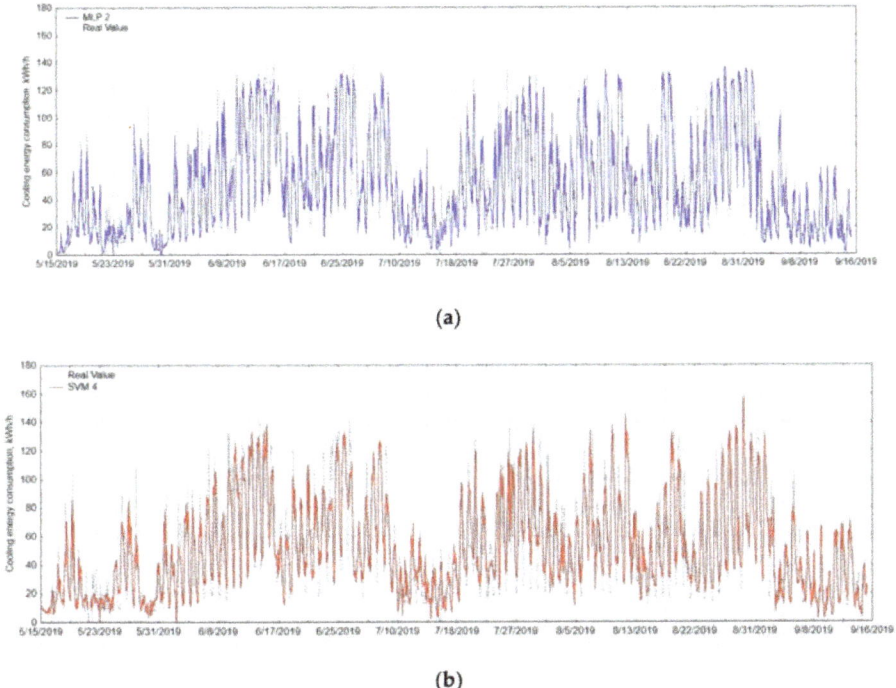

Figure 9. Prediction performance during cooling season of the model with highest accuracy (**a**) neural network MLP 2; (**b**) Support Vector Machine SVM 4.

The figure shows the real values of hourly consumption of cooling energy in the analyzed period with a dashed gray line. The blue color marks the results obtained with the MLP 2 model, which received the best results among the proposed ANN networks, and the red color indicates SVM 4, analogically the best among the presented SVM models. In both of the proposed models, the worst effects present at predicting extreme values, where the cooling load is very high or very low. Comparing the models with each other, it is noticeable that neural networks perform better in this matter, and the difference between the real load and the predicted value is significantly smaller.

4. Discussion

The analysis of the cooling energy load in the hotel building showed that, based on several parameters characterizing the external conditions, building using conditions and time, it is possible to estimate the overall cooling load of the building. The analysis used nine parameters, including hour, day of the week, average temperature, occupancy level, wind direction, average wind speed, maximum wind speed, total hourly precipitation, and relative humidity. The external temperature has an obvious influence on energy consumption, which determines the demand for cooling energy. As shown by the preliminary analysis, relative humidity, wind speed, and direction can also be considered significant factors. Despite the low correlation coefficients of the time parameters (hour and day of the week) and the occupancy level, it was decided that, in the general model, they may be significant; therefore, they were not omitted in further analysis. Based on the assumed parameters, five models of neural networks based on the MLP algorithm and four SVM models with different configurations of parameters defining the models were proposed. Five model evaluation indicators were used to check the quality of the models shown: the mean absolute error (*MAE*), root mean square error (*RMSE*), weighted absolute percentage error (*WAPE*), coefficient of variance (*CV*), and coefficient of determination (*r*). Among all the proposals, a better fit was found for the ANN models, the correlation coefficient of which did not fall below 0.9. As shown by the sensitivity analysis of inputs (Table 4), time and occupancy level played an important role in the model, despite the seemingly low importance of these parameters. Total hourly precipitation had the least impact on the overall result. The selected neural networks are characterized by the *WAPE* coefficient ranging from 19.93 to 22.77% and the *MAE* coefficient of 10.27 to 11.80 kWh/h. The best of the proposed models, i.e., MLP 2, achieved the value of the correlation coefficient *r* at the level of 0.93. For comparison, the mean absolute error for the proposed SVM models varies from 13.13 to 14.46 kWh/h, and the weighted absolute percentage error from 25.45 to 28.04%. The most precise SVM model—SVM 4—is characterized by a correlation coefficient of 0.88. The proposed prediction methods are widely used for energy prediction due to the possibility of their application in nonlinear dependencies of many variables. As the results show, the use of neural networks in energy prediction enables the achievement of better model fit coefficients. Both methods used were based on the nine variables mentioned above. The greatest differences between the proposed models are visible in the extreme values. The SVM model performs much worse at very low and very high cooling loads. In the event of high values of cooling energy consumption, both the ANN and SVM models lower the predicted values compared to the actual values. This is especially noticeable with the SVM model, which directly affects the lower fit factors.

5. Conclusions

Reducing energy consumption is an issue that is becoming more and more popular nowadays. This is related to both the obvious economic aspects as well as the growing awareness of the society regarding the limitation of human impact on the environment, including the exploitation of non-renewable resources and emissions of pollutants into the environment. There are new solutions on the market to reduce the energy demand of buildings. Systems of proper management of energy demand, especially in combined energy economies, are also becoming more and more popular. The key issue in such solutions is the ability to accurately determine the demand for energy at a given moment. As a consequence, machine learning techniques are also becoming more and more popular, used to predict the energy load of a building.

The article presents two popular methods of energy prediction: neural networks and support vector machines. For each of the methods, several models differing in characteristic parameters and selected optimization methods are presented. The subject of the analysis is a building of historical importance, the modernization of which is limited due to the minimization of interference with the cubature and the overall appearance of the facility. In addition to replacing the heat source and internal installations, one of the effective methods of reducing operating costs may be the proper management of energy installations, which requires a building energy consumption forecasting system. The models use parameters that may affect consumption, both characterizing external conditions, time of use, and the number of guests. Based on the collected data, five models of neural networks and four SVM models were presented. Each of them was compared according to the proposed forecasting accuracy coefficients. The analysis showed that, despite the initially insignificant influence of some parameters on the obtained results, they played an important role in the predictive model. The best fit coefficients were obtained for models using artificial neural networks to predict the heating load. The proposed model made it possible to estimate the demand for cooling energy with a matching coefficient of 0.93. The quality of the model can be improved by extending the analysis time to several summer seasons. Hotel buildings are a specific object due to the users who have different preferences as to the conditions inside the rooms. Individual control allows them to maintain thermal comfort by changing the temperature, and thus their behavior directly affects energy consumption for cooling purposes.

Author Contributions: Conceptualization, M.B. and K.Z.; methodology, M.B.; software, K.Z.; validation, M.B.; formal analysis, M.B.; investigation, K.Z.; resources, K.Z.; data curation, K.Z.; writing—Original draft preparation, K.Z.; writing—Review and editing, M.B.; visualization, K.Z.; supervision, M.B.; project administration, M.B. All authors have read and agreed to the published version of the manuscript.

Funding: This research was funded by the European Regional Development Fund, Intelligent Development Program, Grant No. POIR.01.01.01-00-0720/16.

Conflicts of Interest: The authors declare no conflict of interest.

Abbreviations

AHU	Air Handling Unit
ANFIS	Adaptive Neuro-Fuzzy Interference System
ANN	Artificial Neural Network
ARIMA	Autoregressive, Integrated and Moving Average
ARIMAX	Auto-Regressive Integrated Moving Average with eXplanatory variable
BFGS	Broyden–Fletcher–Goldfarb–Shanno
BPNN	Back Propagation Neural Network
BSI	Blind System Identification
CV	Coefficient of Variance
DT	Decision Tree
ELM	Extreme Learning Machine
FDD	Fault Detection and Diagnosis
FFNN	Feed Forward Backpropagation Neural Network
GBDT	Gradient Boosting Decision Tree
GRNN	General Regression Neural Network
HVAC	Heating, Ventilation, Air Conditioning
kNN	k Nearest Neighbors
LS-SVR	Least Squares SVR
LSTM	Long Short-Term Memory Neural Networks
MAE	Mean Absolute Error
MLP	Multilayer Perceptron
MLR	Multiple Linear Regression
MOGA	Multi-Objective Genetic Algorithm
NARX RNN	Nonlinear Autoregressive Exogenous Recurrent Neural Networks
PPD	Predicted Percentage Dissatisfied
RBF	Radial Basis Function
RBFN	Radial Basis Function Network
RF	Random Forest
RMSE	Root Mean Squared Error
SVM	Support Vector Machine
SVR	Support Vector Regression
WAPE	Weighted Absolute Percentage Error
XGBoost	Extreme Gradient Boosting

References

1. International Energy Agency. *Global Status Report for Buildings and Construction. Towards a Zero-Emissions, Efficient and Resilient Buildings and Construction Sector*; United Nations Environment Programme: Paris, France, 2019.
2. International Energy Agency. *Energy Efficiency Indicators*; Highlights, IEA: Paris, France, 2019.
3. Directive 2010/31/EU of the European. *Parliament and of the Council of 19 May 2010 on the Energy Performance of Buildings*; IWA Publishing: London, UK, 2010.
4. ASHRAE. *Handbook of Fundamentals*; ASHRAE: Atlanta, GA, USA, 2013.
5. Lee, H.; Kim, S. Black-Box Classifier Interpretation Using Decision Tree and Fuzzy Logic-Based Classifier Implementation. *Int. J. Fuzzy Log. Intell. Syst.* **2016**, *16*, 27–35. [CrossRef]

6. Wei, Y.; Zhang, X.; Shi, Y.; Xia, L.; Pan, S.; Wu, J.; Han, M.; Zhao, X. A review of data-driven approaches for prediction and classification of building energy consumption. *Renew. Sustain. Energy Rev.* **2018**, *82*, 1027–1047. [CrossRef]
7. Yu, Z.; Haghighat, F.; Fung, B.C.M.; Yoshino, H. A decision tree method for building energy demand modeling. *Energy Build.* **2010**, *42*, 1637–1646. [CrossRef]
8. Chou, J.S.; Bui, D.K. Modeling heating and cooling loads by artificial intelligence for energy-efficient building design. *Energy Build.* **2014**, *82*, 437–446. [CrossRef]
9. Wang, R.; Lu, S.; Li, Q. Multi-criteria comprehensive study on predictive algorithm of hourly heating energy consumption for residential buildings. *Sustain. Cities Soc.* **2019**, *39*, 1016–1023. [CrossRef]
10. Ahmad, T.; Chen, H. Nonlinear autoregressive and random forest approaches to forecasting electricity load for utility energy management systems. *Sustain. Cities Soc.* **2019**, *45*, 460–473. [CrossRef]
11. Sendra-Arranz, R.; Gutiérrez, A. A long short-term memory artificial neural network to predict daily HVAC consumption in buildings. *Energy Build.* **2020**, *216*, 109952. [CrossRef]
12. Wang, R.; Lu, S.; Feng, W. A novel improved model for building energy consumption prediction based on model integration. *Appl. Energy* **2020**, *262*, 114561. [CrossRef]
13. Casteleiro-Roca, J.L.; Gómez-González, J.F.; Calvo-Rolle, J.L.; Jove, E.; Quintián, H.; Gonzalez Diaz, B.; Mendez Perez, J.A. Short-Term Energy Demand Forecast in Hotels Using Hybrid Intelligent Modeling. *Sensors* **2019**, *19*, 2485. [CrossRef]
14. Jaber, A.A.; Saleh, A.M.A.; Ali, H.A.M. Prediction of Hourly Cooling Energy Consumption of Educational Buildings Using Artificial Neural Network. *Int. J. Adv. Sci. Eng. Inf. Technol.* **2019**, *9*, 159–166. [CrossRef]
15. Nasruddin, N.; Sholahudin, S.; Satrio, P.; Mahlia, T.M.I.; Giannetti, N.; Saito, K. Optimization of HVAC system energy consumption in a building using artificial neural network and multi-objective genetic algorithm. *Sustain. Energy Technol. Assess.* **2018**, *35*, 48–57. [CrossRef]
16. Shaa, H.; Xua, P.; Hub, C.; Lib, Z.; Chena, Y.; Chena, Z. A simplified HVAC energy prediction method based on degree-day. *Sustain. Cities Soc.* **2019**, *51*, 101698. [CrossRef]
17. Wei, Y.; Xia, L.; Pan, S.; Wu, J.; Zhang, X.; Han, M.; Zhang, W.; Xie, J.; Li, Q. Prediction of occupancy level and energy consumption in office building using blind system identification and neural networks. *Appl. Energy* **2019**, *240*, 276–294. [CrossRef]
18. Zhong, H.; Wang, J.; Jia, H.; Mu, Y.; Lv, S. Vector field-based support vector regression for building energy consumption prediction. *Appl. Energy* **2019**, *242*, 403–414.
19. Koschwitz, D.; Frisch, J.; van Treeck, C. Data-driven heating and cooling load predictions for non-residential buildings based on support vector machine regression and NARX Recurrent Neural Network: A comparative study on district scale. *Energy* **2018**, *165*, 134–142. [CrossRef]
20. Manjarres, D.; Mera, A.; Perea, E.; Lejarazu, A.; Gil-Lopez, S. An energy-efficient predictive control for HVAC systems applied totertiary buildings based on regression techniques. *Energy Build.* **2017**, *152*, 409–417. [CrossRef]
21. Ahmad, M.W.; Mourshed, M.; Rezgui, Y. Trees vs Neurons: Comparison between random forest and ANN forhigh-resolution prediction of building energy consumption. *Energy Build.* **2017**, *147*, 77–89. [CrossRef]
22. Jovanović, R.Ž.; Sretenović, A.A.; Živković, B.D. Ensemble of various neural networks for prediction of heating energyconsumption. *Energy Build.* **2015**, *94*, 189–199. [CrossRef]
23. Kuo, P.H.; Huang, C.J. A High Precision Artificial Neural Networks Model for Short-Term Energy Load Forecasting. *Energies* **2018**, *11*, 213. [CrossRef]
24. Güngör, O.; Akşanlı, B.; Aydoğan, R. Algorithm selection and combining multiple learners for residential energy prediction. *Future Gener. Comput. Syst.* **2019**, *99*, 391–400.
25. Işik, E.; Inalli, M. Artificial neural networks and adaptive neuro-fuzzy inference systems approaches to forecast the meteorological data for HVAC: The case of cities for Turkey. *Energy* **2018**, *154*, 7–16. [CrossRef]
26. Borowski, M.; Mazur, P.; Kleszcz, S.; Zwolińska, K. Energy Monitoring in a Heating and Cooling System in a Building Based on the Example of the Turówka Hotel. *Energies* **2020**, *13*, 1968. [CrossRef]
27. *European Standard EN-1434:2015 Thermal Energy Meters*; BSI: London, UK, 2015.
28. Pham, D.T.; Liu, X. *Neural Networks for Identification, Prediction and Control*; Springer: London, UK, 1995.
29. Ke-Lin, D.; Swamy, M.N.S. *Neural Networks and Statistical Learning*; Springer: London, UK, 2014.
30. Hong, W.-C. *Hybrid Intelligent Technologies in Energy Demand in Forecasting*; Springer International Publishing: Berlin/Heidelberg, Germany, 2020.

31. Cherkassky, V.; Ma, Y. Practical Selection of SVM Parameters and Noise Estimation for SVM Regression. *Neural Netw.* **2004**, *17*, 113–126. [CrossRef]
32. Chang, C.C.; Lin, C.J. LIBSVM. A Library for Support Vector Machines. 2001 Software. Available online: http://www.csie.ntu.edu.tw/~{}cjlin/libsvm (accessed on 30 September 2020).
33. Schroeter, B.J.E. Artificial Neural Networks in Precipitation Nowcasting: An Australian Case Study. In *Artificial Neural Network Modelling*; Shanmuganathan, S., Samarasinghe, S., Eds.; Springer International Publishing: Cham, Switzerland, 2016; pp. 325–339.

Publisher's Note: MDPI stays neutral with regard to jurisdictional claims in published maps and institutional affiliations.

© 2020 by the authors. Licensee MDPI, Basel, Switzerland. This article is an open access article distributed under the terms and conditions of the Creative Commons Attribution (CC BY) license (http://creativecommons.org/licenses/by/4.0/).

Article

Assessment of the Impact of Occupants' Behavior and Climate Change on Heating and Cooling Energy Needs of Buildings

Gianmarco Fajilla [1,2,*], **Marilena De Simone** [1], **Luisa F. Cabeza** [3] **and Luís Bragança** [2,*]

1. Department of Environmental Engineering, University of Calabria, 87036 Rende, Italy; marilena.desimone@unical.it
2. Department of Civil Engineering, University of Minho, 4800-058 Guimarães, Portugal
3. GREiA Research Group, Universitat de Lleida, 25001 Lleida, Spain; luisaf.cabeza@udl.cat
* Correspondence: gianmarco.fajilla@unical.it (G.F.); braganca@civil.uminho.pt (L.B.); Tel.: +39-393-093-3551 (G.F.); +351-966-042-447 (L.B.)

Received: 9 November 2020; Accepted: 3 December 2020; Published: 7 December 2020

Abstract: Energy performance of buildings is a worldwide increasing investigated field, due to ever more stringent energy standards aimed at reducing the buildings' impact on the environment. The purpose of this paper is to assess the impact that occupant behavior and climate change have on the heating and cooling needs of residential buildings. With this aim, data of a questionnaire survey delivered in Southern Italy were used to obtain daily use profiles of natural ventilation, heating, and cooling, both in winter and in summer. Three climatic scenarios were investigated: The current scenario (2020), and two future scenarios (2050 and 2080). The CCWorldWeatherGen tool was used to create the weather files of future climate scenarios, and DesignBuilder was applied to conduct dynamic energy simulations. Firstly, the results obtained for 2020 demonstrated how the occupants' preferences related to the use of natural ventilation, heating, and cooling systems (daily schedules and temperature setpoints) impact on energy needs. Heating energy needs appeared more affected by the heating schedules, while cooling energy needs were mostly influenced by both natural ventilation and usage schedules. Secondly, due to the temperature rise, substantial decrements of the energy needs for heating and increments of cooling energy needs were observed in all the future scenarios where in addition, the impact of occupant behavior appeared amplified.

Keywords: occupant behavior; climate changes; energy needs; ventilation; residential buildings; DesignBuilder

1. Introduction

In most developed countries, buildings are the major energy consumers, and they may not be able to reach the new energy standards [1,2]. In the EU, most of the buildings have more than twenty years and present low energy performance [1]: The percentage of well-designed buildings is less than 2%, with almost 60% of heating systems inefficient and almost 40% of the windows being single glazed [3]. As recognized by the Energy Performance of Buildings Directive (EPBD) [4], buildings are responsible for 40% of the total energy consumption and 36% of global annual greenhouse gas emissions [3,5–7]; these consumptions could drastically increase double or even triple by 2050 if not faced in the right way [8]. As a consequence, governments worldwide have implemented energy requirements in their building regulations to reduce levels of energy consumed by buildings and to promote more energy-efficient envelopes and systems [9].

2. Literature Review

Nowadays, most researchers agree that occupant behavior plays an increasingly important role in building energy performance [10,11]. Despite the efforts made in improving the envelope of buildings and the efficiency of the systems, reducing energy consumption can be achieved considering also the impact that occupants' behavior (OB) has on buildings consumptions [12–18]. Furthermore, OB is often neglected or too simplified in energy design and assessment, causing large discrepancies between calculated and measured energy performances [12,19,20]. For example, a recent study conducted by Carlucci et al. [19] claimed that occupant behavior related to thermostat control (thermostat setpoints and operation schedules) is often too simplified in the building performance standards and calculation procedures, causing significant uncertainty in the predictions of building energy demand. Moreover, Mora et al. [20] simulated the energy consumption of a residential building considering three occupancy scenarios: Regulations, Current-use, and Statistical. Compared to the Current-use schedules, the Regulation schedules provided a significant underestimation of the heating energy needs, while the statistical schedules led to an overestimation. Different authors [13,14,18] highlighted that OB has an important responsibility in determining the energy consumption of buildings, pointing out that this impact is more significant in the new buildings where the envelope and the systems are optimized. Furthermore, Rouleau et al., in their work [15] claimed that the impact of OB has to be recognized to obtain a reduction in energy consumption. Because OB impacts in many ways on energy consumption (e.g., through heating and cooling systems or the interaction with windows and blinds), they deem that it should be not surprising if there is a huge gap between actual and prevised consumption. Furthermore, Zhang et al. [16] analyzed the role of occupant behavior in building energy performance, concluding that the energy-saving potential of occupant behavior in residential buildings is in the range of 10–25%. Similar results were obtained by [17] that quantify to 20% the achievable energy saving by modifying occupants' behavior using recommendations and feedback. Consequently, occupant behavior in buildings is becoming an increasingly topic so much so that different projects, performed within the framework of the International Energy Agency—Energy in Buildings and Communities Program (IEA-EBC), such as IEA EBC Annex 66 [21] and IEA EBC Annex 79 [22], focused on understanding and studying this issue.

The impact of OB on energy consumption of buildings is also recognized by the Intergovernmental Panel on Climate Change (IPCC) that in the IPCC AR5 [23] reported that factors of 3 to 10 differences can be found worldwide in residential energy use for similar dwellings, due to different usage of natural ventilation and thermal control of the indoor environment.

The reduction of buildings' energy consumption is a growing and global problem, mainly due to the looming threat of climate change. Goal 13 of the 2030 Agenda for Sustainable Development [24] calls for urgent action to tackle climate change and its impacts. Indeed, due to climate change and more frequent extreme events, buildings will have to deal with new climatic conditions for which they were not designed [25]. Thus, an increasing body of literature is now emerging on this topic. A recent work [26] assessed the scientific literature on the energy efficiency of buildings and the climate impact through a comparative analysis of Web of Science and Scopus. It was found that while most of the works focused on technologies for heating, ventilation, air-conditioning, and phase change materials, there is still a knowledge gap in the areas of behavioral changes, circular economy, and some of the renewable energy sources (e.g., geothermal, biomass, wind). The authors in [6] analyzed the impact of climate change on the energy performance of a zero energy building in Valladolid (Spain). Three future weather scenarios (2020, 2050, and 2080) were investigated, and the results showed a drop in the space heating demand and an increase in space cooling. Due to these consumptions' variations, they estimated an increase equal to 25% of the burning biomass to provide more energy to the absorption cooling system. Berardi and Jafarpur [27] assessed the impact of climate change on building heating and cooling energy demand of 16 building prototypes located in Toronto (Canada). Authors estimated for 2070 an average decrease of 18–33% and an average increase of 15–126% for the heating and cooling energy use, respectively. Ciancio et al. [28] simulated the energy performances of a building in three cities (Aberdeen, Palermo, and Prague) considering three climatic scenarios (2020, 2050, and 2080). In general, decreasing trends for heating energy needs and increasing trends for cooling energy needs were obtained. The highest variations were observed for 2080: A reduction of the heating energy needs from −36% to −80% and an increase of cooling energy needs from +142% to 2316%. In another study, Ciancio et al. [29] analyzed the energy needs of a hypothetic building by varying its location in 19 cities with different climate conditions. The simulations performed for 2020, 2050, and 2080 showed, once again, a general decrease in heating energy needs and an increase in cooling energy needs. The authors highlighted that the effects of climate change will be more predominant in the Mediterranean basin than in other European areas. Same results were also found from other studies, such as [25], that argued that Southern Europe will be more vulnerable to climate change than Northern Europe. Furthermore, the authors in [30] studied the climate change-driven increase of energy demand in residential buildings in the area of Qatar, founding an increase equal to around 30%. They stressed how such an increase would cause higher CO_2 emissions, more consumptions of water and fossil fuel, as well as an increase in the impact on the already strained local marine ecosystem. They also suggested renovating the building stocks and substitute fossil fuels with renewable energies (e.g., PV plants, wind farms, and tidal plants) as approaches to reduce the environmental impacts of climate change. Cabeza and Chàfer [8] published a systematic review of the technological options and strategies to achieve zero energy buildings contributing to climate change mitigation. Findings showed that buildings, if properly designed, can help to mitigate the impact of climate change—decreasing both the embodied energy in the materials, used during the construction phase, and the energy demand and use in the operation phase. Moreover, regarding new buildings, authors in [31] proposed an innovative method for designing buildings with robust energy performance under climate change for supporting architects and engineers in the design phase. To the extent of our knowledge, the effect of environmental (climate change) and behavioral variables (such as usage profiles and thermal comfort preferences) on the energy performance of buildings was investigated separately in the literature till now. What is missing are studies that consider both the influencing variables and provide predictions combining the double impact. Table 1 synthesizes the literature review related to this area highlighting: Subject of the study, outcomes and limitations, and considered impacts (occupant behavior/ climate change).

Table 1. Comparison of the Literature review.

Ref.	Subject	Outcomes and Limitations	Considered Impact *	
			OB	CC
[6]	Effect of CC on the energy performance of ZEBs	Heating/cooling demand registered in future scenarios		√
[8]	Technological options to achieve ZEBs	Gaps in the application of different technologies to reduce CC		√
[10]	Occupant-related energy codes and standards	Considerable variations across the occupancy and usage profiles	√	
[11]	Impact of OB on the energy demands	High variability of OB effect	√	
[12]	Influence of OB on natural ventilation	OB is the reason for discrepancies between calculated and measured energy performance and comfort. The characteristics of the local climate are not considered	√	
[14]	Physical and behavioral factors affecting energy performances	The most significant physical parameters are floor area and climate. Age, number of household members, and income are the most important occupancy variables	√	

Table 1. *Cont.*

Ref.	Subject	Outcomes and Limitations	Considered Impact *	
			OB	CC
[15]	Modeling and prediction of the number of occupants, domestic hot water (DHW) use, and non-HVAC electricity use	Acceptable results were obtained from the comparison between simulated and measured values	√	
[16]	Understanding of OB	Gaps: Understand OB in a systematic framework and evaluate its role in building energy policies	√	
[17]	Provide occupants with recommendations to reduce energy consumption	Procedure to develop an energy-efficient Reference Building	√	
[18]	Occupancy patterns on the energy performance of nZEB	Being a nZEB is not related only to the construction and plants, but is also dependent on occupant related factors	√	
[19]	Production and landscape on OPA modeling	Need to develop new studies in climate contexts where models are missing	√	
[20]	Heating and DHW energy consumptions and indoor comfort	Simplified approaches are not suitable to describe adequately the usage scenarios	√	
[22]	Occupant-centric building design and operation	The need of relieving occupants from a passive role in building design	√	
[25]	Impact of CC and variability on thermal comfort	Ventilation and insulation lead to a decrease in internal temperatures		√
[26]	Energy efficiency and CC mitigation	Gaps in the areas of behavioral changes and non-technological measures		√
[27]	Effects of CC on heating and cooling energy demand	Decrease in the heating energy use intensity and increase in the cooling energy use intensity		√
[28]	Resilience to CC of a residential building located in different European cities	CC will affect the heating and cooling energy demands		√
[29]	Impact of CC on heating and cooling energy consumption in different cities	The trends appear more impacting in Southern than Northern Europe		√
[30]	Effect of CC on the residential sector and environmental implications	Importance of renovating the building stocks and use renewable energies		√
[31]	Designing buildings with robust energy performance under CC	OB as a source of variation in combination with CC is indicated as a future work		√

* OB = Occupant behavior, CC = Climate Change.

Aim of the Study

As emerged from the literature review, the impact of occupant behavior, and the effect of climate change on the energy performance of buildings was largely recognized. Their impacts were investigated, highlighting the importance of future scientific contributions to these topics and encouraging more comprehensive studies considering that behavioral variables and climate change were still analyzed separately. Consequently, this paper aims to fill this gap by proposing a study that combines the double effect of these variables on the energy performance of buildings. By considering the information and indications of the available literature, the aim of this study was addressed to assess the impact of both occupants' behavior and climate change on the heating and cooling energy performance of a typical residential unit located in Southern Europe. Here, the energy performance was referred to as the heating and cooling energy needs defined as the heat to be delivered to, or extracted from, a conditioned space to maintain the intended temperature conditions during a given period of time.

Energy needs constitute the base of calculation of the primary energy demand that is determined by the energy supply system and the user types of fuel.

In particular, the authors wanted to answer the following research questions:

- RQ1: How does climate change influence the heating and cooling hours of operation?
- RQ2: How do the daily heating, cooling, and ventilation use profiles affect energy needs?
- RQ3: How does climate change affect the energy performance of buildings in winter and summer?
- RQ4: How do occupants' preferences related to the heating and cooling setpoints temperature affect energy needs in different climate scenarios?

The answers to these research questions can provide useful indications for scientists and policymakers to assess how human factors and environmental conditions can impact the energy consumptions of buildings, and consequently give due weight to them in future regulations and design criteria.

3. Methodology

The general schema and the consecutive steps of the investigation are illustrated in Figure 1.

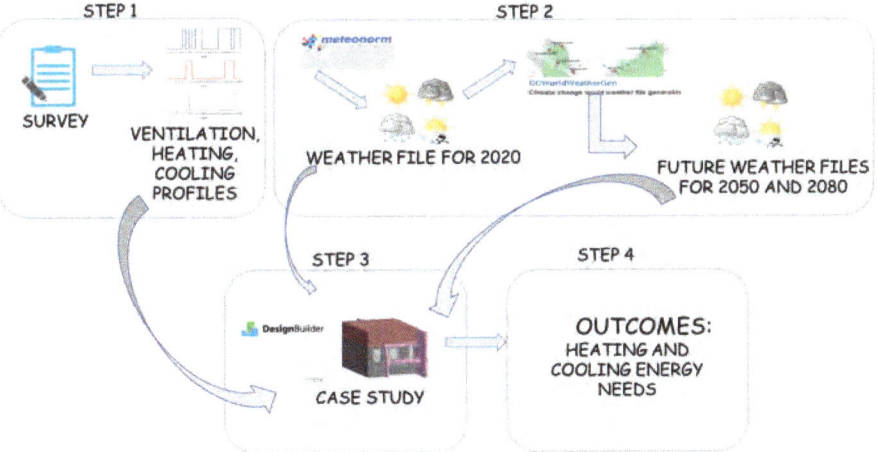

Figure 1. Schema of the adopted methodology.

The research can be summarized in four steps:

- step 1: Survey distribution and data for the creation of heating, cooling, and natural ventilation profiles;
- step 2: Weather file for 2020 was downloaded from METEONORM and then adopted in CCWorldWeatherGen tool to obtain the weather files for the future scenarios;
- step 3: An apartment was chosen as a case study and modeled trough DesignBuilder by considering different usage profiles and climate scenarios;
- step 4: Results in terms of heating and cooling energy needs were obtained to assess the impact of occupant behavior and climate change on the energy performance of buildings.

Furthermore, this section introduces more in detail Step 1 to Step 3: The survey to collect information on the occupants' behavior to be used in energy simulations, the energy model of the residential unit investigated in the study, and the tool adopted to obtain the weather files of future climate scenarios.

3.1. Questionnaire Survey

Data of a questionnaire survey delivered in Southern Italy were used to obtain use profiles to be provided as input in energy simulations. During two survey campaigns conducted from 2017 [9] to 2019, 237 surveys were collected, and among them, 193 were accepted as valid for these analyses. The questionnaire presents a total of 64 questions grouped into three main categories, as shown in Figure 2.

Information about building and equipment

a) Dwelling:
- Year of construction
- Type of house
- Floor area
- Location
- Structure (envelope, window, number of rooms, ..)

b) Cooling, Heating, and DHW:
- Energy source, production system, hourly usage schedule

c) Presence of solar thermal collectors
d) Presence of PV panels
e) Lighting and appliances:
- Lamps information, energy class of major appliances

Family composition and energy consumption

a) Family composition:
- Age
- Education level
- Employment for each of the household's component
- Total annual income of the family
- Presence of smokers
- Presence of pets

b) Energy consumption:
- Annual electricity and fuel consumption and expenditures from bills

Energy-related occupant behavior

a) Occupancy habits:
- Hourly Occupancy schedule for each room of the house

b) DHW consumption habits:
- Hourly DHW usage schedule in the kitchen and bathroom
- Occurrence of shower or bath per week

c) Cooling, Heating consumption habits:
- Hourly cooling usage schedule in each room of the house
- Hourly heating usage schedule in each room of the house

d) Windows and blinds opening schedule:
- Hourly usage schedule of windows for each room of the house in winter and summer
- Hourly usage schedule of blinds for each room of the house in winter and summer

e) Energy consumption habits:
- Hourly usage schedule of lighting in winter and summertime
- Weekly usage schedule of appliances (number of usages per week, duration)

Figure 2. Questionnaire contents.

Consistently with the aim of this paper, the attention was dedicated to the questions regarding the cooling and heating operation habits and the window opening preferences. The responses collected for the buildings located in Rende, characterized by Mediterranean climate conditions and defined as "Csa" according to the Köppen climate classification [32], were considered.

For the selected buildings, the schedules were first subjected to a cleaning process to verify their reliability. After that, the profiles were clustered based on the timing and length of the usage, and typical hourly profiles were obtained for heating, cooling, and natural ventilation.

3.2. Case Study

Among the collected sample, an apartment built in 2008 located on the second floor of a six-story building, with a gross floor area of 80 m² was chosen as a case study. The building structure is made of reinforced concrete, and the external walls consist of double hollow brick layers with an internal air gap partially filled with expanded polystyrene, resulting in a U-value of 0.6 W/m²·K. The windows are double glazing and a frame with thermal break. The heating system, used both for heating and DHW production, is an autonomous wall-mounted gas boiler. A zone thermostat regulates the operating of the heating system, and the heat emitters are aluminum radiators. The cooling system consists of air conditioners installed in the living room and in the bedrooms. METEONORM weather data [33] were used for the dynamic energy simulations conducted by DesignBuilder [34]. The model of the residential unit is shown in Figure 3.

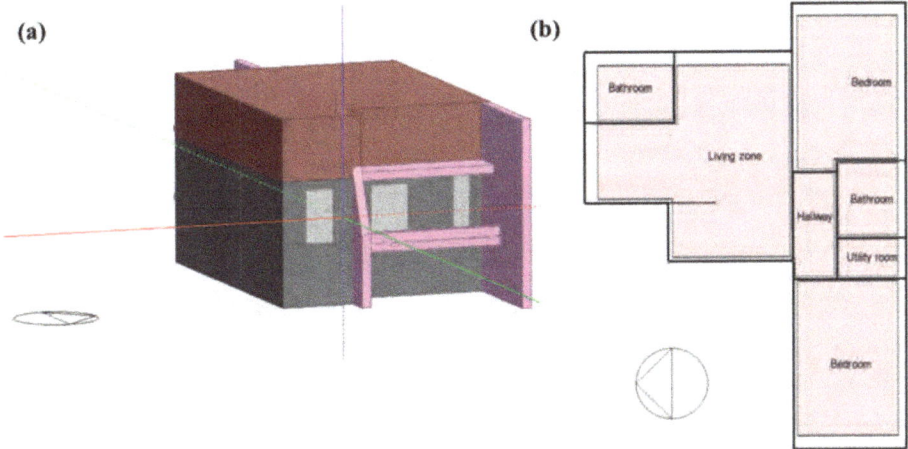

Figure 3. The model of the residential unit: (a) DesignBuilder model of the building; (b) plan of the apartment.

The reliability of the model was verified by the authors in previous work [20] following the ASHRAE Guideline 14-2002 [35]. The predicted results obtained from the simulation of the actual use and the measured data extracted from the energy bills were compared on a monthly scale through the Normalized Mean Bias Error(NMBE) and the Coefficient of Variation of the Root Mean Square Error (CVRMSE). Values lower than the limit values were obtained for both the parameters.

Downstairs there is an unconditioned thermal zone; while upstairs, there is an adiabatic block, due to the presence of another heated dwelling. Horizontal and vertical overhangs were shaped through standard component block considered by the software in shading calculation. Three thermal zones (living area, bedrooms, and bathrooms) were considered, and the characteristic parameters were changed in terms of management of the heating and cooling system, as both activation period and setpoint temperature, and ventilation hourly profiles. The internal heat loads were determined following the indications of the Standard UNI/TS 11300-1 [36] that uses the relation:

$$\phi_{int} = 7.987 \, A_f - 0.0353 \, A_f^2 \quad (1)$$

where A_f is the usable floor area of the house [m²]. The calculated value amounts to 5.56 W/m² and groups all contributions of occupancy, miscellaneous equipment, catering process, and lighting. The dynamic simulations were performed by combining different hourly ventilation profiles with heating and cooling operation schedules and setpoints temperature. In the reference case, energy simulations were conducted for the heating (from 1 October to 30 April) and cooling (from 1 May to 30 September) season by considering the current climate data and a setpoint temperature of 20 °C and 26 °C, respectively. Further energy assessments were obtained by varying the climatic scenarios (2050 and 2080) and the internal setpoints temperature (18 °C and 22 °C in the heating season, 26 °C and 24 °C in the cooling season).

3.3. Climate Scenarios

In this study, the climate change world weather file generator (CCWorldWeatherGen) [37] was used to create the weather files of future climate scenarios. Several studies used this tool to obtain future weather files [6,25,27–30], and the authors in [38] presented a critical analysis of it. Specifically, CCWorldWeatherGen is a Microsoft Excel-based tool commonly used that, employing the morphing procedure [39], provides weather files for future scenarios using outputs from the UK Hadley Centre Coupled Model (version 3, HadCM3) [40].

The future scenarios selected for this study were 2050 and 2080. The three adopted climate weather files were first analyzed in terms of variations in the external air temperature values. Figure 4a shows the monthly average air temperatures of the current climate, while the ΔT between current and future monthly average air temperatures are reported in Figure 4b for 2050 and in Figure 4c for 2080.

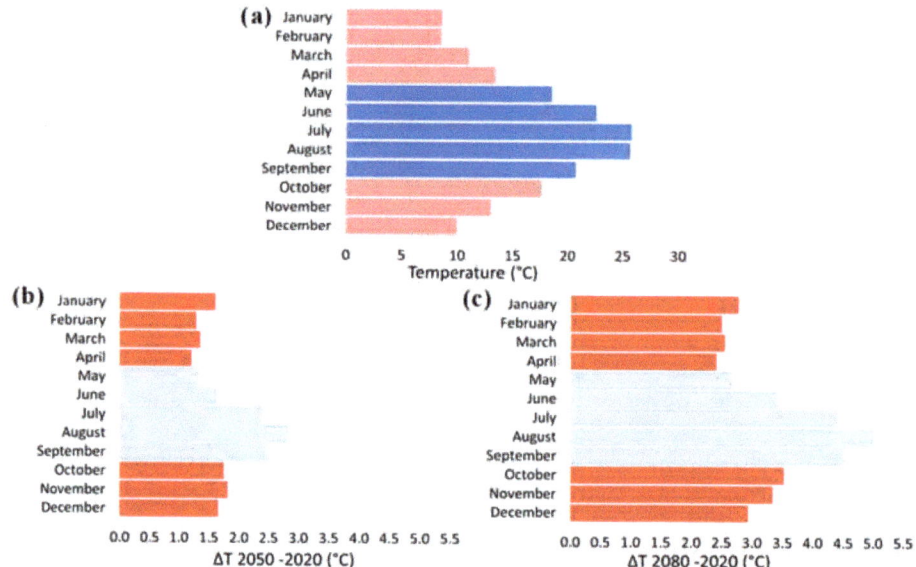

Figure 4. Monthly average air temperature: (**a**) in 2020, and monthly average air temperature increment ΔT (**b**) in 2050 and (**c**) in 2080.

Compared to the current climate, an increase in the monthly average air temperatures for each month of both 2050 and 2080 is projected. In particular, increments from 1.2 °C observed in April to 2.8 °C in August and from 2.4 °C to 5 °C, in the same months, were expected for 2050 and 2080, respectively.

4. Results and Discussion

This section presents the results obtained from the survey and the energy simulations conducted for the heating and cooling season. The results are organized as follow:

- ventilation, heating, and cooling profiles obtained from the survey;

- monthly hours of operation of the heating and cooling systems in 2020, 2050, and 2080 with setpoint temperatures of 20 °C and 26 °C;
- impact of diverse usage schedules of heating, cooling, and natural ventilation on the heating and cooling energy needs in the current climate conditions;
- variations of energy needs in future weather scenarios;
- variations of energy needs by changing the heating and cooling setpoint temperatures of ±2 °C in the different climate conditions.

4.1. Ventilation, Heating, and Cooling Profiles

Tables 2–4 show the typical hourly profiles obtained for heating, cooling, and natural ventilation. Moreover, respondents declared to generally use the heating system from October to April with a typical setpoint temperature of 20 °C, and the cooling system from May to September with a setpoint temperature of 26 °C. Further setpoints temperature ranging from 18 °C to 22 °C in winter, and from 24 °C to 28 °C in summer, were encountered.

Table 2. Daily heating schedules (On = 1, Off = 0).

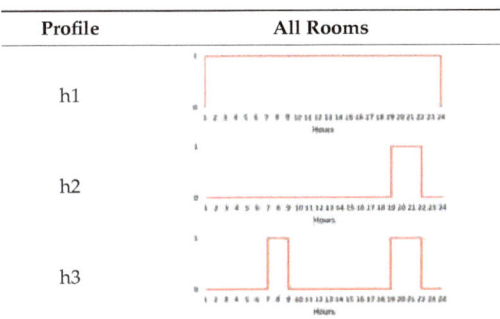

Table 3. Daily cooling schedules (On = 1, Off = 0).

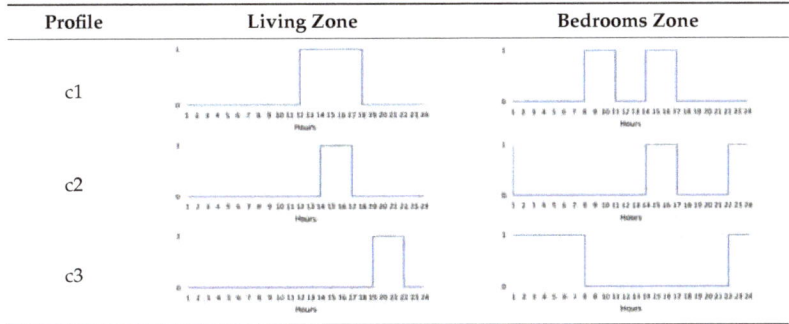

The heating schedules varied in terms of both the duration and time of operation. The heating system could operate for 24 h (profile h1), for three hours during the evening (from 19:00 to 22:00) in profile h2, and during the morning (from 07:00 to 09:00) and in the evening (from 19:00 to 22:00) in profile h3.

Table 4. Daily Natural ventilation schedules (Open = 1, Close = 0).

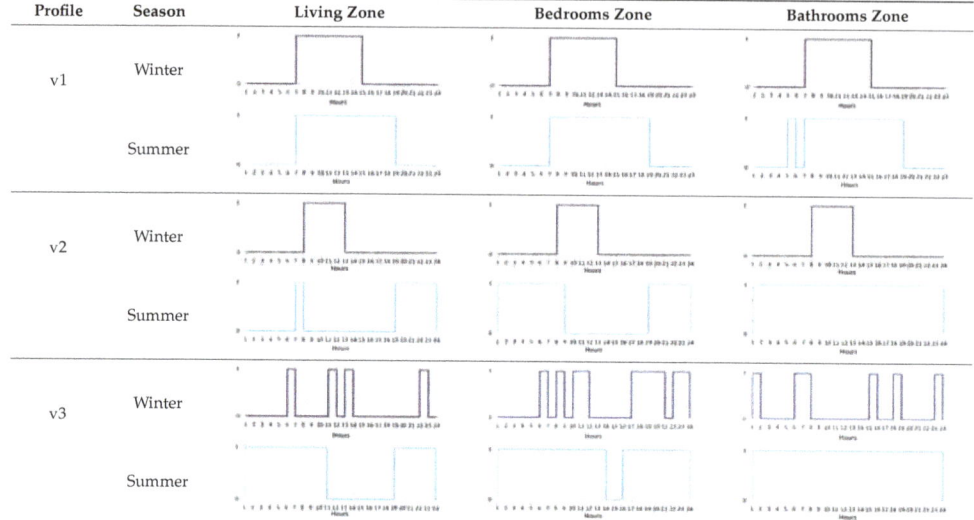

As shown in Table 3, the cooling system was installed only in the living and bedroom zones and used with diverse daily schedules. In profile c1, it was used in the hottest hours of the day (from 12:00 to 18:00) in the living zone, and for two time ranges in the bedrooms (from 08:00 to 11:00, and from 14:00 to 17:00). The schedules of the cooling system were more similar between the zones with profile c2: In the afternoon (from 14:00 to 17:00) in the two zones, and in the late evening (from 22:00 to 01:00) only in the bedrooms. Profile c3 was different from the others because the cooling operation was only activated during the late afternoon: From 19:00 to 22:00 in the living zone, and from 22:00 to 07:00 in the bedrooms.

Usually, occupants welcome natural ventilation more than mechanical ventilation, where they can only passively accept the system operation [41]. On the other hand, natural ventilation impacts negatively on the energy needs of a building when the external air temperature is lower than the internal air temperature in winter, or higher in summer, producing greater values of heat losses. On the other hand, benefits from window openings can be obtained in summer when the external air is used for natural cooling during the late afternoon or at night.

Looking at the graphs, shown in Table 4, it can be seen the variations of the occupants' preferences related to ventilation through the seasons. Profile v1 was typical of families who preferred to use continuous hours of ventilation during the day from the morning to the afternoon. The daily schedules were equal among the rooms, but different in duration between the seasons: From 07:00 to 15:00 and from 07:00 to 19:00 in winter and summer, respectively. Profile v2 showed the use of the natural ventilation limited to the morning hours in winter (from 08:00 to 13:00) and concentrated in the coolest hours in the summer. Finally, profile v3 presented an intermittent, but prolonged use throughout the day in winter, and continuous use in the coolest hours in the summer (from 19:00 to 11:00). Similar habits could be seen in the bathrooms area in both v2 and v3 profiles where people used to leave the windows open for the entire day. Natural ventilation profiles, as well as heating and cooling profiles, are linked to occupancy. Generally, it is noted that in homes with greater hours of daily occupancy, there is a more frequent occupant-window interaction and prolonged use of the heating and cooling systems (e.g., heating schedule h1 with continuous activation).

The heating and cooling schedules were combined with the natural ventilation profiles, and nine profiles, both for winter and summer seasons, were applied to perform the energy simulations of a residential unit.

4.2. Monthly Hours of Operation of the Heating and Cooling System in the Climate Scenarios

Due to the increase of the monthly average air temperature, it is also interesting to analyze how the hours of operation of the heating and cooling systems vary from the current climate to the future scenarios. In this study, "monthly hours of operation" was the sum of the hours in which the heating/cooling system provides the energy necessary to reach and maintain the indoor temperature at the setpoint value.

In particular, Figure 5 shows the monthly hours of operation of the heating system in the current climate (Figure 5a) and the differences (Δh) with respect to 2050 (Figure 5b) and 2080 (Figure 5c), by setting the internal air temperature value equal to 20 °C. The energy simulations were performed by considering all the heating schedules (h1, h2, h3) coupled with the ventilation profiles (v1, v2, v3).

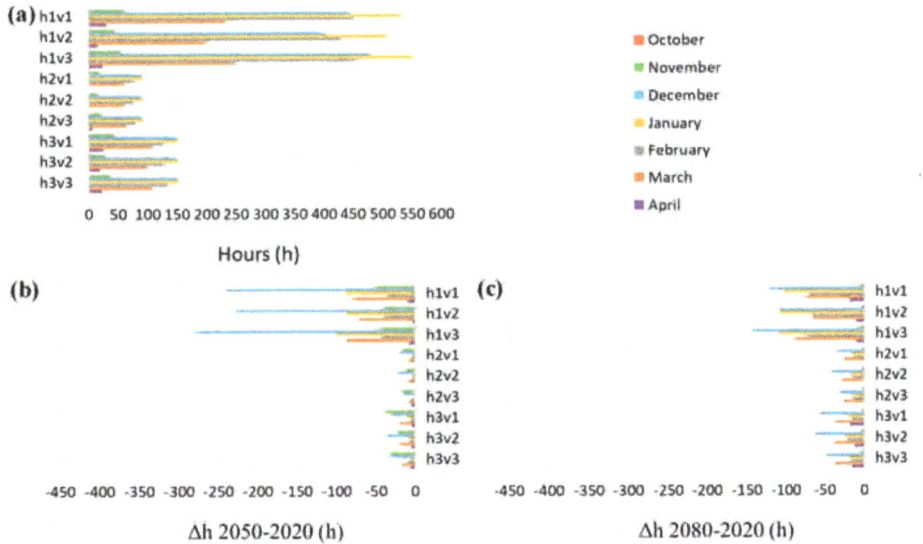

Figure 5. Monthly hours of operation of the heating system: (**a**) in the current climate, and differences Δh in the year (**b**) 2050 and (**c**) 2080.

The study shows a decreasing trend in the operation hours of the heating system for each month of the future climate scenarios. The major differences arose when the three profiles of the natural ventilation were combined with the heating schedule h1 characterized by continuous activation. In general, December was the month where more variations from 2020 to the future scenarios were observed.

The results for the cooling season, in terms of monthly hours of operation, were obtained with a setpoint temperature of 26 °C and are shown in Figure 6.

As a consequence of the external temperature rise, it is possible to observe an increasing trend of the monthly hours of operation of the cooling system in the future climate conditions. May, June, and September registered the main increases with the schedules c1 and c2. This growth was more visible with profile c1 because the cooling system could operate for more hours and mainly in the hottest hours. Considering the schedule c3, the operation of the cooling system was from June to September in 2020, and also needed in May during the future climate scenarios. It mainly happened when the cooling schedule c3 was coupled with the natural ventilation profile v1 because the ventilation occurred in the hottest hours of the day, producing an increase of the internal air temperature, and consequently, a prolonged cooling system operation.

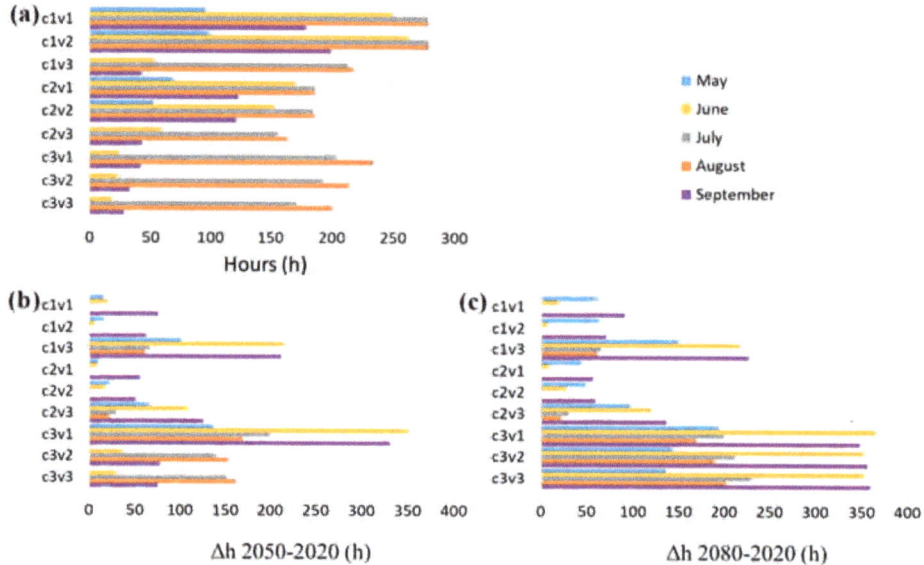

Figure 6. Monthly hours of operation of the cooling system: (**a**) in the current climate, and differences Δh in the years (**b**) 2050 and (**c**) 2080.

4.3. Impact of Occupant Behavior on Energy Needs

Figure 7 shows the heating and cooling energy needs in the current climate with a heating setpoint temperature of 20 °C and a cooling setpoint temperature equal to 26 °C.

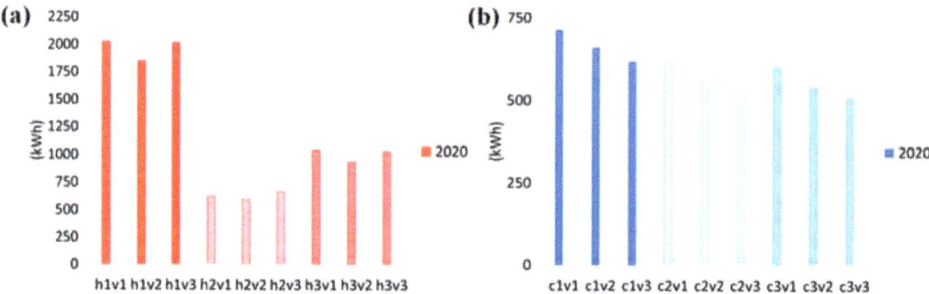

Figure 7. Energy needs in the current climate for: (**a**) the heating season; (**b**) the cooling season.

In winter, the energy requirement (Figure 7a) was more influenced by the heating schedules than the natural ventilation type. In particular, values of the order of 2000 kWh, 1000 kWh, and 700 kWh were registered for the heating schedule h1, h3, and h2, respectively. These differences in energy needs were due to the diverse duration of the heating system operation.

On the other hand, the cooling energy need seems to be affected by both the operation type and natural ventilation schedules. A decreasing trend of the energy requirement from the cooling schedule c1 to c3 and from the natural ventilation schedule v1 to v3 was observed. In more detail, the cooling energy need ranged from 714.8 kWh to 619.7 kWh, from 616.4 kWh to 534.7 kWh, and from 606.4 kWh to 511.7 kWh for c1, c2, and c3, respectively. These results can be explained by analyzing the cooling and ventilation profiles. In fact, the cooling system could operate for more hours and in the hottest hours of the day with the schedule c1.

Also, the natural ventilation with profile v1 mainly occurred in the hours in which the external air temperature can be higher than the internal one leading to greater cooling energy needs. In contrast, the schedules v2 and v3 produced a positive effect the energy balance.

In the current climate, h2v2 and c3v3 were the less heating and cooling energy-demanding profiles, while h1v1 and c1v1 were those with the most heating and cooling energy requirement.

4.4. Impact of Climate Changes on the Energy Needs

The use profiles were also used to assess their impact on future climate scenarios characterized by temperature rise. Figure 8 illustrates the relative differences of the energy needs in 2050 and 2080 compared to 2020.

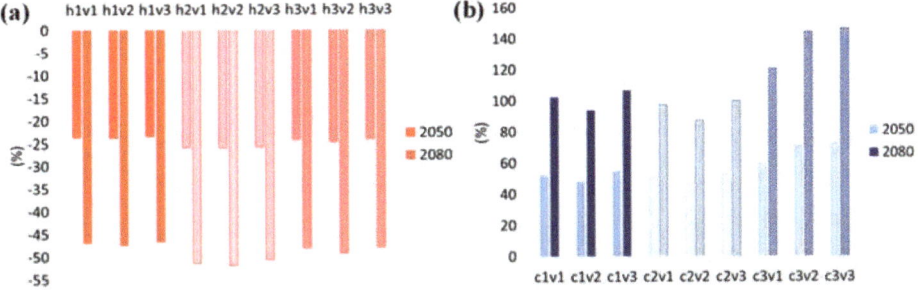

Figure 8. Relative differences between the energy needs calculated in the current climate and in future climate scenarios for: (**a**) the heating season; (**b**) the cooling season.

For all future scenarios, energy needs reductions were observed in the heating season, and energy needs increments in the cooling season. In winter, the impact of climate change was more predominant than the impact of occupant behavior (Figure 8b). In fact, significant variations were found from one year to the next and not in different heating and ventilation profiles. The differences varied from −24% to −26%, and from −47% to −52% in 2050 and 2080, respectively.

In summer, visible variations were observable varying the use profiles and passing from a climatic scenario to another (Figure 8b). In fact, energy requirements increased from +48% to +54%, from +46% to +53%, and from +60% to +73% with the cooling schedule c1, c2, and c3 in 2050, respectively. Moreover, for 2080, cooling need increased from +94% to +107%, from +87% to 100%, and from +121% to +146% with the schedule c1, c2, and c3, respectively.

4.5. Impact of the Heating and Cooling Setpoint Temperatures on Energy Needs

Occupants can impact the energy performance of buildings also by varying the setpoint temperature of the heating and cooling system.

Figures 9–11 present, for the different climate scenarios, the variations of the heating and cooling energy needs when the setpoint temperatures were modified of ±2 °C.

As expected, the decrease in the heating setpoint temperature by 2 °C led to a reduction in energy requirements, and the increase in temperature consequently produced an increase in energy need (see Figure 9a). Opposite trends in thermal behavior were observed by varying the cooling setpoint temperature (see Figure 9b).

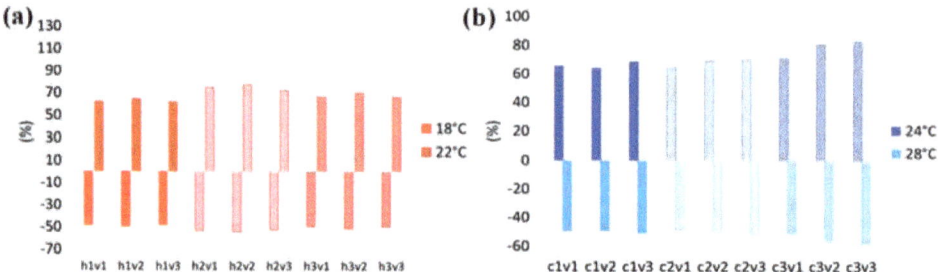

Figure 9. Relative differences of the (**a**) heating and (**b**) cooling energy needs caused by a variation of the setpoint temperature of ±2 °C in 2020.

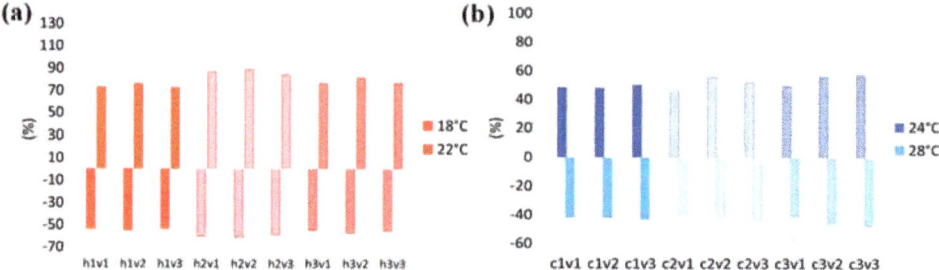

Figure 10. Relative differences of the (**a**) heating and (**b**) cooling energy needs caused by a variation of the setpoint temperature of ±2 °C in 2050.

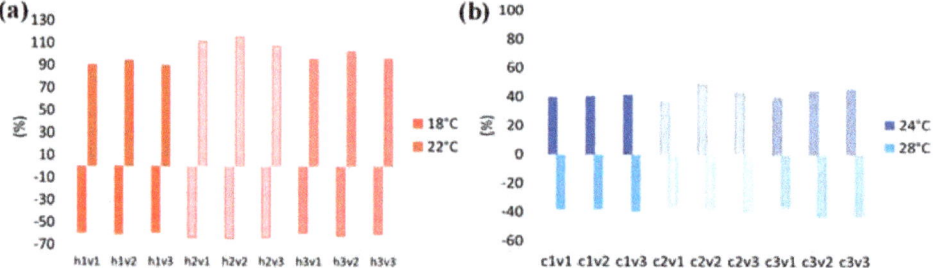

Figure 11. Relative differences of the (**a**) heating and (**b**) cooling energy needs caused by a variation of the setpoint temperature of ±2 °C in 2080.

More in detail, regarding the heating season in the current climate, the energy need decreased from −48% to −54% when the internal air temperature was set equal to 18 °C and increased from +62% to +77% when 22 °C was the selected setpoint. The maximum variation was found for profile h2v2, for which the energy need was 595.2 kWh at 20 °C, and 271 kWh and 1052 kWh at 18 °C and 22 °C, respectively.

In summer, the energy need increased from +65% to +83%, when the setpoint temperature was 28 °C and decreased from −48% to −58% when it was equal to 24 °C, with a maximum variation for profile c3v3 with both 24 °C and 28 °C. In particular, the greatest variations were found for the c3v3 profile that registered an energy requirement of 511.7 kWh at 26 °C, and of 935.5 kWh and 215 kWh at 24 °C and 28 °C, respectively.

The same information as above, but referring to 2050, is shown in Figure 10. For both the heating (Figure 10a) and cooling needs (Figure 10b), the general trends were similar to those noticed in 2020, what changes were the magnitude of the variations.

Specifically, in 2050, the heating energy need encountered higher fluctuations by varying the setpoint temperature (decrement from −53% to −61% and increment from +72% to +89%). The maximum variation, also in 2050, was observed in both cases for profile h2v2. Instead, in summer, the variations due to the occupants' preferences had a minor impact: The energy need increased from +46% to +57% and decreased from −40% to −48%. The results for 2080 are shown in Figure 11.

In 2080, the reduction and increase of the heating setpoint temperature led to remarkable changes in energy requirements (from −59% to −65% for 18 °C, and from +90% to +114% for 22 °C). In the cooling season, the variations of the setpoint temperature determined more limited modifications in terms of energy needs that increased from +36% to +45% and decreased from −35% to −43%. As happened in 2020, also in 2050 and 2080, the maximum variations were observed for profile h2v2 in winter and c3v3 in summer.

4.6. Discussion

Energy simulations were first performed with setpoint temperature equal to 20 °C in winter and 26 °C in summer. In 2020, the heating energy needs were more influenced by heating schedules than ventilation profiles, and values of the order of 2000 kWh, 1000 kWh, and 700 kWh were registered for the continuous and the two intermittent operations, respectively. In summer, the cooling energy needs were affected by both cooling and ventilation operations. They ranged from 511.7kWh to 606.4 kWh, from 534.7 kWh to 616.4 kWh, and from 619.7 kWh to 714.8 kWh in the three operation modes.

In future scenarios, the temperature rise determined the decrement of the heating energy needs and the augmentation of the cooling energy needs, in agreement with the results of the previous studies. Specifically, during the heating season, energy needs reductions from −24% to −26% in 2050, and from −47% to −52% in 2080 were obtained. In summer, energy requirements increased from +48% to +54%, from +46% to +53%, and from +60% to +73% by changing the cooling schedule in 2050. Moreover, the increments obtained in 2080 were around double then those registered in 2050.

In addition to natural ventilation habits and systems operation mode, the occupants' can have different preferences in thermal comfort conditions, thus, variations of the setpoint temperature of ±2 °C were considered.

In particular, in 2020, the heating energy needs decreased from −48% to −54% and increased from +62% to +77% when the setpoint temperature was set equal to 18 °C and 22 °C, respectively. On the other hand, cooling energy needs increased from +65% to +83% and decreased from −48% to −58% with setpoint temperature equal to 28 °C and 24 °C, respectively.

From 2020 to 2080, the variations of energy needs were smaller for the heating and greater for the cooling. In any case, occupants' behavior in controlling and personalizing the indoor thermal conditions had a consistent impact in each climatic scenario.

To the extent of our knowledge, this study was the first that jointly assessed both the impact of occupant behavior and climate change on the energy performance of buildings. The results of this study can be considered indicative of what could be predicted in other Mediterranean countries.

A limitation of this study consists in the fact that energy evaluations were carried out in one location and for a type of building. Thus, the results are contextual and suggest further investigations to address the implication of both occupant behavior and climate change on the heating and cooling energy needs in diverse building typologies and climatic conditions.

Furthermore, this initial study provided informative results for scientists and policymakers as both human factors and environmental conditions can consistently affect the energy consumptions of buildings. Moreover, if the temperature rise determines the reduction of the energy needs in winter and the increment in summer, different preferences and behavior of occupants can lead to better managing of the systems' operation following energy-saving intentions in every season.

Therefore, adequate attention is needed for the aforementioned aspects in future regulations and design criteria.

5. Conclusions

Dynamic simulations were conducted to assess how the heating and cooling energy needs of a residential unit were affected by occupants' behavior and climate change. In particular, the impact of occupants' behavior was investigated by applying nine usage profiles of heating, cooling, and natural ventilation in winter and summer. Moreover, the influence of occupant behavior was taken into account by varying the indoor setpoint temperature. Regarding climate changes, three scenarios were considered—2020, 2050, and 2080.

The heating energy needs in 2020 were more influenced by heating schedules than ventilation profiles, while the cooling energy needs were consistently affected by both cooling and ventilation operations.

As expected, reducing the energy needs in winter and a rise in summer were noticed in future scenarios. In addition, due to the temperature increase, the variations of energy needs in 2080 were doubled than those obtained in 2050. More relevant results were highlighted concerning the impact of the setpoint temperature. In fact, the variations of energy needs registered from 2020 to 2080 were higher for the cooling than those for the heating, indicating that standards and codes should place more attention to future prescriptions about this control parameter.

In general, this study quantified how occupant preferences related to heating, cooling, and natural ventilation affect the energy performance of buildings. It was also demonstrated that due to climate change, buildings could be subjected to more critical climate conditions, which will lead them to have higher energy needs and to emit more CO_2. In future scenarios, the impacts of occupant behavior will be amplified, and especially the preferences related to the cooling system will have a consistent impact in Mediterranean countries.

Author Contributions: Conceptualization, G.F. and M.D.S.; methodology, G.F. and M.D.S.; software, G.F. and M.D.S.; validation, M.D.S.; formal analysis, G.F. and M.D.S.; investigation, G.F. and M.D.S.; data curation, G.F. and M.D.S.; writing—original draft preparation, G.F.; writing—review and editing, M.D.S, L.F.C., and L.B.; visualization, G.F. and M.D.S.; project administration, M.D.S.; funding acquisition, M.D.S. and L.F.C. All authors have read and agreed to the published version of the manuscript.

Funding: This research was funded by the Calabria Region Government with the Gianmarco Fajilla's Ph.D. scholarship (POR Calabria FSE/FESR 2014–2020) grant number H21G18000170006. A part of this publication has received funding from Secretaría Nacional de Ciencia y Tecnología (SENACYT) under the project code FID18-056. This work was partially funded by the Ministerio de Ciencia, Innovación y Universidades de España (RTI2018-093849-B-C31 - MCIU/AEI/FEDER, UE). This work was partially funded by the Ministerio de Ciencia, Innovación y Universidades - Agencia Estatal de Investigación (AEI, RED2018-102431-T). This work is partially supported by ICREA under the ICREA Academia program.

Acknowledgments: The authors would like to thank the Catalan Government for the quality accreditation given to their research group (GREiA 2017 SGR 1537). GREiA is certified agent TECNIO in the category of technology developers from the Government of Catalonia.

Conflicts of Interest: The authors declare no conflict of interest.

References

1. Almeida, M.; Ferreira, M. Ten questions concerning cost-effective energy and carbon emissions optimization in building renovation. *Build. Environ.* **2018**, *143*, 15–23. [CrossRef]
2. Ferreira, M.; Almeida, M.; Rodrigues, A. Impact of co-benefits on the assessment of energy related building renovation with a nearly-zero energy target. *Energy Build.* **2017**, *152*, 587–601. [CrossRef]
3. de Fátima Castro, M.; Colclough, S.; Machado, B.; Andrade, J.; Bragança, L. European legislation and incentives programmes for demand Side management. *Sol. Energy* **2020**, *200*, 114–124. [CrossRef]
4. European Union. Directive (EU) 2018/844 of the European Parliament and of the Council of 30 May 2018 amending Directive 2010/31/EU on the energy performance of buildings and Directive 2012/27/EU on energy efficiency. *Off. J. Eur. Union* **2018**, *156*, 75–91.

5. Araújo, C.; Almeida, M.; Bragança, L.; Barbosa, J.A. Cost-benefit analysis method for building solutions. *Appl. Energy* **2016**, *173*, 124–133. [CrossRef]
6. Rey-Hernández, J.M.; Yousif, C.; Gatt, D.; Velasco-Gómez, E.; San José-Alonso, J.; Rey-Martínez, F.J. Modelling the long-term effect of climate change on a zero energy and carbon dioxide building through energy efficiency and renewables. *Energy Build.* **2018**, *174*, 85–96. [CrossRef]
7. Streimikiene, D.; Balezentis, T.; Alebaite, I. Climate change mitigation in households between market failures and psychological barriers. *Energies* **2020**, *13*, 2797. [CrossRef]
8. Cabeza, L.F.; Chàfer, M. Technological options and strategies towards zero energy buildings contributing to climate change mitigation: A systematic review. *Energy Build.* **2020**, *219*, 110009. [CrossRef]
9. Carpino, C.; Fajilla, G.; Gaudio, A.; Mora, D.; De Simone, M. Application of survey on energy consumption and occupancy in residential buildings. An experience in Southern Italy. *Energy Procedia* **2018**, *148*, 1082–1089. [CrossRef]
10. O'Brien, W.; Tahmasebi, F.; Andersen, R.K.; Azar, E.; Barthelmes, V.; Belafi, Z.D.; Berger, C.; Chen, D.; De Simone, M.; D'Oca, S.; et al. An international review of occupant-related aspects of building energy codes and standards. *Build. Environ.* **2020**, *179*, 106906. [CrossRef]
11. Gaetani, I.; Hoes, P.J.; Hensen, J.L.M. Estimating the influence of occupant behavior on building heating and cooling energy in one simulation run. *Appl. Energy* **2018**, *223*, 159–171. [CrossRef]
12. Roetzel, A.; Tsangrassoulis, A.; Dietrich, U.; Busching, S. A review of occupant control on natural ventilation. *Renew. Sustain. Energy Rev.* **2010**, *14*, 1001–1013. [CrossRef]
13. Balint, A.; Kazmi, H. Determinants of energy flexibility in residential hot water systems. *Energy Build.* **2019**, *188–189*, 286–296. [CrossRef]
14. Mora, D.; Carpino, C.; De Simone, M. Behavioral and physical factors influencing energy building performances in Mediterranean climate. *Energy Procedia* **2015**, *78*, 603–608. [CrossRef]
15. Rouleau, J.; Ramallo-González, A.P.; Gosselin, L.; Blanchet, P.; Natarajan, S. A unified probabilistic model for predicting occupancy, domestic hot water use and electricity use in residential buildings. *Energy Build.* **2019**, *202*. [CrossRef]
16. Zhang, Y.; Bai, X.; Mills, F.P.; Pezzey, J.C. V Rethinking the role of occupant behavior in building energy performance: A review. *Energy Build.* **2018**, *172*, 279–294. [CrossRef]
17. Ashouri, M.; Fung, B.C.M.; Haghighat, F.; Yoshino, H. Systematic approach to provide building occupants with feedback to reduce energy consumption. *Energy* **2020**, *194*, 116813. [CrossRef]
18. Carpino, C.; Mora, D.; Arcuri, N.; De Simone, M. Behavioral variables and occupancy patterns in the design and modeling of Nearly Zero Energy Buildings. *Build. Simul.* **2017**, *10*, 875–888. [CrossRef]
19. Carlucci, S.; De Simone, M.; Firth, S.K.; Kjærgaard, M.B.; Markovic, R.; Rahaman, M.S.; Annaqeeb, M.K.; Biandrate, S.; Das, A.; Dziedzic, J.W.; et al. Modeling occupant behavior in buildings. *Build. Environ.* **2020**, *174*, 106768. [CrossRef]
20. Mora, D.; Carpino, C.; De Simone, M. Energy consumption of residential buildings and occupancy profiles. A case study in Mediterranean climatic conditions. *Energy Effic.* **2018**, *11*, 121–145. [CrossRef]
21. Yan, D.; Hong, T.; Dong, B.; Mahdavi, A.; D'Oca, S.; Gaetani, I.; Feng, X. IEA EBC Annex 66: Definition and simulation of occupant behavior in buildings. *Energy Build.* **2017**, *156*, 258–270. [CrossRef]
22. O'Brien, W.; Wagner, A.; Schweiker, M.; Mahdavi, A.; Day, J.; Kjærgaard, M.B.; Carlucci, S.; Dong, B.; Tahmasebi, F.; Yan, D.; et al. Introducing IEA EBC Annex 79: Key challenges and opportunities in the field of occupant-centric building design and operation. *Build. Environ.* **2020**, *178*, 106738. [CrossRef]
23. Lucon, O.; Ürge-Vorsatz, D.; Zain Ahmed, A.; Akbari, H.; Bertoldi, P.; Cabeza, L.F.; Eyre, N.; Gadgil, A.; Harvey, L.D.D.; Jiang, Y.; et al. *Climate Change 2014: Mitigation of Climate Change*; Contribution of Working Group III to the Fifth Assessment Report of the Intergovernmental Panel on Climate Change Chapter 9, Buildings; Cambridge University Press: Cambridge, UK; New York, NY, USA, 2014; pp. 671–738.
24. UN General Assembly. *Transforming Our World: The 2030 Agenda for Sustainable Development*; A/Res/70/1; Division for Sustainable Development Goals: New York, NY, USA, 2015.
25. Barbosa, R.; Vicente, R.; Santos, R. Climate change and thermal comfort in Southern Europe housing: A case study from Lisbon. *Build. Environ.* **2015**, *92*, 440–451. [CrossRef]
26. Cabeza, L.F.; Chàfer, M.; Mata, É. Comparative analysis of web of science and scopus on the energy efficiency and climate impact of buildings. *Energies* **2020**, *13*, 409. [CrossRef]

27. Berardi, U.; Jafarpur, P. Assessing the impact of climate change on building heating and cooling energy demand in Canada. *Renew. Sustain. Energy Rev.* **2020**, *121*, 109681. [CrossRef]
28. Ciancio, V.; Falasca, S.; Golasi, I.; de Wilde, P.; Coppi, M.; de Santoli, L.; Salata, F. Resilience of a building to future climate conditions in three European cities. *Energies* **2019**, 12. [CrossRef]
29. Ciancio, V.; Salata, F.; Falasca, S.; Curci, G.; Golasi, I.; de Wilde, P. Energy demands of buildings in the framework of climate change: An investigation across Europe. *Sustain. Cities Soc.* **2020**, *60*, 102213. [CrossRef]
30. Andric, I.; Al-Ghamdi, S.G. Climate change implications for environmental performance of residential building energy use: The case of Qatar. *Energy Rep.* **2020**, *6*, 587–592. [CrossRef]
31. Moazami, A.; Carlucci, S.; Nik, V.M.; Geving, S. Towards climate robust buildings: An innovative method for designing buildings with robust energy performance under climate change. *Energy Build.* **2019**, *202*, 109378. [CrossRef]
32. Köppen, W.P.; Geiger, R. *Handbuch der Klimatologie*; Gebrüder Borntraeger: Berlin, Germany, 1930.
33. Meteonorm. *Meteonorm Global Meteorogical Database Version 7.1.8*; METEOTEST: Bern, Switzerland, 2012.
34. DesignBuilder. DesignBuilder Version 6.1.6.008. DesignBuilder Software Ltd. 2020. Available online: http://www.designbuilder.co.uk/ (accessed on 1 February 2020).
35. ASHRAE. *ANSI/ASHRAE Guideline 14-2002 Measurement of Energy and Demand Savings*; ASHRAE: Peachtree Corners, GA, USA, 2002.
36. Italian Standardization Body, UNI/TS 11300-1. *Energy Performance of Buildings, Part 1: Evaluation of Energy Need for Space Heating and Cooling*; Italian Standardization Body: Milan, Italy; Rome, Italy, 2014.
37. Jentsch, M.F.; Bahaj, A.B.S.; James, P.A.B. Climate change future proofing of buildings-Generation and assessment of building simulation weather files. *Energy Build.* **2008**, *40*, 2148–2168. [CrossRef]
38. Moazami, A.; Carlucci, S.; Geving, S. Critical Analysis of Software Tools Aimed at Generating Future Weather Files with a view to their use in Building Performance Simulation. *Energy Procedia* **2017**, *132*, 640–645. [CrossRef]
39. Belcher, S.E.; Hacker, J.N.; Powell, D.S. Constructing design weather data for future climates. *Build. Serv. Eng. Res. Technol.* **2005**, *26*, 49–61. [CrossRef]
40. Met Office. Met Office HadCM3: Met Office Climate Prediction Model. Available online: https://www.metoffice.gov.uk/research/approach/modelling-systems/unified-model/climate-models/hadcm3 (accessed on 15 March 2020).
41. Day, J.K.; McIlvennie, C.; Brackley, C.; Tarantini, M.; Piselli, C.; Hahn, J.; O'Brien, W.; Rajus, V.S.; De Simone, M.; Kjærgaard, M.B.; et al. A review of select human-building interfaces and their relationship to human behavior, energy use and occupant comfort. *Build. Environ.* **2020**, *178*, 106920. [CrossRef]

Publisher's Note: MDPI stays neutral with regard to jurisdictional claims in published maps and institutional affiliations.

© 2020 by the authors. Licensee MDPI, Basel, Switzerland. This article is an open access article distributed under the terms and conditions of the Creative Commons Attribution (CC BY) license (http://creativecommons.org/licenses/by/4.0/).

Article

A Method for Establishing a Hygrothermally Controlled Test Room for Measuring the Water Vapor Resistivity Characteristics of Construction Materials

Toba Samuel Olaoye [1,*], Mark Dewsbury [1] and Hartwig Kunzel [2]

1. Architecture and Design, College of Sciences and Engineering, Inveresk Campus, University of Tasmania, Launceston 7250, Australia; mark.dewsbury@utas.edu.au
2. Fraunhofer Institute for Building Physics IBP, Fraunhoferstr. 10, 83626 Valley, Germany; hartwig.kuenzel@ibp.fraunhofer.de
* Correspondence: toba.olaoye@utas.edu.au; Tel.: +61-4-0627-7304

Citation: Olaoye, T.S.; Dewsbury, M.; Kunzel, H. A Method for Establishing a Hygrothermally Controlled Test Room for Measuring the Water Vapor Resistivity Characteristics of Construction Materials. *Energies* 2021, 14, 4. https://doi.org/10.3390/en14010004

Received: 10 November 2020
Accepted: 19 December 2020
Published: 22 December 2020

Publisher's Note: MDPI stays neutral with regard to jurisdictional claims in published maps and institutional affiliations.

Copyright: © 2020 by the authors. Licensee MDPI, Basel, Switzerland. This article is an open access article distributed under the terms and conditions of the Creative Commons Attribution (CC BY) license (https://creativecommons.org/licenses/by/4.0/).

Abstract: Hygrothermal assessment is essential to the production of healthy and energy efficient buildings. This has given rise to the demand for the development of a hygrothermal laboratory, as input data to hygrothermal modeling tools can only be sourced and validated through appropriate empirical measurements in a laboratory. These data are then used to quantify a building's dynamic characteristic moisture transport vis-a-vis a much more comprehensive energy performance analysis through simulation. This paper discusses the methods used to establish Australia's first hygrothermal laboratory for testing the water vapor resistivity properties of construction materials. The approach included establishing a climatically controlled hygrothermal test room with an automatic integrated system which controls heating, cooling, humidifying, and de-humidifying as required. The data acquisition for this hygrothermal test room operates with the installation of environmental sensors connected to specific and responsive programming codes. The room was successfully controlled to deliver a relative humidity of 50% with ±1%RH deviation and at 23 °C temperature with ±1 °C fluctuation during the testing of the water vapor diffusion properties of a pliable membrane common in Australian residential construction. To validate the potential of this testing facility, an independent measurement was also conducted at the Fraunhofer Institute of Building Physics laboratory (IBP) Holzkirchen, Germany for the diffusion properties of the same pliable membrane. The inter-laboratory testing results were subjected to statistical analysis of variance, this indicates that there is no significant difference between the result obtained in both laboratories. In conclusion, this paper demonstrates that a low-cost hygrothermally controlled test room can successfully replace the more expensive climatic chamber.

Keywords: water vapor resistivity; hygrothermal modeling; condensation; mold; hygrothermal properties; energy efficiency; moisture transport; inter-laboratory testing

1. Introduction

Over the last three decades, the increased expectations for energy efficient buildings combined with greater thermal comfort has established significant differences between the interior and exterior environmental water vapor pressure. This has created the need to manage water vapor diffusion and moisture, and has led to an increased demand for appropriate hygrothermal assessment [1]. Hygrothermal analysis is capable of calculating the dynamic transport of moisture, heat, and air in a building envelope. In most developed nations, this has become an essential part of the production of durable, healthy, comfortable, and energy-efficient buildings [2,3]. The presence of uncontrolled moisture above a critical limits can result in various degrees of deterioration which can include corrosion, rusting, freezing, and swelling of many materials used in the building [2,4]. The most concerning aspect of uncontrolled moisture in a building is the opportunity for mold to grow within interior spaces. This can have serious implications for the health of the occupants [5,6]. In addition, recent research has shown that high levels of moisture can impact the energy performance of a building and the quality of the indoor air [7–10].

In Australia, moisture problems have become apparent in many new buildings. Up to 50% of National Construction Code Class 1 and Class 2 buildings constructed in the last 15 years have a visible internal formation of condensation [11]. The complexity involved in understanding water vapor transport through appropriate hygrothermal calculation is posing significant challenges to the design and construction professionals in Australia especially when considering moisture management and energy efficiency in buildings [12–14].

While hygrothermal assessment, the key scientific approach to managing condensation and mold in buildings, has been deployed to address these challenges in many other developed nations, it is an emerging field in the Australia [13]. This may be because there were no building regulations requiring insulation in building envelopes until 2003, and the first regulations regarding risk of condensation management only came into effect in 2019. The long-term impact of moisture accumulation on building durability and human health has now become a critical aspect of the Australian regulatory agenda for new buildings.

Across other developed nations, hygrothermal analysis has evolved from manual calculation methods to computer simulations [15–17]. In the last two decades, this has moved from a limited focus on condensation risk analysis to a greater understanding of moisture accumulation, energy efficiency, and the drying capacity envelopes. Over the same period of time, the simulation method has advanced from steady state to transient simulation [18–20].

Several elements need to be considered in choosing an appropriate approach to hygrothermal modeling. In addition to precision and accuracy, the flexibility to allow selection from a variety of climatic zones and the quality of the climatic data are important aspects [21]. Other things to consider include the simulation runtime, the size of the material data library, and how the vapor diffusion and moisture absorption data have been sourced and validated. For instance, WUFI Pro [15], which appears to be the most popularly used hygrothermal software in Europe and North America, has been considered to be reliable because of its ability to deliver a realistic transient calculation and also because all the construction materials in its data library have been well validated [15,22].

The most appropriate method to source and validate construction material's vapor diffusion properties is to conduct measurements in the laboratory. For many nations, the laboratory measurement of water vapor diffusion characteristics of individual construction materials is evolving, and robust databases are being created. The internationally accepted method to represent vapor diffusion is material vapor resistivity. Due to Australia's slower adoption of highly insulated envelopes and vapor resistivity material data has not been required. It is inappropriate to adopt internationally available data directly for use in Australia without appropriate empirical evaluation of their applicability to materials used in Australia's envelope systems and the physical properties of Australian manufactured construction materials. As of 2019, the Australian National Construction Code requires hygrothermal calculations [23,24] in order for the design of new buildings to be approved. Early adopters are using non-Australian data from international material databases for hygrothermal modeling; however, these data may not provide a true representation of Australian construction materials. Without empirical information regarding the vapor diffusion properties of Australian construction materials, there is the potential that inappropriate decisions will be made.

Four types of laboratory-based test methods are internationally recognized for the quantification of the water vapor diffusion properties of materials. These include the electron-analytical, sweating guarded hot plate, dynamic moisture permeation cell test, and the gravimetric methods [4,25–31]. The testing process requires the establishment of two environments with different vapor pressures on each side of the material. Increasingly, the most preferred method for establishing the water vapor diffusion properties of most construction materials is the gravimetric method [26,32–36]. This involves the measurement of the mass of moisture that has resulted from water vapor diffusion into or out of a test dish assembly, often referred to as the wet-cup or dry-cup test method, respectively [25,32,37]. Depending on whether it is a wet-cup or dry-cup test, salt solutions, distilled water or a desiccant are used to establish a predetermined relative humidity within the test dish. The material is cut and attached to the test dish and then placed in a temperature and humidity-controlled cabinet or room. The humidity outside the cup, in the room, or cabinet, is controlled so that the desired relative humidity condition outside is achieved [37,38]. The conditions created within the cabinet or test room are designed to replicate the hygrothermal conditions the material may expect to experience as a component of the built fabric. The focus of this paper centers on the establishment of an appropriately hygrothermally controlled test room required for gravimetric vapor diffusion testing.

The general principle for the gravimetric method (shown in Figure 1) is to create two environments with different vapor pressures, by establishing different relative humidities inside and outside the cup, while the temperature remains constant. During the test period, the dish is weighed at regular intervals until the mass does not change, indicating the vapor pressure of the test dish and the room have reached equilibrium. For wet cup gravimetric testing (shown in Figure 2, the vapor flux is expected to go from the cup which has a higher RH through the material being tested to the environment which has a lower RH. The reverse is the case for dry cup gravimetric testing, shown in Figure 3. The process is discontinued after a minimum of four consecutive weighing which shows no change in mass.

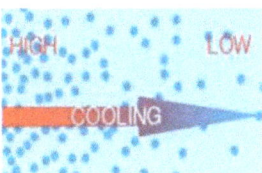

Figure 1. Diagram of water vapor diffusion [13].

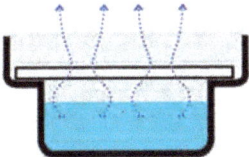

Figure 2. Diagram of wet cup test method [13].

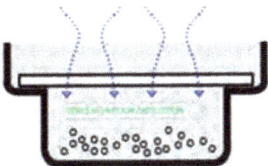

Figure 3. Diagram of dry cup test method [13].

While many research papers have reported different procedures for quantifying the water vapor diffusion of construction materials using the gravimetric method in a climatic cabinet [34,39,40], no research has reported the development of a hygrothermally controlled test room. However, the demand for more hygrothermally controlled test rooms will increase over the coming years both in Australia and internationally. This is because the demand for energy efficient buildings has increased in many jurisdictions as building codes have moved towards the requirement of near-zero energy consumption in buildings. Hence, the need to establish more hygrothermally suitable construction systems will increase and laboratory testing will be required to establish the hygrothermal properties of individual component materials.

The merits of a hygrothermally conditioned test room over the climatic cabinet is the elimination of experimental errors. During the gravimetric weighing, process errors may arise from opening, closing, and transporting test dishes from the cabinet. In a test room, all weighing activities occur within the climatically controlled space. Despite this distinct advantage, little or no research has reported the design, construction, installation of the equipment, and the operations of such a laboratory. This may be because the acquisition and installation of laboratories is not regarded as a research output. In addition, due to commercial reasons, those engineering firms that have built such rooms have never made available the details of the design, construction, and installation of such a facility. This paper describes the methods employed to develop Australia's first hygrothermal laboratory for quantifying the diffusion properties of materials using common appliances, which included a round-robin test conducted between Fraunhofer Institute of Building Physics laboratory (IBP) Holzkirchen Germany, and this hygrothermal testing laboratory at the University of Tasmania (UTAS), Australia.

The approach employed included establishing a climatically controlled hygrothermal test room with an automatic integrated system which allows heating, cooling, humidifying, and de-humidifying as required. The data acquisition for this hygrothermal test room operates with the installation of environmental sensors connected to specific and responsive programming codes. The room reported here, has been used to successfully complete wet and dry cup vapor diffusion material testing for relative humidities RH between 50% with ±1%RH deviation and temperatures between 23 °C with ±1 °C fluctuation. The test results indicate that a hygrothermally controlled test room can successfully replace the more expensive climatic chamber.

2. Materials and Methods

To establish a conditioned hygrothermally controlled test room, it was necessary to design and install environmental equipment that controls the interior temperature and relative humidity within the conditioned room. The accurate control of temperature and relative humidity conditions, within the bandwidths prescribed in ISO 12572, is critical to enable gravimetric based testing of building material vapor resistivity properties. For this research, a test building located at the Newnham campus of the University of Tasmania, was reconfigured to enable the conditioned room to be dynamically controlled. The controls included heating, cooling, humidification, and dehumidification. The second stage involved a round-robin testing of the water vapor resistivity properties of a pliable membranes at Fraunhofer Institute of Building Physics laboratory Holzkirchen Germany, and at this hygrothermal testing laboratory. The following sections discuss the design, installation, operation, and the performance of test room, the inter-laboratory testing that was conducted to compare test facilities and results for measuring vapor resistivity properties.

2.1. Design and Description of the Thermal Test Building

The University of Tasmania has three thermal test buildings at the Newnham campus in Launceston. They include an unenclosed-perimeter platform-floored building, an enclosed-perimeter platform-floored building and a concrete slab-on-ground floored building. Previous research had established that the well-insulated concrete slab-on-ground floored test building demonstrated the most stable interior temperatures without any stratification in both conditioned and unconditioned modes of operation. This building has an internal floor area of 30.03 m^2 (5.48 m by 5.48 m), a ceiling height of 2.44 m and total volume of 73.3 m^3 and has no window, as shown in Figures 4 and 5. The building, constructed in

2006, applied Australian best practice wall and ceiling insulation and air-tightness methods. The combination of the ground keyed concrete slab, external walls with R2.5 in-frame wall insulation, R4.2 ceiling insulation, and a well-installed air barrier system ensured a high-quality test building with minimal internal temperature variability.

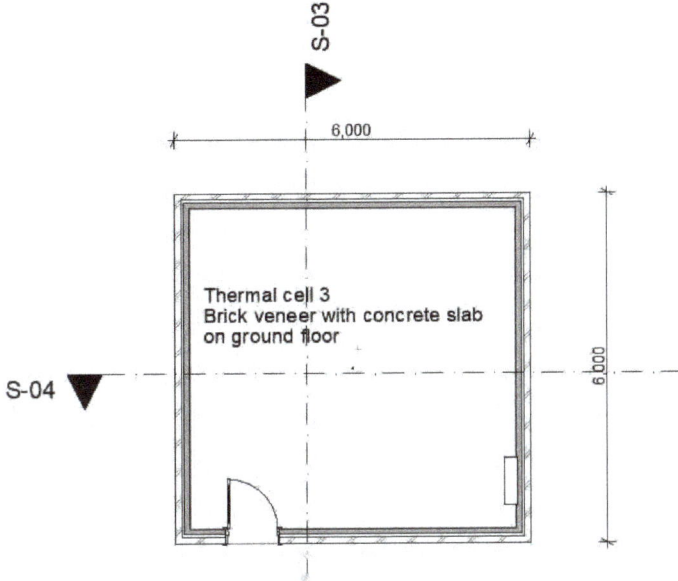

Figure 4. Floor plan of test building.

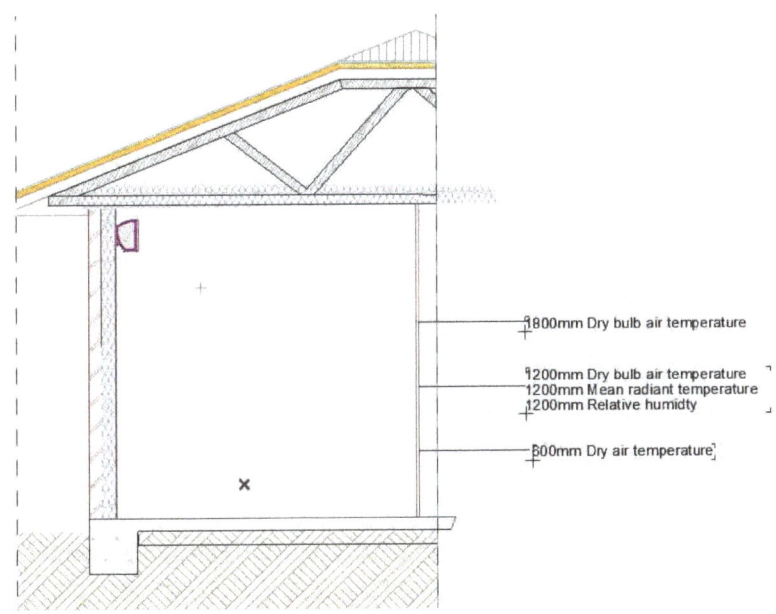

Figure 5. Architectural section of test building.

2.2. Cabling and Installation of Integrated Data Acqusition System

The control of air temperature and relative humidity are critical to the successful operation of a hygrothermally controlled test room. To enable accurate control of the test room interior a data acquisition system was used. Normally, data acquisition requires one or more transducers (sensors) to sense, process, and send signals from a measuring instrument to the system, the data acquired is then stored or logged into the central processing unit of a computer or external memory for later analysis. The data acquisition system generally includes: the sensors; a device that converts the primary signal from the sensors into a compactible form with the information processing systems; a computer by which the overall system is able to be managed and on which data from sensors are stored. For this research, DataTaker DT500 dataloggers with a channel extension module (CEM) (see Figure 6) were used. Connection between the Datataker and Dell PC was established via a RS232 communication cable (Figure 7). The De Transfer interface software was used for communication between the DT500 data logger and the Dell PC. Two DT 500 DataTaker data-loggers were used, one for temperature sensors and the second for the relative humidity sensors. An array of four wire PT100 sensors were used to measure temperature. An array of two wire Vaisala HMW40U relative humidity sensors were used to measure relative humidity. Due to the number of terminals required for the array of four wire PT100 sensors, they were connected to both the data-logger and the CEM. The second DT500 DataTaker was used to connect the array of relative humidity sensors used for this project. The primary sensor location was on a pole located in the center of the room (see Figure 8). The need for at least three sensors in each location was based on previous research, which queried the reliability of single sensors and when two sensors had varied measured values [41]. The sensors and other apparatus used to control the room are described in Table 1.

Figure 6. Data acquisition system (DT 500 datalogger).

Figure 7. Desk control.

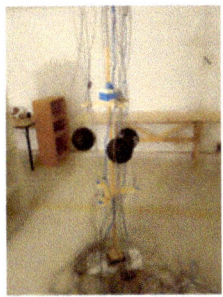

Figure 8. Environmental control equipment.

Table 1. Summary of sensors and other equipment.

Sensor/Equipment	Type	Location	Function
Dry bulb air temperature (V1)	Four wire Platinum RTD	Version 1–Center of room, three sensors at each reference height of 600 mm, 1200 mm, and 1800 mm	To measure test room air temperature and to inform the control of the air conditioner
Dry bulb air temperature (V2)	Four wire Platinum RTD	Version 2–same as Stage 1 plus air-conditioner supply air	Same as above
Mean radiant temperature	Four wire Platinum RTD within 150 mm diameter copper globes	Center of room, 3 sensors at 1200 mm	Information only
Relative Humidity	Two wire Vaisala HMW40U	Center of room, 3 sensors at 1200 mm	To measure test room relative humidity and to inform the control of the humidifier and de-humidifier
Air-conditioner	Daikin split system	South east corner	To heat or cool the room
Humidifier	6 L Air Humidifier Ultrasonic Cool Mist Steam Nebulizer Diffuser Purifier E	South east corner	To provide additional water vapor to the test room air
De-humidifier	Breville The Smart Dry Dehumidifier	Center of room	To remove water vapor from the test room air
Data Acquisition	Datataker DT500 with Channel expansion module		To continuously collect measured room temperature and relative humidity data
Relay	Solid state	Relay board	To control and switch humidifier and de-humidifier operation with alarm programming code
Silicone DC relays		South east wall connected to air-conditioner	To control and switch heating and cooling with switch alarm programming code

2.3. Cooling and Heating System

Automated heating and cooling were essential for the control of this hygrothermally conditioned test room. Figure 9 shows the position of the air-conditioner within the test room. This equipment is a reverse-cycle heat pump and can heat up to 30 °C. When heating above 30 °C was required for the room, the wall mounted electric heater shown in Figure 10 was turned on. Silicone DC relays (Figure 11) was used as the power switching interface between the data-logger and the appliances.

Figure 9. Air-conditioner.

Figure 10. Wall-mounted heater.

Figure 11. Silicone DC relays.

2.4. Humidity and Pressure Control System

The capability to control humidity was essential for this hygrothermally controlled room. For this research, this was achieved through the installation of humidity equipment which enabled water vapor to either be added or removed as required. The power switching for the humidity equipment utilized two solid-state relays shown in Figure 12. The first method to add water vapor to the air was to use a fishpond with a water heater. However, after preliminary testing and discussions with other research collaborators, it was established that there would be a significant water vapor lag with this method. This led to an analysis of quick response humidifiers. This resulted in the selection of a 6 L Ultrasonic Cool Mist Steam Nebulizer Diffuser Purifier (shown in Figure 13). This humidifier quickly demonstrated a very fast response to add extra water vapor to the room. Similarly, a Breville Smart dry de-humidifier (Figure 14), was installed to remove excessive water vapor from the room. The power supply for the humidifier and dehumidifier was controlled by a solid-state relay, which in turn was controlled by the DT500 data-logger. In practical terms, when the relative humidity in the room was too high the programmed data logger alarm switched the relay, thus providing power to the dehumidifier. When the desired relative humidity value was achieved, the programmed data logger alarm switched the relay off. Conversely, when the relative humidity was too low, the data logger alarm switched the relay to provide power to the humidifier, thus adding water vapor into the room until the required relative humidity setpoint was reached.

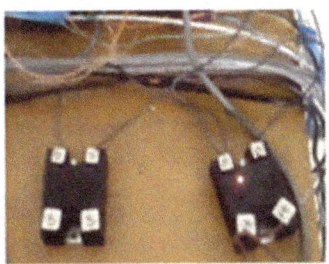

Figure 12. Solid state relay.

Figure 13. 6 litres Ultrasonic Humidifier.

Figure 14. Dehumidifier.

Additionally, a household fan was installed to provide circulation of the air in the room to minimize water vapor stratification.

2.5. Calibration of the Environmental Instruments

Calibration of the temperature and relative humidity sensors was completed to avoid intrinsic error that may have existed in the devices or data logging equipment. In the first instance, all sensors were carefully chosen for their level of accuracy and long-term reliability. A diagnostic procedure was established to ensure that wiring from the data logger to each sensor did not cause errors in measurements. The on-site calibration utilized pre-calibrated NATA certified temperature and relative humidity sensors provided by Industrial Technik. The calibration of the temperature sensors included zero degrees, room temperature and near boiling temperature. This was to ensure that there were no linear or non-linear errors. Any sensor that had erroneous outputs was replaced. The output from the relative humidity sensors was compared to a certified and pre-calibrated sensor, whilst the relative humidity was increased and decreased

2.6. Monitoring and Controlling Environmental Conditions

As previously mentioned, the DataTaker DT500 data logger was used for data acquisition. This system relied on programming code for data acquisition from the sensors and to control the switching relays for the heating, cooling, humidifying, and de-humidifying appliances. The acquisition systems collected temperature and relative humidity data from the sensors and simultaneously stored the data in the memory of Datataker for later use. Figure 15 shows a snapshot of an example of the programming code use to operate and collect temperature data from the PT100 sensors. This code was written according to the sensor type. Similarly, the programming code for acquiring the relative humidity data within the hygrothermal room is shown in Figure 16. In this research, temperature and relative humidity data was collected every 10 min. The examples of the programing code also show alarm codes. The coding shows minimum and maximum values for temperature and relative humidity. The alarms required the data logger to continuously monitor the relative humidity and temperature conditions in the test room. The alarm-controlled power supply to the digital switches on the data loggers. In turn, the digital switches controlled the power supply to the silicone and solid-state relays, which controlled the appliances. The combination of continuous measurement and the control of the four appliances, enabled the room temperature and relative humidity to be adequately controlled by the heating, cooling, humidifying, and dehumidifying appliances.

```
U
SCHEDULE
CM
CDATA
D=15/01/2020
T=2:30pm
' stage 1 - reset action
H
CLEAR
\W5
CDATA
\W5
RESET
\W5

' stage 2 - 'switches, parameters
/h
/e
/R
/S
S1=0,100,400,2000"%"'relative humidity'

' stage 3 - date, time
D=\d
T=\t
BEGIN
RA10M
D       'DAY
T       'TIME
1PT385(4W,"PT100-1800-1")
3PT385(4W,"PT100-1800-2")
4PT385(4W,"PT100-1800-3")
5PT385(4W,"PT100-1200-4")
6PT385(4W,"PT100-1200-5")
7PT385(4W,"PT100-1200-6")
8PT385(4W,"PT100-1200GLOBE-7")
9PT385(4W,"PT100-1200GLOBE-8")
10PT385(4W,"PT100-1200GLOBE-9")
1:1PT385(4W,"PT100-600-10")
1:2PT385(4W,"PT100-600-11")
1:3PT385(4W,"PT100-600-12")
1:4PT385(4W,"PT100-North_wall")
1:5PT385(4W,"PT100-East_wall")
1:6PT385(4W,"PT100-roofspace-top-insul")
RZ1S
ALARM1(8PT385(4W,"PT100")<23.2)1:1DSO
ALARM2(8PT385(4W,"PT100")>23.5)1:1DSO
END
LOGON

G
```

Figure 15. Example of temperature programming code.

```
STATUS
UM    'unload memory
U     'unload
Q     'quit unload
/H    'csv format
/h    'text format

' stage 1 - reset action
H
CLEAR
\W5
CDATA
D=15/12/2019
T=02:04pm
\W5
RESET
\W5

' stage 2 - 'switches, parameters
/h
/e
/R
/S

S1=0,100,400,2000"%"'relative humidity'
\W2
' stage 3 - date, time

D=\d
T=\t

BEGIN
RA5M
/D       'DAY
/T       'TIME

6*V(S1,"RH roomorange")
7+V(S1,"RH roombrown")
7-V(S1,"RH roomgreen")
8*V(S1,"RH roofblue")
8+V(S1,"RH orange sth wall mid")
8-V(S1,"RH brown sth wall base")
9-V(S1,"RH green sth wall base")

RZ1S
ALARM1(6*V(S1,"RH roomorange")>35.0)1:7DSO
ALARM2(6*V(S1,"RH roomorange")<35.5)1:6DSO

END

LOGON

G
```

Figure 16. Example of relative humidity programming code.

2.7. Inter-Laboratory Testing of Wet-Cup and Dry-Cup Dishes

The procedure for the interlaboratory testing involved the selection of a pliable membrane classified as permeable material in clause AS 4200:1 and carrying out a standard test as referred to in ISO 12572. The independent testing of water vapor resistivity properties was completed on a pliable membrane commonly used in Australian external envelope construction systems. The same material was tested under the same climatic condition of 23 °C/50%RH at both the hygrothermal laboratory at Fraunhofer IBP Germany, and UTAS, Australia. Table 2 shows the comparison of the important testing parameters that were used.

Table 2. Summary of testing parameters.

Parameter	At IBP, Laboratory	At UTAS
Dishes	Round glass dish (80 × 200 mm)	Round glass dish (60 × 195 mm)
Air space	20 mm	20 mm
Average barometric pressure	933.26 hPa	1030.5 hPa
Water vapor permeability of air	2.12×10^{-10} kg/(m·s·Pa)	1.92×10^{-10} kg/(m·s·Pa)

It was necessary to employ very similar round glass dishes with diameter of 200 mm. While the depth of the dishes at IBP is 80 mm, at UTAS, the dept is 60 mm. For accuracy, three dishes were used for wet-cup and another three were used for dry-cup gravimetric measurement both in Germany and in Australia. To achieve the desired humidity testing condition within wet-cup dishes, ammonium dihydrogen phosphate solution was placed in the dish, by both laboratories during the testing. This achieved a dish relative humidity of 93% (Figure 17). Similarly, to achieve the desired testing humidity condition within the dry-cup test dishes, silica gel beads were used at both laboratories, as shown in Figure 18. This achieved relative humidity of 3% within the dishes. Both laboratories employed a 20 mm air space between the top surface of the substrates and the bottom surface of the test specimen. The pliable membrane specimens were then glued to the top edge of the dishes. To avoid water vapor leakages between the dishes and test specimens, the edges between the materials were taped and sealed with molten paraffin wax at 100 °C. The dishes were then placed on shelving within these test rooms, as shown in Figures 19 and 20.

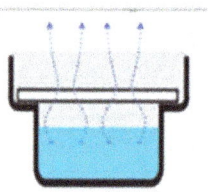  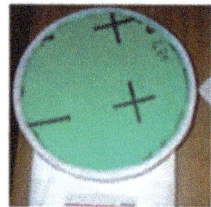

Figure 17. Wet-cup test method.

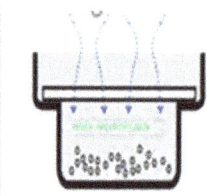  

Figure 18. Dry-cup test method.

Figure 19. Shelving in the interior of test room at IBP Germany.

Figure 20. Shelving in the interior of test room at UTAS, Australia.

Regular weighing measurements of the test dishes were taken every two hours until equilibrium was achieved. The measurements were in milligrams and all weighing data were recorded. The calculations of the water vapor resistivity properties were obtained mathematically (see Tables 3 and 4). Microsoft Excel 365 was used to complete a statistical analysis of variance to establish if there was any significant difference between the result obtained from the laboratory at Fraunhofer IBP and UTAS.

Table 3. Water vapor diffusion properties measured at IBP.

	Wet cup @ 23°C 93/50% Test @IBP Germany								
Specimen	Mean thickness d (m)	Area m^2	Mass of specimen (g)	Water vapour flux g = G/A in kg/(s*m^2)	Water vapour permeance W = g/dp in kg/(s*m^2*Pa)	Water vapour resistance Z = 1/W in (s*m^2*Pa)/kg	Water vapour resistance factor μ	Diffusion-equivalent air layer thickness Sd (m)
TA1	0.00082	0.0293	7.40	3.53×10^{-6}	2.68×10^{-9}	3.74×10^{8}	71.86	0.0590
TA2	0.00080	0.0290	7.31	3.44×10^{-6}	2.61×10^{-9}	3.83×10^{8}	76.17	0.0610
TA3	0.00084	0.0287	7.82	2.84×10^{-6}	2.15×10^{-9}	4.65×10^{8}	93.84	0.0790
Mean value	0.00082	0.0290	7.51	3.27×10^{-6}	2.48×10^{-9}	4.07×10^{8}	80.62	0.0663
Standard deviation	0.00002	0.0003	0.27	3.77×10^{-7}	2.86×10^{-10}	5.01×10^{8}	11.65	0.0110
	Dry cup @ 23°C 3/50% Test @IBP Germany								
Specimen	Mean thickness d (m)	Area m^2	Mass of specimen (g)	Water vapour flux g = G/A in kg/(s*m^2)	Water vapour permeance W = g/dp in kg/(s*m^2*Pa)	Water vapour resistance Z = 1/W in (s*m^2*Pa)/kg	Water vapour resistance factor μ	Diffusion-equivalent air layer thickness Sd (m)
TA4	0.00079	0.0284	7.27	4.06×10^{-6}	3.08×10^{-9}	3.25×10^{8}	62.18	0.0490
TA5	0.00081	0.0281	7.09	4.24×10^{-6}	3.21×10^{-9}	3.11×10^{8}	57.04	0.0460
TA6	0.00082	0.0278	7.56	3.95×10^{-6}	2.99×10^{-9}	3.34×10^{8}	61.63	0.0510
Mean value	0.00081	0.0281	7.30	4.08×10^{-6}	3.09×10^{-9}	3.24×10^{8}	60.28	0.0487
Standard deviation	1.53×10^{-5}	0.0003	0.24	1.44×10^{-7}	1.09×10^{-10}	1.13×10^{7}	2.82	0.0025

Table 4. Water vapor diffusion properties measured at UTAS.

Specimen	Mean thickness d (m)	Area (m²)	Mass of specimen grammes (g)	Water vapour flux g = G/A in kg/(s*m²)	Water vapour permeance W = g/dp in kg/(s*m²*Pa)	Water vapour resistance Z = 1/W in (s*m²*Pa)/kg	Water vapour resistance factor μ	Diffusion-equivalent air layer thickness Sd (m)
Wet cup @ 23 °C 93/50% Test @University of Tasmania, Australia								
TA1	0.000819	0.0275	7.05	3.08×10^{-6}	2.33×10^{-9}	4.28×10^{8}	76.02	0.0623
TA2	0.000794	0.0266	6.95	3.03×10^{-6}	2.30×10^{-9}	4.35×10^{8}	80.09	0.0636
TA3	0.000784	0.0260	7.21	3.98×10^{-6}	3.01×10^{-9}	3.32×10^{8}	55.76	0.0437
Mean	0.000799	0.0267	7.07	3.36×10^{-6}	2.55×10^{-9}	3.99×10^{8}	70.62	0.0565
Standard deviation	1.80×10^{-5}	0.00076	0.13114877	5.32×10^{-7}	4.03×10^{-10}	5.78×10^{7}	13.03	0.0111
Dry cup @ 23 °C 3/50% Test @University of Tasmania, Australia								
TA4	0.000824	0.0275	7.43	3.34×10^{-6}	2.76×10^{-9}	3.62×10^{8}	60.99	0.0503
TA5	0.000804	0.0278	7.40	3.55×10^{-6}	2.94×10^{-9}	3.40×10^{8}	57.15	0.0459
TA6	0.000805	0.0275	7.17	3.40×10^{-6}	2.82×10^{-9}	3.55×10^{8}	60.81	0.0490
Mean	0.000811	0.0276	7.33	3.43×10^{-6}	2.83×10^{-9}	3.52×10^{8}	59.65	0.0484
Standard deviation	1.13×10^{-5}	0.000160728	0.142243922	1.11×10^{-7}	9.28×10^{-11}	1.13×10^{7}	02.17	0.0023

3. Results

3.1. Hygrothermal Control of the Test Room

This section discusses the result from the climatic control of the hygrothermal test room which was used to quantify the water vapor diffusion properties of the permeable pliable membrane, when the test room was maintained at 50% relative humidity and the temperature remained at 23 °C (±1 °C) for the material testing periods. It was found that the room would take up to 72 h to initially reach and stabilize at the desired temperature and relative humidity.

During the establishment of the test room, sensors which controlled the operation of heating, cooling, humidifying, and dehumidifying appliances were moved until adequate control of the room was established. The final two versions of the sensor locations are shown in Table 1. The principle reason for the change in sensor location between Version 1 and Version 2 was a measured, and significant time lag for room temperature control. The time lag issues were addressed by the Version 2 configuration.

To demonstrate the potential of this hygrothermally controlled room at UTAS, the temperature and relative humidity during the material testing period was retrieved for analysis. Figure 21 shows the temperature profile of test room for the period of six weeks, while Figure 22 shows the relative humidity profile for this same period which required the relative humidity be kept at 50%. The blue box plot (Figure 23) shows the observations from three temperature sensors located 1800 mm above the floor, the orange box plot shows the observations from three temperature sensors located 1200 mm above the floor, the grey box plot shows the observations from three globe temperature (mean radiant) sensors located 1200 mm above the floor, and the yellow box plot shows the observations from three temperature sensors located 600 mm above the floor. Summarily the box plot observation indicates that aside from occasional outliers, the temperature in the room was maintained between 23.2 °C and 22.6 °C, with an average of 22.9 °C (±1 °C). Figure 24 shows the results from the three relative humidity sensors for the corresponding period, and the box plots show that aside from occasional outliers, the relative humidity was maintained between 49.8% and 50.8%, with an average humidity of 50.4% (±1%).

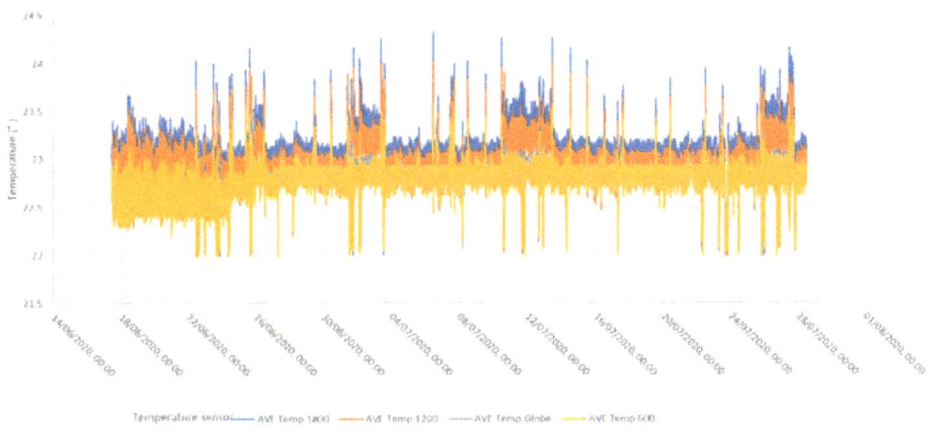

Figure 21. Temperature profile of the room aimed at 23 °C (+/−0.5 °C) for the testing period 2.

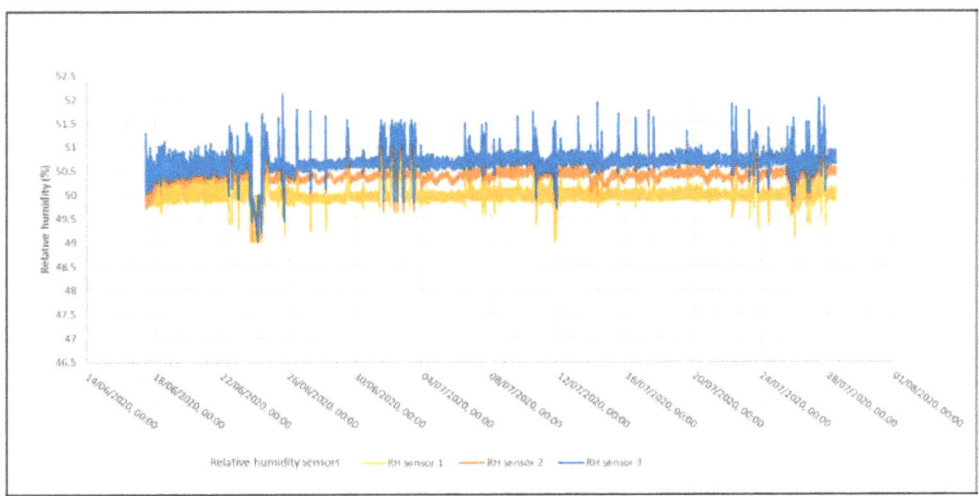

Figure 22. Relative humidity profile of the room aimed at 50% for the testing period 2.

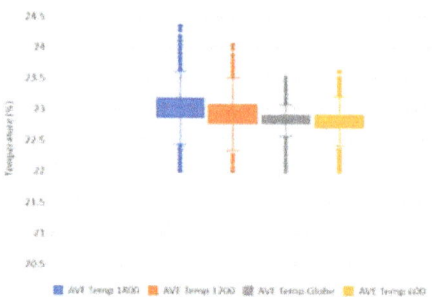

Figure 23. Box and whisker plot of temperature observations during test 2.

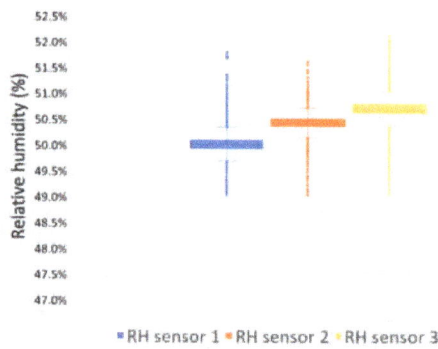

Figure 24. Box and whisker plot of relative humidity observations during test 2.

3.2. Comparison of the Interlaboratory Results for the Water Vapor Diffusion Properties

The gravimetric measurement of change in mass over a particular period commenced as soon as the dishes were placed in the test room. Initially, weighing was completed at two hourly intervals. This was to establish if the dish gained or lost weight (depending on the dry-cup or wet-cup substrate). Tables 3 and 4 show the water vapor resistivity properties measured for the permeable pliable membrane commonly used for Australian construction system.

The analysis of variance that was completed shows that there was no significant difference ($p = 0.38$) between the results of the water vapor resistance factor (Table 5) for the wet-cup test obtained in both IBP and UTAS. Similarly, for the dry-cup test, the there was no significant difference ($p = 0.77$) between the results of the test obtained in both laboratories. Table 6 also indicates that there was no significant difference ($p = 0.34$) between the result of the wet-cup test obtained in both IBP and UTAS for the diffusion-equivalent air layer thickness, and there was no significant difference ($p = 0.89$) between the results of the dry-cup test obtained in both laboratories.

Table 5. Inter-laboratory comparison of the ANOVA result for the resistance factor (μ) of wet-cup test.

Water Vapour Resistance Factor (μ)

Anova: Single Factor
SUMMARY

Groups	Count	Sum	Average	Variance	Standard deviation
Wet-cup test IBP	3	241.87	80.62333	135.6542	11.64706973
Wet-cup test UTAS	3	211.87	70.62333	169.8302	13.03189293

ANOVA

Source of Variation	SS	df	MS	F	p-value	F crit
Between Groups	150	1	150	0.982047	0.377789806	7.708647
Within Groups	610.9689	4	152.7422			
Total	760.9689	5				

Groups	Count	Sum	Average	Variance	Standard deviation
Dry-cup test IBP	3	180.85	60.28333	7.965033	2.822239064
Dry-cup test UTAS	3	178.954	59.65133	4.685605	2.164625911

ANOVA

Source of Variation	SS	df	MS	F	p-value	F crit
Between Groups	0.599136	1	0.599136	0.09472	0.77361956	7.708647
Within Groups	25.30128	4	6.325319			
Total	25.90041	5				

Table 6. Inter-laboratory comparison of the ANOVA result for the diffusion-equivalent air layer thickness Sd(m) of dry-cup test.

Diffusion-Equivalent Air Layer Thickness Sd(m)

Anova: Single Factor

SUMMARY

Groups	Count	Sum	Average	Variance	Standard deviation		Groups	Count	Sum	Average	Variance	Standard deviation
Wet-cup test IBP	3	0.199	0.066333	0.000121	0.011015141		Dry-cup test IBP	3	0.146	0.048667	6.33×10^{-6}	0.002516611
Wet-cup test UTAS	3	0.1696	0.056533	0.000124	0.011132984		Dry-cup test UTAS	3	0.14515	0.048383	5.08×10^{-6}	0.00225407

ANOVA

Source of Variation	SS	df	MS	F	p-value	F crit		Source of Variation	SS	df	MS	F	p-value	F crit
Between Groups	0.000144	1	0.000144	1.174673	0.339403454	7.708647		Between Groups	1.20×10^{-7}	1	1.2×10^{-7}	0.02109951	0.891533715	7.708647
Within Groups	0.000491	4	0.000123					Within Groups	2.28×10^{-5}	4	5.70×10^{-6}			
Total	0.000635	5						Total	2.29×10^{-5}	5				

4. Discussion

Firstly, the set-up and configuration of the test room followed many practices common for the establishment of environmentally controlled spaces. The points of interest were the challenges in controlling the room temperature and the configuration and operation of the humidifier and de-humidifier. The ability to keep the temperature and relative humidity within specific bandwidths was critical. The temperature was kept within +/−1 °C and the relative humidity was kept within +/− 1% RH. Table 1 makes note of Version 1 and Version 2 for the measurement of dry bulb air temperature. The data logger combined with relay switches demonstrated a simple mechanism to control room temperature. However, there was a recognized time lag and regular over-heating of the test room. After several iterations of data logger programming and the co-location of additional sensors around the air-conditioning appliance, localized temperature stratification near the appliance was identified. An additional PT100 temperature sensor was installed close to the air-conditioner thermostat to establish the step difference that was occurring. This extra data allowed for a more informed approach to the data-logger alarm bandwidths, which controlled the air-conditioner power supply.

Secondly, the result of the inter-laboratory measurement of the water vapor resistance factor and the diffusion equivalent air layer thickness of a permeable membrane was investigated to validate the performance of the UTAS laboratory. Under the same experimental procedure and parameters, similar results were obtained, while experimental procedural error was minimized. Recent research [42] had indicated that irrespective of the material to be tested or the test procedure, discrepancies in results may normally occur during any inter-laboratory measurement to determine the water vapor diffusion properties of material through gravimetric cup test. The ANOVA test for this research has demonstrated that discrepancies in the result of interlaboratory measurement of pliable membrane is insignificant. This implies that the hygrothermally controlled room at UTAS can be used for the same experimental purposes obtained at IBP.

The results of the water vapor diffusion properties from the interlaboratory testing with the world leading IBP laboratory indicates that the operation of this laboratory is promising, as this method can be employed to set up a low-cost hygrothermal testing facility.

5. Conclusions and Recommendations

Essentially, the equipment in the test cells, comprised of an all-embracing range of temperature and relative humidity sensors, and an integrated data acquisition system, which enable flexible monitoring and control of heating, cooling, humidifying, and dehumidifying appliance. This combination of equipment enabled the stabilization of temperature and relative humidity which are key parameters for construction material wet-cup and dry-cup water vapor diffusion testing. The integrated system enabled the stabilization of the temperature and the relative humidity through the use of simple data-logger programming code. The current configuration, operation, and performance of the test room temperature and humidity indicated that the precise profiles required for the vapor diffusion measurement were achieved and maintained for test room conditions of 23 °C with a 50% RH.

This paper reports the establishment of Australia's first precisely controlled hygrothermal room for measuring the water vapor diffusion properties of building materials via the use of a conditioned test room. As a key component of this research is to provide national guidance and methods for the establishment of vapor diffusion properties of Australian Construction materials, this is a positive outcome. The use of an environmentally controlled test room for measuring water vapor diffusion properties of building materials is considered more appropriate than other published methods. This is because the process of taking test dishes in and out of conditioned cabinets for weighing allows for the possibility of intrinsic errors. In summary, this research has demonstrated that the establishment of a conditioned hygrothermal test room may not be financially onerous for prospective researchers seeking to establish a hygrothermally controlled laboratory, that can be used to quantify water vapor diffusion properties for locally made construction materials.

Author Contributions: T.S.O.—Main author carried out the experiments both at UTAS and IBP, involved in the conceptualization; collects data; analyze data; graphs and visualization; provide the original draft manuscript, and revised manuscript. M.D.—Second author provides guidance to experiment, source for funding to procure equipment; project administration; contribute to data analyze; data curation; contribute to graphs and visualization, edit and provided revision to the manuscript. H.K.—Provides guidance for the laboratory operation, supervision and validation at IBP, and revision of manuscript. All authors have read and agreed to the published version of the manuscript.

Funding: This research received no external funding.

Acknowledgments: These authors acknowledge the Commonwealth Scientific Industrial Research Organization (CSIRO) for co-funding this research project.

Conflicts of Interest: The authors declare no conflict of interest.

References

1. Künzel, H.M.; Holm, A.; Zirkelbach, D.; Karagiozis, A.N. Simulation of Indoor Temperature and Humidity Conditions Including Hygrothermal Interactions with the Building Envelope. *Sol. Energy* **2005**, *78*, 554–561. [CrossRef]
2. Künzel, H.M.; Zirkelbach, D. Advances in Hygrothermal Building Component Simulation: Modeling Moisture Sources Likely to Occur Due to Rainwater Leakage. *J. Build. Perform. Simul.* **2013**, *6*, 346–353. [CrossRef]
3. Hall, M.R.; Casey, S.P.; Loveday, D.L.; Gillott, M. Analysis of Uk Domestic Building Retrofit Scenarios Based on the E. On Retrofit Research House Using Energetic Hygrothermics Simulation–Energy Efficiency, Indoor Air Quality, Occupant Comfort, and Mold Growth Potential. *Build. Environ.* **2013**, *70*, 48–59. [CrossRef]
4. Nilsson, L. *Methods of Measuring Moisture in Building Materials and Structures: State-of-the-Art Report of the Rilem Technical Committee 248-Mmb*; Springer: Cham, Switzerland, 2018.
5. You, S.; Li, W.; Ye, T.; Hu, F.; Zheng, W. Study on Moisture Condensation on the Interior Surface of Buildings in High Humidity Climate. *Build. Environ.* **2017**, *125*, 39–48. [CrossRef]
6. Dewsbury, M.; Law, T. Temperate Climates, Warmer Houses and Built Fabric Challenges. *Procedia Eng.* **2017**, *180*, 1065–1074. [CrossRef]
7. Fořt, J.; Šál, J.; Kočí, J.; Černý, R. Energy Efficiency of Novel Interior Surface Layer with Improved Thermal Characteristics and Its Effect on Hygrothermal Performance of Contemporary Building Envelopes. *Energies* **2020**, *13*, 2012. [CrossRef]
8. Moon, H.J.; Ryu, S.H.; Kim, J.T. The Effect of Moisture Transportation on Energy Efficiency and Iaq in Residential Buildings. *Energy Build.* **2014**, *75*, 439–446. [CrossRef]
9. Wang, Y.; Ma, C.; Liu, Y.; Wang, D.; Liu, J. Effect of Moisture Migration and Phase Change on Effective Thermal Conductivity of Porous Building Materials. *Int. J. Heat Mass Transf.* **2018**, *125*, 330–342. [CrossRef]
10. Dong, W.; Chen, Y.; Bao, Y.; Fang, A. A Validation of Dynamic Hygrothermal Model with Coupled Heat and Moisture Transfer in Porous Building Materials and Envelopes. *J. Build. Eng.* **2020**, *32*, 101484. [CrossRef]
11. Dewsbury, M.; Law, T.; Potgieter, J.; Fitzgerald, D.; McComish, B.; Chandler, T.; Soudan, A. *Scoping Study of Condensation in Residential Buildings: Final Report*; Australian Building Codes Board, Department of Industry Innovation and Science Canberra Australia: Canberra, Australia, 2016.
12. Nath, S.; Dewsbury, M.; Orr, K. Is New Housing Health Hazard? In Proceedings of the Engaging Architectural Science: Meeting the Challenges of Higher Density, 52nd International Conference of the Architectural Science Association, RMIT University, Melbourne, Australia, 28 November–1 December 2018.
13. Olaoye, T.S.; Dewsbury, M. Establishing an Environmentally Controlled Room to Quantify Water Vapor Resistivity Properties of Construction Materials. In *Revisiting the Role of Architecture for 'Surviving' Development, Proceedings of the 53rd International Conference of the Architectural Science Association, Roorkee, India, 28–30 November 2019*; Agrawal, A., Gupta, R., Eds.; Architectural Science Association (ANZAScA): Roorkee, India, 2019; pp. 675–684.
14. Dewsbury, M.; Soudan, A.; Su, F.; Geard, D.; Cooper, A.; Law, T. *Condensation Risk Mitigation for Tasmanian Housing*; Department of Justice Tasmania, Hobart: Hobart, Australia, 2018.
15. Woloszyn, M.; Rode, C. Tools for Performance Simulation of Heat, Air and Moisture Conditions of Whole Buildings. *Build. Simul.* **2008**, *1*, 5–24. [CrossRef]
16. International Standard Organization. Hygrothermal Performance of Building Components and Building Elements—Internal Surface Temperature to Avoid Critical Surface Humidity and Interstitial Condensation—Calculation Methods (ISO 13788:). In *EVS-EN ISO 13788:2012*; Estonian Center for Standardization: Brussels, Belgium, 2012.
17. Ramos, N.M.; Delgado, J.Q.; Barreira, E.; de Freitas, V.P. Hygrothermal Properties Applied in Numerical Simulation: Interstitial Condensation Analysis. *J. Build. Apprais.* **2009**, *5*, 161–170. [CrossRef]
18. Roels, S.; Depraetere, W.; Carmeliet, J.; Hens, H. Simulating Non-Isothermal Water Vapor Transfer: An Experimental Validation on Multi-Layered Building Components. *J. Therm. Envel. Build. Sci.* **1999**, *23*, 17–40. [CrossRef]

19. Hagentoft, C.; Kalagasidis, A.S.; Adl-Zarrabi, B.; Roels, S.; Carmeliet, J.; Hens, H.; Grunewald, J.; Funk, M.; Becker, R.; Shamir, D.; et al. Assessment Method of Numerical Prediction Models for Combined Heat, Air and Moisture Transfer in Building Components: Benchmarks for One-Dimensional Cases. *J. Therm. Envel. Build. Sci.* **2004**, *27*, 327–352. [CrossRef]
20. Glass, S.V.; TenWolde, A.; Zelinka, S.L. *Hygrothermal Simulation: A Tool for Building Envelope Design Analysis*; Wood Design Focus: LaGrange, GA, USA, 2013; Volume 23, Number 3; Fall Issue 2013; pp. 18–25.
21. Libralato, M.; Saro, O.; de Angelis, A.; Spinazzè, S. Comparison between Glaser Method and Heat, Air and Moisture Transient Model for Moisture Migration in Building Envelopes. *Appl. Mech. Mater.* **2019**, *887*, 385–392. [CrossRef]
22. Pallin, S.; Boudreaux, P.; Shrestha, S.; New, J.; Adams, M. *State-of-the-Art for Hygrothermal Simulation Tools*; US Department of Energy: Springfield, VA, USA, 2017.
23. ABCB. *The National Constrcution Code: Volume 1*; Australian Building Codes Board: Canberra, Australia, 2019.
24. ABCB. *The National Construction Code: Volume 2*; Australian Building Codes Board: Canberra, Australia, 2019.
25. ASTM. Standard Test Methods for Water Vapor Transmission of Materials. In *E96/E96M*; ASTM International: ASTM: West Conshohocken, PA, USA, 2010.
26. Borjesson, F. An Investigation of the Water Vapor Resistance-the Humidity Detection Sensor Method in Versmaperm Mkiv Compared to the Gravimetric Method. Master's Thesis, Chalmers University of Technology, Gothenburg, Sweden, 2013.
27. McCullough, E.A.; Kwon, M.; Shim, H. Comparison of Standard Methods for Measuring Water Vapor Permeability of Fabrics. *Meas. Sci. Technol.* **2003**, *14*, 1402–1408. [CrossRef]
28. Gibson, P.; Rivin, D.; Berezin, A.; Nadezhdinskii, A. Measurement of Water Vapor Diffusion through Polymer Films and Fabric/Membrane Laminates Using a Diode Laser Spectroscope. *Polym. Plast. Technol. Eng.* **1999**, *38*, 221–239. [CrossRef]
29. Huang, J.; Qian, X. A New Test Method for Measuring the Water Vapor Permeability of Fabrics. *Meas. Sci. Technol.* **2007**, *18*, 3043–3047. [CrossRef]
30. Richter, J.; Staněk, K. Measurements of Water Vapor Permeability—Tightness of Fibreglass Cups and Different Sealants and Comparison of M-Value of Gypsum Plaster Boards. *Procedia Eng.* **2016**, *151*, 277–283. [CrossRef]
31. ISO, International Standard Organization EN. Hygrothermal Performance of Building Materials and Products—Determination of Water Vapor Transmission Properties—Cup Method Iso 12572. In *EVS-EN ISO 12572:2016*; Estonian Center or Standardization: Brussels, Belgium, 2016.
32. Janz, M. Methods of Measuring the Moisture Diffusivity at High Moisture Levels. Ph.D. Thesis, Building Materials LTH, Lund Unversity, Lund, Sweden, 1997.
33. Li, K.; Xu, Z.; Jun, G. Experimental Investigation of Hygrothermal Parameters of Building Materials under Isothermal Conditions. *J. Build. Phys.* **2009**, *32*, 355–370.
34. Olaoye, T.S.; Dewsbury, M. Australian Building Materials and Vapor Resistivity. In *Building Physics Forum*; AIRAH: Wollogong, Australia, 2018.
35. Bomberg, M.; Pazera, M. Methods to Check Reliability of Material Characteristics for Use of Models in Real Time Hygrothermal Analysis. In *Research in Building Physics—Proceedings of the First Central European Symposium on Building Physics, Cracow–Lodz, Poland, 13–15 September 2010*; Gawin, D., Kisielewicz, T., Eds.; Technical University of Cracow: Cracow-lodz, Poland, 2010.
36. Galbraith, G.H.; Kelly, D.J.; McLean, R.C. Alternative Methods for Measuring Moisture Transfer Coefficients of Building Materials. In Proceedings of the 2nd International Conference on Building Physics, Antwerpen, Belgium, 14–18 September 2003; pp. 249–254.
37. Couturier, M.; Boucher, C. Dynamic Water Vapor Permeance of Building Materials and the Benefits to Buildings. In Proceedings of the 26th RCI International Convention and Trades Show, Reno, NV, USA, 29 April 2011.
38. Wu, Y. Experimental Study of Hygrothermal Properties for Building Materials. Master's Thesis, Concordia University, Montreal, QC, USA, 2007.
39. Rafidiarison, H.; Rémond, R.; Mougel, E. Dataset for Validating 1-D Heat and Mass Transfer Models within Building Walls with Hygroscopic Materials. *Build. Environ.* **2015**, *89*, 356–368. [CrossRef]
40. Dewsbury, M.; Fay, M.R.; Nolan, G.; Vale, R.J.D. The Design of Three Thermal Performance Test Cells in Launceston. In *Towards Solutions for a Liveable Future: Progress, Practice, Performance, People*; Dirk, J.S., Coulson, T.R., Eds.; Deakin University: Geelong, Australia, 2007; pp. 91–100.
41. Dewsbury, M.; Fay, R.; Nolan, G. Thermal Performance of Light-Weight Timber Test Buildings. In Proceedings of the World Congress of Timber Engineering, Miyazaki, Japan, 2–5 June 2008.
42. Feng, C.; Guimarães, A.S.; Ramos, N.; Sun, L.; Gawin, D.; Konca, P.; Hall, C.; Zhao, J.; Hirsch, H.; Grunewald, J. Hygric Properties of Porous Building Materials (Vi): A Round Robin Campaign. *Build. Environ.* **2020**, *185*, 107242. [CrossRef]

Article

Non-Intrusive Measurements to Incorporate the Air Renovations in Dynamic Models Assessing the In-Situ Thermal Performance of Buildings

María José Jiménez [1,*], José Alberto Díaz [1], Antonio Javier Alonso [2], Sergio Castaño [1] and Manuel Pérez [2]

[1] Energy Efficiency in Buildings R&D Unit, CIEMAT, 28040 Madrid, Spain; alberto.diaz@ciemat.es (J.A.D.); sergio.castano@psa.es (S.C.)
[2] CIESOL Research Center on Solar Energy, Joint Center University of Almería-CIEMAT, 04120 Almería, Spain; a.javialbox@gmail.com (A.J.A.); mperez@ual.es (M.P.)
* Correspondence: mjose.jimenez@psa.es; Tel.: +34-950-38-7900

Citation: Jiménez, M.J.; Díaz, J.A.; Alonso, A.J.; Castaño, S.; Pérez, M. Non-Intrusive Measurements to Incorporate the Air Renovations in Dynamic Models Assessing the in-Situ Thermal Performance of Buildings. *Energies* **2021**, *14*, 37. https://dx.doi.org/10.3390/en14010037

Received: 13 November 2020
Accepted: 21 December 2020
Published: 23 December 2020

Publisher's Note: MDPI stays neutral with regard to jurisdictional claims in published maps and institutional affiliations.

Copyright: © 2020 by the authors. Licensee MDPI, Basel, Switzerland. This article is an open access article distributed under the terms and conditions of the Creative Commons Attribution (CC BY) license (https://creativecommons.org/licenses/by/4.0/).

Abstract: This paper reports the analysis of the feasibility to characterise the air leakage and the mechanical ventilation avoiding the intrusiveness of the traditional measurement techniques of the corresponding indicators in buildings. The viability of obtaining the air renovation rate itself from measurements of the concentration of the metabolic CO_2, and the possibilities to express this rate as function of other climatic variables, are studied. N_2O tracer gas measurements have been taken as reference. A Test Cell and two full size buildings, with and without mechanical ventilation and with different levels of air leakage, are considered as case studies. One-month test campaigns have been used for the reference N_2O tracer gas experiments. Longer periods are available for the analysis based on CO_2 concentration. When the mechanical ventilation is not active, the results indicate significant correlation between the air renovation rate and the wind speed. The agreement between the N_2O reference values and the evolution of the metabolic CO_2 is larger for larger initial values of the CO_2 concentration. When the mechanical ventilation is active, relevant variations have been observed among the N_2O reference values along the test campaigns, without evidencing any correlation with the considered boundary variables.

Keywords: building energy; building envelope; performance assessment; air renovation; non-intrusive measurements; on-board monitoring

1. Introduction

Buildings use about 40% of the total energy produced globally and have a relevant potential in terms of energy savings and reducing the pollutant emissions to the atmosphere [1]. These issues are driving an increasing interest to foster the energy efficiency in buildings leading to the elaboration and incorporation of related regulations, stressing the demand to broaden the knowledge related to the energy performance of the buildings, and motivating many research initiatives in this area. Presently, the majority of the checks of compliance and energy performance labelling of buildings rely on design values and theoretical assessments or simulations. Nevertheless, many researches have demonstrated that the actual performance of a building can be very different from the one theoretically evaluated [2,3]. The readiness of reliable enough test procedures applicable to as built buildings for assessing their thermal performance, would contribute to eliminate the problems related to the performance gap. The need for tools identifying the sources of the performance gaps, and providing feedback to different stakeholders, is included among the research themes considered by the Energy in Buildings and Communities (EBC) Technology Collaboration Programme (TCP) of the International Energy Agency (IEA) [1]. One of the elements having a significant influence on the energy behaviour of the buildings is the building envelope. The identification of the intrinsic thermal properties characterising the as built building envelope from on board monitoring system is recently attracting the attention of many research groups in the context of international collaboration initiatives [4]. In this context, those monitoring systems with a limited set of non-intrusive measurement devices, embedded in the building, as those typically used for billing or for controlling the Heating, Ventilating and Air Conditioning (HVAC) systems are considered as on board monitoring systems. The energy performance assessment of the building envelope can be carried out through data analysis techniques that require the measurement (that can be direct or indirect) of all the effects that contribute to the energy balance in the space that is confined by the building envelope being characterised [5]. One of the contributions to this energy balance is the one from air renovations, either by mechanical or natural ventilation, or by infiltrations as consequence of cracks or material porosity [6].

There are several procedures for the experimental assessment of the air renovation rate in rooms. Some of these procedures are based on pressurisation and others are based on tracer gas techniques [7]. These traditionally applied methods that could give precise results are complex, expensive and highly intrusive for the building users and inhabitants. Additionally, these traditional techniques characterise the air renovations by a constant parameter. Some standardised procedures obtain this parameter under a pressure that is raised regarding the pressure of the building in use [8]. These constant values can introduce some degree of uncertainty on the data based dynamic modelling techniques that are applied for the thermal performance assessment of the building envelope from on-board monitoring systems [5,9,10]. Part of this uncertainty can be driven by the use of the air renovation rate as a constant parameter when actually it is a variable. A review paper that has been recently published identifies the dynamic behaviour of the air renovation rate as an issue contributing to the uncertainty in tracer gas-based methods [11]. Other authors have analysed the uncertainties due to wind in building pressurisation tests [12]. They identified errors in the rage 6–12% for wind speed in the range 6–10 ms^{-1} for test carried out under a standard pressure of 50 Pa, while the errors raised up to 35% and 60% for wind speeds of 6 ms^{-1} and 10 ms^{-1}, respectively under a pressure of 10 Pa. When the air renovation rate is obtained according the standardised building pressurisation tests, the transformation of the pressurised value to the non-pressurised one, can introduce also certain degree of uncertainty in the dynamic models that are used for the energy performance assessment of in-use buildings. The presence of some uncertainty and variability in the air renovation rate due to infiltrations as well as mechanical ventilation, can contribute to understand and explain the behaviour of the Heat Loss Coefficient (HLC) experimentally assessed and its uncertainties [13,14].

The work reported in this paper is focused on the experimental assessment of the air renovation rate analysing the reliability of cheaper and more cost effective techniques regarding the traditional techniques based on tracer gas. The feasibility to characterise air leakage and mechanical ventilation avoiding the intrusiveness of the traditional measurement techniques is analysed. The viability to obtain the air renovation rate itself, as well as the possibilities to express it as function of other variables (such as wind speed, atmospheric pressure, etc.), are studied extending some preliminary studies [15]. Tracer gas measurements based on N_2O have been used as reference. Experimental relations between the air renovations and the wind speed, the indoor-outdoor air temperature difference, and the atmospheric pressure have been analysed. The reliability of an alternative method based on the evolution of the metabolic CO_2 using wall mounted sensors of CO_2 concentration is evaluated. A PASLINK Test Cell [16,17] and two full size buildings are considered as case studies. First the Test Cell and a very simple single zone building, without mechanical ventilation, are considered. Afterwards, a room in an office building has been studied with and without mechanical ventilation. One-month test campaigns have been used for the reference study based on tracer gas measurements using N_2O, in both buildings and the Test Cell. Longer periods are available for the analysis based on CO_2 concentration.

The next sections are organised as follows: Section 2 presents the considered case studies, and briefly describes the experiment set up and the methodology applied for data analysis, Section 3 presents and discusses the results that have been obtained for the different case studies, and finally Section 4 summarises the conclusions regarding the behaviour of the air renovation rate, discusses the effect of this behaviour on the Heat Loss Coefficient (HLC) and suggest further research on this issue.

2. Materials and Methods

The next subsections included under this section describe the three considered case studies, the experiment set up, the tests carried out, and finally the methodology applied for data analysis.

2.1. Case Studies

A PASLINK Test Cell and two full size extensively monitored buildings are considered as case studies [16,17]. These buildings and the Test Cell, briefly described in Section 2.1.1, Section 2.1.2, Section 2.1.3 are at the CIEMAT's Plataforma Solar de Almería (PSA), in Tabernas (37.1° N, 2.4° W), Almería (Spain). They are in a rural area where the climate is semi-arid, with large day-night temperature variations.

2.1.1. PASLINK Test Cell

The PASLINK Test Cell consists in a test facility with a high-thermal-insulation test room and an auxiliary room (Figure 1a). The test room has a surface of 4.825×2.48 m^2 and its high is 2.47 m. The Test Cell is placed in a large open area without any shading. It has an air conditioning system and measurement devices for testing full-scale building components. Its test room envelope is highly insulated by 40 cm of polystyrene and it is equipped with the Pseudo-Adiabatic Shell (PAS) Concept. This system is based on a thermopile that detects if there is heat flux through the envelope of the test room, and cancels it by means of a heating foil. The interior surface of the test room is finished with an aluminium sheet giving it thermal uniformity. The Test Cell is over a rotating device that enables it for testing in any orientation.

Figure 1. Buildings considered as case studies: (**a**) PASLINK Test Cell; (**b**) Single-zone building; (**c**) Office building.

The south wall and the roof of the test chamber are interchangeable, which permits any vertical or horizontal building component to be installed for testing. The tests of air renovations considered in this work correspond to a reference experiment. In this case, the Test Cell incorporates a homogeneous and opaque wall in its replaceable façade.

This test was conducted in the framework of a series of tests that included several photovoltaic modules and electrochromic windows replacing a piece of the component taken as reference. The Heat Loss Coefficients of these components are obtained by subtracting the Heat Loss Coefficient obtained with the photovoltaic modules or the electrochromic windows, from the Heat Loss Coefficient obtained from the reference component. The Test Cell is designed to be very airtight. Typical air renovation rates during testing are between 0.02 and 0.05 renovations per hour [18]. The assessment of its air renovation rate is important in order to check the achieved level of air tightness and to assess the deviations from this level due to the climatic variables.

2.1.2. Single-Zone Building

This building is a small workshop with just one room, and its area is 31.83 m² (Figure 1b) [17]. It can give experimental support to diverse research activities maintaining it empty or with low occupancy rates. It was built in 2002. It is near another twin building that is placed 2 m from its east wall. Both are built in an open area without any other obstacles around that could shade them.

This building was designed to reduce the energy demand incorporating the following passive strategies: South orientation, shading elements avoiding the solar gains in summer and maximising them in winter, the windows are double-glazed to reduce heat losses, and diagonally aligned (north-south) to facilitate the natural ventilation, thermal mass incorporated in the building envelope, external insulation and high ceilings.

2.1.3. Office Building Prototype

The so called C-DdI ARFRISOL at PSA is a one floor building with most of the regularly occupied offices facing south (Figure 1c). Its net floor area is 1007.40 m². It was constructed in 2007 in the framework of the PSE-ARFRISOL project [19]. It is a prototype of a new plant, built on one floor longitudinal plan.

A double-wing structure, that is installed on the roof along the main axis of the building, protects it from the solar radiation. This structure integrates two different types of solar collectors. Uncovered collectors which are designed to operate as radiant coolers by night are over the wing facing north. Flat plate collectors that are designed to supply hot water for the heating, cooling and DHW systems are over the south facing wing. Small solar chimneys that provide night ventilation of the offices are constructed on the central part of this structure. The south windows are protected by an overhang that provides shade during the summertime and facilitates passive heating in winter.

This building is in use, but it must be taken into account that the experiments used for this work were carried out when the considered room was positively empty; at lunch time and also once the working day is finalised (identified every day as test 1, and test 2, respectively).

2.2. Experiment Set Up

A tracer gas device combined with a gas analyser have been used to carry out Decay experiments based on the evolution of N_2O concentration in both buildings and the Test Cell.

The Test Cell and the two buildings are extensively monitored. The monitoring system records minutely read measurements of the following variables:

- N_2O concentration when the Decay experiments are being conducted.
- Indoor and outdoor air temperatures, relative humidity, and concentration of CO_2. Two sensors are installed to measure this variable. An accurate and expensive sensor used as reference, and a cheaper and less accurate sensor (Identified as CO_{2_ref} and CO_2 respectively in this document).
- Temperature of walls, floor and glass surfaces.
- Energy delivered by the heating system (radiant floor).
- Electric consumption due to computers and lighting
- Whether doors and windows are closed or "not closed".
- Ground temperature.
- Beam, diffuse, global horizontal, global vertical south and global vertical north solar irradiance.
- Longwave radiation.

One-month test campaigns for each building were considered for the analysis. These campaigns were conducted under different conditions: Dynamic heating sequence in the Test Cell maintaining a large indoor to outdoor air temperature difference, free running test in the single-zone building, and space heating maintaining the indoor air temperature in a comfort range in the office building.

2.3. Methodology

2.3.1. Analysis of the Relations between the Air Renovation Rate and Climate Variables

- For both buildings and the Test Cell, for infiltrations and mechanical ventilation, tracer gas measurements based on N_2O have been used as reference. The air renovation rate has been obtained using the Decay method [7]. Experimental relations between the air renovation rate and the following variables have been analysed.
- The difference between the indoor and outdoor air temperatures ($T_i - T_e$).
- The wind speed (W).
- The product of the wind speed and the difference between the indoor and outdoor air temperatures ($W(T_i - T_e)$).
- The product of the wind speed raised to two and the difference between the indoor and outdoor air temperatures ($W^2(T_i - T_e)$).
- The atmospheric pressure (P_{atm}).
- The absolute value of the variation of wind speed per unit of time ($|dW/dt|$).

2.3.2. Analysis of Feasibility to Obtain Air Renovation Rate from Wall Mounted CO_2 Sensors

Additionally, the reliability of an alternative method based on the evolution of the metabolic CO_2 using wall mounted sensors of CO_2 concentration is evaluated in a room of the office building. A reference value ($CO_{2infinite}$) has been used, such that the variable used for the Decay method is the $CO_2-CO_{2infinite}$. This value was obtained as the average of the CO_2 concentration in a period when the room is positively non-occupied (from 9 pm to 7 am), starting when the Decay curve has reached its asymptotic value. An error obtained as the percentage of deviation regarding the reference value (based on N_2O), has been represented as function of the maximum value of the CO_2 concentration at the beginning of the decay method curve.

3. Results and Discussion

A reference value has been obtained for each of the considered case studies. These reference values have been obtained using a N_2O tracer gas applying the Decay method. The measurements carried out for the different case studies, presented in Figure 2, evidence that the air renovation rates are different for the different case studies.

The air renovation rates obtained from these tests are:

- PASLINK Test Cell: 0.056 renovations/hour.
- Single-zone building: 0.308 renovations/hour.
- Office building without mechanical ventilation: 0.825 renovations/hour.
- Office building with the mechanical ventilation active: 2.12 renovations/hour.

The dependence of these infiltration rates on the considered climate variables, and the feasibility to obtain them from the concentration of the metabolic CO_2, is discussed in the next subsections.

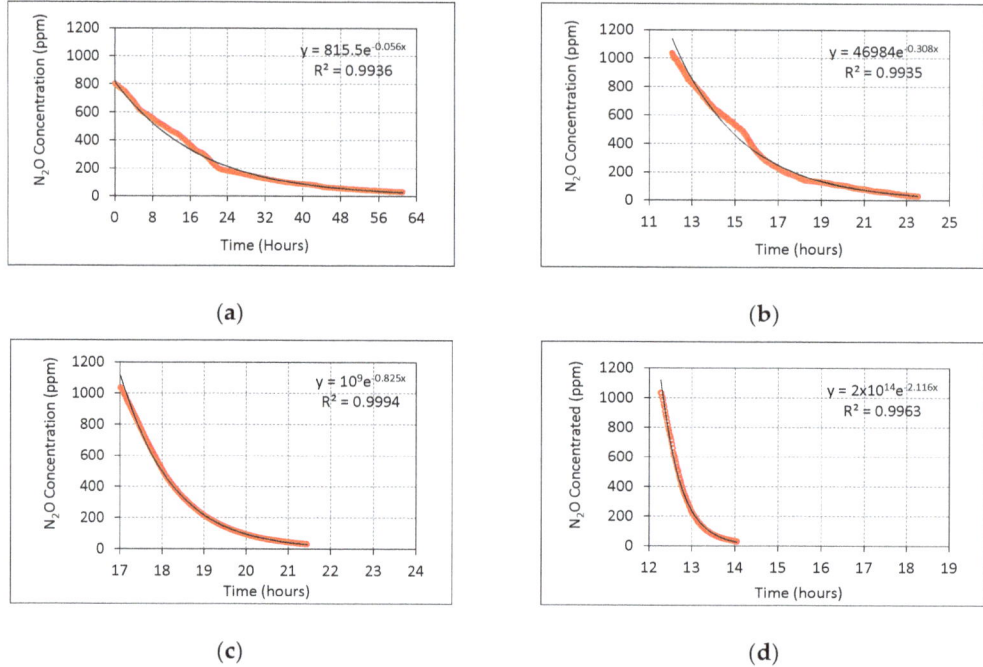

Figure 2. Decay method based on N$_2$O as tracer gas, applied to the three case studies: (**a**) PASLINK Test Cell (08/10/2018–11/10/2018); (**b**) Single-zone building 24/02/2016; (**c**) Office building without mechanical ventilation (10/02/2017); (**d**) Office building with mechanical ventilation (02/02/2017).

3.1. PASLINK Test Cell. Infiltrations

As expected, very low infiltration rates have been obtained for all the tests carried out in the PASLINK Test Cell. These results are shown in Figure 3 and Table 1. In this case, the infiltration rate does not show any relevant correlation with the indoor to outdoor air temperature difference (Figure 3a). This correlation also is not relevant with the atmospheric pressure (Figure 3d). However, the air infiltration rate presents some correlation with other considered variables. It shows significant linear dependency on the wind speed (Figure 3e), and the dependency is remarkable on the absolute value of the variation of the wind speed per unit of time (Figure 3f).

3.2. Single-Zone Building. Infiltrations

The results obtained for the single zone building are summarised in Table 2. This table shows that the air renovation rate (n) presents a large variation in the range 0.16–0.97 renov/hour. Its average is 0.37 renov/hour, and its standard deviation is 0.26 renov/hour. Figure 4a,c,e,g,i) shows that the n value has evident correlation with all the considered boundary variables except the atmospheric pressure (Figure 4g). The most relevant correlation detected is regarding the wind speed (Figure 4c). The absolute value of the variation of the wind speed per unit of time is also relevant (Figure 4i).

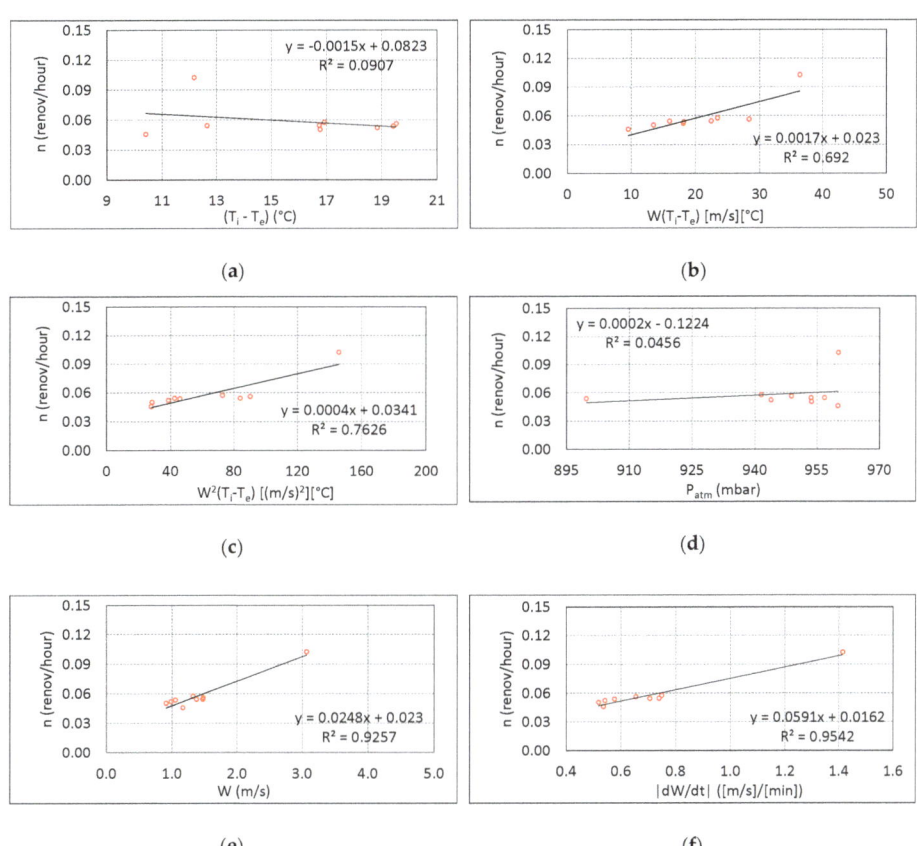

Figure 3. PASLINK Test Cell. Relations between the air renovation rate and the climatic variables. N_2O tracer gas measurement taken as reference. (**a**) Indoor to outdoor air temperature difference; (**b**) product of the indoor to outdoor air temperature difference and the wind speed; (**c**) product of the indoor to outdoor air temperature difference and the wind speed raised to two; (**d**) atmospheric pressure; (**e**) wind speed; (**f**) absolute value of the variation of wind sped per unit of time.

Table 1. PASLINK Test Cell. Experimentally determined air infiltration rates and climate variables.

Days In 2018	n (N_2O) [1] (ren/h)	r^2 (N_2O) [1] (·)	$T_i - T_e$ (°C)	W (m/s)	$W(T_i - T_e)$ [m/s] [°C]	$W^2(T_i - T_e)$ [(m/s)^2] [°C]	P_{atm} (mbar)
24/09–26/09	0.1022	0.9834	12.2	3.07	36.39	145.7	960
27/09–29/09	0.0543	0.9778	16.7	1.48	22.46	83.9	957
02/10–04/10	0.0536	0.9917	19.4	1.05	18.20	45.9	900
05/10–07/10	0.0503	0.9983	16.8	0.91	13.45	28.3	954
08/10–11/10	0.0560	0.9936	19.5	1.48	28.35	90.3	949
12/10–14/10	0.0543	0.9829	12.7	1.38	15.94	42.4	954
23/10–25/10	0.0458	0.9367	10.4	1.17	9.52	27.8	960
26/10–29/10	0.0574	0.9882	16.9	1.33	23.45	72.9	941
29/10–01/11	0.0520	0.9979	18.8	0.98	18.11	38.6	944

[1] The (N_2O) indicates that the values included in the column were obtained using the N_2O tracer gas.

3.3. Office Building Prototype

3.3.1. Infiltrations

The results obtained for the studied room are summarised in Figure 4b,d,f,h,j and Table 3. Considering the analysis based on N_2O, the air renovation rate (n) presents some variation. However, the observed variation is not so large as in the single-zone building. The n value is between 0.61 and 0.75 renov/hour. Its average is 0.67 renov/hour, and it standard deviation is 0.05 renov/hour. Figure 4b,d,f,h,j shows that the n value has relevant correlation with all the considered boundary variables except the indoor to outdoor air temperature difference and the atmospheric pressure. The most relevant correlation detected is regarding the wind speed (Figure 4d).

It is noticeable the different behaviour observed for the dependence of the n value with the indoor-outdoor temperature difference in this heated room regarding the single zone free running building. The n value for the heated office does not show relevant dependence with this variable (Figure 4b). This behaviour is also observed in the Test Cell, also heated during the test campaign, that does not show relevant dependence with this variable (Figure 3a). However, a linear tendency is seen for the free running single-zone building (Figure 4a). This different behaviour could be explained by the different ranges of indoor-outdoor air temperature differences in the case studies (Figures 3a and 4a,b).

Acceptable agreement is observed for the values obtained using the metabolic $CO_{2\,ref}$ concentration, measured with the wall-mounted sensors, regarding the reference n values based on N_2O (Table 3 and Figure 5a). The agreement is very poor when the less accurate CO_2 sensor is used (Table 3 and Figure 5a). This behaviour is explained by taking into account that the office has just one user, and consequently, the level of CO_2 concentration produced by the metabolic activity is very low, which is leading to relevant uncertainties in the estimations of the n values if the used sensor does not have enough resolution. These uncertainties show a decreasing tendency when the CO_2 concentration increases (Figure 5b). Taking into account this behaviour a better performance of this sensor is foreseen for larger CO_2 concentrations that would be present in rooms with more occupants. This issue will be further investigated.

Table 2. Single-zone building. Experimentally determined air infiltration rates and climate variables.

Day	n (N_2O) [1] (ren/h)	r^2 (N_2O) [1] (·)	$T_i - T_e$ (°C)	W (m/s)	$W(T_i - T_e)$ [m/s] [°C]	$W^2(T_i - T_e)$ [(m/s)^2] [°C]	P_{atm} (mbar)
09/02/2016	0.74	0.9756	−2.49	9.20	−21.88	−214.9	956
10/02/2016	0.60	0.9125	−0.79	10.27	−9.58	−121.6	954
11/02/2016	0.50	0.9708	0.28	9.27	2.66	25.8	951
12/02/2016	0.97	0.9942	−0.61	11.55	−7.07	−87.0	951
15/02/2016	0.22	0.9776	9.44	3.75	32.07	157.8	952
16/02/2016	0.19	0.9874	12.07	2.45	25.39	78.9	960
17/02/2016	0.31	0.9550	9.54	4.48	39.96	201.3	955
18/02/2016	0.16	0.9965	9.57	3.04	29.71	107.1	955
19/02/2016	0.22	0.9936	9.68	4.44	42.93	213.0	958
22/02/2016	0.16	0.9666	6.58	2.84	18.40	78.8	958
23/02/2016	0.17	0.9976	10.42	2.21	18.26	46.1	959
24/02/2016	0.31	0.9935	6.30	4.73	26.75	131.4	952
25/02/2016	0.22	0.9608	10.12	3.82	29.83	141.6	952

[1] The (N_2O) indicates that the values included in the column were obtained using the N_2O tracer gas.

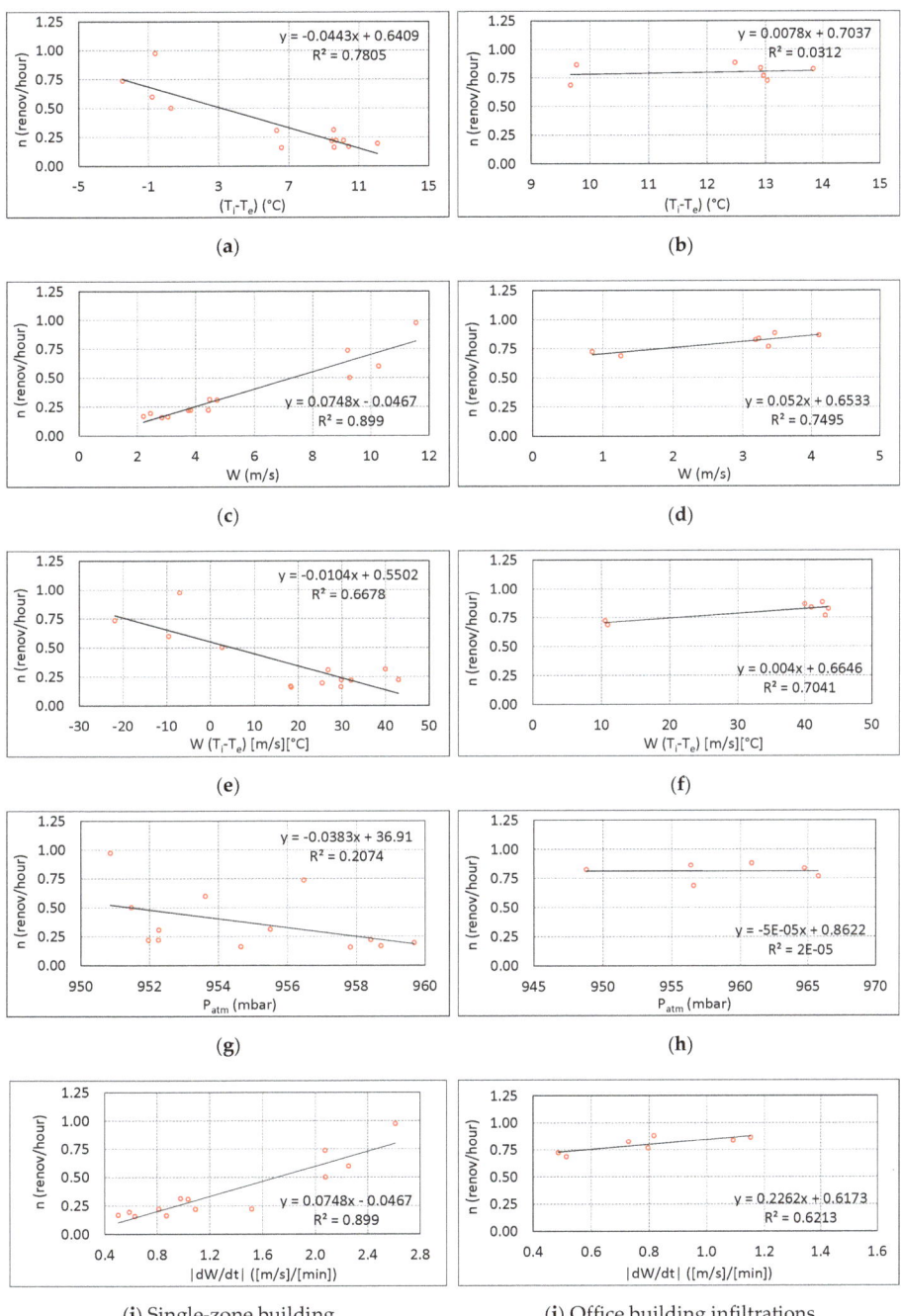

Figure 4. Relations between the air renovations and the climatic variables. Left: single-zone building. Right: Room of the office building. (**a,b**) Indoor to outdoor air temperature difference; (**c,d**) wind speed; (**e,f**) product of the indoor to outdoor air temperature difference and the wind speed; (**g,h**) product of the indoor to outdoor air temperature difference and the wind speed raised to two; (**i,j**) absolute value of the variation of the wind sped per unit of time.

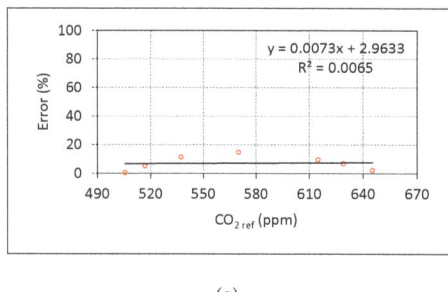
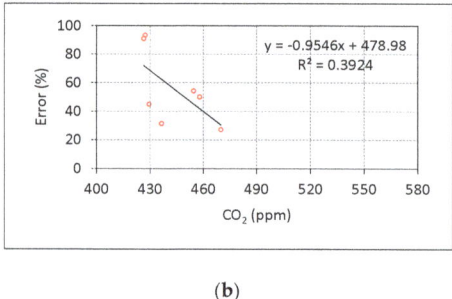

(a)   (b)

Figure 5. Office number 1, analysis of infiltrations. Percentage of error of the results obtained from the Decay method using the metabolic CO_2 concentration and considering as reference the value obtained from the N_2O tracer gas. (**a**) Using the CO_2 reference sensor; (**b**) using the cheaper CO_2 sensor.

3.3.2. Mechanical Ventilation

The results obtained for the studied room are summarised in Tables 4 and 5. Considering the analysis based on N_2O, the air renovation rate (n) presents a large variation. It is between 0.95 and 3.08 renov/hour. Its average is 1.98 renov/hour which is very close to the design value (2 renov/hour), and it standard deviation is 0.59 renov/hour. However, the n value does not show relevant correlation with any of the considered boundary variables. The observed large spread could be caused by the instability of the electricity that powers the mechanical ventilation system that transmits such instability to the ventilation rate. Other effects, such as hysteresis of the mechanical components of the ventilation system could contribute to produce the detected variations. The causes of the detected large spread will be further investigated in future research works.

Large uncertainties are observed for the values obtained using the metabolic CO_2 concentration measured with the wall-mounted sensors (Tables 4 and 5 and Figure 6). These uncertainties are remarkably larger than those observed for the same room without mechanical ventilation (Figure 5). This high uncertainty is attributed the low level of metabolic CO_2 concentration produced by just one user. This issue also leads to large uncertainties in the air renovation rate obtained for the same room without mechanical ventilation using the less accurate sensor (Figure 5b). However, such uncertainty is worsened in the case of mechanical ventilation taking into account that the time interval available for each calculation of the n value is shortened regarding the case of not using mechanical ventilation.

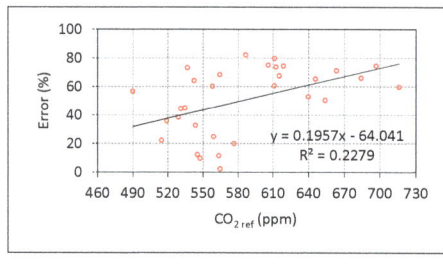
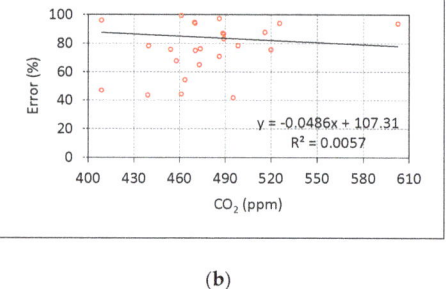

(a)   (b)

Figure 6. Office number 1, tests with mechanical ventilation active. Percentage of error of the results obtained from the Decay method using the metabolic CO_2 concentration and considering as reference the value obtained from the N_2O tracer gas. (**a**) Using the CO_2 reference sensor; (**b**) using the cheaper CO_2 sensor.

Table 3. Office number 1. Experimentally determined air infiltration rates, climate variables and deviations between the results obtained using the metabolic CO_2 concentration and the N_2O tracer gas.

Day	n (N_2O)[1] (ren/h)	n (CO_2)[1] (ren/h)	n (CO_{2_ref})[1] (ren/h)	r^2 (N_2O)[1] (-)	r^2 (CO_2)[1] (-)	r^2 (CO_{2_ref})[1] (-)	$T_i - T_e$ (°C)	W (m/s)	$W(T_i - T_e)$ [m/s][°C]	$W^2(T_i - T_e)$ [(m/s)^2][°C]	P_{atm} (mbar)	$CO_{2ref.max}$ (ppm)	Error (CO_{2_ref})[1] (%)	CO_{2max} (ppm)	Error (CO_2)[1] (%)
09/02/2017	0.72	0.07	0.72	0.9994	0.0334	0.9639	13.03	0.85	10.5	13.5	800	506	0.2	426	91.0
10/02/2017	0.83	0.57	0.78	0.9994	0.7564	0.9864	13.83	3.18	43.5	161.5	949	517	5.1	436	31.4
14/02/2017	0.88	0.64	0.82	0.9997	0.7520	0.9923	12.47	3.46	42.6	161.7	961	629	6.9	470	27.3
15/02/2017	0.77	0.38	0.78	0.9991	0.7900	0.9909	12.96	3.37	43.0	156.5	966	645	2.2	458	50.1
16/02/2017	0.84	0.38	0.74	0.9996	0.8052	0.9913	12.91	3.22	40.9	149.3	965	538	11.3	454	54.3
21/02/2017	0.86	0.48	0.78	0.9990	0.6157	0.9888	9.77	4.10	40.0	178.0	956	615	9.5	430	44.8
01/03/2017	0.69	0.05	0.59	0.9941	0.0446	0.9677	9.67	1.26	10.9	19.8	957	570	14.7	427	93.3

[1] The (N_2O), (CO_2) and (CO_{2_ref}) indicate that the values included in the column refer to the measurements using the N_2O tracer gas, the CO_2 or the CO_{2_ref} devices respectively.

Table 4. Office number 1, test 1 for each day. Experimentally determined air infiltration rates when the mechanical ventilation is active, climate variables and deviations between the results obtained using the metabolic CO_2 concentration and the N_2O tracer gas.

Day	n (N_2O) [1] (ren/h)	n (CO_2) [1] (ren/h)	n (CO_{2_ref}) [1] (ren/h)	r^2 (N_2O) [1] (-)	r^2 (CO_2) [1] (-)	r^2 (CO_{2_ref}) [1] (-)	$T_i - T_e$ (°C)	W (m/s)	$W(T_i - T_e)$ [m/s][°C]	$W^2(T_i - T_e)$ [(m/s)²][°C]	P_{atm} (mbar)	$CO_{2ref.max}$ (ppm)	Error (CO_{2_ref}) [1] (%)	CO_{2max} (ppm)	Error (CO_2) [1] (%)
31/01/2017	2.41	0.41	0.65	0.9983	0.7064	0.9819	2.12	1.06	2.3	3.2	953	537	73	489	83
01/02/2017	2.41	1.10	0.76	0.9982	0.8000	0.9872	2.08	2.19	4.5	11.6	955	564	68	463	54
02/02/2017	2.12	0.26	0.73	0.9963	0.5889	0.9852	6.21	1.64	10.1	18.6	953	646	65	516	88
03/02/2017	2.70	−0.98	0.97	0.9984	0.2534	0.9405	1.97	5.04	9.9	54.5	956	543	64	433	136
07/02/2017	2.61	0.63	1.05	0.9971	0.8246	0.9941	0.30	4.61	0.8	3.7	958	716	60	520	76
08/02/2017	2.45	0.07	0.50	0.9959	0.0427	0.9694	5.74	3.91	22.3	98.9	800	612	80	486	97
09/02/2017	2.50	0.61	0.81	0.9973	0.7653	0.9795	7.63	3.89	29.7	128.6	800	615	68	454	76
10/02/2017	2.31	0.30	0.91	0.9954	0.6379	0.9789	9.50	1.76	16.9	44.5	949	611	61	488	87
13/02/2017	3.08	−0.11	1.23	0.9993	0.0015	0.9823	6.44	3.81	24.5	111.8	947	558	60	436	104
14/02/2017	2.61	0.14	0.68	0.9966	0.2700	0.9487	7.32	2.92	21.4	71.9	959	613	74	470	95
15/02/2017	2.41	0.14	0.69	0.9973	0.2015	0.9894	7.77	4.32	33.6	159.1	964	663	71	525	94
16/02/2017	2.39	−0.04	0.61	0.9959	0.0133	0.9682	9.06	5.74	52.0	323.7	965	619	74	446	102
17/02/2017	1.51	0.33	0.71	0.9965	0.7170	0.9889	9.60	2.07	19.9	50.8	962	640	53	498	78
21/02/2017	2.41	1.34	0.60	0.9935	0.6193	0.9612	8.13	8.48	68.9	605.3	957	606	75	461	44
22/02/2017	2.26	0.54	0.58	0.9955	0.7501	0.9738	5.63	5.94	33.4	207.4	954	697	74	473	76
23/02/2017	1.57	−0.65	0.77	0.9984	0.0542	0.9751	6.17	4.87	29.9	166.5	948	654	51	433	142
24/02/2017	1.54	0.09	0.52	0.9995	0.0588	0.9776	6.92	1.39	9.5	18.8	950	684	66	603	94
02/03/2017	2.64	0.02	0.47	0.9980	0.0014	0.9794	1.47	1.57	2.3	4.4	955	587	82	461	99

[1] The (N_2O), (CO_2) and (CO_{2_ref}) indicate that the values included in the column refer to the measurements using the N_2O tracer gas, the CO_2 or the CO_{2_ref} devices respectively.

Table 5. Office number 1, test 2 for each day. Experimentally determined air infiltration rates when the mechanical ventilation is active, climate variables and deviations between the results obtained the using the metabolic CO_2 concentration and the N_2O tracer gas.

Day	n (N_2O)[1] (ren/h)	n (CO_2)[1] (ren/h)	n (CO_{2_ref})[1] (ren/h)	r^2 (N_2O)[1] (-)	r^2 (CO_2)[1] (-)	r^2 (CO_{2_ref})[1] (-)	$T_i - T_e$ (°C)	W (m/s)	$W(T_i - T_e)$ [m/s][°C]	$W^2(T_i - T_e)$ [(m/s)²][°C]	P_{atm} (mbar)	$CO_{2ref.max}$ (ppm)	Error (CO_{2_ref})[1] (%)	CO_{2max} (ppm)	Error (CO_2)[1] (%)
31/01/2017	2.11	0.29	1.42	0.9979	0.7698	0.9911	6.75	2.12	14.1	32.6	954	544	33	489	86
01/02/2017	1.96	−0.29	1.21	0.9949	0.3562	0.9823	5.47	1.11	5.5	8.0	955	529	38	443	115
02/02/2017	2.00	0.12	1.10	0.9960	0.5396	0.9894	6.08	1.03	6.2	7.8	954	535	45	470	94
06/02/2017	2.25	0.56	0.98	0.9949	0.7154	0.9835	2.36	3.37	6.4	24.7	960	490	56	470	75
07/02/2017	1.34	−0.17	0.86	0.9988	0.1874	0.9901	4.10	3.30	13.7	53.2	957	519	36	461	113
08/02/2017	1.33	0.43	1.04	0.9997	0.6881	0.9916	8.91	1.65	14.0	30.0	800	515	22	458	68
13/02/2017	0.95	0.04	0.76	0.9826	0.0970	0.9128	8.90	2.55	22.4	69.2	951	577	20	409	96
17/02/2017	1.22	0.71	1.07	0.9767	0.8458	0.9208	11.91	2.65	31.2	94.4	961	545	12	495	42
20/02/2017	0.95	0.51	1.38	0.9992	0.7741	0.9104	11.29	7.22	81.3	618.8	958	531	44	409	47
22/02/2017	1.12	0.39	0.99	0.9844	0.7994	0.9958	8.21	3.54	29.0	121.8	952	563	11	473	65
23/02/2017	1.42	0.41	1.07	0.9983	0.7682	0.9951	9.87	1.74	16.9	34.1	948	559	25	486	71
24/02/2017	1.36	0.30	1.33	0.9800	0.1849	0.9935	9.48	1.02	9.3	12.4	951	565	2	440	78
02/03/2017	1.46	0.83	1.60	0.9965	0.3075	0.9893	6.54	1.33	7.9	13.2	954	547	9	439	43

[1] The (N_2O), (CO_2) and (CO_{2_ref}) indicate that the values included in the column refer to the measurements using the N_2O tracer gas, the CO_2 or the CO_{2_ref} devices respectively.

4. Conclusions

This section summarises the conclusions regarding the behaviour of the air renovation rate and discusses the effect of this behaviour on the experimental assessment of the Heat Transfer Coefficient (HLC).

The following conclusions are extracted regarding the air renovation rate from the different tests carried out:

- When the mechanical ventilation is not active: Significant correlation between air renovation rate and the wind speed has been observed in both buildings and the Test Cell. The agreement between the values obtained using N_2O and the evolution of metabolic CO_2 increases when the starting value of CO_2 concentration increases.
- When the mechanical ventilation is active: Large variations have been observed among the different values obtained along the test campaign using N_2O tracer gas. However, these values do not show any correlation with any of the considered climate variables. Consequently, the observed spread has been used to estimate an uncertainty of the air renovations rate. The measurements based on CO_2 concentrations do not show good agreement to the values obtained using N_2O tracer gas. This issue will be further investigated, but in principle it is attributed to the low level of CO_2 measured along the analysed test campaign when the mechanical ventilation is active. This explanation is in agreement with previous works carried out regarding the air renovation in the same building [15].

The behaviour observed in the air renovation rate, showing large variability considering infiltrations and also considering mechanical ventilation, contributes to understand the behaviour of the HLC experimentally assessed and its uncertainties. The following text summarises the conclusions extracted from this work and some ideas for further research, regarding the influence of the air renovation rate on the behaviour of the Heat Loss Coefficient (HLC):

- Regarding infiltrations, the dependencies of the n value with the wind speed and its variation per unit of time in absolute value, can explain some variability of the HLC and some uncertainty when it is assumed as a constant value. Further analysis of this wind dependence is an interesting issue regarding future research works that could lead to a wind dependent HTC reducing the uncertainties of this coefficient in experimental assessments.
- The behaviour observed in the n value for the case of mechanical ventilation leads to conclude that the experimental assessment of an HLC assuming n as constant could lead to some degree of uncertainty. The work presented in this paper has not identified any variable that could contribute to model such variability reducing the associated uncertainty. This issue is identified as a relevant topic regarding future research.

Author Contributions: Measurements, S.C.; Data curation, A.J.A. and J.A.D.; Data analysis, elaboration of graphs and synthesis of results, A.J.A., J.A.D. and M.P.; writing—review and editing M.J.J. and M.P., Methodology and writing—original draft preparation, M.J.J. All authors have read and agreed to the published version of the manuscript.

Funding: This research was funded by the Spanish National Research Agency (Agencia Estatal de Investigación) through the In-Situ-BEPAMAS project, reference PID2019-105046RB-I00. Additionally, the operation of the test facilities that supported this study was partially funded by the Spanish Ministry of Economy, Industry and Competitiveness through ERDF funds (SolarNOVA-II project Ref. ICTS-2017-03-CIEMAT-04).

Conflicts of Interest: The authors declare no conflict of interest.

References

1. EBC Executive Committtee. *Strategic Plan 2019–2024. Energy in Buildings and Communities Technology Collaboration Programm*; International Energy Agency: Paris, France, 2019.
2. Tian, W.; Heo, Y.; de Wilde, P.; Li, Z.; Yan, D.; Park, C.S.; Feng, X.; Augenbroe, G. A review of uncertainty analysis in building energy assessment. *Renew. Sustain. Energy Rev.* **2018**, *93*, 285–301. [CrossRef]
3. Kampelis, N.; Gobakis, K.; Vagias, V.; Kolokotsa, D.; Standardi, L.; Isidori, D.; Cristalli, C.; Montagnino, F.M.; Paredes, F.; Muratore, P.; et al. Evaluation of the performance gap in industrial, residential & tertiary near-Zero energy buildings. *Energy Build.* **2017**, *148*, 58–73. [CrossRef]
4. Annex 71 of the Programme "(EBC)" of the IEA on EBC Annex 71. Building Energy Performance Assessment Based on In-Situ Measurements. 2016–2021. Available online: http://www.iea-ebc.org/projects/project?AnnexID=71 (accessed on 4 December 2020).
5. Jiménez, M.J. IEA, EBC annex 58, report of subtask 3, part 1. In *Thermal Performance Characterization Based on Full Scale Testing–Description of the Common Exercises and Physical Guidelines*; Jiménez, M.J., Ed.; KU Leuven: Leuven, Belgium, 2016; ISBN 9789460189876. Available online: http://www.iea-ebc.org/Data/publications/EBC_Annex_58_Final_Report_ST3a.pdf (accessed on 3 November 2020).
6. Marin, M.; Vlase, S.; Paun, M. Considerations on double porosity structure for micropolar bodies. *AIP Adv.* **2015**, *5*, 037113. [CrossRef]
7. Sherman, M.H. On the estimation of multizone ventilation rates from tracer gas measurements. *Build. Environ.* **1989**, *24*, 355–362. [CrossRef]
8. ISO 9972:2015. *Thermal Performance of Buildings–Determination of Air Permeability of Buildings–Fan Pressurization Method*; ISO: Geneva, Switzerland, 2015.
9. Olazo-Gómez, Y.; Herrada, H.; Castaño, S.; Arce, J.; Xamán, J.P.; Jiménez, M.J. Data-based RC dynamic modelling to assessing the in-situ thermal performance of buildings. Analysis of several key aspects in a simplified reference case toward the application at on-board monitoring level. *Energies* **2020**, *13*, 4800. [CrossRef]
10. Díaz-Hernández, H.P.; Torres-Hernández, P.R.; Aguilar-Castro, K.M.; Macias-Melo, E.V.; Jiménez, M.J. Data-based RC dynamic modelling incorporating physical criteria to obtain the HLC of in-use buildings: Application to a case study. *Energies* **2020**, *13*, 313. [CrossRef]
11. Remion, G.; Moujalled, B.; El Mankibi, M. Review of tracer gas-based methods for the characterization of natural ventilation performance: Comparative analysis of their accuracy. *Build. Environ.* **2019**, *160*, 106180. [CrossRef]
12. Carrié, F.R.; Leprince, V. Uncertainties in building pressurisation tests due to steady wind. *Energy Build.* **2016**, *116*, 656–665. [CrossRef]
13. Farmer, D.; Johnston, D.; Miles-Shenton, D. Obtaining the heat loss coefficient of a dwelling using its heating system (integrated coheating). *Energy Build.* **2016**, *117*, 1–10. [CrossRef]
14. Marshall, A.; Fitton, R.; Swan, W.; Farmer, D.; Johnston, D.; Benjaber, M.; Ji, Y. Domestic building fabric performance: Closing the gap between the in situ measured and modelled performance. *Energy Build.* **2017**, *150*, 307–317. [CrossRef]
15. Enríquez, R.; Bravo, D.; Díaz, J.A.; Jiménez, M.J. Mechanical ventilation performance assessment in several office buildings by means of Big Data techniques. In Proceedings of the 36th AIVC Conference "Effective Ventilation in High Performance Buildings", Madrid, Spain, 23–24 September 2015; ISBN 2-930471-45-X.
16. Baker, P.H.; van Dijk, H.A.L. PASLINK and dynamic outdoor testing of building components. *Build. Environ.* **2008**, *43*, 127–128. [CrossRef]
17. Castaño, S.; Guzmán, J.D.; Jiménez, M.J.; Heras, M.R. LECE-UiE3-CIEMAT. In *Report of Subtask 1a: Inventory of Full Scale Test Facilities for Evaluation of Building Energy Performances. IEA EBC Annex 58*; Janssens, A., Ed.; KU Leuven: Leuven, Belgium, 2016; ISBN 9789460189906.
18. Jiménez, M.J.; Porcar, B.; Heras, M.R. Estimation of UA and gA values of building components from outdoor tests in warm and moderate weather conditions. *Solar Energy* **2008**, *82*, 573–587. [CrossRef]
19. Olmedo, R.; Sánchez, M.N.; Enríquez, R.; Jiménez, M.J.; Heras, M.R. ARFRISOL Buildings-UIE3-CIEMAT. In *Report of Subtask 1a: Inventory of Full Scale Test Facilities for Evaluation of Building Energy Performances. IEA EBC Annex 58*; Janssens, A., Ed.; KU Leuven: Leuven, Belgium, 2016; ISBN 9789460189906.

Article

On the Retrofit of Existing Buildings with Aerogel Panels: Energy, Environmental and Economic Issues

Paola Marrone [1]**, Francesco Asdrubali** [2,*]**, Daniela Venanzi** [3]**, Federico Orsini** [1]**, Luca Evangelisti** [2]**, Claudia Guattari** [2]**, Roberto De Lieto Vollaro** [2]**, Lucia Fontana** [1]**, Gianluca Grazieschi** [2]**, Paolo Matteucci** [3] **and Marta Roncone** [2]

[1] Department of Architecture, Roma TRE University, via Madonna dei Monti 40, 00184 Rome, Italy; paola.marrone@uniroma3.it (P.M.); federico.orsini@uniroma3.it (F.O.); lucia.fontana@uniroma3.it (L.F.)

[2] Department of Engineering, Roma TRE University, via Vito Volterra 62, 00146 Rome, Italy; luca.evangelisti@uniroma3.it (L.E.); claudia.guattari@uniroma3.it (C.G.); roberto.delietovollaro@uniroma3.it (R.D.L.V.); gianluca.grazieschi@uniroma3.it (G.G.); marta.roncone@uniroma3.it (M.R.)

[3] Department of Economics, Roma TRE University, via Silvio D'Amico 77, 00145 Rome, Italy; daniela.venanzi@uniroma3.it (D.V.); paolo.matteucci@uniroma3.it (P.M.)

* Correspondence: francesco.asdrubali@uniroma3.it

Citation: Marrone, P.; Asdrubali, F.; Venanzi, D.; Orsini, F.; Evangelisti, L.; Guattari, C.; De Lieto Vollaro, R.; Fontana, L.; Grazieschi, G.; Matteucci, P.; et al. On the Retrofit of Existing Buildings with Aerogel Panels: Energy, Environmental and Economic Issues. *Energies* **2021**, *14*, 1276. https://doi.org/10.3390/en14051276

Academic Editor: Paulo Santos

Received: 19 January 2021
Accepted: 22 February 2021
Published: 25 February 2021

Publisher's Note: MDPI stays neutral with regard to jurisdictional claims in published maps and institutional affiliations.

Copyright: © 2021 by the authors. Licensee MDPI, Basel, Switzerland. This article is an open access article distributed under the terms and conditions of the Creative Commons Attribution (CC BY) license (https://creativecommons.org/licenses/by/4.0/).

Abstract: Among the super insulating materials, aerogel has interesting properties: very low thermal conductivity and density, resistance to high temperatures and transparency. It is a rather expensive material, but incentives in the field can improve its economic attractiveness. Starting from this, the thermal behavior of a test building entirely insulated with aerogel panels was investigated through an extended experimental campaign. A dynamic simulation model of a case study building was generated to better comprehend the energy savings obtained through aerogel in terms of energy demand over a whole year. The investigation was completed by computing the carbon and energy payback times of various retrofit strategies through a life cycle assessment approach, as well as by a cost-benefit analysis through a probabilistic financial framework. Compared to conventional insulation materials, aerogel is characterized by a higher energy and carbon payback time, but it guarantees better environmental performance in the whole life cycle. From an economic-financial perspective, the aerogel retrofit is the best in the current tax incentive scenario. However, due to its higher lump-sum investment, aerogel's net present value is very sensitive to tax deductions, and it is riskier than the best comparable materials in less favorable tax scenarios.

Keywords: aerogel; thermal behavior; dynamic simulation; retrofitting; LCA; economic analysis

1. Introduction

Climate-changing gases (GHG), mainly produced by anthropogenic activities, are now considered to be the main responsible factor for the global warming; in fact, the global average temperature has increased by about 1 °C compared to the pre-industrial era [1]. Consequently, due to global warming and climate change (CC), large and densely populated areas risk becoming inhospitable [2]. To avoid, or at least reduce, the negative effects of climate change, it is necessary to globally modify the development model aiming at reducing GHG emissions [3,4].

This objective can be pursued by promoting renewable resources [5], inspiring economic development to the principles of the circular economy [6–9], producing low-carbon materials [10,11] and reducing energy consumption [12]. In this context, cities and buildings play a fundamental role. They are among the main responsible factors for energy consumption and GHG emissions (over 30% of the total amount), mainly caused by urban and extra-urban transport, buildings' energy needs (electrical appliances, heating and cooling) and the production of construction materials [13].

The European Green Deal set the target of reducing GHG emissions by at least 50–55% below the levels of 1990 by 2030 [14].

Among the possible actions encouraged by this agreement, strategies to reduce buildings' energy consumption can be listed. In particular, policies targeted at improving the thermal performance of buildings' envelopes [15,16] in terms of reducing heat loss and increasing thermal lag can be implemented.

Alongside the traditional insulator materials [17] made of inorganic constituents (for example, rock wool, expanded polystyrene, etc.), or organic ones (cork, wood fibers, etc.), today non-traditional materials, defined as super-insulating, made with innovative production processes and/or materials, are available [18].

The super-insulating materials are characterized by high performance with a thermal conductivity value lower than 0.020 W/mK, compared to traditional materials (rock wool or glass wool) whose thermal conductivity is equal to 0.035–0.040 W/mK. Comparison can also be made with transition materials, such as expanded polyurethane or propylene, characterized by thermal conductivity values ranging between 0.02 and 0.03 W/mK [19,20]. Furthermore, the high thermal performance of non-traditional insulation materials is characterized by a significant reduction in their thicknesses compared to the traditional ones. Different kinds of innovative and high-insulating materials have already been studied by researchers: reflective multilayer insulation [21–23], vacuum insulation panels (VIPs) [24] and gas-filled panels [25].

Among these innovative materials, aerogel appears to be of great interest, ranking among the most interesting innovative products for the near future [26]. Discovered in the early 1930s [27,28], aerogel is a porous synthetic product, in which the gel's liquid component is replaced with a gas. This solution allowed the creation of a highly performing material in terms of thermal insulation, with a thermal conductivity of about 0.013 W/mK. In fact, several studies have highlighted its excellent thermal performance for opaque envelope applications, integrated in panels [29,30] or mortars [31], and for translucent applications, integrated in panels and frames [32,33]. Cuce et al. [34] presented a comprehensive review on aerogel utilization in buildings: the applications range from energy insulation purposes, to sound insulation, fire retardation and air purification. The use of aerogel in retrofits of historical buildings is very competitive since it permits saving inner space, maintaining the external façades unaltered [35]. Karim et al. [36] proposed a super-insulated plaster made with aerogel particles mixed in the matrix. Finally, their optical properties permit the integration of aerogels in different types of glazing systems [37].

If, on one hand, the high performance of aerogel is nowadays well-known owing to several studies, on the other hand, these studies have also focused on the high cost of this material [35]. Nowadays, this aspect is considered as a very strong limit to its widespread application to the construction sector [38].

However, in Italy this limit can now be partially overcome thanks to the introduction in Italian law of a tax credit of 90% (so-called "bonus façades") for the costs incurred for the retrofit of the building façades (see Budget Law 2020). The standard also includes energy retrofit interventions that meet the so-called minimum requirements and the thermal transmittance limit values of the building envelope [39]. Another incentive that is today guaranteed by the Italian legislation is the so called Superbonus 110% (a tax reduction of the 110% of the expenditure sustained for the works aiming at deep energy retrofits of existing buildings [40]). The insulation of building envelopes is one of the driving interventions that are promoted by Superbonus 110%.

Starting from this, it seems important to evaluate the possibility of employing aerogel for the energy retrofit of existing buildings in order to define its effectiveness in terms of both thermal performance and economic feasibility.

This paper has the following structure: Section 2 provides the aim and scope of the research; Section 3 provides some information about the test rooms and the case study, the experimental campaign in the test rooms, data post-processing, simulations in a case study building and the cost–benefit analysis; Section 4 presents the results; finally, Section 5 draws conclusions.

2. Aim and Scope

The thermal performance of aerogel is well-known in the literature. It is a super insulating material able to improve the thermal performance of a wall with reduced thicknesses. On the other hand, aerogel is a rather expensive material, and its use needs a comparison between energy savings and installation costs in order to identify costs and benefits.

Thus, the aim of this study is a wide-ranging analysis, examining and comparing aerogel performance to that of other diffused insulation materials employed as an external insulation layer in regions characterized by a mild climate (central Italy). From an economic-financial point of view, the analysis here conducted applies a complete financial approach, based on a probabilistic method used to measure both the most probable value of the Net Present Value (NPV) of each retrofit alternative and its probability distribution (however, limited to the monetizable costs and benefits). This approach assumes optimistic and pessimistic estimates (defined in a subjective manner) of the uncertain variables and measures the corresponding range of NPV. Therefore, it derives (under some hypotheses) the variance of the NPV that allows obtaining its probability distribution.

The whole analysis was carried out in order to compare and quantify the advantages/disadvantages of employing aerogel instead of other insulation materials, also in the light of the Italian tax credit.

3. Materials and Methods

This work integrates four evaluation fields to assess the competitiveness of aerogel in comparison with other insulation materials. After an experimental campaign aiming at studying the real thermal performance of an aerogel coating insulation (described in Section 3.1), a simulation was carried out to evaluate the energy savings achievable by building retrofits using aerogel or other insulation materials (introduced in Section 3.2); the evaluation of the related environmental benefits in the life cycle (see Section 3.3) and the estimation of the achievable economic benefits (see Section 3.4) were finally performed.

3.1. The Experimental Campaign

The experimental measurement campaigns took place in the external area of the CEFME CTP school for construction, located in Pomezia, a small city close to Rome. According to Italian legislation, the climatic zone of Pomezia is D (on a scale from A to F, with A corresponding to the warmest places and F the coldest), with a degree day value equal to 1536. The experimental investigations involved two test rooms characterized by the same geometry, walls and roof stratigraphy, and the same orientation. One of them is not thermally insulated; the other is insulated with aerogel panels. It is worth noting that the two test rooms, despite their close proximity, do not cast shadows on one another. Figure 1a provides an aerial view of the construction site, and Figure 1b shows the geometrical characteristics of the investigated test rooms. Original vertical walls are characterized by brick construction technique, with plastered tuff blocks, reinforced concrete slabs and internal and external cladding with cement plaster. Table 1 lists the stratigraphy of the test rooms' components.

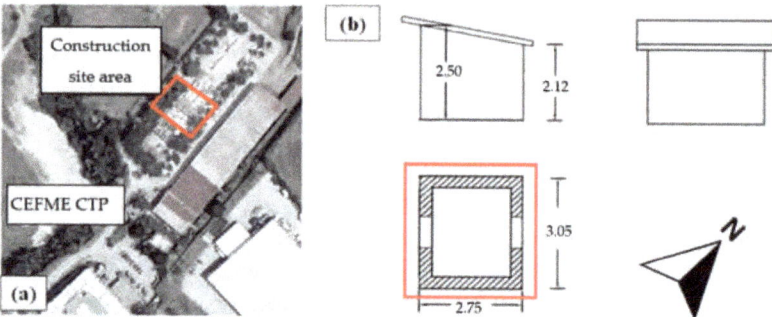

Figure 1. Aerial view of the construction site (**a**), geometrical characteristics of the monitored test rooms (**b**).

Table 1. Test Rooms' Components Stratigraphy.

Component	Material	Thickness [m]
External wall	External cement plaster	0.04
	Tuff blocks	0.26
	Internal cement plaster	0.04
Roof	Reinforced concrete slab	0.14
Ground floor	Reinforced concrete slab	0.12
Door	Oak wood	0.04

Sample images of the analyzed test rooms are reported in Figure 2, where it is possible to observe the external insulation system during installation (aerogel panels characterized by a thickness equal to 0.01 m) and after installation. The external insulation layer was realized with semi-rigid panels [41] (dimensions equal to 1400 × 720 mm), realized by means of a layer of silica aerogel reinforced with PET (Polyethylene terephthalate) fibers (felt), water-repellent and breathable, with mass density equal to 230 kg/m^3, thermal conductivity equal to 0.015 W/mK and specific heat capacity equal to 1000 J/kgK. The external finish of the coat was realized with cement fiber panels, which are also mounted with dowels.

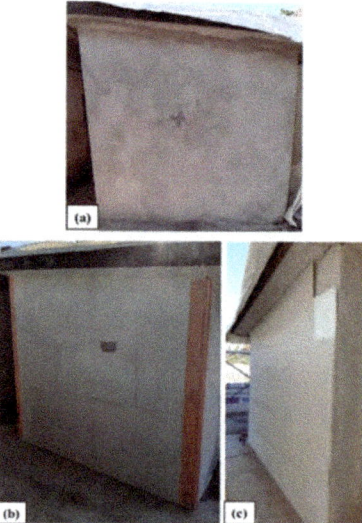

Figure 2. Selected test rooms in their original state (**a**) and during aerogel panel installation (**b**,**c**).

As already mentioned, one of the test rooms was monitored as a reference structure for measurements, without any thermal insulation. The other was fully insulated with 0.01-m-thick aerogel panels.

In order to assess the thermal behavior of the reference test room and the insulated one, a heat flow meter sensor and internal and external surface and air temperature probes were installed on the walls [42–45] facing north-west. In particular, heat flow meter sensors were installed on the inner side of the walls, and surface temperature probes were installed on the inner and outer sides of the walls. In addition, internal and external air temperatures were monitored. The experimental campaign was carried out during the winter, specifically during January and February 2020. The schematic representation of the experimental setup is shown in Figure 3. Table 2 lists the technical specifications of the measuring instruments.

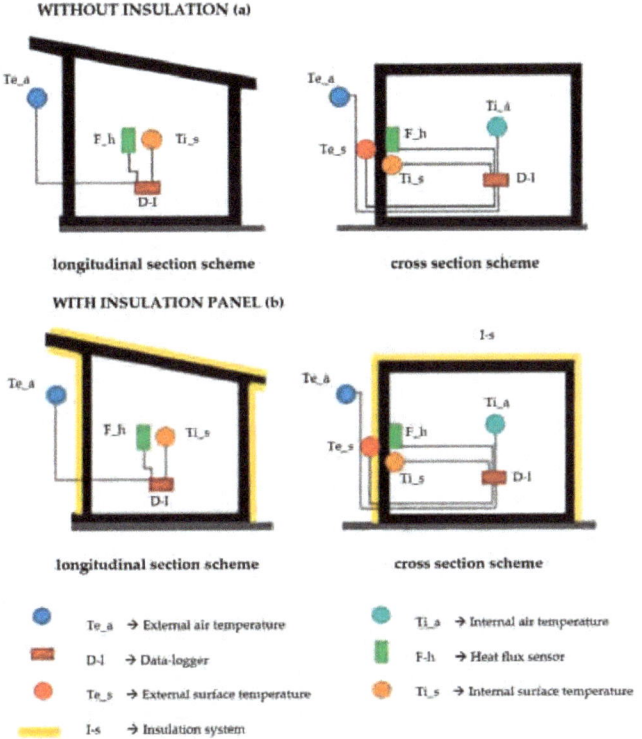

Figure 3. Schematic representation of the experimental setup in the reference test room (**a**) and in the insulated one (**b**).

Table 2. Technical Specifications of the Measuring Instruments.

Measuring Instrument	Manufacturer	Model	Measuring Range	Resolution	Accuracy
Heat-flow meter	Hukseflux	HFP01	$-2000 \div 2000$ W/m^2	0.01 W/m^2	5% on 12 h
Thermometer	LSI	Pt100	$-40 \div 80$ °C	0.01 °C	0.10 °C (0 °C)
Surface temperature probe	LSI	EST124	$-40 \div 80$ °C	0.01 °C	0.15 °C (0 °C)

The measurements of the thermal transmittances of the walls were carried out in compliance with the ISO 9869-1 Standard [46]. The acquired data were processed using the progressive averages method, applying the following formula:

$$U = \frac{\sum_{j=1}^{n} q_j}{\sum_{j=1}^{n} (T_{ai} - T_{ae})_j} \quad (1)$$

where q is the heat flow density, T_{ai} and T_{ae} are the temperature of the air inside and outside the analyzed test room, respectively.

The phase shift (briefly defined **PS**) of the thermal waves can be determined as the time difference between the recording time of the highest external surface temperature value ($h\_{Ts\,max_e}$) compared to that which corresponds to the highest internal surface temperature ($h\_{Ts\,max_i}$).

$$PS = h\_{Ts\,max_e} - h\_{Ts\,max_i} \quad (2)$$

The thermal wave attenuation (briefly defined **DF**) can be calculated as the ratio of the difference between the maximum internal surface temperature ($T_{s\,max_i}$) and the average one ($T_{s\,avg_i}$), and the difference between the maximum external surface temperature ($T_{s\,max_e}$) and the average one ($T_{s\,avg_e}$) [47]:

$$DF = \left[\frac{T_{s\,max_i} - T_{s\,avg_i}}{T_{s\,max_e} - T_{s\,avg_e}} \right] \quad (3)$$

In order to carry out a complete and reliable measurements campaign, the thermal behaviors of the two test rooms were analyzed taking into account different scenarios in terms of operational times of the heating system (made with electric fan heaters properly shielded to avoid direct disturbing effects to the sensors).

The first analyzed scenario took five days; during this time, the heating systems was always switched off (this first scenario is defined in the following as *Free-Floating*).

In the second scenario (the so-called *On*), the heating systems were switched on for four days continuously, and at the end of this time, the cooling phase of the two structures was evaluated during the 3 following days.

Finally, in the third scenario (the so-called *On-Off*), the thermal behavior of the two structures was studied by switching on the heating systems for nine hours per day (switching on the fan heaters in the morning at 9.00 a.m. and switching them off at 06.00 p.m.).

3.2. Energy Simulation Model

The data obtained from the experimental campaign were employed to build a dynamic energy simulation model of an ideal building. The test rooms, in fact, are too small and not representative of a real residential building. The ideal building, which was used by the authors for simulations in previous works, has the same envelope thermal performance (thermal transmittance, phase shift, wave attenuation, etc.) as that of the monitored test rooms, but it is more representative of an actual building since it has transparent surfaces and plants that are essential in residential spaces.

An hourly energy simulation was performed using Design Builder software [48], a computational code based on Energy Plus as an internal simulation engine.

Design Builder was used for modeling a building larger than the actual buildings where measurements were carried out. A simple building with a square shape of a 6 m side, similar to a two-storey detached house already used in other studies [49], was considered as a case study (Figure 4). Each wall has a surface area equal to 36 m^2, characterized by the stratigraphy listed previously in Table 1. The fifteen windows adopted in the model are double glazed windows (6 mm–6 mm filled with air in the gap and with a solar factor of 0.7), with a thermal transmittance of 3.094 W/m^{2}K and an area of 1.44 m^2 for each one; the frame is made of painted wood and is characterized by a thickness of 8 cm. The shadings of the windows are composed by shutters that are simulated as external systems.

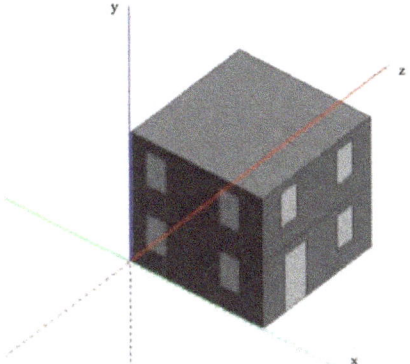

Figure 4. 3D view of the case study used for simulation.

The following settings were adopted:

- The internal gains were not considered, excluding the ones linked to people's metabolic rate that varies between 110 and 180 W/person depending on the activity performed in the different rooms; the employed metabolic rate factor was equal to 0.9.
- The clothing was equal to 1 clo in winter and 0.5 clo in summer.
- The heating system schedule was 5–9 a.m. and 5–12 p.m.
- The infiltration rate considered was equal to 0.7 1/h.
- The ventilation was natural and set to 1 1/h.
- The internal set point temperature was set as equal to 20 °C for winter.

As Italian buildings are usually equipped with only heating system, a natural gas boiler was supposed for the heat generation, and the global efficiency of the system was set as equal to 0.83. In the energy model, an occupancy value of 0.02 people/m² has been defined.

The energy need of the building was simulated. Later, different insulating materials were tested, taking always into account a thickness of the insulating layer equal to 0.01 m (equal to the thickness of the aerogel panel tested during the in situ campaign). This choice allows the comparison of different insulation materials with equal saving of inner space in the case of internal application; the use of aerogel is, in fact, a competitive solution in the retrofit of historical buildings when the intervention on the external façade is not possible for architectural conservatory constraints [35]. In particular, the simulated insulating materials are: Expanded PolyStyrene (EPS), rock wool, kenaf and aerogel (whose thermophysical properties are shown in Table 3).

Table 3. Thermal characteristics, duration and decay rate of the aerogel and other materials.

Insulating	Conductivity [W/mK]	Specific Heat [J/kgK]	Density [kg/m³]	Duration [Years]	Decay Rate [%]
Expanded PolyStyrene (EPS)	0.040	1400	15	20	0.20
Rock Wool	0.038	840	40	25	0.25
Kenaf	0.040	1700	30	15	0.17
Aerogel	0.015	1000	230	45	0.21 [1]

[1] Obtained as the mean of other materials' decay rate.

For the materials, a useful life of 45, 20, 25 and 15 years was considered, respectively, for aerogel, EPS, rock wool and kenaf. However, the materials may not be removed from the walls, and they could continue to partially carry out their task for the whole duration of the building. Therefore, for these kinds of interventions, a linear compound decay rate was estimated as equal to 0.21%, 0.20%, 0.25% and 0.17% per year, respectively, for aerogel, EPS, rock wool and kenaf (Table 3). As far as the duration is concerned, a duration of 50 years was considered for the building.

These insulating materials modified the walls' thermal transmittances, as reported in Table 4 (the insulating material is installed on the outer side of the wall, before plaster). According to this, an energy analysis was carried out to quantify the energy savings obtained by means of different insulating materials.

Table 4. U-Value of the Walls Considering Different Insulating Materials.

Wall	U-Value [W/m^{2}K]
Original wall	1.647
Insulated with Expanded PolyStyrene (EPS)	1.167
Insulated with Rock Wool	1.149
Insulated with Kenaf	1.167
Insulated with Aerogel	0.785

3.3. Environmental Assessment Based on LCA

Following the quantification of the energy savings obtained after the implementation of different retrofits, a life cycle assessment (LCA) was performed to determine the effectiveness of the intervention when considering the environmental burdens embodied in the building materials installed. The LCA is an interesting methodology that permits the comparison of the energy requirements of the buildings and the related environmental burdens from a more comprehensive perspective that takes into account the whole life cycle stage of the constructions (production, installation, operation, end-of-life). In fact, different authors have already warned about the burden shifting that characterizes every retrofit intervention [50,51]: the reduction of the operational energy requirement and related environmental burdens is followed by an increase in embodied components linked to the installation of new building materials and systems. Two indicators were introduced to describe the environmental performances of the different external insulation coatings supposed: the Energy Payback Time (***EPBT***) and the Carbon Payback Time (***CPBT***). The first one can be defined as the ratio between the variation of the Embodied Energy (***EE***) of the building following the retrofit and the annual Energy Savings (***ESa***) achieved through the retrofit (see Equation (4)). The latter is similarly the ratio between the variation of the Embodied Carbon (***EC***) of the building and the annual emissions avoided (***CSa***) through the retrofit (see Equation (5)).

$$EPBT = \frac{\Delta EE}{ESa} \quad (4)$$

$$CPBT = \frac{\Delta EC}{CSa} \quad (5)$$

The LCA analysis was carried out using Ecoinvent (Ecoinvent, Zurich, Switzerland) data, and when this was not possible, Environmental Product Declaration datasheets were consulted [52]. The EE was calculated using the single-issue indicator Cumulative Energy Demand (***CED***), while the Global Warming Potential (***GWP***) (100 years) was employed to determine the EC of the retrofit. As shown in Figure 5, a "cradle to site" approach was employed for the life cycle assessment. Since the application of external insulation coatings in a low-height building does not imply an energy intensive installation process, stage A5 can be considered negligible.

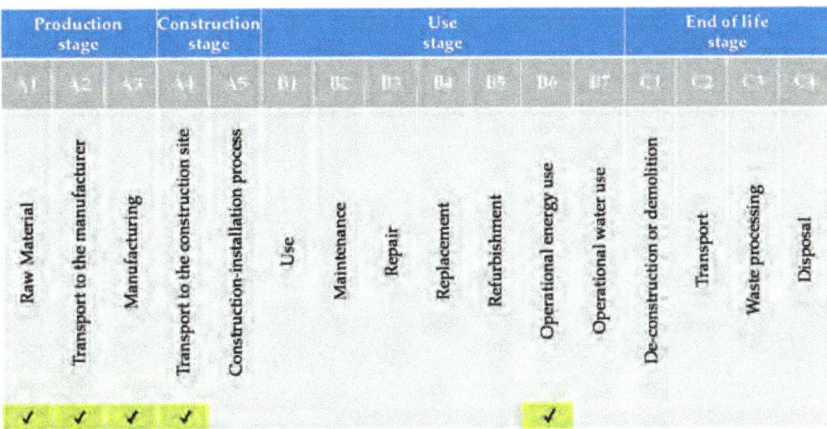

Figure 5. Life cycle stages considered in this study (green marked).

On the other hand stages B1–B5 were not included in the calculation of the payback times since, generally, they are much lower than the useful life adopted in this work for the insulation materials (see Table 3) [53].

3.4. The Cost-Benefit Analysis

From an economic-financial perspective, the international literature estimates the convenience of retrofit intervention in a partly incomplete manner.

Table 5 provides a systematic review of some typical studies.

The international literature on this topic is very ample and an in-depth analysis of it goes beyond the objective of this study. Therefore, only some studies are analyzed, which can be considered representative of different approaches.

Most studies consider only the energy savings resulting from retrofit interventions and some related items (initial investment, maintenance costs, running and replacement costs, etc.) which are measurable in monetary terms. They do not consider the environment benefits of a retrofit, nor, generally speaking, its impacts on the Internal Environmental Quality (IEQ), due to the difficulty and subjectivity of their economic measurement.

Table 5. Studies on the Economic Convenience of a Retrofit Intervention.

Studies	Main Objective		Decision Scenario Considered		Performance Measure					Costs and Benefits Considered		Uncertainty Explicitly Considered	Uncertain Variables	
	Methodological (+Example Case)	Real Specific Application	Stand-Alone Investment Convenience	Comparison among Alternative Interventions	Payback Period	Discounted Payback Period (1)	Present Value of Differential Costs (Pseudo NPV or PI) (1) (2)	NPV (3)	Energy Efficiency	LCC (4)	IEQ (5)		Technical	Economic
Almeida et al. [54]	x							x		x		x	x	
Almeida-De Freitas [55]	x			x	x		x			x	thermal comfort			
Ballarini et al. [56]	x			x	x		x		x	x		x	x	
Burhenne et al. [57]	x		x				x		x			x	x	x
Gustaffson [58]	x		x				x							
Hasan et al. [59]		x		x			x			x				
Hopfe-Hensen [60]	x											x	x	
Niemela et al. [61]		x		x	x		x			x		x		
Ortiz et al. [62]		x		x			x			x		x	x	
Ozel [63,64]		x		x		x			x			x	x	
Verbeeck-Hens [65]		x		x	x		x		x			x	x	x

(1) free-risk rate as a discount rate (corrected by inflation or not). (2) PI = Profitability Index: the ratio between the present value of net benefits and initial investment. (3) NPV= Net Present Value, that uses cost of capital (risk-free rate + risk premium) as a discount rate. (4) LCC = Life-Cycle Costs. (5) IEQ = Indoor Environmental Quality.

Some of them choose the intervention which guarantees the quickest recovery of the investment or the shortest payback time [55,56,61,63–65]. This method is quite easy to apply, but it shows two elements of weakness:
- It does not discount, with an appropriate cost of capital, the costs and benefits of the investment which occur in different years, often over very long periods, and that are estimates (i.e., uncertain values).
- It does not provide a threshold value with which to compare the recovery period of individual interventions (for a stand-alone evaluation of their convenience).

When a more complete approach is provided [54,57–59,62], the present value of differential costs/benefits is calculated (in [55,56,61,65] as a further method) by using a free-risk rate for discounting (often corrected by the expected inflation rate), which does not take into account a premium for the risk of the discounted cash flows.

The analysis here conducted applies a financial approach consistent with the modern financial theory. The net present value (NPV) is used (however, only considering the monetary costs and benefits of a retrofit, in line with the international literature), which measures today's monetary value of the intervention, and it discounts the net cash flows by a rate which considers the time value of the money and the risk premium, calculated with reference to the main risk drivers of the investment. The retrofit is convenient if the NPV is non-negative, and it is the more convenient the higher its value.

Furthermore, a probabilistic approach was used to measure the risk of the NPV of each retrofit alternative. Many studies [54,56,57,60–64] explicitly consider the uncertainty, more often with regard to the technical variables than the economic ones. Some studies consider different possible values of technical input variables (rarely of economic variables, as for example the discount rate in [57,61] and the initial investment and gas price in [57]) and estimate the resulting range of outcome measure, others [54,57,60] use very complex methods to deal with the uncertainty (various sensitivity analysis methods and Monte Carlo techniques), but they are quite methodological exercises: in fact, these techniques are very difficult to apply in a real context, since many of the necessary data cannot be realistically provided, and the approach is quite difficult for the decision-maker to understand.

In this paper, optimistic and pessimistic estimates of the uncertain drivers of NPV were assumed and defined in a subjective manner (i.e., on the basis of the analyst/decision-maker's forecasts), and the corresponding range of NPV was measured. This analysis provides two useful results for a decision-maker:
- To identify which variable, that influences the investment's NPV, most affects its variability (sensitivity analysis).
- To derive an approximate measure of the risk of retrofit under some hypotheses [66]; in fact, the sensitivity analysis allows the estimation of the probability distribution of the NPV, which enables the decision-maker to choose better than in the case of a single value: the decision-maker can translate his/her risk aversion into a minimum acceptable percentage of non-negative values of the NPV and compare the percentage emerging from the NPV probability distribution with this threshold value [67].

Finally, as far as the IEQ aspects are concerned, a multi-criteria methodology (MCDA) is being developed that would measure for the different retrofit interventions the main descriptors of the IEQ in relative terms, with respect to the acceptable ranges defined by EU regulations. This approach would use linear optimization models in order to allow the decision-maker to compare the retrofit alternatives with each other and with the current state of a building.

4. Results and Discussion

4.1. Experimental Campaign Results

As mentioned before, the monitoring campaign was carried out during the winter, specifically during the months of January and February 2020, and it was focused on the assessment of the thermal behavior of the studied test rooms in three different scenarios:

Free-Floating conditions (no heating in the two test rooms), the so-called On scenario (heating system always on) and the so-called On-Off scenario (heating system on only during a specific daily time interval). The obtained results during winter can be summarized as follows:

- Free-floating conditions: Data processing in this phase mainly focused on defining the thermal waves' phase-shift and attenuation according to Equations (2) and (3). In particular, the surface temperature values were analyzed, and their trend over time is reported in Figure 6, where the internal and external surface temperatures for the reference test room are called *Tsi_ref* and *Tse_ref*, respectively.

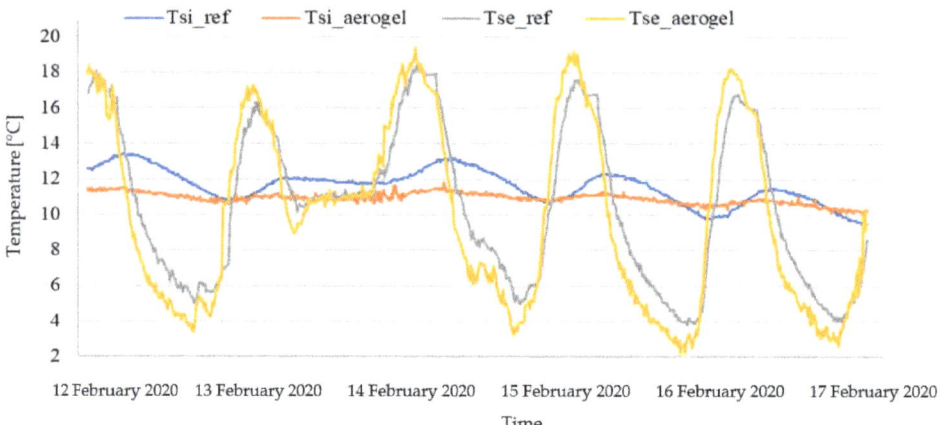

Figure 6. Internal and external surface temperatures of north-west walls registered for the insulated and the reference test rooms in the free-floating conditions.

On the other hand, the internal and external surface temperatures for the insulated test room are called *Tsi_aerogel* and *Tse_aerogel*, respectively.

It is clear that the internal surface temperatures immediately present a stabilized periodic regime. In particular, the internal surface temperature data measured on the thermally insulated test room provide almost constant values along time, mostly lower than those registered on the reference test room.

Both attenuation and phase shift values were calculated with respect to a daily interval, while the final average value (shown in Table 6) was calculated as the average of the daily attenuation and phase shift values.

Table 6. Attenuation and Phase-Shift Average Values Obtained under Free-Floating Conditions.

	Attenuation	Phase Shift
Reference Test Room	0.124	4 h 07 min
Thermally Insulated Test Room	0.044	4 h 58 min

By comparing these data, it is possible to observe that applying a thin layer of aerogel does not cause a significant variation of the thermal inertia of the wall. In fact, the thermally insulated test room is characterized by an average phase shift just 20.6% higher than the reference one. On the contrary, one centimeter of aerogel, due to its high insulating performance, produced a decrease in the average attenuation of about 64.5% compared to that calculated for the reference test room, with a better indoor air temperature steadiness. Thus, it is possible to affirm that the application of a thin layer of aerogel can improve the dynamic behavior of the structure. However, this is not surprising. In a steady state regime, the layer arrangement makes no difference. On the contrary, under dynamic boundary conditions, the layer arrangement becomes fundamental, and by interchanging the layers the wall properties change. Hence, this aspect needs to be considered for improving the inertial behaviour of a wall.

- On scenario: The second phase was related to the investigation of the thermal behavior of the two test rooms with the heating always on. In this case, the progressive increase in the air temperature of the two different test rooms was focused, as shown in Figure 7 (before the vertical black dotted line).

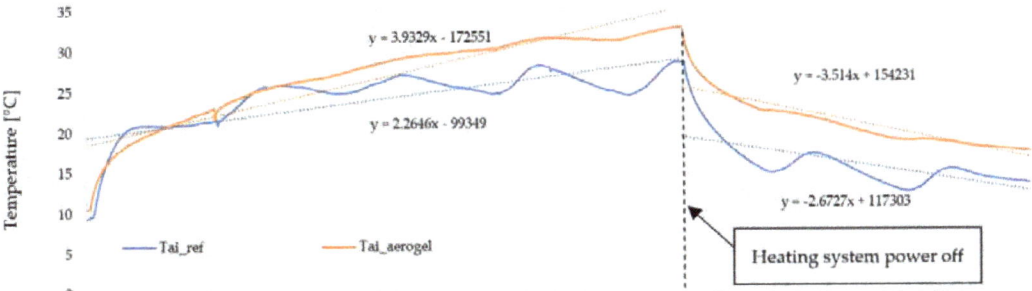

Figure 7. Indoor air temperatures registered in the test rooms during the second scenario.

Taking into account the thermally insulated test room, it is possible to observe a faster internal air temperature rise if compared with the reference test room. At the end of the always-on heating period, a stabilized regime was not achieved in the thermally insulated test room, as the internal temperature gradually increased. The use of the aerogel led to an internal air temperature of about 33 °C, compared to about 27 °C (average value) obtained in the reference test room, where an almost periodic regime was identified after about two days.

The absence of the external insulating coat made the reference test room more sensitive to the typical variations of the outdoor air temperatures also during the heating system shutdown. Figure 7 (after the vertical black dotted line) shows a more rapid decrease in the values of the internal air temperature, as expected.

- On-Off scenario: The last part of the winter monitoring was aimed at evaluating the thermal behavior of the two test rooms, assuming that the heating system was switched on and off; i.e., switching on the fan heaters in the morning and switching them off in the evening, thus simulating the irregular working of an actual heating system. The acquired data were employed for evaluating the thermal transmittance of the walls facing north-west. Figure 8 shows the thermal transmittances as a result of the data post-processing based on the progressive average method. The thermal insulation of the test room through the thin layer of aerogel allowed obtaining a thermal transmittance reduction equal to -28.3%.

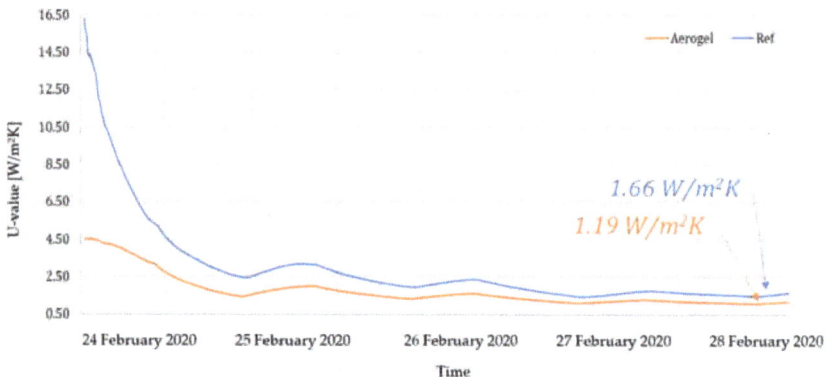

Figure 8. Thermal transmittances obtained from the U-value measurements.

4.2. Energy and LCA Results

Table 7 and Figure 9 show the results obtained through the software Design Builder. In the reference non-insulated building an energy demand for heating equal to 11,621.5 kWh/year of natural gas was obtained. The energy analysis shows how the installation of different insulating materials allows the reduction in the energy need of the building. Quite similar percentage reductions were obtained when the structure was insulated with EPS, rock wool and kenaf panels. Similar energy savings of about 6% are not surprising, because of similar thermal conductivity values among the different insulating materials. Aerogel, due to its reduced thermal conductivity, allowed achieving a heating energy need of 10,313.7 kwh/year. A percentage difference in terms of heating energy requirement equal to −11.3% was obtained—almost double that obtained using the other insulating materials.

Table 7. Data on the Energy Need and Energy Saving of Insulated Buildings Compared with the Reference.

	Reference	EPS	Rock Wool	Kenaf	Aerogel
Energy need [kWh]	11,621.5	10,945	10,917.7	10,944.9	10,313.7
Energy saving [%]	-	−5.8	−6.1	−5.8	−11.3

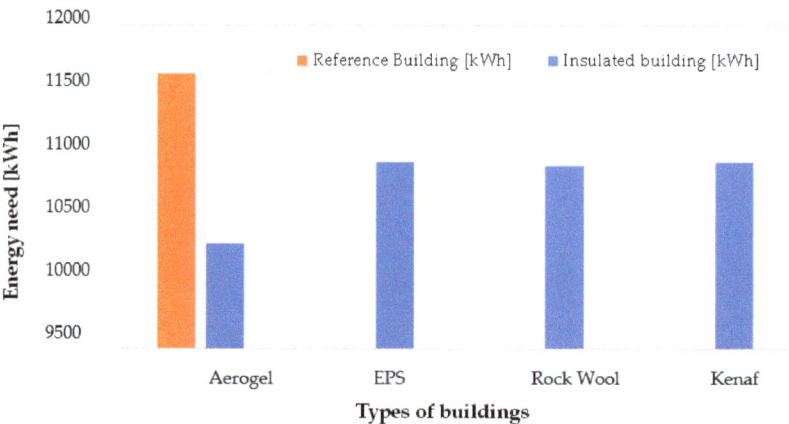

Figure 9. Energy needs of the reference building and the insulated one.

The results on the EPBT and CBPT are reported in Table 8. The retrofit with aerogel is characterized by a higher EPBT and CPBT in comparison to traditional insulation materials (e.g., rock wool or EPS). As previously supposed in Section 3.3, the values obtained are lower than the supposed useful life of the insulation materials installed. This means that the burden shifting on embodied components is only temporary and that every intervention is characterized by a positive environmental effect on its life cycle. Figures 10 and 11 show the total CED and the cumulative GWP versus time considering as positive values the energy saved and the emissions avoided: the coating with traditional insulation materials has a lower payback time in comparison with the scenario considering aerogel as insulation material, but the latter results guarantee, after about fourteen years from the installation, a higher energy saving and carbon emission reduction potential.

Table 8. Energy and carbon payback times of the various retrofit solutions.

	ΔEE (kWh)	ΔEC (kgCO$_2$eq)	EPBT (Years)	CPBT (Years)
EPS	1341	213	1.98	1.57
Rock Wool	1110	227	1.58	1.61
Kenaf	1793	323	2.65	2.38
Aerogel	9073	1682	6.94	6.40

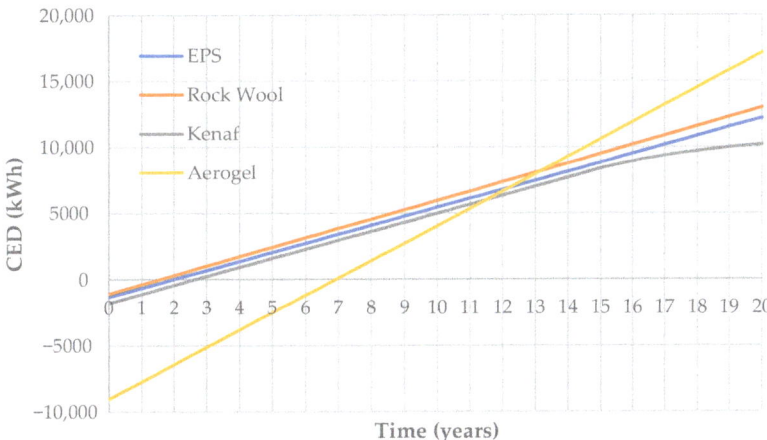

Figure 10. Cumulative energy demand for every retrofit intervention (positive values stand for energy saved).

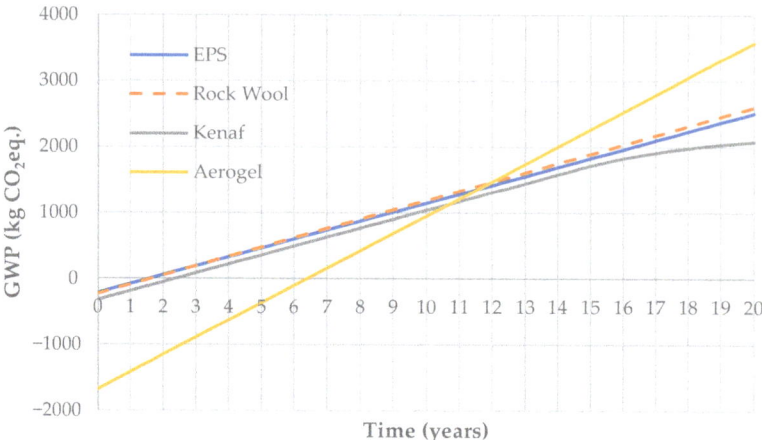

Figure 11. Cumulative GWP for every retrofit intervention (positive values stand for emissions avoided).

4.3. Economic Analysis Results

In order to calculate the NPV, the differential costs and benefits of the four analyzed materials (compared to the current state) were estimated.

In order to estimate benefits, the annual savings related to the methane gas consumption and the tax incentives of "Bonus Facciate" [68] were considered.

The price of methane gas was obtained by the average of prices applied by different suppliers in 2020, equal to 0.0985 euro per kWh. From the historical price series of methane gas over the last 10 years (Eurostat data [69]), an annual growth rate of methane gas of 1.77% was obtained, applied during the whole duration of the retrofit (as a trend estimate).

For the duration and decay rates of the insulating materials, the values reported in Table 3 were adopted.

To estimate the tax incentives, based on the current regulation, a tax deduction of 90% of the total expenditure was assumed over 10 years.

To estimate the cost of capital, the Capital Asset Pricing Model approach has been used, with the following parameters:

- Risk free rate equal to 1.18%, obtained from the average of the 10-year BTP returns during the last 12 months (investing data [70]); the rate includes both inflation expectations and country risk premium perceived by the market.
- Market risk premium equal to 5% (IBES consensus estimate).
- Beta equal to 0.65, estimated on the basis of the systematic variability of the methane gas price (source: Eurostat), referred to the Italian GDP (sources: Bank of Italy [71] and Istat [72]) from 1991 to today. The 1991–2019 time series of Italian GDP and methane gas price were considered, obtaining a variation coefficient (i.e., their normalized volatility) of 22.36% and 18.76%, respectively; their Pearson coefficient of correlation is 0.77. Beta was calculated as Equations (6) and (7) show:

$$\text{beta} = \frac{\text{correlation}_{\text{gas price, GDP}} \times \text{volatility}_{\text{gas price}}}{\text{volatility}_{\text{GDP}}} \quad (6)$$

$$\text{beta} = \frac{0.77 \times 0.1876}{0.2236} \cong 0.65 \quad (7)$$

A cost of capital of 4.42% was obtained as Equations (8) and (9) show:

$$\text{cost of capital} = \text{risk free rate} + \text{beta} \times \text{market risk premium} \quad (8)$$

$$\text{cost of capital} = 1.18 + 0.65 \times 5 = 4.42\% \tag{9}$$

which was used as a discount rate of the energy savings. The tax incentives were discounted by the risk-free interest rate since they are relatively certain. Table 9 summarizes the costs, benefits and NPV of the four materials.

Table 9. Calculation of the Net Present Value (NPV) of materials (data in euros).

Cash Flows and NPV	Aerogel	EPS	Rock Wool	Kenaf
Present value (energy savings)	3510.50	1795.03	1871.13	1790.53
Present value (tax deduction)	12,266.58	4605.13	4708.38	5555.07
Lump-sum investment [1]	14,359.52	5390.86	5511.74	6502.88
NPV	1417.55	1009.30	1067.78	842.71

[1] Estimated for a surface of 121 m^2.

It is possible to conclude that in this scenario aerogel gives the best economic benefits, with a positive expected NPV of EUR 1417.55.

As far as the intervention risk is concerned, a sensitivity analysis was implemented for retrofit interventions of aerogel and rock wool (on the basis of the results of the previous analysis, the latter is the best among the alternatives to aerogel). Optimistic and pessimistic estimates of the main uncertain drivers of the NPV have been forecasted: duration, methane gas price, cost of capital and tax incentive. The assumptions were the following:

- The duration was included in the range of 45–50 years for aerogel and 20–30 for rock wool. The decay rate during the building residual duration was estimated as a linear compound decay rate from material duration to building duration (in contrast, in the case of aerogel, the average decay rate has been used: due to its longer duration, this hypothesis is more realistic).
- The methane gas price's change is equal to ±13% (compared to 2020), measured on the basis of the price semiannual time series (Eurostat data). Gas price is assumed to be normally distributed, and the values corresponding to 5° and 95° percentiles of probability distribution are considered (this variation is added to the growth trend, hypothesized above).
- The cost of capital was included in the range 3.83–5.26%, calculated as follows: (i) as an optimistic estimate, a risk-free rate equal to 1.59% and a beta of 0.53 were considered (the average beta of listed producers from Datastream [73] dataset was used); (ii) as a pessimistic estimate, the average beta of the gas industry (but including both gas producers and related service providers) and a risk-free rate equal to 1.56% were used. In this scenario, the risk-free rate was measured by adopting a more conservative approach; in fact, the German Bund 10-year returns were corrected by means of the inflation differential between Germany and Italy, and a country risk premium was added by using the differentials of credit default swap (CDS) spreads over 10 years (Bloomberg data [74]).
- Furthermore, in order to provide a more general assessment of the convenience of the different materials here considered, the current tax deduction of 90% has been assumed as an optimistic estimate: 50% and 65% are assumed as pessimistic and average tax incentives, respectively (all over 10 years). In this hypothesis (which is different from the current scenario, adopted in the above NPV calculation), given the most probable values of the other uncertain drivers discussed before, the most probable NPV of the two retrofits are negative, EUR −1989.83 and −240.11, respectively.
- Table 10 shows the NPVs corresponding to the above estimates (changing a driver at a time) and the related NPV range.

Table 10. Sensitivity Analysis of the NPV.

Uncertain Drivers	Input Data		NPV (Optimistic Estimate) [1]		NPV (Pessimistic Estimate) [1]		NPV Range [1]		Coefficient of Sensitivity	
	Optimistic Estimate	Pessimistic Estimate	Aerogel	Rock Wool	Aerogel	Rock Wool	Aerogel	Rock Wool	Aerogel	Rock Wool
duration (years)	aerogel 50 rock wool 30	aerogel 40 rock wool 20	−1988.71	−235.83	−1992.36	−244.79	3.65	8.96	0.0%	0.0%
methane price (kwh)	0.1113	0.0857	−1533.46	3.14	−2446.19	−483.35	912.73	486.49	2.6%	3.9%
cost of capital	3.83%	5.26%	−1547.46	665.68	−2495.91	−563.10	948.44	1228.78	2.9%	24.6%
tax incentive	90%	50%	1417.55	1067.78	−4034.26	−1024.84	5451.81	2092.61	94.5%	71.5%

[1] Data in euros.

The last two columns measure the coefficients of NPV sensitivity; i.e., how much each driver variability influences the NPV variability.

Figure 12 shows the cumulative probability distribution of NPV of each retrofit, where NPV variance is measured following Equation (10) (by simplifying, the uncertain drivers are assumed to be independent of each other and linearly related to NPV):

$$\sigma^2_{NPV} = \sum k_i^2 \times S_i^2 \qquad (10)$$

where S_i = NPV range between the optimistic estimate U_i and the pessimistic one L_i of uncertain driver i (see columns 8–9 in Table 9) and, where σ_i is its volatility (in this case, k_i is equal to 0.3).

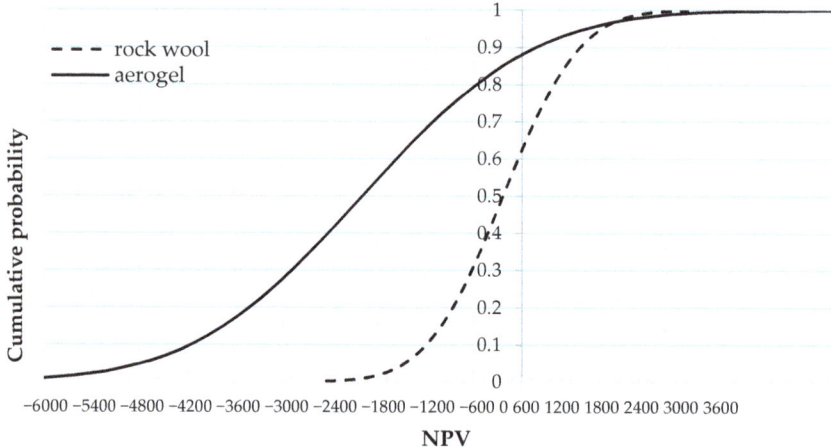

Figure 12. NPV cumulative probability distributions of retrofit with aerogel and rock wool.

The main results are the following:

- NPV volatility mainly depends on the change of tax incentive for both retrofits, which is the most important driver of performance and risk of the two retrofits considered here. The cost of capital affects NPV variance of rock wool retrofit more than aerogel retrofit (25% versus 3%); the methane gas price volatility similarly affects NPV variance of two retrofits; the duration variability has no impact on both retrofits.
- Aerogel NPV assumes non-negative values in only 12% of the cases (rock wool in 38% of the cases, instead) and outperforms the competing material only in the right tail of the NPV probability distribution.

Due to the crucial effect of tax incentives on NPV of both retrofits here assessed, we further analyzed how NPV changes, depending on tax deduction (the most probable values of the other drivers were considered). Figure 13 shows that the aerogel NPV: (i) becomes the more advantageous, in comparison to rock wool, the higher the tax deduction; (ii) is positive for tax deduction larger than 80% (70% for rock wool, instead); and (iii) beats the competing material when tax deduction is larger than 87%. This analysis is important, since it shows that aerogel material is more convenient than competitors only in the fiscal framework here considered (or in a more favorable one, as for example in the case of the Superbonus 110%), while in other scenarios it is not, due to its higher lump-sum investment (even though it provides double energy savings than the alternative materials).

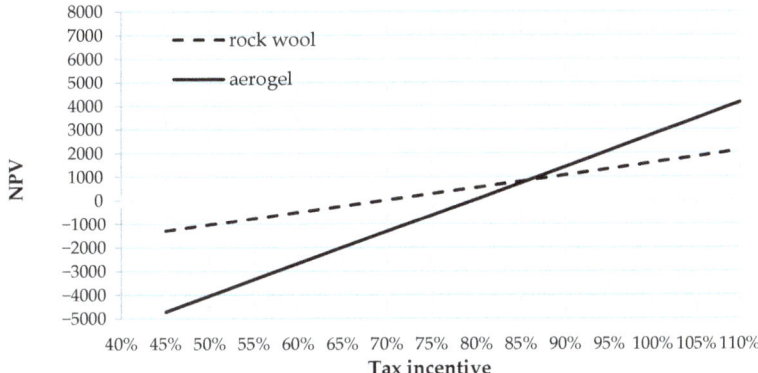

Figure 13. NPV as a function of tax incentives.

5. Conclusions

A small test room, totally insulated with aerogel panels, was investigated by an experimental point of view. The thermal behavior of the aerogel insulated test room was compared with a non-insulated identical test room. Heat transfers across walls were assessed by installing heat-flow meters and air and surface temperature sensors. Experimental data verified the well-known aerogel capability to improve the thermal performance of test room envelopes, even if reduced thicknesses of thermal insulation were applied. In particular, the thermally insulated test room showed an average phase shift 20.6% higher than the reference one. On the other hand, the small layer of aerogel allowed obtaining an average attenuation decrease of about 64.5% when compared to the reference test room. Moreover, 1 cm of aerogel allowed to obtain a thermal transmittance reduction of −28.3%.

Successively, a dynamic simulation model was generated to better comprehend the energy savings obtained through aerogel across a whole year, in terms of energy demand. By comparing aerogel with other commonly used insulation materials, a heating energy demand reduction of −11.3% was found.

Subsequently, the investigation was completed by computing the environmental and energy payback times of this retrofit strategy as well as by a cost-benefit analysis through a probabilistic financial framework. In sum, it is possible to conclude that in the current tax incentive scenario the aerogel retrofit gives the best positive expected NPV, and from the whole LCA perspective it also guarantees both the highest energy saving and emissions avoidance. However, due to its higher lump-sum investment, aerogel's NPV is very sensitive to tax deductions and it is riskier than the best comparable material (roof wool) in less favorable tax scenarios: for example, if a 65% tax deduction is assumed (given the probability distributions of the uncertain variables considered here), aerogel gives a non-negative NPV only in 12% of the cases (roof wool in 38% of the cases, instead).

Therefore, the proposed interdisciplinary study aimed to investigate the environmental, energy and economic impacts associated with the application of aerogel compared to other insulating materials. Furthermore, the proposed methodological approach could also be replicated in other countries characterized by different climatic conditions.

Future developments will address a comparison of the different energy, environmental and economic benefits of aerogel under different climatic conditions, also applying thin aerogel panels to different wall stratigraphy. In addition, the analysis could be performed also considering the energy, environmental and economic benefits of aerogel by analyzing walls thermally insulated with materials of different thicknesses but characterized by the same U-value.

Author Contributions: Conceptualization, P.M. (Paola Marrone), F.A. and D.V.; methodology, P.M. (Paola Marrone), F.A., D.V., L.F. and R.D.L.V.; thermal measurements, L.F., L.E., C.G., M.R. and F.O.; energy simulations, G.G. and M.R.; economic-financial analysis: P.M. (Paolo Matteucci) and D.V.; data curation, L.E., C.G. and M.R.; writing—original draft preparation, L.E., C.G., M.R., F.O., P.M. (Paolo Matteucci) and D.V.; writing—review and editing, F.A., M.R., G.G., L.E., P.M. (Paolo Matteucci) and D.V.; supervision, P.M. (Paola Marrone), F.A., D.V., L.F. and R.D.L.V.; project coordination, P.M. (Paola Marrone). All authors have read and agreed to the published version of the manuscript.

Funding: This research received no external funding.

Institutional Review Board Statement: Not applicable.

Informed Consent Statement: Not applicable.

Data Availability Statement: Data available on request due to restrictions. The data presented in this study are available on request from the corresponding author. The data are not publicly available due to privacy reasons.

Acknowledgments: The authors would like to thank Eng. Alfredo Simonetti and all the staff at CEFME CTP for the valuable collaboration.

Conflicts of Interest: The authors declare no conflict of interest.

References

1. IPCC. Global Warming of 1.5 °C. Available online: https://www.ipcc.ch/sr15 (accessed on 27 September 2020).
2. United Nations UNSD. Environmental Indicators. Available online: https://unstats.un.org/unsd/envstats/qindicators.cshtml (accessed on 14 November 2020).
3. European Commission. Paris Agreement. Available online: https://ec.europa.eu/clima/policies/international/negotiations/paris_it (accessed on 14 November 2020).
4. United Nations. Environment Programme The Emissions Gap Report 2018. Available online: https://www.unenvironment.org/resources/emissions-gap-report-2018 (accessed on 14 November 2020).
5. Sanson, A.; Giuffrida, L.G. Decarbonizzazione Dell'economia Italiana, il Catalogo Delle Tecnologie Energetiche. Available online: https://www.dsctm.cnr.it/images/Eventi_img/de_carbonizzazione_3_ottobre_2017/catalogo-tecnologie-energetichevers.ultima.pdf (accessed on 14 November 2020).
6. Pomponi, F.; Moncaster, A. Circular economy for the built environment: A research framework. *J. Clean. Prod.* **2017**, *143*, 710–718. [CrossRef]
7. Geissdoerfer, M.; Savaget, P.; Bocken, N.M.P.; Hultink, E.J. The Circular Economy—A new sustainability paradigm? *J. Clean. Prod.* **2017**, *143*, 757–768. [CrossRef]
8. De los Rios, I.C.; Charnley, F.J.S. Skills and capabilities for a sustainable and circular economy: The changing role of design. *J. Clean. Prod.* **2017**, *160*, 109–122. [CrossRef]
9. Winans, K.; Kendall, A.; Deng, H. The history and current applications of the circular economy concept. *Renew. Sustain. Energy Rev.* **2017**, *68*, 825–833. [CrossRef]
10. Orsini, F.; Marrone, P. Approaches for a low-carbon production of building materials: A review. *J. Clean. Prod.* **2019**, *241*, 118380. [CrossRef]
11. Orsini, F.; Marrone, P. Prodotti a basse emissioni di carbonio: Potenzialità e limiti della manifattura della regione Lazio. In *XIX Congresso Nazionale CIRIAF—Energia e Sviluppo Sostenibile*; Morlacchi Editore University Press: Perugia, Italy, 2019; pp. 173–186.
12. Chitnis, M.; Sorrell, S.; Druckman, A.; Firth, S.; Jackson, T. Estimating Direct and Indirect Rebound Effects for U. S. Households. Available online: http://www.sustainablelifestyles.ac.uk/sites/default/files/publicationsdocs/slrg_working_paper_01-12.pdf (accessed on 8 November 2020).
13. IEA. World Energy Outlook. 2018. Available online: https://www.iea.org/reports/world-energy-outlook-2018 (accessed on 8 November 2020).
14. European Commission. *Secretariat-General Communication from the Commission to the European Parliament, the European Council, the Council, the European Economic and Social Committee and the Committee of the Regions—The European Green Deal Green Deal*; European Commission: Brussels, Belgium, 2019.

15. McKinsey & Company Pathways to a Low-Carbon Economy: Version 2 of the Global Greenhouse Gas Abatement Cost Curve 2009. Available online: https://www.mckinsey.com/~{}/media/mckinsey/dotcom/client_service/sustainability/costcurvepdfs/pathways_lowcarbon_economy_version2.ashx (accessed on 8 November 2020).
16. Evangelisti, L.; Guattari, C.; Asdrubali, F. On the sky temperature models and their influence on buildings energy performance: A critical review. *Energy Build.* **2019**, *183*, 607–625. [CrossRef]
17. Papadopoulos, A.M. State of the art in thermal insulation materials and aims for future developments. *Energy Build.* **2005**, *37*, 77–86. [CrossRef]
18. Cuce, E.; Riffat, S.B. A state-of-the-art review on innovative glazing technologies. *Renew. Sustain. Energy Rev.* **2015**, *41*, 695–714. [CrossRef]
19. Aegerter, M.A.; Leventis, N.; Koebel, M.M. *Aerogels Handbook*; Springer: New York, NY, USA, 2011; ISBN 978-1-4419-7477-8.
20. Hüsing, N.; Schubert, U. Aerogels—Airy Materials: Chemistry, Structure, and Properties. *Angew. Chem. Int. Ed.* **1998**, *37*, 22–45. [CrossRef]
21. Lee, S.W.; Lim, C.H.; Salleh, E.; Ilias, B. Reflective thermal insulation systems in building: A review on radiant barrier and reflective insulation. *Renew. Sustain. Energy Rev.* **2016**, *65*, 643–661. [CrossRef]
22. Tenpierik, M.J.; Hasselaar, E. Reflective multi-foil insulations for buildings: A review. *Energy Build.* **2013**, *56*, 233–243. [CrossRef]
23. Aditya, L.; Mahlia, T.M.I.; Rismanchi, B.; Ng, H.M.; Hasan, M.H.; Metselaar, H.S.C.; Muraza, O.; Aditya, H.B. A review on insulation materials for energy conservation in buildings. *Renew. Sustain. Energy Rev.* **2017**, *73*, 1352–1365. [CrossRef]
24. Karami, P.; Al-Ayish, N.; Gudmundsson, K. A comparative study of the environmental impact of Swedish residential buildings with vacuum insulation panels. *Energy Build.* **2015**, *109*, 183–194. [CrossRef]
25. Baetens, R.; Jelle, B.P.; Gustavsen, A.; Grynning, S. Gas-filled panels for building applications: A state-of-the-art review. *Energy Build.* **2010**, *42*, 1969–1975. [CrossRef]
26. Richter, K.; Norris, P.M.; Tien, C.-L. Aerogels: Applications, structure and heat transfer phenomena. *Annu. Rev. Heat Transf.* **1995**, *6*, 61–114. [CrossRef]
27. Kistler, S.S. Coherent Expanded Aerogels and Jellies. *Nature* **1931**, *127*, 741. [CrossRef]
28. Kistler, S.S.; Caldwell, A.G. Thermal Conductivity of Silica Aërogel. *Ind. Eng. Chem.* **1934**, *26*, 658–662. [CrossRef]
29. Koebel, M.; Rigacci, A.; Achard, P. Aerogel-based thermal superinsulation: An overview. *J. Sol Gel Sci. Technol.* **2012**, *63*, 315–339. [CrossRef]
30. Baetens, R.; Jelle, B.P.; Gustavsen, A. Aerogel insulation for building applications: A state-of-the-art review. *Energy Build.* **2011**, *43*, 761–769. [CrossRef]
31. Garay Martinez, R.; Goiti, E.; Reichenauer, G.; Zhao, S.; Koebel, M.; Barrio, A. Thermal assessment of ambient pressure dried silica aerogel composite boards at laboratory and field scale. *Energy Build.* **2016**, *128*, 111–118. [CrossRef]
32. Jensen, K.I.; Kristiansen, F.H.; Schultz, J.M. Highly Insulating and Light Transmitting Aerogel Glazing for Super-Insulating Windows: HILIT+ Public Final Report 2005. Available online: http://www.vinduesvidensystem.dk/Artikler-Rapporter-Noter-mm/Rapporter/HighlyInsulatingandLIghtTransmittingAerogelGlazingforSuperInsulatingWindows.pdf (accessed on 8 November 2020).
33. Buratti, C.; Moretti, E. Glazing systems with silica aerogel for energy savings in buildings. *Appl. Energy* **2012**, *98*, 396–403. [CrossRef]
34. Cuce, E.; Cuce, P.M.; Wood, C.J.; Riffat, S.B. Toward aerogel based thermal superinsulation in buildings: A comprehensive review. *Renew. Sustain. Energy Rev.* **2014**, *34*, 273–299. [CrossRef]
35. Orsini, F.; Marrone, P.; Asdrubali, F.; Roncone, M.; Grazieschi, G. Aerogel insulation in building energy retrofit. Performance testing and cost analysis on a case study in Rome. *Energy Rep.* **2020**, *6*, 56–61. [CrossRef]
36. Karim, A.N.; Johansson, P.; Kalagasidis, A.S. Super insulation plasters in renovation of buildings in Sweden: Energy efficiency and possibilities with new building materials. *IOP Conf. Ser. Earth Environ. Sci.* **2020**, *588*, 042050. [CrossRef]
37. Buratti, C.; Belloni, E.; Merli, F.; Zinzi, M. Aerogel glazing systems for building applications: A review. *Energy Build.* **2021**, *231*, 110587. [CrossRef]
38. Riffat, S.B.; Qiu, G. A review of state-of-the-art aerogel applications in buildings. *Int. J. Low Carbon Technol.* **2013**, *8*, 1–6. [CrossRef]
39. Italian Revenue Agency. Circolare n. 2/2020—Detrazione per gli interventi finalizzati al recupero o restauro della facciata esterna degli edifici esistenti prevista dall'articolo 1, commi da 219 a 224 della legge 27 dicembre 2019 n. 160 (Legge di bilancio 2020). Available online: https://www.agenziaentrate.gov.it/portale/documents/20143/2338359/Circolare+n.+2+del+14+febbraio+2020.pdf/cee7f814-8750-9d6d-05ca-46d485c6470f (accessed on 8 November 2020).
40. President of Italian Republic Law Decree of 19 May 2020, Number 34—Article 119. Available online: https://www.gazzettaufficiale.it/eli/id/2020/05/19/20G00052/sg (accessed on 8 November 2020).
41. Ama Composites Aeropan Nanotech Thermal Insulation. Available online: http://www.aeropan.it/it/prodotti/aeropan/ (accessed on 8 November 2020).
42. Evangelisti, L.; Guattari, C.; Grazieschi, G.; Roncone, M.; Asdrubali, F. On the energy performance of an innovative green roof in the mediterranean climate. *Energies* **2020**, *13*, 5163. [CrossRef]
43. Evangelisti, L.; Guattari, C.; Gori, P.; De Lieto Vollaro, R.; Asdrubali, F. Experimental investigation of the influence of convective and radiative heat transfers on thermal transmittance measurements. *Int. Commun. Heat Mass Transf.* **2016**, *78*, 214–223. [CrossRef]

44. Evangelisti, L.; Guattari, C.; Asdrubali, F. Comparison between heat-flow meter and Air-Surface Temperature Ratio techniques for assembled panels thermal characterization. *Energy Build.* **2019**, *203*, 109441. [CrossRef]
45. Evangelisti, L.; Guattari, C.; Asdrubali, F. Influence of heating systems on thermal transmittance evaluations: Simulations, experimental measurements and data post-processing. *Energy Build.* **2018**, *168*, 180–190. [CrossRef]
46. International Organization for Standardization. *ISO 9869-1:2014 Thermal Insulation—Building Elements—In-Situ Measurement of Thermal Resistance and Thermal Transmittance Heat Flow Meter Method*; ISO: Geneva, Switzerland, 2014.
47. Kontoleon, K.J.; Bikas, D.K. The effect of south wall's outdoor absorption coefficient on time lag, decrement factor and temperature variations. *Energy Build.* **2007**, *39*, 1011–1018. [CrossRef]
48. DesignBuilder Software Ltd. DesignBuilder, Version 6.1.5.004. Available online: http://designbuilderitalia.it (accessed on 16 October 2020).
49. Grazieschi, G.; Gori, P.; Lombardi, L.; Asdrubali, F. Life cycle energy minimization of autonomous buildings. *J. Build. Eng.* **2020**, *30*, 101229. [CrossRef]
50. Ramesh, T.; Prakash, R.; Shukla, K.K. Life cycle energy analysis of buildings: An overview. *Energy Build.* **2010**, *42*, 1592–1600. [CrossRef]
51. Chastas, P.; Theodosiou, T.; Bikas, D. Embodied energy in residential buildings-towards the nearly zero energy building: A literature review. *Build. Environ.* **2016**, *105*, 267–282. [CrossRef]
52. Aspen Aerogels Inc. Environmental Product Declaration SPACELOFT®Aerogel Insulation. Available online: https://www.environdec.com/Detail/?Epd=10953 (accessed on 8 November 2020).
53. Asdrubali, F.; Ballarini, I.; Corrado, V.; Evangelisti, L.; Grazieschi, G.; Guattari, C. Energy and environmental payback times for an NZEB retrofit. *Build. Environ.* **2019**, *147*, 461–472. [CrossRef]
54. Almeida, R.M.S.F.; Ramos, N.M.M.; Manuel, S. Towards a methodology to include building energy simulation uncertainty in the Life Cycle Cost analysis of rehabilitation alternatives. *J. Build. Eng.* **2015**, *2*, 44–51. [CrossRef]
55. Almeida, R.M.S.F.; De Freitas, V.P. An Insulation Thickness Optimization Methodology For School Buildings Rehabilitation Combining Artificial Neural Networks And Life Cycle Cost. *J. Civ. Eng. Manag.* **2016**, *22*, 915–923. [CrossRef]
56. Ballarini, I.; Corrado, V.; Madonna, F.; Paduos, S.; Ravasio, F. Energy refurbishment of the Italian residential building stock: Energy and cost analysis through the application of the building typology. *Energy Policy* **2017**, *105*, 148–160. [CrossRef]
57. Burhenne, S.; Tsvetkova, O.; Jacob, D.; Henze, G.P.; Wagner, A. Uncertainty quantification for combined building performance and cost-benefit analyses. *Build. Environ.* **2013**, *62*, 143–154. [CrossRef]
58. Gustafsson, S.-I. Optimisation of insulation measures on existing buildings. *Energy Build.* **2000**, *33*, 49–55. [CrossRef]
59. Hasan, A.; Vuolle, M.; Sirén, K. Minimisation of life cycle cost of a detached house using combined simulation and optimisation. *Build. Environ.* **2008**, *43*, 2022–2034. [CrossRef]
60. Hopfe, C.J.; Hensen, J.L.M. Uncertainty analysis in building performance simulation for design support. *Energy Build.* **2011**, *43*, 2798–2805. [CrossRef]
61. Niemelä, T.; Kosonen, R.; Jokisalo, J. Cost-effectiveness of energy performance renovation measures in Finnish brick apartment buildings. *Energy Build.* **2017**, *137*, 60–75. [CrossRef]
62. Ortiz, J.; Fonseca i Casas, A.; Salom, J.; Garrido Soriano, N.; Fonseca i Casas, P. Cost-effective analysis for selecting energy efficiency measures for refurbishment of residential buildings in Catalonia. *Energy Build.* **2016**, *128*, 442–457. [CrossRef]
63. Ozel, M. Cost analysis for optimum thicknesses and environmental impacts of different insulation materials. *Energy Build.* **2012**, *49*, 552–559. [CrossRef]
64. Ozel, M. Determination of optimum insulation thickness based on cooling transmission load for building walls in a hot climate. *Energy Convers. Manag.* **2013**, *66*, 106–114. [CrossRef]
65. Verbeeck, G.; Hens, H. Energy savings in retrofitted dwellings: Economically viable? *Energy Build.* **2005**, *37*, 747–754. [CrossRef]
66. Hull, J.C. *The Evaluation of Risk in Business Investment*; Elsevier: New York, USA, 1980; ISBN 9780080240749.
67. Berk, J.; De Marzo, P.; Venanzi, D. *Capital Budgeting*; Pearson-Addison Wesley: Milan, Italy, 2009; ISBN 9788871925875.
68. Italian Parliament Law 27 December 2019, n. 160—Bilancio di Previsione Dello Stato per L'anno Finanziario 2020 e Bilancio Pluriennale per il Triennio 2020–2022. Available online: https://www.gazzettaufficiale.it/eli/id/2019/12/30/19G00165/sg (accessed on 8 November 2020).
69. Eurostat Energy Database. Available online: https://ec.europa.eu/eurostat/web/energy/data/database (accessed on 8 November 2020).
70. Fusion Media Limited. Rates Italian Bonds: 10 Year Bond Yield. Available online: https://it.investing.com/rates-bonds/italy-10-year-bond-yield (accessed on 8 November 2020).
71. Bank of Italy. Historical Series of the Bank of Italy—Statistical Series. Available online: https://www.bancaditalia.it/pubblicazioni/collana-storica/statistiche/index.html (accessed on 8 November 2020).
72. Italian National Institute of Statistics (ISTAT). Italian National Gross Domestic Product. Available online: https://www.istat.it/it/archivio/pil (accessed on 8 November 2020).
73. Refinitiv Datastream. Available online: https://eikon.thomsonreuters.com/index.html (accessed on 8 November 2020).
74. Bloomberg Finance, L.P. Credit Default Swap: 10 Years Spreads. Available online: https://bba.bloomberg.net/ (accessed on 8 November 2020).

MDPI
St. Alban-Anlage 66
4052 Basel
Switzerland
Tel. +41 61 683 77 34
Fax +41 61 302 89 18
www.mdpi.com

Energies Editorial Office
E-mail: energies@mdpi.com
www.mdpi.com/journal/energies

www.ingramcontent.com/pod-product-compliance
Ingram Content Group UK Ltd.
Pitfield, Milton Keynes, MK11 3LW, UK
UKHW052121230426
12049UKWH00010BA/147